a survey of
GENERAL MATHEMATICS

GENERAL

a survey of
MATHEMATICS

JAMES E. SHOCKLEY

**Virginia Polytechnic Institute
and State University**

HOLT, RINEHART AND WINSTON

New York   Chicago   San Francisco   Atlanta
Dallas   Montreal   Toronto   London   Sydney

Copyright © 1976 by Holt, Rinehart and Winston
All Rights Reserved

Library of Congress Cataloging in Publication Data
Shockley, James E
A survey of general mathematics.

1. Mathematics—1961–   I. Title.
QA39.2.S495   510   74-20171
ISBN 0-03-079230-4

9 8 7 6 5 4 3 2 1   032   9 8 7 6

This book was set in Laurel by York Graphic Services, Inc.
Editor: Holly Massey   *Designer:* Scott Chelius
Production Manager: Paul Holder   *Printer and Binder:* Von Hoffmann Press
Drawings: Vantage Art   *Cover:* Edmée Froment
Cover illustration courtesy of California Computer Products, Inc.,
Anaheim, Calif.

to
EVELYN

preface

This text develops a new approach to the college level general mathematics course. Designed for a one-year terminal course at the freshman level, it is a unique blend of technical and cultural topics. Elementary work is included on mathematics, computer science, probability, and statistics—the mathematical sciences that touch our daily lives.

The main thrust of the book is towards achieving mathematical literacy rather than detailed technical knowledge. I have attempted to develop a fresh approach based on two parallel themes: (1) the utilization of mathematics in the solution of problems in the real world; and (2) the development of mathematics over the centuries in an attempt to solve these problems.

Cultural Enrichment. I firmly believe that all educated persons should have at least a rudimentary knowledge of how mathematics developed and an elementary understanding of how it is applied. A large part of this book is devoted to a cultural grounding in elementary mathematics with emphasis on those two subjects. For the most part the cultural material is separated from the main body of the text. It is contained in a series of short notes, supplemental figures, pictures from museums, and so on, which are located in the margins near the relevant portions of the text. Topics requiring more space are discussed in short essays spliced, magazine-style, in the body of the text. This arrangement allows the individual instructor to decide the amount of the cultural material to incorporate into the course and to leave the remainder for the interested student.

Computer Science, Probability, and Statistics. Many of the modern applications of mathematics are based on computer science, probability, and statistics rather than on the more traditional mathematical topics. This book contains a strong introduction to these subjects in Chapters 6 (an optional chapter), 9, and 10.

Because the content of Chapter 6 is unusual for a book of this type, a few additional remarks will be made about it. The chapter is designed to acquaint the students with the basic operations that the computer can perform, thereby helping them to understand which problems can be solved and which cannot be solved by its use. The chapter is optional. With the exception of a few sample programs, the remainder of the book is independent of this work.

If a computer is not available, the first two sections can be taught as a cultural introduction to computer science. A few sample programs in BASIC can then be illustrated by having the students work through the steps as the computer would do.

Chapter Outline. Chapter 1 (*Flow Charts*) introduces the flow chart as a device for breaking complicated problems into simpler component problems. This chapter is fundamental for the book.

Chapters 2 (*The Whole Numbers*) and 3 (*The Rational Numbers*) contain a review of the basic operations of arithmetic. Much of this work is remedial and can be omitted if the students have good mathematical backgrounds.

Chapters 4 (*Elements of Geometry*) and 8 (*Topics in Geometry*) introduce the student to the practical side of geometry using a descriptive, rather than an axiomatic, approach. Basic work is included on similarity, right-angle trigonometry, coordinate geometry, and graphing, as well as a discussion of "area" and "length." Supplemental topics, such as the classical construction problems, are included for the interested student.

Chapters 5 (*Linear Equations*) and 7 (*Nonlinear Relationships*) introduce the student to basic algebraic techniques. The emphasis is on the solution of stated problems rather than on formal manipulation.

Chapter 6 (*The Computer*), an optional chapter, gives an introduction to computer science with programming in the BASIC language.

Chapter 9 (*Introduction to Statistics*) develops the elementary topics in statistics needed by literate persons in the modern world. The emphasis is on frequency distributions, and on the calculation and interpretation of the mean and the standard deviation for normally distributed data.

Chapter 10 (*Probability*) introduces both empirical and *a priori* probability theory. The motivation is based largely on gambling games, but is quickly extended to more practical problems. The chapter closes with a discussion of Pascal's triangle and its relation to the binomial distribution.

Examples and Exercises. The exercises are the most important part of any mathematics course. It is while working them that the student masters the material. The first few exercises of most sections are designed to imprint the basic techniques on the mind of the student. A few discuss topics related to those in the text and a few are included just because they are fun to work. Most of the exercises, however, involve the solution of simple word problems, similar to those the student might try to solve in the world. I have tried to keep the exercises relevant to the needs of the modern student and have avoided problems that illustrate particular techniques but are never encountered outside of the mathematics classroom.

All techniques and theory are illustrated by worked-out examples in the text. Although no new techniques need to be learned for the exercises, the student is frequently challenged to use a little ingenuity.

Acknowledgements. The technical portions of this book closely follow the CUPM "Math E" guidelines ("A Course in Basic Mathematics for Colleges").

I am indebted to a number of persons for their help in the development of this book. Valuable assistance was given by Professors Joseph Dorsett, Robert Hoyer, Harold Schoen, Michael Sullivan, and June Wood, who read all or parts of the manuscript in its preliminary versions and made many excellent suggestions that resulted in improvements. Mr. A. E. Montague typed the entire manuscript, making sense out of scribblings in one of the world's worst hand writings. Editors Jane Ross, Robert Linsenman, and Holly Massey of Holt, Rinehart and Winston worked closely with me for several years on the book. Their desire to publish a beautiful, as well as useful, book has contributed greatly to the project. Special thanks go to Evelyn Rubin, to whom the book is dedicated, who worked closely with me in all stages of the project and who did much of the research for the marginal notes.

J. E. S.

Blacksburg, Va.
August, 1975

contents

3

THE RATIONAL NUMBERS

4

ELEMENTS OF GEOMETRY

5

LINEAR EQUATIONS

6

THE COMPUTER

7

NONLINEAR RELATIONSHIPS

8

TOPICS IN GEOMETRY

9

INTRODUCTION TO STATISTICS

10

PROBABILITY

° Optional section.

a survey of
GENERAL MATHEMATICS

1

FLOW CHARTS

1-1 INTRODUCTION

During the past 20 years, electronic computers have caused a major upheaval in our way of life. Thousands of tasks that formerly were done by men are now either done or controlled by computers. These machines predict orbits of spacecraft, compute dosages of medicine, process checks for banks, take inventories, send orders to repossess automobiles, and perform countless other tasks that affect us daily.

The computer handles its thousands of jobs in an impersonal, coldly efficient, automatic way. Very little trouble is caused when informed men use the results of the computer to decide on courses of action—after all, the machine does its calculations as it was instructed by men. A great deal of trouble can be caused, however, when the computer is allowed to make decisions affecting policy or to interact directly with the public. For example, there are records of companies losing business because computers sent threatening letters to customers who had made trivial errors in paying their bills.

In the modern world, when all of our lives are touched daily by computers, it is important for us to have some knowledge of how these machines do their work. If nothing else, this knowledge can help us have a healthy respect for the usefulness of computers, coupled with an understanding of some of their very real limitations.

How does the computer work? This is one of the questions that we attempt to answer in this book. As with most other meaningful questions, we need practical experience, not just words of explanation. Consequently,

Figure 1-1 A modern computing center. Photo courtesy of IBM Corp.

a large part of the book is concerned with the preparation of "flow charts" and "programs" for computers and with the solution of problems by steps similar to those used by computers.

The modern computer consists of three basic machines (*components*) wired together into one elaborate complex. These components are the *input–output unit*, the *memory unit*, and the *arithmetic unit*.

(1) The *input–output unit* is used by the computer's operator to store information and programs in the computer. It is used by the computer to print out solutions to problems. It is connected to teletype terminals, card readers, tape machines, high-speed printers, and other devices.

(2) The *memory unit* contains millions of storage locations, each capable of storing a 10-digit number. (Think of the memory unit as a wall of post office boxes, each containing one slip of paper on which a single number can be written.) During the execution of a program, the computer constantly works with the memory unit—storing numbers in certain locations, reading numbers, changing numbers, and so on.

(3) The *arithmetic unit* is the heart of the computer. It works out all of the computations, controls the input–output and memory units, and makes simple decisions.

We have used the word "program" several times. A *program* is a detailed set of instructions that the computer must follow in order to solve a particular problem. The computer must be instructed at each step. This includes instructions on how to read the data, how and when to print out the answers, where to store numbers in memory, when to read numbers previously stored in memory, as well as how to perform every tedious calculation. Frequently it takes longer to prepare a program for the computer than it takes a human to solve the original problem.

After reading the last comment, the reader may wonder why we bother with computers at all. After all, if it takes longer to instruct the machine to solve a problem than to solve it ourselves, why not just solve the problem and forget about the computer? To be quite frank, the computer is superior to the human in solving certain types of problems. For example, the computer is better at solving problems that involve great amounts of computation in which a single error could cause a wrong answer. Other capabilities of the computer contribute to its versatility in solving problems. Three of these are listed below:

(1) *Repetition.* Many problems, such as the computation of department store bills, occur over and over again in the same form. One standard program can be stored permanently in the computer and used each time the problem arises.

(2) *Speed.* Many problems, such as those connected with the reentry of a spacecraft, require almost instantaneous solutions. Programs can be worked out in advance and stored in the computer until needed. When the need actually arises, the solutions can be obtained within seconds.

(3) *Reliability.* Not all of us can solve difficult problems. In general, we need a good intuition or advanced training in a subject to solve any more

Contrary to the impression given by certain television shows, computers and human beings cannot communicate directly with each other. All input to the computer must be on punched cards, magnetic tape, or some other device.

The individual instructions in a program are quite simple. This is true even when the programs are written to solve difficult and complex problems. Such programs are very complicated, containing hundreds of interrelated simple instructions.

than the most trivial problems. Once an expert works out a computer program for the solution of a problem, however, anyone able to follow simple instructions can load the program and the data into a computer, push the buttons that activate it, and read the solution on the high-speed printer.

Furthermore, once a program has been tested and proved to be correct, we can depend on the answers. This is not the case when a human works on a problem, because he can make a mistake at any step. The computer, unlike the human, works through the steps of the program exactly as written without making errors.

Programs are usually written to be as general as possible so that they can be used in many different applications. Specific information needed to complete the program for a particular problem is given the computer in *data* statements. For example, the programmer may prepare a program to add two numbers and print their sum. If "5" and "3" are then written on the data statements and read into the computer, it will print "8." If "17" and "194" are on the data statements, it will print "211."

There are three basic steps in solving a problem on the computer:

(1) The *program* and the *data* are read into the computer's memory unit by means of the input–output unit. (The *program* is a detailed plan for the solution of the problem. The *data* statements tell the computer the specific numbers that are to be used with the program.)

(2) The computer executes the program exactly as it is written, using the data provided.

(3) The computer prints out the solution to the problem in a form intelligible to the user. (The controlling statements for the printout are part of the program.)

Figure 1-2

If we write out the three basic steps for the solution of a problem in schematic form, we have the diagram, known as a *flow chart*, shown in Figure 1-2. The arrows indicate the order in which the basic steps are performed. Obviously, these steps can be broken down into smaller steps as in the flow chart of Figure 1-3.

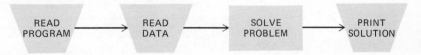

Figure 1-3

In most cases, a detailed flow chart is made for each program. This flow chart shows each step that the computer goes through in reading the data, solving the problem, and printing the answers. In the remainder of this chapter, we shall learn the basic principles used in making flow charts. Later we shall prepare programs that can be run on the computer.

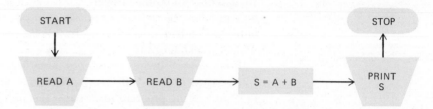

Figure 1-4

1-2 FLOW CHARTS

The computer does its work in a special language using only numbers. Consequently, each instruction in the program is stored as a number in the computer's memory. In the early days of programming, a programmer had to write all instructions in this special language used by the computer. He first had to design a flow chart for the solution of the problem, then make a program based on the flow chart, and then translate the program into the number language used by the computer. This last difficulty was removed later by the invention of *translator languages*. These allow instructions to be written in a modified form of English according to a strict format. The computer recognizes these instructions as symbols written in a special code. It translates them into specific instructions in the number language that it "understands."

Some of the common translator languages are BASIC, FORTRAN (used in scientific and mathematical work), and COBAL (used in business). A number of refinements have been made in these languages over the years, resulting in various dialects. In the later chapters of this book, we shall write programs in BASIC, a simple translator language that can be used on most computers.

Example 1. The flow chart in Figure 1-5 instructs the computer to add three numbers A, B, C and print out the sum.

The steps in the flow chart of Figure 1-5 are quite simple. The computer is instructed to

(1) read the numbers A, B, C from the data statement,
(2) add the three numbers it just read, calling the total "S,"
(3) print the value of S, and
(4) stop work on this program and proceed to the next one.

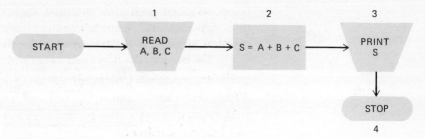

Figure 1-5

The following computer program, written in BASIC, is based on the flow chart of Figure 1-5:

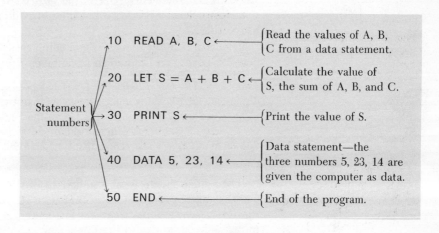

The computer works through the program exactly as it is written, types out the single number

42

which is the sum of the three numbers 5, 23, and 14, figures the bill for its work, then stops work on this program and goes on to the next one.

Several standard features are illustrated in the flow chart for the above program:

(1) *The flow chart begins with a* START *box and ends with a* STOP *box.* This is a convention that helps us when we examine an unfamiliar flow chart.

(2) *If we trace through the steps of a flow chart, we must eventually get to a* STOP *box.* This convention causes us no trouble in our simple example. We must be careful, however, when we construct more complicated flow charts to make sure that this always occurs. Otherwise, the

computer might finish work on a program, but might not know it. It would then stop all work except for the calculation of the bill, which would continue to grow until the operator stopped the machine.

(3) The boxes in the flow chart have different standard shapes to distinguish their functions. *Oval* boxes contain (START) and (STOP) instructions. *Trapezoidal* boxes contain \READ/ and \PRINT/ instructions. *Rectangular* boxes contain | Arithmetic Instructions |. Boxes of other special shapes will be discussed later.

Decision Boxes

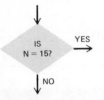

One of the most useful boxes in a flow diagram is the *decision box*. This is a diamond-shaped box that contains a question that can be answered YES or NO. If the answer is YES, the flow chart branches in one direction to an instruction; if it is NO, it branches in a different direction to a different instruction.

In most flow charts the questions in decision boxes involve the size of numbers.

The next example, which is nonmathematical in nature, illustrates how decision boxes can be used to advantage.

Example 2. The Car Thief. Big Louie has been ordered to steal a late model expensive car. He wants a car that is less than two years old, that cost at least $5000 when new, and that has the key in the ignition. At 5:00 P.M. he will quit work and go home to his wife and children. Figure 1-6 shows the portion of the flow chart outlining the decisions that Big Louie will make as he examines a typical car.

The reader should work through all of the flow charts with examples. These should be chosen so as to illustrate all of the possible paths through the flow chart. For the fragment in Figure 1-6, we should choose at least four examples similar to the following:

(1) Car is three years old.
(2) Car is one year old and costs $2000.
(3) Car is one year old, costs $8000, and does not have the key in the ignition.

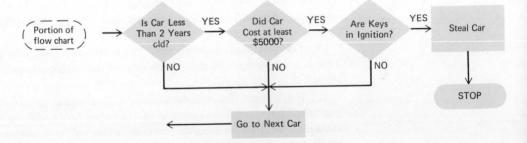

Figure 1-6   Portion of flow chart (Example 2).

(4) Car is one year old, costs $6000, and does have the key in the ignition. Other combinations of age, price, and availability of keys would be informative.

At this point, we can examine one of the major shortcuts in flow-charting. Figure 1-6 illustrates the decisions for *each* car. Thus, we can draw an arrow from the box "Go to Next Car" to the first decision box, indicating that Louie continues on through the decision process until he finds a car to steal. This provides us with the first approximation to our complete flow chart (Figure 1-7).

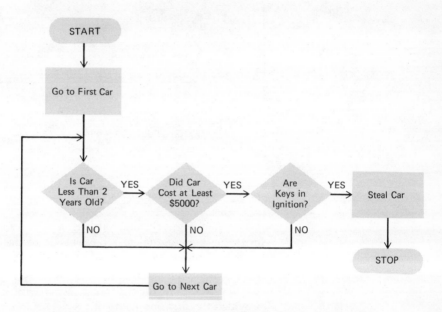

Figure 1-7   First approximation to the complete flow chart in Example 2.

If we work through the flow chart in Figure 1-7, we see that everything works out properly if Louie eventually finds a car to steal. If not, however, it instructs him to stay at his task forever. We must modify our chart so that he will stop at 5:00 P.M. if he does not find a suitable car. Obviously, another decision box is needed: "Is It 5:00 Yet?" If this box is placed just before the box "Go to Next Car!" we get the final flow chart shown in Figure 1-8.

If we now check out all possible combinations of age, cost, availability of keys and time, we see that we must eventually get to a STOP box. Thus, Big Louie is not doomed to an eternal search for a car.

A set of instructions and decision boxes that can be repeated over and over in a flow chart is known as a *loop*. The flow chart in Figure 1-8 contains a loop that begins with the question "Is Car Less Than 2 Years Old?" and ends with the instruction "Go to Next Car." Notice that there are three ways we can progress through this loop.

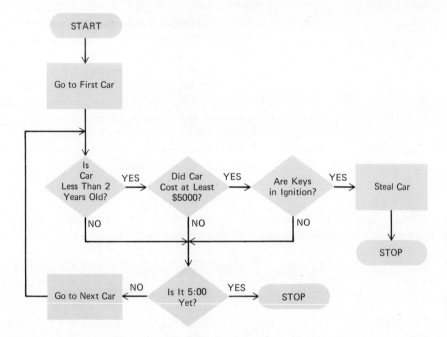

Figure 1-8 The car thief: complete flow chart for Example 2.

Every loop must contain at least one decision box that will eventually break the loop and go to a box outside of the loop. The loop in Figure 1-8 can be broken in two ways: (1) by a YES answer to the question "Are Keys in Ignition?" and (2) by a YES answer to "Is It 5:00 Yet?" Observe that the second of these decision boxes must eventually break the loop, provided the first one does not.

EXERCISES

1. Construct a flow chart for your morning routine including (a) getting dressed, (b) eating breakfast, and (c) getting to work or school. Include at least two decision boxes. (For example, the weather conditions determine whether or not you will take a raincoat or umbrella.)
2. Construct a flow chart for the operation of changing a tire. Include at least two decision boxes. (For example, "Is the spare tire inflated?")
3. Modify the flow chart in Figure 1-8 (the car thief) according to one of the following sets of instructions:
   (a) The age of the car does not matter.
   (b) The keys are not necessary (Big Louie can cross the ignition wires), but the car must be *unlocked.*
   (c) The car must be *at least* two years old, cost *no more* than $4000 when new, and must be *unlocked.*
4. According to Greek mythology, as punishment for giving fire to man,

Prometheus was sentenced to be chained to a mountain until an immortal would consent to take his place. Furthermore, each day a vulture was sent to the mountain to eat his liver. Make a flow chart for the activities of the vulture. (Include a loop for the vulture's daily routine.)

Must there be a way out of the daily loop *as the problem is posed?* If not, what modifications should we make in the statement of the problem to guarantee that the poor vulture eventually gets a break from his steady diet of liver?

5. Machine AUTOWAY-1734-92A at the Break-a-Jaw Bubble Gum factory automatically picks up and weighs each piece of bubble gum as it comes down a conveyor belt. If a piece weighs less than $\frac{1}{10}$ of an ounce, it is rejected. Otherwise, it is put back on the conveyor belt. If 10 pieces are rejected in any 1-minute period of time, the machine stops the assembly line and rings a bell to summon the supervising engineer. Construct a flow chart for the activities of this machine. Explain why a STOP box is *not* necessary in this particular flow chart (contrary to our general instructions).

1-3 REPLACEMENT BOXES

In the preceding section, we saw how loops can be set up to handle operations that are repeated several times. In our previous flow charts, these loops were incidental. We now consider some situations where loops can be planned in advance in order to simplify the flow charts.

Many problems involve operations that must be repeated over and over many times. Imagine that you are a clerk in a discount store and you must stock a counter with 1500 bottles of El Kisso mouthwash. You must go through a simple set of operations "Take Bottle from Box" and "Put Bottle on Shelf" 1500 times in a row. In flow-charting the operations we could repeat these instructions 1500 times. It is much easier, however, to set up a loop to program the actions.° The fragment in Figure 1-9 illustrates the loop.

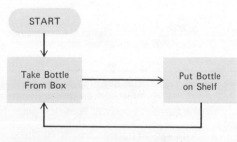

Figure 1-9

°This assumes, of course, that the cartons have been opened, the price stamped on the bottles, and so on.

This is not a complete flow chart because we have no way out of the loop. We must find a way to stop the process after 1500 bottles have been put on the shelf. Obviously, we need to count the number of times we go through the loop and stop after we have gone through it 1500 times. We let N denote a quantity that can vary during the program, start with $N = 0$ and raise the value of N each time we go through the loop. (See Fig. 1-10.)

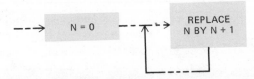

Figure 1-10

At the start N has the value 0. As we go through the loop each time we replace the value of N by $N + 1$. Thus, we give N the value 1 when we go through the loop the first time, the value 2 when we go through the second time, and so on, as shown in the flow chart of Figure 1-11. To get out of the loop, we put a decision box that will stop the program after 1500 bottles have been put on the shelf.

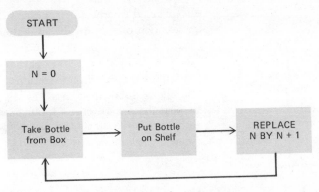

Figure 1-11 Incomplete flow chart for stock clerk.

Example 1. The flow chart for the stock clerk is shown in Figure 1-12.

The key step in this flow chart involves the *replacement box*. This is a box that instructs us to *replace* one number with another. In this particular program, we start with $N = 0$, replace N by 1, then by 2, then by 3, and so on.

The number N in the flow chart is an example of a *variable*. It is a symbol that can stand for different numbers in different parts of the program. A variable that is used to count the number of times we go through a loop is called a *counter*.

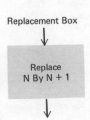

Replacement Box

Replace N By N + 1

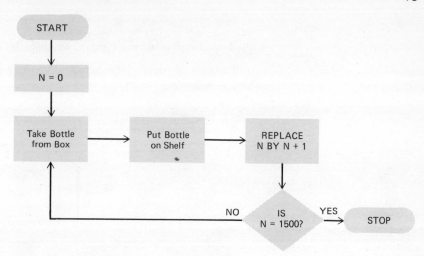

Figure 1-12 Complete flow chart for the stock clerk.

In the remainder of this section, we shall consider several flow charts that illustrate the types of operations we can use in computer programming.

Example 2. The Assembly Line. George, a fifth grade dropout, is the last man on the assembly line at the Wonder Wicket Company, the world's largest manufacturer of croquet wickets. George puts the wickets in cartons, one at a time. When he gets 200 wickets in a carton, he pushes it down a chute, gets a new carton, and repeats the process. Every two hours he stops for a break.

Figure 1-13 pictures the main steps in George's work. Obviously, these are (1) get a carton, (2) fill it with 200 wickets, (3) push it down a chute, (4) return to step (1). At some point George must also check his watch and see if he has worked two hours without a break. If so he will take a break; if not he will continue work. From past experience he knows that it is best to stop work after he has pushed a full carton down the chute and before he gets a new one. (This avoids a lot of confusion on George's part.) Before proceeding further, the reader should work through the steps of Figure 1-13.

We now make a more detailed flow chart. The only part of our chart that needs a more detailed analysis is the one box "Fill Carton with 200 Wickets."

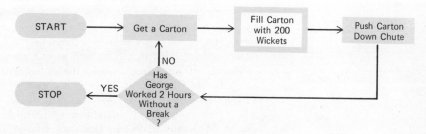

Figure 1-13 Flow chart outlining main steps in Example 2.

Recall that George puts the wickets in the carton one at a time. Thus, we must set up a loop for this operation and count the number of times we go through the loop. The fragment in Figure 1-14 illustrates the loop.

The task of piecing the two flow charts together is left for the reader (see Exercise 2).

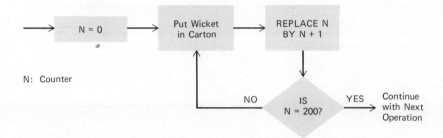

N: Counter

Figure 1-14   Portion of flow chart corresponding to the instruction "Fill Carton with 200 Wickets" in Figure 1-13 (Example 2).

Observe that many of the steps in Figure 1-14 can be broken down into still smaller steps. For example, the box "Get a Carton" can be broken down as shown in Figure 1-15. This breakdown into smaller steps is always possible. The good programmer lists only the operations that are essential, combining the small steps into larger ones. When we work with the computer we must learn which operations can be programmed and which must be broken down into simpler operations.

Figure 1-15   Detailed breakdown of the step "Get a Carton" in Figure 1-13.

Example 3. The Date   John, typical male college student, wants a date for the evening. He has the names of 27 girls in his address book. He plans to call all of them on the telephone, going in order from the front to the back of the book. If he does not complete a call, he will forget about that girl and go on to the next one. If he gets a date, he will stop calling and get dressed. If he does not get a date, he will toss a coin. If it comes up *heads*, he will either study mathematics or drink beer. If *tails*, he will go to a movie.

Figure 1-16 was constructed by the author. The reader should decide if it is adequate. He should work through the steps of the flow chart using several different examples. ("John completes first, second, fifth calls, gets date on fifth call," "John does not get a date, coin comes up *heads*," and so on.) The reader also should identify all decision boxes, all counters, and locate all loops. He should make sure that every loop must eventually terminate and that a STOP box must eventually be reached (see Exercise 3).

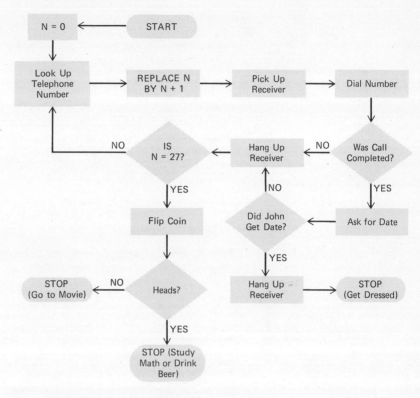

Figure 1-16 Flow chart for Example 3.

We close with a flow chart of a more mathematical nature. In fact, this flow chart can be translated directly into a computer program.

Example 4. Professor Mindboggle has been recognized as an outstanding teacher by the college administration and, as a reward, has been assigned to teach a class of 297 freshmen students. Obviously, he needs all of the help he can get, so he plans to have the computer average the grades on each test. The computer will add the grades and divide the sum by the number of individuals who took the test. Since several students usually manage to miss any given test, the first data

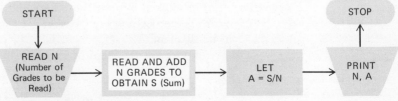

N: Number of grades
S: Sum of grades
A: Average of grades (A = S/N)

Figure 1-17 The main steps in Example 4.

card will tell the computer how many cards to read. (Otherwise, it might run out of cards to read and become frustrated.)

The flow chart should have a *variable* N for the *total number of grades* to be read, a *variable* G to represent *each grade* as it is read, a *counter* K for the *number of grades* read up to that point, and *variables* S and A for the *sum of* the grades and the *average* of the grades. The flow chart in Figure 1-17 lists the main steps.

The detailed flow chart in Figure 1-18 is the one devised by the good professor. The reader should test the flow chart by working through it with several examples (see Exercises 4 and 5).

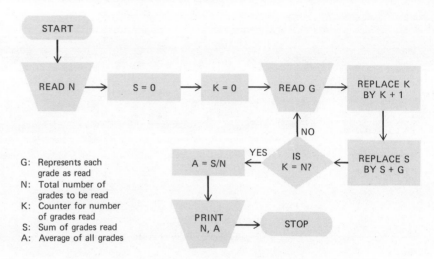

Figure 1-18   Flow chart constructed by Professor Mindboggle (Example 4).

EXERCISES

1. Identify all decision boxes and loops in the flow charts in the four examples of this section.
2. Combine the two incomplete flow charts (Figs. 1-13 and 1-14) into one master flow chart for Example 2. Are any further modifications necessary (or desirable)?
3. Examine Figure 1-16 (Example 3).
    (a) Are the two boxes "Hang Up Receiver" necessary? Could they be combined into one box? Explain!
    (b) Is the decision box "Heads?" a necessary part of the flow chart? Explain!
4. Work through the steps of Figure 1-18 using the following data (Example 4):
    (a) N = 1, grade 87
    (b) N = 3, grades 62, 71, 50
    (c) N = 7, grades 21, 99, 58, 100, 76, 87, 41
5. Make the following modification in Figure 1-18: Erase the "NO" line

that runs from the decision box "IS K = N?" to the box "READ G" and *replace* it with a "NO" line from the decision box to the box "K = 0." What effect does this change have on the program? What happens when you work through the new flow chart? Test it by attempting to work through the new flow chart with the data in Exercise 4.

1-4 ENGLISH INTO MATHEMATICS

Many of us find it quite difficult to solve a mathematical problem in a step-by-step method. We do, however, possess the ability to solve the practical problems that face us every day. We simply lack the experience and the knowledge of techniques necessary to solve problems posed in mathematical terms.

Before we discuss any problems in detail, it will be useful to consider a few general principles that may help us. First, we must learn to *avoid getting bogged down with unnecessary details*. It is essential in mathematics that we strip each problem of its nonessential details, ignoring isolated facts that appear to be part of the problem but, in actuality, are not.

As an example, consider the problem of estimating the value of a rectangular tract of land. If we know the area and the price per acre, we can calculate the value. The fact that the land is in the suburbs, that it fronts on a main street, or is covered with artificial turf has no bearing on the problem.

At this point the author can almost hear the reader protest, "But if these factors are considered, the land will be increased in value." Think again. The location of the land has already been considered in the price per acre. The artificial turf should be considered separately—after all, it is not part of the land. Thus, the price per acre and the number of acres are sufficient. (The author will concede the point that the land may *vary* in value. The part of the land near the street may be worth more than the remainder of the land. Again, this can be dealt with by knowing the price(s) per acre.)

We now examine our second principle. It is frequently the case that simple problems have many of the same features as related complicated problems. Usually we can do a better job of solving a problem if we have had experience with related problems. Consequently, if we are faced with a difficult problem that we cannot solve, a helpful strategy is to first *solve a simpler related problem*. The experience gained with the simple problem may be sufficient to enable us to solve the original difficult one.

This is the idea behind most of the "textbook" problems in mathematics and, in fact, the idea behind much of formal classroom instruction. Most mathematics textbooks contain sets of problems designed to impart specific skills—factoring, for example. The first few problems in a set are usually quite simple, enabling the student to get needed experience before tackling the more difficult problems at the end of the set.

As an example of how we can use this idea of replacing difficult problems with simple ones, consider the *rate-time-distance* problems encountered in high school algebra. Many students have difficulty with these problems, because they can never remember when to multiply and when to divide. They know that they must do one or the other, but get confused about which. These same students would know immediately that if they traveled 150 miles (*distance*) in two hours (*time*), they had traveled at an average speed of 75 miles per hour (*rate*). This would indicate that

$$\frac{\text{distance}}{\text{time}} = \text{rate}$$

or, equivalently, that

$$\text{time} \cdot \text{rate} = \text{distance}$$

and

$$\frac{\text{distance}}{\text{rate}} = \text{time}$$

These common sense rules could then be used to help solve the original difficult problem.

There is a third idea that is a companion to the one above. Try to *replace the original problem with a series of simple ones.* That is, break the original complicated problem into a number of simple problems in such a way that solving the simple problems in order actually solves the original problem for us.

If we think about it, we see that this is the method we use in our everyday lives to solve most problems. The problem of studying for an examination, for example, may be partially solved by (1) organizing the material into basic units, (2) mastering the subject matter of each of the units, and (3) studying the interrelationships between the units. Once we have worked out the three simple problems, we have also solved the main one.

In summary, we have three principles that can help us solve a mathematical problem:

(1) Strip the problem of unimportant details and concentrate on the pertinent facts.

(2) Solve a simpler related problem to get valuable experience that can be used to solve the original complicated problem.

(3) Break the original problem into a series of simple problems that can be solved in a definite order.

It is with this last idea that flow-charting can help. A simple flow chart can outline the components of a problem, showing their relationships to one another and to the original problem. Each of the main steps in this flow chart can then be broken down into the steps needed to solve the original problem.

We now turn our attention to some specific examples.

Example 1. Mr. MacGregger is due at an important meeting in his cabbage patch at 11:00 A.M. He has been driving back from a business trip for the appointment, but has been delayed by highway construction. To make a long story short, he started from a point 300 miles away from the patch at 6:30 A.M. and averaged 75 miles per hour for the first two hours. Because of the highway construction, he traveled only 25 miles during the next hour. He has now passed the construction and has a clear road ahead. How fast must he travel in order to reach the cabbage patch by 11:00 A.M.?

Solution. If we think about it, we see that we need to know the distance that remains to be traveled and the amount of time that is left. Both of these quantities can be calculated from the data. (See Fig. 1-19.) We then complete the problem by calculating that

$$\text{required rate} = \frac{\text{remaining distance}}{\text{remaining time}}$$

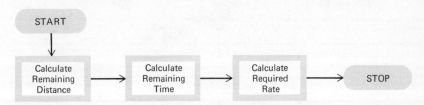

Figure 1-19 Basic flow chart showing the breakdown into component problems (Example 1).

In Figure 1-19 we have broken the original problem into three component problems. Detailed flow charts can now be made for each of these problems. For example, the problem "Calculate Remaining Distance" can be partially solved by the steps in Figure 1-20.

Figure 1-20 Flow chart for the component problem "Calculate Remaining Distance" (Example 1).

The final solution of the problem is left for the reader (see Exercise 1).

Example 2. Ophelia, a character from a Victorian novel, is terrified of thunderstorms. At the present time she is sitting on the front porch of her mansion watching with dread as a thunderstorm approaches. To calm her nerves, she is attempting to measure the rate at which the storm is moving toward her. A few minutes ago she timed the interval between a lightning flash and the associated thunder as 15 seconds. Ten minutes later she timed the interval as 6 seconds. She knows that sound travels at the rate of 1100 feet per second. How fast is the storm approaching?

Solution. This is a rate-time-distance problem. In order to calculate the *rate* that the storm is approaching, we need to

(1) calculate the *distance* that the storm traveled over the 10-minute period that Ophelia timed it, and

(2) divide the distance by the *time* (10 minutes).

If we let D1 denote the distance between Ophelia and the storm at the beginning of the 10-minute period and D2 denote the distance at the end of the period, then the storm has moved a total distance of D1 − D2. Then the *rate* is

$$\text{rate} = \frac{\text{distance}}{\text{time}} = \frac{\text{D1} - \text{D2}}{10}$$

Figure 1-21 shows the relationships between the component problems.

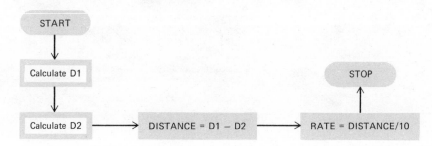

Figure 1-21 Flow chart showing the breakdown into component problems (Example 2).

To calculate D1, the distance from Ophelia to the storm at the beginning of the period of time, we recall that sound travels at the rate of 1100 feet per second. Since the sound traveled for 15 seconds it traveled a total distance of

$$\text{D1} = (1100) \cdot (15) = 16{,}500 \text{ feet}$$

D2 can be calculated similarly. The final solution of the problem is left for the student (see Exercise 2).

Example 3. Mr. Moneybags wishes to put a 3-foot wide cement sidewalk around the rectangular reflecting pool he constructed in his garden at a cost of $14,300. The pool is 50 feet long and 10 feet wide. The sidewalk is to be 6 inches thick. Calculate the amount of cement, in cubic yards, that the contractor should order. (27 cubic feet = 1 cubic yard.)

Solution. The problem involves the calculation of a volume. Since the sidewalk is to be 6 inches thick, the volume is equal to the area A of the land covered by the sidewalk multiplied by the thickness ($\frac{1}{2}$ foot = 6 inches). Consequently, the problem involves first calculating the area. (See Fig. 1-22.) To calculate the

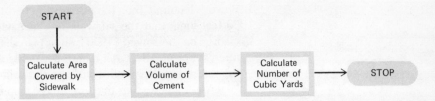

Figure 1-22 Flow chart showing the breakdown into component problems (Example 3).

area covered by the sidewalk, we first draw pictures of the reflecting pool and the sidewalk (Fig. 1-23).

Obviously, the area covered by the sidewalk is equal to the total area covered by the pool and sidewalk together *less* the area covered by the pool. We see from Figure 1-23 that the total area covered by the pool and sidewalk together is $16 \cdot 56 = 896$ square feet. The area covered by the pool is $10 \cdot 50 = 500$ square feet. Consequently, the area covered by the sidewalk is

$$\text{area} = 896 - 500 = 396 \text{ square feet}$$

The *volume* is equal to the *area* multiplied by the thickness. The volume of cement needed for the sidewalk is

$$\text{volume} = \text{area} \cdot \text{thickness} = 396 \cdot \tfrac{1}{2} = 198 \text{ cubic feet}$$

Now each cubic yard contains 27 cubic feet. If we divide 27 into 198, we get $7\frac{1}{3}$ cubic yards. Thus, the contractor should order about $7\frac{1}{2}$ or 8 cubic yards to allow for minor variations in depth, loss due to spillage, waste, and so forth.

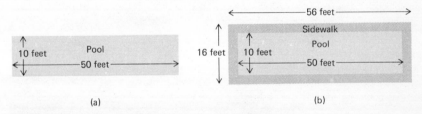

(a)   (b)

Figure 1-23 (a) Area covered by pool = 500 square feet; (b) area covered by pool and sidewalk together = $16 \cdot 56 = 896$ square feet (Example 2).

EXERCISES

1. (a) Make a detailed flow chart for the complete solution to Example 1.

   (b) Work out the numerical solution to Example 1.

2. (a) Make a detailed flow chart for the complete solution to Example 2.

   (b) Work out the numerical solution to Example 2.

3. Rework Example 3 on the assumption that the pool is 20 feet wide and 30 feet long, that the sidewalk is 4 feet wide and 4 inches thick.

4. George has borrowed $1000 from a loan shark at the rate of 10 percent interest per month. Each month he makes a payment of $250 to cover the interest and pay off a little of the principal.

(a) Make a flow chart to compute the amount owed after one payment.

(b) Make a flow chart to compute the amount owed after N payments, where N is a fixed integer given at the beginning of the program. Incorporate the flow chart from (a) if possible. (Assume that the interest is computed only on the unpaid balance.)

2

THE WHOLE NUMBERS

2-1 INTRODUCTION

Prehistoric tally-stick made from a piece of bone.

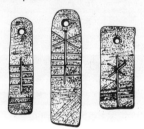

These tally-sticks were used by the Swiss peasants to record debts.

Twentieth-century tally sticks.

Man has used the whole numbers, the numbers *one, two, three, four,* and so forth, since the beginning of history. At first he probably used these numbers only for counting. In fact, in the most primitive periods, it is likely that he only could tally—that is, he restricted his concept of number to marks on a stick. These marks corresponded to the number of objects he wished to trade, he owned, or had killed.

Tallying was sufficient for a primitive society, but was inadequate for even simple commerce. As civilization became more complicated, tallying gave way to finger counting, which in turn evolved into the use of counting frames and written systems of enumeration.

Finger counting quickly became more sophisticated than we might imagine. Observe that there are two natural units of measure, the basic unit (*one*) represented by one finger, and the large unit (*ten*) represented by one man. It was quite natural to represent both units by holding up fingers. Thus, a man could hold up seven fingers followed by four fingers to represent the number *seventy-four* (seven large units and four small units). The system is similar to the way we count money using the cent as the basic unit and the dollar as the large unit.

This primitive type of finger counting was limited in that only numbers up to 110 could be represented conveniently. (There is no simple way to represent a number larger than 10 large units and 10 small units.) Because of this limitation, some early genius invented a *counting board* or *abacus*. (See Fig. 2-2.)

Most of us are familiar with the bead-on-wire type of abacus used in the Orient. This type of counting board is a modern invention, and a number

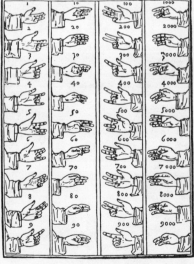

Figure 2-1 Sophisticated finger counting. (From the book *Summa de Arithmetica,* published in Venice in 1494. Written by the Italian mathematician Luça Pacioli, the *Summa de Arithmetica* was the first important book on mathematics to be published in the western world.) Courtesy of the British Library.

Left hand   Right hand

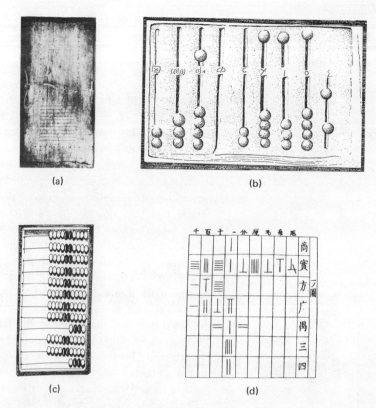

(a)

(b)

(c)

(d)

Figure 2-2 Four types of abacus. (a) Greek counting board for use with movable counters—fourth century B.C.; (b) small Roman abacus designed to be held in one hand; (c) Russian ščёt; (d) early Chinese counting board for use with movable counting rods. The figure shows several numbers represented by the rods.

of other types have been used over the centuries. The ancient Chinese used "counting rods" that they manipulated with lightning speed. Other societies used boards marked with columns on which sand was sprinkled. Various numbers could be represented by marking in the sand in the proper columns. The ancient Greeks and Romans used boards marked with rows or columns on which movable counters could be placed.

Because the bead-on-wire type of abacus is the most familiar, we shall restrict our discussion to that type, keeping in mind, however, that most of the mathematical properties are common to all types of abacus. The prototype that we shall consider is modified from a Russian model and has 6 wires with 10 beads per wire (Fig. 2-3).

Our abacus allows a direct extension of the process of finger counting. The numbers from 1 to 10 are represented by pushing up beads on the right-hand wire. Each bead on that wire represents one "basic" unit just as one finger represented one "basic" unit in finger counting. Each bead

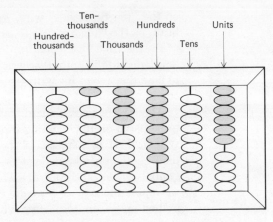

Figure 2-3 Prototype of the abacus. The number *four-teen thousand eight hundred six* is represented.

on the second wire represents one "large" unit (10 small units). Thus, the number *seventy-four* is represented by pushing up *seven* beads on the "tens" wire and *four* beads on the "units" wire.

We now can go one step further than we could with finger counting. Each bead on the third wire represents the number *one hundred* (10 of the "tens" units), each bead on the fourth wire represents *one thousand,* (10 of the "hundreds" units), and so on, For example, we can represent the number *fourteen thousand eight hundred six* by pushing up *one* bead

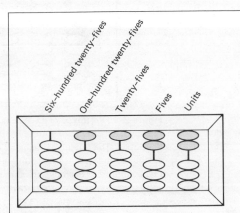

The number 10 plays a special role in our system. Originally our ancestors used all 10 of their fingers in counting, resulting in an abacus with 10 counters in each column. If the fingers of only one hand had been used, the resulting abacus probably would have had only *five* counters in each column. The figure illustrates such an abacus with the number *162* represented.

The symbols in the Roman numeral system indicate that a primitive system based on 5 may have been incorporated into a system based on 10.

on the "ten thousands" wire, *four* beads on the "thousands" wire, *eight* beads on the "hundreds" wire, and *six* beads on the "units" wire (Fig. 2-3).

By the time that counting frames began to be used, civilization had developed to the point that written records were necessary. It was natural to base the system for writing numbers on the methods developed for the counting frame.

There were two standard ways in which the writing of numbers came to be based on the counting frame. The Egyptians and Romans used one system, the Babylonians and the Hindus, the other. The method used by the Egyptians can be understood most easily if we imagine an abacus with symbols written at the top of each column (Fig. 2-4). To represent a number, we repeat the symbol for each column as many times as beads would be pushed up on the respective wire when representing the number on the abacus.

For example, the number *four thousand one hundred thirty-two* would be written as 𝟤𝟤𝟤𝟤𝟤9 ∩∩∩ ‖ indicating four beads on the "thousands" wire, *one* on the "hundreds" wire, *three* on the "tens," and *two* on the "units" wire.

The Hindus and the Babylonians used a completely different method to represent numbers by symbols. The Hindus invented nine symbols to

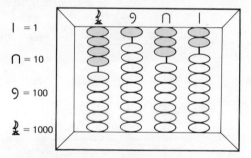

Figure 2-4 Prototype of an abacus showing how the Egyptian system for writing numbers may have developed. The Egyptians repeated the "column" symbol as many times as counters were used in that column. For example, the number *four thousand one hundred thirty-two* was written

𝟤𝟤𝟤𝟤9 ∩∩∩ ‖

represent the number of beads pushed up on any wire. (A table of the Hindu symbols is shown in Fig. 2-5.) To represent a number, they wrote the symbols that represented the number of beads on the various wires, the symbols being strung out in the same order as the wires on the abacus. For example, the number *three hundred forty-six* was written ३ ४ ६ , indicating *three* (३) beads on the "hundreds" wire, *four* (४) beads on the "tens" wire, and *six* (६) beads on the "units" wire.

(a)

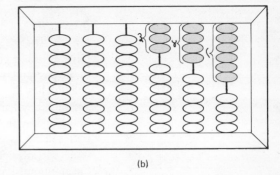

(b)

Figure 2-5 (a) The Hindu numerals—600–800 A.D.; (b) in the Hindu system, the number *three hundred forty-six* was written 𝟥 𝟪 𝟨 indicating *three* (𝟥) hundreds, *four* (𝟪) tens, and *six* (𝟨) units.

The primitive Hindu system had one defect when compared to the Egyptian system. There was no simple way to represent "no beads pushed up on a wire." If an Egyptian wanted to write *three hundred seven*, for example, he simply wrote 9 9 9 ᵢᵢᵢ , indicating *three* hundreds and *seven* ones. A Hindu could only write 𝟥 𝟩 or 𝟥 𝟩 with a note appended explaining that the first digit represented 𝟥 hundreds rather than 𝟥 tens.

Eventually some genius got the idea of using a special symbol—a small circle *○* —to represent an "empty" column on the abacus. From that time on the Hindus wrote *three hundred seven* as 𝟥 ○ 𝟩—*three* hundreds, no tens, *seven* ones.

Later the Arabs adopted the Hindu system, along with simple modifications in the written symbols. About 1200 A.D. the system was introduced into Europe, where the symbols were known as *Arabic* numbers. Today they are known as *Hindu-Arabic* numbers.

Leonardo of Pisa (Fibonacci)

Leonardo of Pisa (1180–1250), better known as *Fibonacci* (the son of Bonaccio), was the man who introduced the Hindu-Arabic numerals into the European world.

The son of a businessman who traveled widely in North Africa, Leonardo learned as much of the Arabian science and mathematics as possible when a young man. In 1202 he published the influential book *Liber* *abaci*, which contained much of what he had learned about Arabian mathematics. Part of the book was devoted to a systematic development of the Hindu-Arabic number system. In spite of minor setbacks, this system was firmly established in Italy within a century. Not until the sixteenth century, however, was the system universally used in all of Europe.

EXERCISES

1. Write the following numbers in both the Egyptian and the Hindu systems of numeration:

(a) 749
(b) 4381
(c) 2040

2. Give at least two original examples each of how tallying has been used in the past and the present. (For example, in the old West certain outlaws cut notches in the handles of their pistols to represent the number of persons they had killed.)

3. Make a drawing of an abacus for the United States monetary system that can be used to represent amounts from 1 cent to $9.99. Include wires for all of the "standard" units from 1 cent to $5.00. Have only the minimum number of beads on each wire necessary for the representations of numbers. (For example, only four beads are needed on the "cents" wire because 5 cents can be represented by one nickel.)

4. Make a drawing of an abacus for the Roman numeral system that can be used to represent all numbers from I to M. Include wires for each of the "standard" units I, V, X, and so on. Have only the minimum number of beads on each wire necessary to represent the numbers. (Keep the abacus simple; do not attempt to incorporate the "subtraction principle" of the Roman numeral system. For example, write *nine* as VIIII, not as IX.)

5. The drawings of the abacus in the text show 10 beads on each wire. Explain why 10 beads are not really necessary—why 9 beads on each wire will suffice.

6. *Puzzle Problem.* The following is the most famous problem in the book *Liber abaci*, published by Leonardo of Pisa in 1202: A pair of new born rabbits is given to a man. Assuming that each pair of rabbits at least two months old reproduces a new pair every month, and that no rabbits die, how many pairs of rabbits will be present at the end of 12 months?

I	1
V	5
X	10
L	50
C	100
D	500
M	1,000
$\bar{X}$	10,000
$\bar{C}$	100,000
$\bar{M}$	1,000,000

The Roman numeral system is similar in principle to the Egyptian system for writing numerals. The Roman system is more sophisticated, however, in that *subtraction* was incorporated. For example, the number *four* can be written as IIII, indicating *four* units, or as IV, indicating *one* unit less than *five*.

End of Month Number	Number of Pairs of Old Rabbits	Number of Pairs of New Rabbits	Total Number of Pairs of Rabbits
1	1	0	1
2	1	1	2
3	2	1	3
4	3	2	5
5	5	3	8
6	8	5	13
7	...	8	...
8	...	...	...
9	...	...	...
10	...	34	...
11	89	...	...
12	...	...	...

2-2 ADDITION OF WHOLE NUMBERS

In the ancient civilizations, only the cleverest men became scribes—learning to read and write, to keep accounts, and to calculate with numbers. These men had quick minds and looked for improvements and shortcuts in their work. Probably the most important shortcut was found in addition.

Consider the problem of adding 281 to 413. What this addition really means is that we successively add *one* unit to 413 two hundred eighty-one times. That is, we form the numbers $413 + 1$, $413 + 1 + 1$, and so on, eventually obtaining

$$413 + \underbrace{1 + 1 + 1 + \cdots + 1}_{281 \text{ terms}}$$

This gives us the sequence of numbers

$$414, 415, 416, 417, 418, \ldots, 694$$

The process stops when we reach the 281st *number, which is* 694.

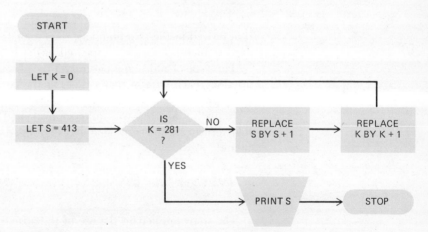

Figure 2-6 Flow chart for addition of 413 and 281 by repeated addition of 1.

The scribes realized that addition could be performed more quickly and efficiently by adding digits in columns. For example, in the Egyptian notation (see Fig. 2-7),

$$\text{99999} \cap \begin{smallmatrix} | & | \\ | \end{smallmatrix} \qquad (413)$$

plus $\quad \text{99} \begin{smallmatrix} \cap\cap\cap\cap \\ \cap\cap\cap\cap \end{smallmatrix} | \qquad (281)$

_____

equals $\quad \begin{smallmatrix} \text{999} \cap\cap\cap\cap\cap\cap | | \\ \text{999 } \cap\cap\cap\cap \text{ } | | \end{smallmatrix} \qquad (694)$

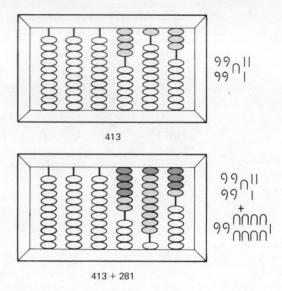

413

413 + 281

Figure 2-7 Addition on the abacus.

That is, they repeated the symbol for the "units" digit as many times as it appeared in the two numbers, then repeated the symbol for the "tens" digit as many times as it appeared in the two numbers, and so on.

This type of addition is natural for the abacus. We simply slide the beads up along each wire as many times as the corresponding beads appear in the representations of the two numbers.

Actually, this step was quite sophisticated in that it involved the fact that terms can be rearranged and regrouped. Using the modern notation + and =, we have

$$\left(\text{9999}\cap\text{|}\text{|}\right) + \left(\text{99}\cap\cap\cap\cap\text{|}\right) = (\text{999999}) + \left(\cap\cap\cap\cap\cap\right) + \left(\text{|}\text{|}\text{|}\right)$$

The steps implicit in the use of the abacus and the Egyptian notation are the same as the ones we use with the Hindu-Arabic numerals. To add 413 and 281 we write one number over the other and add columns. Just as with the Egyptian notation, we are essentially rearranging terms and regrouping:

$$(413) + (281) = (400 + 200) + (10 + 80) + (3 + 1)$$
$$= 600 + 90 + 4 = 694$$

413
281
  4
 90
600
———
694

The addition of 413 and 281 in Hindu-Arabic notation—expanded form of the algorithm

The standard technique for addition, illustrated in the margin, is called the *addition algorithm*. It ultimately rests on two basic principles. The first principle is that the order of writing two numbers is not important in addition. For example,

An *algorithm* is a simple arithmetical process that can be repeated several times to obtain a desired result. An example is the algorithm for the addition of two numbers by repeated addition of digits in columns.

$$5 + 7 = 7 + 5$$
$$8 + 2 = 2 + 8$$
$$183 + 147 = 147 + 183$$

and so on. It is customary to state general principles in mathematics by the use of symbols. In order to state this particular principle, we let A and B represent arbitrary integers. Then we can write this law as

$$A + B = B + A$$

This principle is known as the *commutative law for addition.*

The second principle is that the grouping of terms is not important for addition. For example,

$$3 + (2 + 5) = (3 + 2) + 5$$
$$12 + (9 + 13) = (12 + 9) + 13$$

and so on. In each example we have written the numbers in the same order, but have grouped them differently. If we let A, B, and C represent arbitrary integers, then we can state this principle in compact form as

$$A + (B + C) = (A + B) + C$$

This principle is known as the *associative law for addition.* The reader may be interested in working out all of the steps in the addition of 413 and 281, seeing when each of these laws was used. For a start, note that

$$
\begin{aligned}
413 + 281 &= (400 + 10 + 3) + (200 + 80 + 1) \\
&= (400 + 10) + (3 + 200) + (80 + 1) \quad \text{(regrouping)} \\
&= (400 + 10) + (200 + 3) + (80 + 1) \quad \text{(commuting the middle} \\
&\qquad\qquad\qquad\qquad\qquad\qquad\qquad\qquad \text{two terms)}
\end{aligned}
$$

The algorithm for addition by columns needs to be modified slightly from the way it was described above. If the numbers in any column add to 10 or more, it is necessary to "carry" the "tens" digit. Figure 2-8(a) shows the complete algorithm in expanded form. Figure 2-8(b) shows the standard abbreviated form of the algorithm. Figure 2-9 is a flow chart which can be used to verify the commutative law of addition in particular examples.

Figure 2-8 (a) Addition by repeated addition of digits in columns—expanded form of the algorithm; (b) addition by repeated addition of digits in columns—standard form of the algorithm.

(a)

(b)

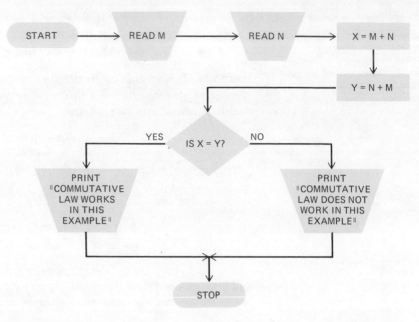

Figure 2-9 Flow chart to verify the commutative law for addition in particular examples.

EXERCISES

1. Make drawings of the "Egyptian" abacus, showing all of the basic steps in the addition of 791 and 632. Write out the equivalent steps using the Egyptian symbols.
2. Follow the instructions of Exercise 1 for the "Hindu" abacus.
3. Perform the following additions, using both the "expanded" form and the "standard" form of the addition algorithms as illustrated in Figure 2-8.
   (a) 18 + 41   (c) 738 + 159
   (b) 311 + 86   (d) 8294 + 6987
4. Work through all of the steps behind the standard algorithm for the addition of 79 and 63, showing exactly when the commutative and associative laws are used.
5. Give examples of two "standard" algorithms different from addition.
6. (a) Make a flow chart to verify the associative law for addition. [Check if $(A + B) + C = A + (B + C)$.]
   (b) Why can such a flow chart only be used to "verify" the associative law in examples? Why can't it be used to show that the associative law always works?
7. Work through the steps of the flow chart in Exercise 6 with the following examples:
   (a) $A = 2$, $B = 7$, $C = 13$
   (b) $A = 1$, $B = 3$, $C = 8$
8. Explain why $a + 0 = a$ for any number a.

2-3 MULTIPLICATION OF WHOLE NUMBERS

Basically, multiplication is repeated addition. When we multiply 6 by 7, we are adding 6 to itself 7 times. One of the shortcuts first noticed by the ancients was that the order makes no difference in multiplication. That is, 6 added to itself 7 times yields the same answer as 7 added to itself 6 times, in symbols

$$6 \cdot 7 = 7 \cdot 6$$

5 × 6, 5 · 6, (5)(6)
$a \times b, a \cdot b, (a)(b), ab$

In elementary arithmetic the symbol × is used to indicate a product. Because of its similarity to the symbol x, which is commonly used in algebra to indicate an unknown, the × is not generally used in more advanced courses. We indicate products by *raised dots,*

5 · 6 $a \cdot b$

by *grouping terms within parentheses,*

(5)(6) $(a)(b)$

or by simple *juxtaposition,*

ab

With these two numbers, there is no saving in time by multiplying in one order rather than the other. However, if we need to multiply 2 by 135 using repeated addition, we find that it is much easier to reverse the order and double 135 rather than adding 2 to itself 135 times; that is,

$$135 \cdot 2 = 2 \cdot 135$$

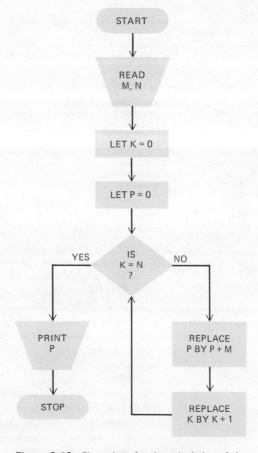

Figure 2-10 Flow chart for the calculation of the product MN by repeated addition.

Similarly,

$$17 \cdot 8 = 8 \cdot 17$$
$$1143 \cdot 82 = 82 \cdot 1143$$
$$8 \cdot 24 \cdot 9 = 9 \cdot 8 \cdot 24$$

and so on. Using the letters A and B to represent arbitrary integers, we have

$$A \cdot B = B \cdot A \qquad \text{(commutative law)}$$

This law is known as the *commutative law* of multiplication.

It also was noticed rather early that the grouping is unimportant when multiplying three or more numbers together:

$$8 \cdot (7 \cdot 3) = (8 \cdot 7) \cdot 3$$
$$142 \cdot (87 \cdot 92) = (142 \cdot 87) \cdot 92$$

and so on. In general, we have

$$A \cdot (B \cdot C) = (A \cdot B) \cdot C \qquad \text{(associative law)}$$

for any three integers A, B, and C. This principle is known as the *associative law* of multiplication.

The use of these laws of multiplication, along with one other rule that relates addition and multiplication—the distributive law, made it easy for

---

"Gelosia" Multiplication

"Gelosia" multiplication is one of several algorithms developed by the Hindus. The method is illustrated by the calculation of the product $728 \cdot 549$. One number is written across the top of the diagram and the other down the right side. Each digit of the first number is multiplied, in turn, by each digit of the second number, and the product is written in the appropriate square, the "tens" digit above the diagonal, the "units" digit below. For example, $4 \cdot 8 = 32$ is written in the third square of the middle row. Finally, the numbers are added along the diagonals, the first digit of a two-digit sum being carried to the next diagonal to the left. For example, the "bottom" diagonal contains only the number 2. The next diagonal contains 2, 7, and 8, which sum to 17. The 7 is written down, and the 1 is carried to the next diagonal.

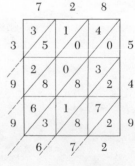

$$728 \cdot 549 = 399{,}672$$

the ancient Hindus to develop simple and efficient algorithms for multiplication. The *distributive law* states that we get the same result if we multiply a fixed number A by a sum $(B + C)$ or if we multiply A by B and by C separately and then add; that is,

$$A \cdot (B + C) = A \cdot B + A \cdot C$$

For example,

$$7 \cdot (8 + 2) = 7 \cdot 8 + 7 \cdot 2 = 70$$
$$12 \cdot (4 + 3) = 12 \cdot 4 + 12 \cdot 3 = 84$$
$$3 \cdot (3 + 1) = 3 \cdot 3 + 3 \cdot 1 = 12$$

and so on.

The Hindu algorithms eventually evolved into the one that we use today, which is illustrated in Figure 2-11. Figure 2-11(a) shows the modern algorithm as carried out by most people. Figure 2-11(b) shows the numbers that we actually get at each step. For example, in (a), when we get *4328 shifted one column to the left*, we actually have the number 43280 (the product of 541 and 80).

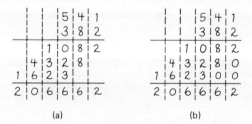

(a)   (b)

Figure 2-11 Two forms of the standard Hindu-Arabic algorithm for multiplication.

The algorithm essentially involves the following steps:

$$541 \cdot 382 = 541 \cdot (2 + 80 + 300)$$
$$= 541 \cdot 2 + 541 \cdot 80 + 541 \cdot 300$$
$$= 1082 + 43,280 + 162,300$$
$$= 206,662$$

Observe that the numbers that we add in order to get the final answer are precisely those that appear in the steps of the algorithm:

$$1,082 \quad 43,280 \quad 162,300$$

The Rhine papyrus (named after Henry Rhine, its first modern owner) is one of the oldest surviving mathematical documents. Written about 1650 B.C. by Ahmos, the scribe, it promised knowledge of all existing things and the clearing up of all mysteries. Actually, however, the papyrus was a practical textbook of elementary mathematics. It shows that the Egyptian scribes were able to develop ingenious methods to solve simple mathematical problems in spite of their cumbersome notation for writing numbers.

Egyptian Multiplication

Our standard multiplication algorithm depends heavily on the Hindu-Arabic place-number system for writing numbers. The following example illustrates the technique used by the ancient Egyptians for multiplication before the place-number system was invented. The method depends on the two operations that can be done easily on the abacus—addition and doubling. Variations of this method were used in parts of Europe until modern times.

The Egyptian method is illustrated by the calculation of 12 times 12. This illustration is taken from the Rhine papyrus. The first step is to write the numbers 1, 2, 4, 8, and so on, in successive lines. From this list pick the numbers that add to 12 (4 and 8) and mark them with oblique lines. Now successively double the other given number (also equal to 12) and write the numbers obtained (12, 24, 48, 96, and so on) opposite the corresponding numbers in the first list. Finally, add the multiples of 12 written opposite the oblique lines, obtaining

$$48 + 96 = 12 \cdot 4 + 12 \cdot 8$$
$$144 = 12 \cdot 12$$

Egyptian notation

Modern notation

12	1	
24	2	
48	4	/
96	8	/

$$48 + 96 = 144$$
$$4 \cdot 12 + 8 \cdot 12 = 144$$
$$12 \cdot 12 = 144$$

Adapted from the Rhine papyrus: written about 1750 B.C. by Ahmos, the scribe.

EXERCISES

1. Multiply
   (a) $17 \cdot 82$ (d) $104 \cdot 276$
   (b) $23 \cdot 46$ (e) $562 \cdot 831$
   (c) $181 \cdot 63$ (f) $114 \cdot 707$

2. Verify in the following examples that multiplication can be considered as repeated addition. For example, in (a) verify that the product $5 \cdot 4$ is equal to 5 added to itself 4 times. Is this verification sufficient to establish the result for all pairs of integers? Explain!
   (a) $5 \cdot 4$ (c) $7 \cdot 6$
   (b) $3 \cdot 2$ (d) $20 \cdot 10$

3. (a) Construct a flow chart that can be used to verify the associative law of multiplication in particular examples; that is, verify that

$$A(BC) = (AB)C$$

    (b)   Work through the steps of your flow chart with the following examples:
        (1) $A = 3$, $B = 7$, $C = 9$
        (2) $A = 8$, $B = 0$, $C = 8$
        (3) $A = 1$, $B = 1$, $C = 10$

4. (a)   Construct a flow chart outlining all of the steps in the algorithm needed to multiply two two-digit numbers A and B.

   (b)   Work through the steps of your flow chart with the following examples:
        (1) $A = 23$, $B = 17$
        (2) $A = 10$, $B = 68$

5. Multiplication can be applied in various ways. For example, the *area of a rectangle* is the product of the *base length* and the *height*. The *distance* traveled at a constant rate of speed is the product of the *rate* and the *time*. Think of two other meaningful applications of multiplication.

6. The extract entitled "Egyptian Multiplication" shows by example how the Egyptians were able to do multiplication without the convenience of the Hindu-Arabic notation. Test your skill by using the method to multiply 13 by 19.

2-4 THE NUMBER LINE; ORDER; ADDITION AND MULTIPLICATION REVISITED

Most of us know how to associate whole numbers with points on a line. The standard way to do this is by marking off equally spaced points on the line. One point, chosen at random, is called "0." The first point to the right is called "1," the next is "2," the next "3," and so on. This gives us a pictorial representation of the positive integers and zero. (See Fig. 2-12.)

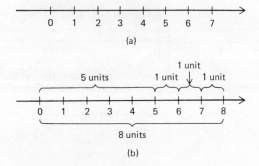

Figure 2-12 (a) The number line. The arrow indicates the direction in which the numbers increase (the *positive* direction). The opposite direction is called the *negative* direction; (b) distances on the number line.

This expanded set of whole numbers is called the system of *nonnegative integers*.

When we use the number line, we work with distances as well as points. Any two consecutive points that we have marked are one unit apart. The distance from the point "0" to any other marked point, say "A," is exactly A units. Thus, the point marked "5" is 5 units from "0," the point "3" is 3 units from "0," and so on.

Addition

Addition can be interpreted in terms of distances on the number line. To add 3 and 2, for example, we first measure *three* units from "0" in the positive direction (obtaining the point "3"), and then measure *two* units further, stopping at the point "5." (See Fig. 2-13.)

The general problem of adding two or more arbitrary numbers can be handled similarly. If *a* and *b* are any two nonnegative integers, we obtain the point represented by *a* + *b* by first measuring a distance of *a* units from "0" and then measuring *b* units further.

Ren. Cartesius

Figure reprinted with the permission of Open Court Publishing Company. *The Geometry of René Descartes*, tr. Latham and Smith, 1925, p. 3.

René Descartes (1596–1650) was the first person to consider the properties of the number line. In *La geometrie*, published in 1637, he developed his main thesis: "Any problem in geometry can easily be reduced to such terms that a knowledge of lengths of certain lines is sufficient for its construction."

Descartes' number line was quite different from the one that we use. As stated above, he was mainly interested in finding general algebraic methods for solving problems in geometry. His first step was to draw two reference lines in the plane and locate points by measuring distances parallel to the two fixed lines. Sophisticated use of these numbers enabled Descartes to solve a number of very difficult problems in geometry. Over a hundred years of refinements were necessary before the number line was used in its present simple form.

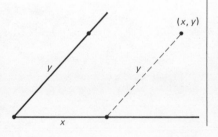

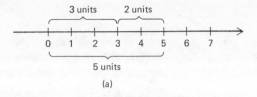

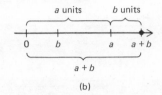

(b)

Figure 2-13 Addition on the number line. (a) Special case: $3 + 2 = 5$. To add 3 and 2, we first measure a distance of 3 units, then a distance of 2 units, resulting in a total distance of 5 units; (b) general case: To add a and b, we first measure a distance of a units, then a further distance of b units, resulting in a total distance of $a + b$ units.

Over the centuries one of the major difficulties in mathematics has been the lack of good symbolism and notation needed to make general arguments. The ancient Greeks made very general arguments in geometry by allowing letters to represent arbitrary points. To us it seems natural to take the next step and use letters to represent arbitrary numbers. This step was not taken, however, until the sixteenth century, 2000 years after the Greeks.

When we use letters such as a and b to represent numbers as in the discussion above, we mean that these symbols can stand for any members of the set under consideration. Thus, in the discussion of addition, a and b represent arbitrary nonnegative integers. It is understood, of course, that the numbers represented by a and b do not change during the discussion. That is, a is an arbitrary but fixed nonnegative integer; b is an arbitrary but fixed nonnegative integer that may be the same as a or may be different. The use of letters to represent arbitrary numbers allows us to get to the heart of mathematical arguments without getting involved with properties of specific numbers.

The commutative property of addition has a natural interpretation in terms of distance. Recall that $a + b$ is obtained by first moving a units, then b units, while $b + a$ is obtained by first moving b units, then a units. It is obvious to anyone who has ever traveled that the same total distance is obtained in either case (Fig. 2-14); that is, $a + b = b + a$.

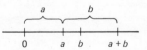

Figure 2-14 Geometrical interpretation of the commutative law of addition: $a + b = b + a$.

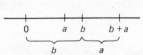

The associative property of addition also can be interpreted geometrically. Observe that $(a + b) + c$ is the number obtained by first adding a to b and then adding c to the result. The number $a + (b + c)$, on the other hand, is obtained by adding a to the sum of b and c. The reader should make his own diagram, based on these remarks, to illustrate that $(a + b) + c = a + (b + c)$. (See Exercise 2.)

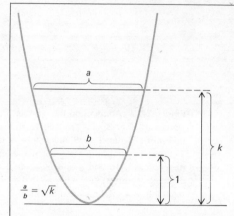

The basic idea of the number line was anticipated by the ancient Greeks (400–200 B.C.), but was never fully developed by them. The great

Greek geometer *Apollonius* was aware of many relationships of the type shown in the figure.

The "cup-shaped" curve is a *parabola*, a curve that can be drawn by mechanical methods. Suppose we mark off vertical distances of 1 unit and k units as shown, then measure across the curve at right angles, obtaining the distances a and b. The Greeks were able to show that the ratio a/b is equal to the square root of k. Such methods enabled the Greeks to estimate numbers, such as square roots, that they needed in their engineering work but could not calculate exactly by arithmetical methods.

Equality

Two numbers a and b are equal only in case they are represented by the same point on the number line. Thus, $2 + 3$ and 5 are equal since they are represented by the same point (Fig. 2-15).

Order

If two numbers a and b are not equal, then one of them must be to the right of the other on the number line. If b is to the right of a (that is, b is on the positive side of the number line from a), we say that b is *greater than* a and that a is *less than* b. We indicate this by either of the symbols

$$a < b \qquad (a \text{ is less than } b)$$

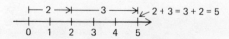

Figure 2-15 Numbers are equal only in case they are represented by the same point.

or

$$b > a \qquad (b \text{ is greater than } a)$$

For any numbers a and b, exactly one of the following must hold: Either a is to the left of b, or a and b are represented by the same point, or a is to the right of b. Thus, we have

$$a < b \qquad \text{or} \quad a = b \qquad \text{or} \quad a > b$$

The above argument is typical of those that can be used to establish simple properties of order. Actually there are even simpler interpretations of order that can be used to verify these properties and to help us remember them. Since monetary transactions are common to everyone, an interpretation in terms of money may be helpful.

Let a and b represent the amounts of money possessed by two rich men, say Aaron and Bernard. The rule

$$a < b \qquad \text{or} \quad a = b \qquad \text{or} \quad a > b$$

can be interpreted as saying that either Aaron is poorer than Bernard, or they are equally wealthy, or Aaron is wealthier than Bernard. Suppose we now give the same amount of money to each man. It is easy to see that the same order relationship must hold. For example, if

$$a < b \qquad \text{(amounts before gift)}$$

and c is the amount given each man then

$$a + c < b + c \qquad \text{(amounts after gift)}$$

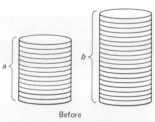

Before

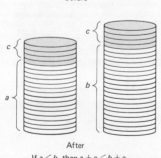

After

If $a < b$, then $a + c < b + c$

A number of properties of order can be remembered by arguments of this type. Three of the most important are

(1) Either $a = b$ or $a < b$ or $a > b$.
(2) If $a < b$, then $a + c < b + c$.
(3) If $a < b$ and c is a *positive* number, then $ac < bc$.

On occasion we need to specify that b is at least as large as a. In other words, we need to indicate that b may be equal to a or greater than a, but definitely is not smaller than a. If this is the case, we write

$$a \leq b \qquad \text{or} \quad b \geq a$$

and we say that *a is less than or equal to b* (*b is greater than or equal to a*).

Example.
(a) $7 \geq 5$ since $7 > 5$.
(b) $3 \geq 3$ since $3 = 3$.

Multiplication

There are several methods that can be used to represent multiplication geometrically on the number line, but they are not very easy to understand. Actually, we can understand multiplication easiest in terms of area. We draw a second number line through "0" perpendicular to the original one. The same scale is used on both lines, and the zero points coincide. The two lines are called *axes*. It is standard to draw the second axis with the positive numbers above the original axis.

Recall that if each side of a square is equal in length to *one unit*, then the *area* of the square is *one square unit*. Similarly, if one side of a rectangle is a units and the other side is b units, then the *area* is *ab square units*. Thus, we can interpret a product ab to be the area of a rectangle with base a and height b (Fig. 2-16).

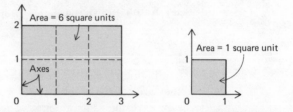

Figure 2-16 The unit square has an area of 1 square unit. The rectangle with sides of length 2 and 3 units has an area of 6 square units.

Observe that the commutative law for multiplication, $ab = ba$, now has a simple interpretation. We start with a rectangle of base a and height b having an area of ab. We rotate the rectangle one quarter turn so that it has base b and height a (Fig. 2-17). The area is unchanged, but is now seen to be equal to ba. Consequently,

$$ab = ba$$

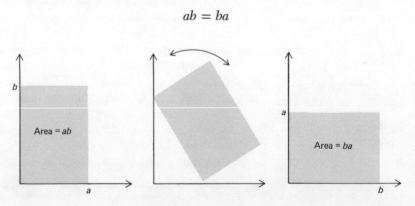

Figure 2-17 Geometrical interpretation of the commutative law for multiplication: $ab = ba$.

Similarly, the distributive law

$$a(b + c) = ab + ac$$

has a simple geometrical idea behind it. Start with a rectangle with base a and height equal to $b + c$. The area is equal to $a(b + c)$. Now decompose the large rectangle into two small rectangles, one of area ab and the other of area ac, as in Figure 2-18. The total area has not changed, so that the total area $= a(b + c) =$ sum of the areas of the small rectangles $= ab + ac$.

The ancient Greeks (500 B.C.–500 A.D.) were the first people to develop a clear philosophical concept of number. Unfortunately, their concept was so limited that it only applied to whole numbers and their ratios, making it impossible to deal with many basic problems of algebra. To get around this restriction, they artificially "geometrized" their algebra and arithmetic, expressing concepts that are arithmetical, such as the distributive law, in terms of geometrical concepts, such as area. The diagram in Figure 2-18 is similar to one used by Euclid to illustrate the distributive law.

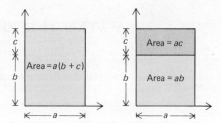

Figure 2-18 Geometrical interpretation of the distributive law: $a(b + c) = ab + ac$.

EXERCISES

1. Draw a diagram similar to Figure 2-13 to illustrate geometrically the process of adding 3 and 4.
2. Figure 2-14 illustrates the commutative law of addition. Draw a diagram to illustrate the associative law of addition: $a + (b + c) = (a + b) + c$.
3. Let a, b, c, and d represent the weights of Aaron, Bernard, Claude, and Darius, respectively. Interpret the following statements (which may or may not be true) in terms of this concept. Then decide if the statements must necessarily be true. (*Hint:* $a < b$ says that Aaron weighs less than Bernard; $a + c < b + d$ says that the combined weight of Aaron and Claude is less than the combined weight of Bernard and Darius, and so on.)
   (a) If $a < b$ and $b < c$ and $c < d$, then $a < d$.
   (b) If $a < b$, then $2a < 3b$.
   (c) If $a < b$ and $c < d$, then $a + c < b + d$.
4. Multiplication of three numbers a, b, and c can be considered geometrically as the calculation of the volume of a rectangular box with sides of length a, b, and c (Fig. 2-19). Show how the associative law $a(bc) = (ab)c$ can be interpreted in terms of volume.

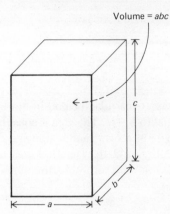

Volume = abc

Figure 2-19 The *volume* of a rectangular solid is equal to the product of the *length, width,* and *height.*

2-5 SUBTRACTION OF WHOLE NUMBERS

Addition and subtraction are like the opposite sides of a single coin. In the process of *addition*, we find the sum of two given numbers. In *subtraction* we find the number we must add to one number in order to get another number for the sum.

In other words, given two numbers a and b, with $a \leq b$, we ask, "What number x do we add to a in order to get b?" This number x is called the *difference* of b and a, written

$$x = b - a$$

Thus, $b - a$ is the number we add to a in order to get b. We also say that x is obtained by *subtracting a from b*. For example, since $5 + 3 = 8$, then

$$8 - 5 = 3$$

and

$$8 - 3 = 5$$

In the physical world, subtraction of whole numbers is equivalent to removing objects from a given set. If b is the number of objects in the original collection and we take a objects away, then $x = b - a$ is the number of objects left over.

We can represent subtraction geometrically on the number line by reversing some of the steps in addition. Recall that to subtract a from b, we find the number x that must be added to a in order to get the sum b. We can do this geometrically by first locating the point b and then measuring back a distance of a units in the *negative* direction. This gives us the point $x = b - a$ (Fig. 2-20).

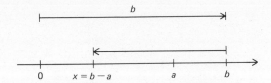

Figure 2-20 Subtraction of the number line. To find the number, $b - a$, we measure a units in the negative direction from b.

For example, to calculate $7 - 5$ geometrically, we measure 7 units in the positive direction, then back 5 units in the negative direction, obtaining the number 2 as a final answer.

Subtraction does not obey most of the laws that hold for addition. For example, subtraction is not commutative. That is, in general $a - b$ is not equal to $b - a$. This is quite easy to see in our present setting, where we deal only with nonnegative integers, because in most cases only one of the numbers $a - b$ and $b - a$ even exists as a nonnegative integer. (As far as we are concerned at this point, if $b > a$, then $b - a$ exists, but $a - b$ does not. When we consider negative numbers in the next chapter, we will see that even in this expanded system $a - b$ and $b - a$ are usually different numbers.)

Subtraction is not associative either. In general, $a - (b - c)$ and $(a - b) - c$ are not equal. This can be seen by calculating both $a - (b - c)$ and $(a - b) - c$ using almost any three numbers a, b, and c. (See Fig. 2-21.)

Among the familiar rules for addition, only the distributive property holds for subtraction; that is,

$$a(b - c) = ab - ac \qquad \text{if} \quad b \geq c$$

This can easily be verified in particular examples by calculating $a(b - c)$ and $ab - ac$ separately.

Example. Verify that

$$a(b - c) = ab - ac \qquad \text{when} \quad a = 7, \quad b = 3, \quad c = 1$$

We calculate

$$b - c = 3 - 1 = 2$$
$$a(b - c) = 7 \cdot 2 = 14$$

Also

$$ab - ac = 21 - 7 = 14$$

Thus, $a(b - c) = ab - ac$ in this case.

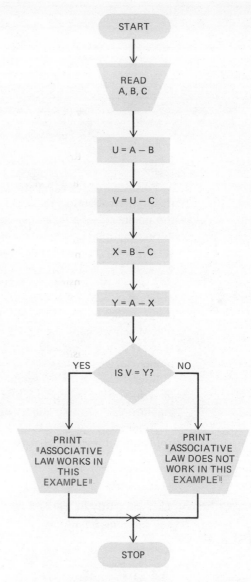

Figure 2-21 Flow chart to check if subtraction is associative. Need A $\geq$ B + C, B $\geq$ C.

To see why the distributive property holds in general, we consider multiplication in terms of area. First, we construct a large rectangle with base a and height c. Now mark off a small rectangle interior to this one with base a and height b as in Figure 2-22. If we let $x = c - b$, then x is the height of the other small rectangle.

The total area = ac = sum of areas of small rectangles = $ab + ax$.

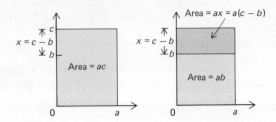

Figure 2-22 $a(c - b) + ab = ac;\ a(c - b) = ac - ab$.

But this means that ax is the number we add to ab in order to get ac so that ax is the difference of ac and ab,

$$ax = ac - ab$$

If we now recall that $x = c - b$, we have

$$ax = a(c - b) = ac - ab$$

which is what we wanted to establish.

STEP BY STEP IN
SUBTRACTION
(borrowing not necessary)

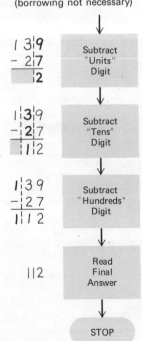

Algorithm for Subtraction

There is a simple algorithm for subtraction, which, essentially, enables us to reverse the steps in the algorithm for addition. Recall that when we add two numbers, we add the digits in columns, progressing from right to left, "carrying" the first digit to the next column whenever the column sum is a two-digit number.

To subtract one number from another, we write the smaller number under the larger, lining up the digits in columns from right to left. We then subtract the digits in columns, progressing from right to left. The figures in the margin illustrate the three steps needed to subtract 27 from 139.

The algorithm becomes more complicated when we need to subtract a large digit from a smaller one. To help with the discussion, we shall at times separate the digits of a number, putting them in boxes. Thus, we write 19 as $\boxed{1}\ \boxed{9}$ and 35 as $\boxed{3}\ \boxed{5}$. Similarly, we also can write $\boxed{2}\ \boxed{15}$ to indicate the number $2 \cdot 10 + 15 = 20 + 15 = 35$.

Consider the problem of subtracting $\boxed{1}\ \boxed{9}$ from $\boxed{3}\ \boxed{5}$. Recall that $\boxed{3}\ \boxed{5}$ is $30 + 5$. The first step is to rewrite $\boxed{3}\ \boxed{5}$ as $20 + 15$, or in our new notation, $\boxed{2}\ \boxed{15}$. We first subtract the 9 from the 15 and then the 1 from the 2, obtaining $\boxed{1}\ \boxed{6} = 16$ for the final answer:

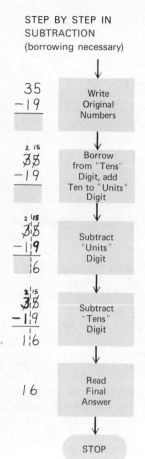

STEP BY STEP IN
SUBTRACTION
(borrowing necessary)

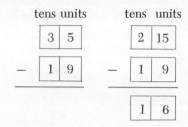

This technique of adding 10 to one digit and subtracting 1 from the digit to the left is known as "borrowing."

Subtraction of larger numbers is handled by the same procedure. We subtract the digits in columns, progressing from right to left, borrowing where we need to.

EXERCISES

1. Subtract
   (a) $46 - 22$ (d) $182 - 83$ (g) $6843 - 5854$
   (b) $84 - 51$ (e) $287 - 198$ (h) $3702 - 3583$
   (c) $72 - 37$ (f) $16,342 - 8534$ (i) $4003 - 1995$

2. (a) By example show that subtraction is not associative; that is, find an example where $a - (b - c)$ and $(a - b) - c$ are different numbers.
   (b) Why is one single example sufficient to show that this law does not generally hold for subtraction?

3. Construct a flow chart that can be used to verify the distributive law for subtraction in particular examples. [Show that $A \cdot (B - C) = AB - AC$.]

4. (a) Construct a flow chart outlining the steps in the algorithm needed for the subtraction of one two-digit number from a different two-digit number.
   (b) Work through the steps of your flow chart with the following examples:
       (1) $63 - 47$
       (2) $84 - 74$
       (3) $53 - 51$

5. Decide which of the following relationships hold:
   (a) If $a < b$, then $a - c < b - c$.
   (b) If $a < b$ and $c < d$, then $a - c < b - d$.
   (c) If $a < b$ and $c < d$, then $a - d < b - c$.

6. Explain why $a \cdot 0 = 0$ for any number a. [*Hint*: $a \cdot 0 = a(1 - 1) = a \cdot 1 - a \cdot 1 = a - a$.]

2-6 REASONABLE ESTIMATES; ACCURACY

In our everyday work with mathematics, we usually estimate quantities that we need rather than calculate their exact values. In many cases these estimates are as close to the actual quantities needed as the amounts that we could compute. As a simple application, consider the following example.

Example 1. A chef wishes to modify a recipe for 12 people in order to feed 15 people. Among other ingredients the original recipe called for 3 beaten eggs and $\frac{3}{4}$ teaspoon of salt. How many eggs and how much salt should be included in the modified recipe?

Solution. The modified recipe must serve $\frac{5}{4}$ times as many persons as the original one. Thus, he must change the ingredients of the recipe by a factor of $\frac{5}{4}$. This means that every quantity listed in the original recipe must be multiplied by $\frac{5}{4}$.

If we multiply 3 by $\frac{5}{4}$, we get $\frac{15}{4} = 3\frac{3}{4}$ for the number of beaten eggs in the new recipe. It is ridiculous to try to put exactly $3\frac{3}{4}$ beaten eggs into a recipe. First of all, the measurement "one egg" is only approximate since eggs vary in size. Thus, $3\frac{3}{4}$ beaten eggs of one size may be equal to 5 beaten eggs of a smaller size or perhaps 3 beaten eggs of a larger size. Consequently, the cook probably will put in 4 beaten eggs and forget about the exact measurements.

The problem is a little different when it comes to the salt. If we multiply $\frac{3}{4}$ by $\frac{5}{4}$, we get $\frac{15}{16}$ teaspoon of salt. This can be measured accurately, but probably it is not worth the trouble to do so. Instead, our cook will put in a little less than one teaspoon of salt and then turn his attention to the other ingredients.

This type of situation in which reasonable estimates are as good as computed values occurs often in our practical work with mathematics. This is quite different from scientific work in which accurate measure is essential. Even scientists, however, ultimately rely on estimates. They simply set up rather strict requirements for the accuracy of their estimates.

Notation: We write $a \approx b$ if a and b are numbers that are approximately equal.

Geometrically, $a \approx b$ if the points representing a and b on the number line are close to each other. (Usually, we take this to mean that the two points are within some preassigned distance of each other.)

For example,

$$27 \text{ feet}, \tfrac{1}{4} \text{ inch} \approx 27 \text{ feet}$$
$$51.003 \approx 51$$
$$1.99997 \approx 2$$

This notation is convenient, but may be misleading at times. Unfortunately, we have no clear-cut standard that can be used to decide if two numbers are approximately equal or not. It depends on our choice of scale and the accuracy needed for the problem. For example, consider the problem of whether New York and Los Angeles are approximately the same distance from Chicago. A motorist would certainly say *no*—they are not. An astronomer observing Earth from another planet, on the other hand, might argue that the two cities are approximately the same distance from New York, each distance being a tiny part of a light year.

Often we need *large* and *small* estimates for quantities. We say that *a* is a *lower (small) estimate* for *b* if

$$a \approx b \quad \text{and} \quad a < b$$

Under the same conditions, we say that *b* is an *upper (large) estimate* for *a*.

For example, π represents the ratio of the circumference of a circle to the diameter. It is known that π is between 3.14 and 3.15. Thus, 3.14 is a lower estimate for π, while 3.15 is an upper estimate. Similarly, 2 is an upper estimate for 1.997; 51 is a lower estimate for 51.3, and so on.

One way to make the concept of approximation precise is to set up a "tolerance" range about the number *a*; that is, we mark off an interval of a fixed, predetermined size on the number line with *a* at its center. If *b* is in this interval, then $a \approx b$.

For example, if we mark off an interval of length $\frac{1}{5}$ of a unit with the number π at its center, then 3.1 and 3.2 are both in the interval. Thus, it makes sense to write

$$\pi \approx 3.1 \quad \text{(small estimate)}$$

and

$$\pi \approx 3.2 \quad \text{(large estimate)}$$

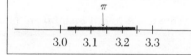

In many problems we only need quick and reasonable estimates, but we must make sure that the estimated values are either larger or smaller than the actual quantities we could compute. A contractor preparing to order cement, for example, may need a quick estimate that is larger than the amount he actually needs. He can easily dispose of the excess cement, but may be in trouble if he does not have enough. On the other hand, a cook putting a potent seasoning, such as curry powder, in a dish might do well to estimate on the small side and put in too little rather than too much.

Example 2. Find upper and lower estimates for the product $(49.7) \cdot (51.6)$.

Solution. $49.7 \approx 50$ (upper estimate) and $51.6 \approx 52$ (upper estimate). When we multiply together two numbers that are larger than the original numbers their product is larger than the product of the original numbers. Thus,

$$(49.7)(51.6) \approx 50 \cdot 52 = 2600 \qquad \text{(upper estimate)}$$

On the other hand,

$$49.7 \approx 49 \qquad \text{(lower estimate)}$$

and

$$51.6 \approx 51 \qquad \text{(lower estimate)}$$

so that

$$(49.7)(51.6) \approx 49 \cdot 51 = 2499 \qquad \text{(lower estimate)}$$

Thus, $(49.7)(51.6)$ is a number between 2499 and 2600.

We can obtain a better estimate than the one in Example 2 by noting that the product will not change too much if we increase one quantity and decrease the other. Thus,

$$49.7 \approx 50 \qquad \text{(upper estimate)}$$

and

The number of objects in a large collection can be estimated by partitioning the collection into small "typical" subsets of approximately the same size, counting the number of objects in one of the subsets, then multiplying this number by the number of subsets.

The small rectangle in the upper right-hand corner of the large rectangle has 12 dots in it. Consequently, the total number of dots in the figure is approximately equal to $9 \cdot 12 = 108$.

This idea is used by managers of wildlife preserves to estimate animal populations, by doctors to make blood-cell counts, and so on.

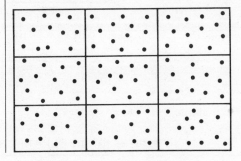

$$51.6 \approx 51 \qquad \text{(lower estimate)}$$

so that

$$(49.7)(51.6) \approx 50 \cdot 51 = 2550$$

If we use this method, it is not apparent whether the estimate is larger or smaller than the actual product. It usually yields a closer approximation, however, than the method using upper and lower estimates.

Powers of Ten

Powers of 10 can be quite useful in estimating products. Recall that we define powers of 10 by

$$10^0 = 1$$

$$10^1 = 10$$

$$10^2 = \underbrace{10 \cdot 10}_{2 \text{ factors}} = 100$$

$$10^3 = \underbrace{10 \cdot 10 \cdot 10}_{3 \text{ factors}} = 1000$$

$$10^4 = \underbrace{10 \cdot 10 \cdot 10 \cdot 10}_{4 \text{ factors}} = 10{,}000$$

and so on.

Thus, 10^n is 10 multiplied by itself n times, which is a number written as a 1 followed by n zeros. It is standard to define 10^0 to be 1.

Now $10^m \cdot 10^n = 10^{m+n}$ for any two nonnegative integers m and n. This is obvious if we write out the product in expanded form:

RULES OF EXPONENTS

$a^m \cdot a^n = a^{m+n}$

$(a^m)^n = a^{mn}$

$(a \cdot b)^m = a^m \cdot b^m$

$$10^m \cdot 10^n = \underbrace{10 \cdot 10 \cdot \,\cdots\, \cdot 10}_{m \text{ factors}} \cdot \underbrace{10 \cdot 10 \cdot \,\cdots\, \cdot 10}_{n \text{ factors}}$$

$$= \underbrace{10 \cdot 10 \cdot \,\cdots\, \cdot 10 \cdot 10 \cdot 10 \cdot \,\cdots\, \cdot 10}_{m + n \text{ factors}}$$

$$= 10^{m+n}$$

For example,

$$10^3 \cdot 10^4 = \underbrace{10 \cdot 10 \cdot 10}_{3 \text{ factors}} \cdot \underbrace{10 \cdot 10 \cdot 10 \cdot 10}_{4 \text{ factors}}$$

$$= \underbrace{10 \cdot 10 \cdot 10 \cdot 10 \cdot 10 \cdot 10 \cdot 10}_{7 \text{ factors}} = 10^7$$

or, using the usual notation,

$$\underbrace{1{,}000}_{3 \text{ zeros}} \cdot \underbrace{10{,}000}_{4 \text{ zeros}} = \underbrace{10{,}000{,}000}_{7 \text{ zeros}}$$

When we need to estimate products of large numbers or numbers that are quite different in size, we frequently replace the numbers by one- or two-digit numbers multiplied by powers of 10.

Example 3. Use powers of 10 to estimate the product (8,017,318)(6,981).

Solution.
$$8{,}017{,}318 \approx 8 \cdot 10^6$$
$$6{,}981 \approx 7 \cdot 10^3$$

Thus,

$$(8{,}017{,}318)(6{,}981) \approx (8 \cdot 10^6)(7 \cdot 10^3)$$
$$= 8 \cdot 7 \cdot 10^6 \cdot 10^3 = 56 \cdot 10^9$$
$$= 56{,}000{,}000{,}000$$

There is no such thing as a ruler or other scale that is absolutely "true" in a mathematical sense. All such devices have errors built into them. For this reason all measurements involve approximations.

When we measure a quantity, there usually is an implicit understanding that the measurement is correct to some standard of accuracy. We want to measure correct to the closest whole number, or the closest multiple of 10, or the closest 16th of an inch, or some other quantity. Also we always choose our estimates so that they are as simple as possible to write and to use. Thus, we might estimate a quantity that is exactly $1208\frac{1}{2}$ inches to be 100 feet, but we would never estimate a quantity that is exactly 100 feet to be $1208\frac{1}{2}$ inches. In all cases we desire simplicity.

There is one standard convention for writing approximations that should be mentioned. Normally *we use the last nonzero digit to determine the degree of accuracy.* Thus, if we write that a weight is 80 pounds, we mean that it is greater than or equal to 75 pounds and less than 85 pounds. If we want to specify that it is between 79.5 and 80.4 pounds, we use a decimal point and write it as 80. pounds. Similarly, a distance written as 17,000 feet indicates a distance greater than or equal to 16,500 feet and less than 17,500 feet.

Example 4. A measurement of 763 indicates a quantity between $762\frac{1}{2}$ and $763\frac{1}{2}$. A measurement of 763. indicates the same degree of accuracy. A measurement of 800 indicates a quantity between 750 and 850. A measurement of 800. indicates a quantity between $799\frac{1}{2}$ and $800\frac{1}{2}$.

EXERCISES

1. Calculate the upper and lower estimates for the product; then approximate the product by using a small estimate for one factor and a large one for the other.

    (a) $9 \cdot 11$ (d) $(17.31)(14.67)$

    (b) $38 \cdot 52$ (e) $(49.5)(80.5)$

    (c) $101 \cdot 44$ (f) $(200.3)(100.2)$

2. One method of estimating quantities is to take the average of an upper and a lower estimate. If we use the figures of Example 2, we find that the product $(49.7)(51.6)$ is a number between 2499 and 2600. Thus,

$$(49.7)(51.6) \approx \frac{2499 + 2600}{2} \approx 2550$$

Use this method to estimate the products in Exercise 1.

3. Use powers of 10 to approximate the product.

    (a) $301 \cdot 699$ (c) $(50,132)(417,863)$

    (b) $511 \cdot 989$ (d) $(22,864)(893,654,277)$

4. George Poorpauper is constructing a concrete patio in his backyard. Since he is doing all of the work himself, the dimensions of the patio are not as accurate as they should be. The patio is approximately rectangular with length about $23\frac{1}{2}$ feet and width about $17\frac{1}{2}$ feet. Before pouring the concrete, he needs to spread gravel 4 inches thick inside the constructed forms. How much gravel should he order? Should he order too much in preference to not ordering enough?

5. Professor Crankcase, a mediocre scholar, has no understanding of approximate numbers. He has just published a book in which he casually mentions that the distance from the Earth to the star Alpha Centauri is 25,262,972,783,280 miles. To obtain this result, he first multiplied the distance 4.3 light years (the distance from the Earth to the star as quoted in an almanac) by the number of seconds in a year $(60 \cdot 60 \cdot 24 \cdot 365\frac{1}{4})$, and then multiplied this number by the distance light travels in 1 second (almanac value: 186,171 miles). Discuss the accuracy of his result. (*Hint:* What does it really mean to say that the distance is 4.3 light years? What accuracy is implied by this number? What accuracy is implied by the professor's number?)

2-7 EXACT DIVISION OF WHOLE NUMBERS

Let a represent a fixed nonnegative integer. The numbers

$$a, 2a, 3a, 4a, 5a, \cdots$$

are called the *multiples of a*. We also include 0 as a multiple of a, because $0 = 0 \cdot a$. If b is one of these multiples, we say that *a divides into b exactly* and that *a is a factor of b*. (See Fig. 2-23.)

We indicate that b is a multiple of a, say $b = na$, by writing

$$b \div a = n$$

Figure 2-23 The multiples of *a*. If *b = na*, then *a* divides *b* exactly.

or

$$\frac{b}{a} = n$$

For example, the multiples of 5 are

$$0, 5, 10, 15, 20, 25, \cdots$$

One of these multiples is $10 = 2 \cdot 5$. Thus,

$$10 \div 5 = 2 \quad \text{and} \quad \frac{10}{5} = 2$$

In general, the symbol b/a represents the number we must multiply by *a* in order to get *b* as the product; that is,

$$\frac{b}{a} = n \quad \text{only in case} \quad na = b$$

The practical problem, of course, is to figure the value of $n = b/a$ when we are given *b* and *a*. In this chapter we consider only the case where b/a is a whole number. In the next chapter we shall extend the concept to include more general fractions. The process of calculating b/a when it is an integer is called *exact division*.

One way of working out exact division is by repeated subtraction. We repeatedly subtract *a* from *b* until we get a difference that is smaller than *a*. If the final difference is zero, then b/a is the number of times we have subtracted *a*. If the final difference is not zero, then *a* does not divide *b* exactly. The flow chart in Figure 2-24 shows exact division of B by A by repeated subtraction.

Example 1. Calculate 70/14 by repeated subtraction.

Exact division reverses the process of multiplication. In multiplication we are given *a* and *b* and are asked, "What is the value of the product $c = ab$?" In exact division we are given *a* and *b* and are asked, "What number *x* do we multiply by *a* in order to get *b*?"

Solution.

$$
\begin{array}{ll}
70 - 14 = 56 & \text{(one subtraction)} \\
56 - 14 = 42 & \text{(two subtractions)} \\
42 - 14 = 28 & \text{(three subtractions)} \\
28 - 14 = 14 & \text{(four subtractions)} \\
14 - 14 = 0 & \text{(five subtractions)}
\end{array}
$$

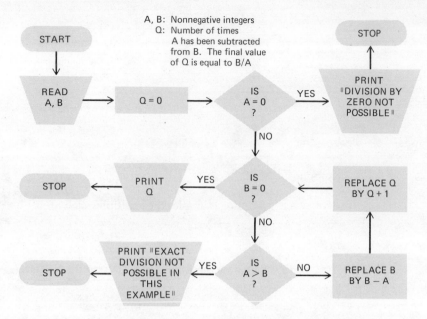

Figure 2-24   Flow chart for exact division of B by A by repeated subtraction.

Since $70 - 5 \cdot 14 = 0$, then

$$70 = 5 \cdot 14$$

and so

$$\frac{70}{14} = 5$$

Example 2.   Use repeated subtraction to show that 8 does not divide 59 exactly.

Solution.

$8 \cdot 1 = 8$
$8 \cdot 2 = 16$
$8 \cdot 3 = 24$
$8 \cdot 4 = 32$
$8 \cdot 5 = 40$
$8 \cdot 6 = 48$
$8 \cdot 7 = 56$
$\longleftarrow 59$
$8 \cdot 8 = 64$
$8 \cdot 9 = 72$

$59 - 8 = 51$	(one time)
$51 - 8 = 43$	(two times)
$43 - 8 = 35$	(three times)
$35 - 8 = 27$	(four times)
$27 - 8 = 19$	(five times)
$19 - 8 = 11$	(six times)
$11 - 8 = 3$	(seven times)

Division by small numbers is usually done with multiplication tables. Since 59 is between two consecutive multiples of 8, then 8 does not divide 59 exactly.

Since the smallest difference is not zero, then 8 does not divide 59 exactly.

In Example 2, even though 8 does not divide 59 exactly, we do have

$$59 - 7 \cdot 8 = 3$$

so that

$$59 = 7 \cdot 8 + 3$$

In this case we say that 7 is the *quotient,* and 3 is the *remainder* when 8 is divided into 59.

Even when a does not divide into b exactly we shall find that a divides into b with a *quotient* and a *remainder.* For example, 8 divides into 59 with a quotient of 7 and a remainder of 3. This can be indicated by writing

$$59 = 7 \cdot 8 + 3$$

In general, if a and b are nonnegative integers, we can find a quotient q and a remainder r such that

$$a = bq + r, \qquad \text{where} \quad 0 \leq r < b$$

We can find the quotient and remainder by the process of repeated subtraction as in Example 2. For example, since

$$59 - 7 \cdot 8 = 3$$

then 7 is the quotient, and 3 is the remainder when 8 is divided into 59. This process will be considered in more detail in Section 2-8.

Although exact division may not be possible in a particular problem, it can help us make estimates. The following example illustrates how we make practical estimates every day.

Example 3. Chloe wishes to try a recipe for beef stew obtained from an old boy friend, an assistant chef in a cafeteria. The full recipe makes 200 servings, but Chloe needs only 7 servings. How should she modify the recipe? Among other ingredients, the recipe calls for 93 pounds of potatoes. How many pounds should she use?

Solution. $200/7 \approx 210/7 = 30$. If she divides all the amounts by 30, she will have about the right amount of stew.

To estimate the potatoes, note that

$$\frac{93}{30} \approx \frac{90}{30} = 3$$

and that 93/30 is a little more than 3. Thus, she should put in a little more than 3 pounds of potatoes.

Powers of 10 can be used to advantage in division. Recall that we add powers of 10 when multiplying:

$$10^x \cdot 10^y = 10^{x+y}$$

Similarly, we subtract powers of 10 when dividing:

$$\frac{10^7}{10^4} = 10^{7-4} = 10^3$$

POWERS OF TEN USED IN DIVISION

$$\frac{10^a}{10^b} = 10^{a-b} \qquad \text{if } a > b$$

$$\frac{10^a}{10^b} = \frac{1}{10^{b-a}} \qquad \text{if } a < b$$

$$\frac{10^a}{10^b} = 1 \qquad \text{if } a = b$$

It is easy to see why this holds. Since

$$10^3 \cdot 10^4 = 10^{3+4} = 10^7$$

then 10^3 is the number we multiply by 10^4 in order to obtain 10^7. But this is just what we mean by $10^7/10^4$. Consequently,

$$\frac{10^7}{10^4} = 10^3$$

More generally,

$$\frac{10^a}{10^b} = 10^{a-b}, \qquad \text{provided that} \quad a \geq b$$

For example,

$$\frac{10^7}{10^3} = 10^{7-3} = 10^4$$

and

$$\frac{10^6}{10^6} = 10^{6-6} = 10^0 = 1$$

Example 4. Use powers of 10 to estimate the following quotients:
(a) 240,010/1,198
(b) 47,782/1,601

Solution. In each case we replace the numerator and denominator with approximate values which can be used for exact division.

$$\text{(a)} \quad \frac{240,010}{1,198} \approx \frac{240,000}{1,200} = \frac{24 \cdot 10^4}{12 \cdot 10^2}$$

$$= \frac{24}{12} \cdot 10^{4-2} = 2 \cdot 10^2 = 200$$

$$\text{(b)} \quad \frac{47,782}{1,601} \approx \frac{48 \cdot 10^3}{16 \cdot 10^2} = \frac{48}{16} \cdot 10^{3-2} = 3 \cdot 10^1 = 30$$

Example 5. Roscoe owns a tract of land that is approximately rectangular. He has measured the width at one point to be 876 feet and the length at another point to be 1980 feet. He knows that similar land has sold for $700 per acre (1 acre = 43,560 square feet). Estimate the value of the land.

Solution. We first estimate the number of square feet, then divide this by 43,560 to get the number of acres. Finally, we multiply the number of acres by $700 to get the value of the land.

The area is

$$\text{area} = \underbrace{(1980)}_{\text{length}} \underbrace{(876)}_{\text{width}} \approx (2000)(900)$$

$$= 2 \cdot 10^3 \cdot 9 \cdot 10^2 = 18 \cdot 10^5 = 1{,}800{,}000$$

$$\text{Number of acres} = \frac{\text{area}}{43{,}560} = \frac{1{,}800{,}000}{43{,}560}$$

$$\approx \frac{1{,}800{,}000}{45{,}000} \approx \frac{180 \cdot 10^4}{45 \cdot 10^3} \approx \frac{180}{45} \cdot 10^{4-3} \approx 4 \cdot 10 \approx 40$$

Since the area is approximately 40 acres, the value is

$$\text{value} = (\text{no. of acres}) \cdot (\text{value per acre})$$
$$\approx (40) \cdot (\$700) \approx \$28{,}000$$

Division by zero is not allowed in mathematics because it leads to results that are not consistent with the basic properties of numbers. Thus, expressions such as 0/0, 5/0, and so on, are considered by mathematicians to be meaningless symbols.

EXERCISES

1. Use the basic concept of exact division, as explained at the beginning of this section, to establish the following facts. (*Hint: a/b* is the number we multiply by *b* in order to get *a*.)

   (a) $\dfrac{7}{7} = 1$ (c) $\dfrac{9 \cdot 6}{6} = 9$

   (b) $\dfrac{0}{13} = 0$ (d) $\dfrac{na}{a} = n$ $(a \neq 0)$

2. Estimate the following numbers

   (a) $\dfrac{64}{31}$ (c) $\dfrac{760{,}010}{3{,}799}$

   (b) $\dfrac{992}{101}$ (d) $\dfrac{2501}{12}$

3. Use powers of 10 to estimate the following numbers:

   (a) $\dfrac{2{,}001{,}003}{40{,}001}$

   (b) $\dfrac{760{,}015}{187}$

   (c) $\dfrac{1{,}439{,}001}{751}$

4. The following information is obtained from a record book kept by George Poorbody in which he lists the number of miles driven on trips, number of gallons of gasoline used, and so on. Use the information to

estimate the number of miles per gallon that he obtained on the three trips.
(a)   187 miles, 16.3 gallons
(b)   212 miles, 18.4 gallons
(c)   491 miles, 26.9 gallons   (George thinks that this last record is an error.)

5. Those exact divisors of a number that are smaller than the number itself are called the *proper divisors* of the number. For example, the proper divisors of 12 are 1, 2, 3, 4, and 6; the proper divisors of 8 are 1, 2, and 4; the only proper divisor of 7 is 1. Find all of the proper divisors of the following numbers:

(a)   15   (d)   18
(b)   16   (e)   19
(c)   17   (f)   20

2-8   THE LONG-DIVISION ALGORITHM

The standard algorithm for long division, learned by most of us in elementary school, is based on repeated subtraction. We shall work through a simple example using repeated subtraction, then modify the method to get the standard algorithm.

Example 1.   Use repeated subtraction to divide 4 into 657, obtaining a quotient and a remainder.

Solution.   We could take the direct approach and successively subtract 4 from 657 until we get a remainder less than 4. We can save a great amount of time and trouble, however, by taking some shortcuts. We shall first successively subtract 400, then successively subtract 40 and then 4.

We can subtract 400 from 657 *one* time:

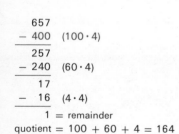

$$657 - 400 = 257$$
$$657 = 100 \cdot 4 + 257$$

We can subtract 40 from 257 *six* times:

$$257 - 6 \cdot 40 = 17$$

This is equivalent to subtracting 4 from 256 *sixty* times. Therefore,

$$257 = 60 \cdot 4 + 17$$

We now can subtract 4 from 17 *four* times:

$$17 - 4 \cdot 4 = 1$$
$$17 = 4 \cdot 4 + 1$$

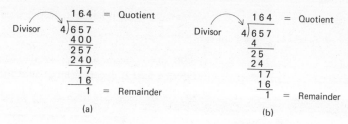

Figure 2-25 The standard algorithm for long division. (a) First form; (b) second form.

Observe that we have subtracted 4 from 657 a total of

$$100 + 60 + 4 = 164$$

times obtaining a remainder of 1:

$$657 = 164 \cdot 4 + 1$$

Thus, 164 is the *quotient* and 1 is the *remainder* when 4 is divided into 657.

The standard algorithm for long division involves the same basic steps that we used above (Fig. 2-25). In the first step, we divide 4 into the first digit, writing $6 = 1 \cdot 4 + 2$. This is equivalent to dividing 400 into 600 one time with a remainder of 200, or to dividing 4 into 600 one hundred times with a remainder of 200. We then "bring down" the 57, add it to 200, and repeat the process, dividing by 40. In the final stage, we divide by 4.

An ability to estimate exact multiples of numbers pays off in long division. For example, suppose we wish to divide 237 into 58,467. We first ask how many times 237 can be subtracted from 584. Observe that

$$\frac{584}{237} \approx \frac{600}{200} = 3 \quad \text{and} \quad \frac{584}{237} < \frac{584}{200} < 3$$

Thus, we would judge that 237 can be subtracted from 584 twice. We verify that this is correct, obtaining 11,067 for the first partial remainder. At the next step, we divide 237 into 1106.

$$\frac{1106}{237} \approx \frac{1100}{200} = 5\frac{1}{2}$$

If we multiply 5 by 237, we find that the product is a little larger than 1106. Thus, we get 4 rather than 5 for the next digit of the quotient. Finally, we divide 237 into 1587:

$$\frac{1587}{237} \approx \frac{1500}{237} < 8$$

In this case we find that 6 is the correct digit. The quotient is 246; the remainder is 165.

The division algorithm always works. If we start with two positive integers a and b, we can keep subtracting multiples of b from a until we get a difference, say r, that is less than b and is nonnegative. At this point we have

$$a - bq = r, \qquad 0 \leq r < b$$

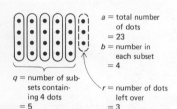

a = total number of dots
= 23
b = number in each subset
= 4
q = number of subsets containing 4 dots
= 5
r = number of dots left over
= 3

The problem of dividing a by b, obtaining a quotient and a remainder, can be considered as the problem of decomposing a set of a objects into subsets, each containing b objects. The *quotient* is the total number of complete subsets formed. The *remainder* is the number of objects "left over" after all of the subsets have been formed.

In the diagram, $a = 23$ (the total number of objects), $b = 4$ (the number in each subset), $q = 5$ (the total number of four-object subsets), $r = 3$ (the number of objects remaining after the subsets have been formed).

where q is the number of times we have subtracted b from a. This can be rewritten as the equation

$$a = bq + r, \qquad \text{where} \quad 0 \leq r < b$$

Furthermore, the quotient q and the remainder r are unique; that is, these are the only numbers that can satisfy the relationship

$$a = bq + r, \qquad 0 \leq r < b$$

For example, if we continually subtract 4 from 23, we could get the following relationships:

$$23 - 1 \cdot 4 = 19, \qquad 23 - 2 \cdot 4 = 15, \qquad 23 - 3 \cdot 4 = 11,$$
$$23 - 4 \cdot 4 = 7, \qquad 23 - 5 \cdot 4 = 3$$

Only the last of these expressions has the final answer in the range $0 \leq r < 5$. Thus,

$$23 = 5 \cdot 4 + 3$$

The *quotient* is 5; the *remainder* is 3.

Galley Division

The standard algorithm for long division has been used extensively since about 1600. Before that time the most common algorithm was the *Galley* (or *Scratch*) *Method*. (The name refers to the final shape of the number arrangement, which resembles a sail. At one time teachers actually required their students to draw sails around the finished calculation.) The

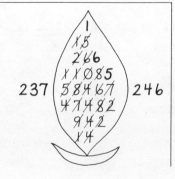

Galley algorithm is an ancient one, invented by the Hindus, used by the Arabs, and eventually introduced into Europe.

The figure illustrates the division of 58,467 by 237. The quotient is 246; the remainder is 165. The multiples of the divisor are written under the number to be divided; the differences are written *above* the number. Subtraction of the digits proceeds from left to right, scratching out the old digits at each step. The first few steps of the calculation are shown below:

Step 1.	237	58467	2
		474	
Step 2.	237	110 ƀ̷8̷4̷67 4̷7̷4̷	2
Step 3.	237	110 ƀ̷8̷4̷67 4̷7̷4̷8 94	24
Step 4.	237	15 2̷6̷ 1̷1̷0̷8 58467 4̷7̷4̷8 9̷4̷	24

One of the oldest classifications of whole numbers is according to whether they are "odd" or "even." *Even* numbers are exactly divisible by 2, while *odd* numbers leave a remainder of 1 when divided by 2. Thus, the first seven odd numbers are 1, 3, 5, 7, 9, 11, and 13, and the first seven even numbers are 2, 4, 6, 8, 10, 12, and 14.

odd · odd = odd
even · even = even
even · odd = even

odd + odd = even
even + even = even
even + odd = odd

This classification is due to the ancient Pythagoreans (fifth and fourth centuries B.C.). The Pythagoreans noted that the sum of two odd numbers must be even; the sum of an odd and an even number must be odd; the product of two odd numbers must be odd, and so forth. These results of the Pythagoreans can be established easily by using the fact that the quotient and the remainder are uniquely determined in division.

Example 2. Show that the product of two odd integers must be odd.

Solution. Let a and b be the two odd integers. If we divide either of these numbers by 2, we get a remainder of 1. Thus, we can write

$$a = 2m + 1$$
$$b = 2n + 1$$

where m and n are the quotients when a and b, respectively, are divided by 2. Then

$$ab = (2m + 1)(2n + 1) = 4mn + 2m + 2n + 1$$
$$= 2(2mn + m + n) + 1$$

By the uniqueness of the quotient and the remainder, this means that $2mn + m + n$ is the quotient, and 1 is the remainder when ab is divided by 2. Since the remainder is 1, then ab is odd.

The Pythagoreans

Pythagoras (about 580–500 B.C.), a Greek mystic, founded the *Society of the Pythagoreans,* a secret society founded along both mathematical and mystical lines. Among other tenets, the Pythagoreans believed that ultimate harmony resides in numbers, which are the basis of all existence. Their mathematical studies, which were kept secret, were partly an attempt to discover this harmony, partly an attempt to exploit the magical properties of numbers.

The Pythagoreans knew many of the basic results in geometry, such as the Pythagorean Theorem, classified numbers according to odd and even, found many of the important proper-

ties of prime numbers, and discovered the mathematical properties of musical harmony. They also assigned magical properties to numbers, attaching significance to such meaningless concepts as male numbers and female numbers.

Eventually the Society lost much of its secret character and developed into an association of professional mathematicians and teachers of mathematics. To a large extent, the Pythagoreans set the stage for the development of theoretical mathematics by the ancient Greeks, which, in turn, has influenced the European development since the Renaissance.

EXERCISES

1. Divide the first number into the second, obtaining a quotient and a remainder. Use both the repeated subtraction method and the long-division algorithm.
   (a) 9, 372   (d) 631, 4689
   (b) 17, 5621   (e) 172, 102,684
   (c) 24, 1109   (f) 264, 24,903

2. Construct a flow chart for the calculation of the quotient and the remainder when the positive integer B is divided into the nonnegative integer A. Use the process of *repeated subtraction.* (Keep the flow chart simple; do not attempt to utilize the positional notation of the numbers. See Figure 2-24.)

3. Work through the steps of the flow chart in Exercise 2 with the following examples:
   (a) $A = 29$, $B = 8$
   (b) $A = 38$, $B = 17$
   (c) $A = 29$, $B = 30$

4. Suppose that $b \geq a$. What are the quotient and the remainder when b is divided into a? (Recall that we want $a = bq + r$, where $0 \leq r < b$.)

5. Modify the technique illustrated in Example 2 to establish the following properties of odd and even numbers.
   (a) The sum of two odd numbers is even.
   (b) The sum of two even numbers is even.
   (c) The sum of an odd number and an even number is odd.
   (d) The product of two even numbers is even.

(e)   The product of an odd number and an even number is even.

6. The Pythagoreans considered a positive integer to be *perfect* if the sum of its proper divisors is the number itself. For example, 6 is perfect, for the sum of its proper divisors is $1 + 2 + 3 = 6$. (See Exercise 5, Section 2-7, for the definition of "proper divisor.")

Find one other perfect number.

7. The Pythagoreans were the first to classify "prime numbers." A *prime number* is an integer greater than 1 that has the number 1 as its only proper divisor. For example, the number 3 is a prime because 1 is its only proper divisor. (See Exercise 5, Section 2-7, for the definition of "proper divisor.")

Find the first 10 prime numbers.

8. Explain why an even integer greater than 2 cannot be a prime number.

SUGGESTIONS FOR FURTHER READING

1. Bergamini, David. *Mathematics,* Chap. 3, "An Alphabet for Deciphering the Unknown." New York: Time-Life, 1963.

2. Dantzig, Tobias. *Number, the Language of Science,* 4th ed. Garden City, N.Y.: Doubleday, 1954.

3. Gies, Joseph, and Gies, Frances. *Leonardo of Pisa and the New Mathematics of the Middle Ages.* New York: T. Y. Crowell, 1969.

4. Kline, Morris. *Mathematics for Liberal Arts,* Chap. 2, "An Historical Introduction." Reading, Mass.: Addison-Wesley, 1967.

3

THE RATIONAL NUMBERS

3-1 FRACTIONS

The earliest forms of society required only the positive integers for the calculations involved with simple commerce. As civilization developed, however, men more and more felt the need for symbols to represent parts of objects and proportions of magnitudes. This need became critical in the Bronze Age, when workers learned to combine various proportions of different metals to form alloys. By 2400 B.C. the Sumerians in the land between the Tigris and Euphrates rivers (present-day Iraq) had developed a very sophisticated system for writing numbers that included fractions as well as whole numbers.

Actually, fractions are the most natural numbers to use other than integers. They measure how many times large units can be split into smaller units. In the practical world, we give names to most of the standard units that we use. For example, since we have taken one-twelfth of a foot as a standard unit, we give it a name, *one inch*. Similarly, one-sixteenth of a pound is *one ounce*. We use combinations of the various units in the most convenient way possible—2 pounds 3 ounces, for example, in preference to $2\frac{3}{16}$ pounds or 35 ounces.

What are fractions? To define $\frac{1}{7}$, for example, we divide an interval measuring one unit in length into seven equal parts. Each of these parts has a measure of $\frac{1}{7}$ of a unit. Essentially, this gives us two "standard" units—the "large" unit of length *one* and the "small" unit of length $\frac{1}{7}$. We write $\frac{5}{7}$ to indicate 5 of these "small" units, $\frac{13}{7}$ to indicate 13 of them, and so on. (See Fig. 3-1.)

The Egyptian system of enumeration did not lend itself to fractions. The "unit" fractions $\frac{1}{2}$, $\frac{1}{3}$, $\frac{1}{4}$, and so on, were indicated by placing the symbol ⌒ over the corresponding integer. For example,

$$\frac{1}{3} = \overset{\frown}{III}$$

$$\frac{1}{12} = \overset{\frown}{\cap II}$$

$$\frac{1}{123} = \overset{\frown}{9\cap II}$$
$$\cap I$$

The Egyptians never developed a general theory of fractions. A fraction such as $\frac{2}{5}$ had to be represented as the sum of unit fractions.

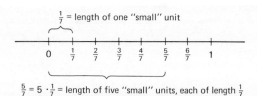

Figure **3-1** If the unit interval is divided into seven equal parts, then each of these small intervals ("small" units) has length $\frac{1}{7}$. Five of these "small" units have a combined length of $\frac{5}{7}$.

A similar process can be carried out for any positive integer n. For example, if $n = 15$, we split the unit interval into 15 equal parts, each having length $\frac{1}{15}$. Then $\frac{7}{15}$ is the measure of 7 of these "small units."

In the fraction $\frac{7}{15}$, the number 7 is called the *numerator* and 15 is called the *denominator*. Similarly, in a more general setting, we can denote a fraction by m/n, where m is the numerator and n is the denominator. The only restriction is that the denominator cannot be equal to zero.

Fractions, such as $\frac{1}{15}$, $\frac{2}{15}$, $\frac{3}{15}$, and so on, can be represented as points

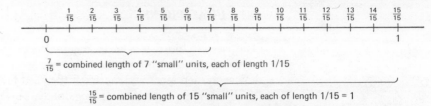

$\frac{7}{15}$ = combined length of 7 "small" units, each of length 1/15

$\frac{15}{15}$ = combined length of 15 "small" units, each of length 1/15 = 1

Figure 3-2   If the unit interval is divided into 15 equal parts ("small" units), then each of the "small" units has length $\frac{1}{15}$. The combined length of 7 of these "small" units is $\frac{7}{15}$. The combined length of 15 of them is $\frac{15}{15} = 1$. Similarly, $a/a = 1$ for any positive integer a.

on the number line as in Figure 3-2. This geometrical representation can help us explain several important properties of fractions. First, observe that $n/n = 1$ for every integer n. In our example, since we originally split the unit interval (length 1) into 15 parts, then the total length of these parts must be 1 unit. Thus,

$$\frac{15}{15} = \text{length of 15 parts each of length } \frac{1}{15} = 1$$

A similar argument can be used to establish that

$$\frac{3}{3} = 1$$
$$\frac{7}{7} = 1$$
$$\frac{183}{183} = 1$$

and so on.

Simplification of Fractions:

$$\frac{an}{bn} = \frac{a}{b}$$

for any positive integer n.

The second important property of fractions that we consider is the fact that

$$\frac{an}{bn} = \frac{a}{b}$$

This property can be explained by a different type of argument. Suppose, for example, we split the unit interval into 3 equal parts so that each part has length $\frac{1}{3}$. Suppose next that we split each of these 3 parts into 5 equal parts. (See Fig. 3-3.) This process splits the original unit interval into 15 equal intervals, each of length $\frac{1}{15}$. It is easy to see that the combined length of 5 of these small intervals is equal to the length of one of the intervals of length $\frac{1}{3}$. Thus,

$$\frac{5}{15} = \frac{1}{3}$$

This type of argument can be extended to cover any three positive integers, say a, b, and n. It can be shown that, in general,

$$\frac{an}{bn} = \frac{a}{b}$$

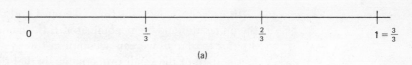

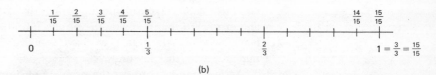

Figure 3-3 $\frac{5}{15} = \frac{1}{3}$; $an/bn = a/b$. (a) Split the unit interval into three equal parts, each of length $\frac{1}{3}$; (b) split each of the intervals of length $\frac{1}{3}$ into five equal parts, each of length $\frac{1}{15}$. Then five of these small intervals have a combined length of $\frac{1}{3}$. Thus, $\frac{5}{15} = \frac{1}{3}$. Similarly, $an/bn = a/b$.

We frequently indicate this relationship by "canceling" the "n" from the fraction by drawing a slash through it. Thus, we write

$$\frac{an}{bn} = \frac{a\cancel{n}}{b\cancel{n}} = \frac{a}{b}$$

For example,

$$\frac{28}{30} = \frac{14 \cdot 2}{15 \cdot 2} = \frac{14 \cdot \cancel{2}}{15 \cdot \cancel{2}} = \frac{14}{15}$$

$$\frac{7}{21} = \frac{1 \cdot 7}{3 \cdot 7} = \frac{1 \cdot \cancel{7}}{3 \cdot \cancel{7}} = \frac{1}{3}$$

$$\frac{25}{65} = \frac{5 \cdot 5}{13 \cdot 5} = \frac{5 \cdot \cancel{5}}{13 \cdot \cancel{5}} = \frac{5}{13}$$

and so on.

Simplification of Fractions

The relationship $an/bn = a/b$ is used to simplify fractions. Whenever possible, we prefer to write fractions in the simplest possible form—with no common factor (other than 1) in both the numerator and the denominator. For example,

$$\frac{6}{4} = \frac{3 \cdot 2}{2 \cdot 2} = \frac{3 \cdot \cancel{2}}{2 \cdot \cancel{2}} = \frac{3}{2}$$

In the original form, the fraction $\frac{6}{4}$ had a common factor of 2 in both the numerator and the denominator. In the reduced form, the fraction $\frac{3}{2}$ has no such common factor.

In order to reduce fractions, we must find the common factors in both the numerator and the denominator. If the numerator and the denominator are not too large, we can accomplish this in a few minutes of work. We first find all of the small prime numbers that divide the numerator, check to see which of them also divide the denominator, then cancel out the common factors. We can find the primes that divide the numerator by trial and error. We simply divide the primes into the numerator and see which leave no remainder. Once we have found a divisor, we can reduce the numerator by that factor.

As an example, consider the problem of simplifying the fraction $\frac{90}{42}$. We first factor the numerator. If we divide 90 by 2, we get

$$90 = 2 \cdot 45$$

so that 2 is a factor. Furthermore,

$$45 = 3 \cdot 15$$

and

$$15 = 3 \cdot 5$$

so that

$$90 = 2 \cdot 45 = 2 \cdot 3 \cdot 15 = 2 \cdot 3 \cdot 3 \cdot 5 \qquad \text{(product of prime numbers)}$$

If we now check to see which of these primes divide into the denominator, we find that

$$42 = 2 \cdot 3 \cdot 7 \qquad \text{(product of prime numbers)}$$

Then

$$\frac{90}{42} = \frac{2 \cdot 3 \cdot 3 \cdot 5}{2 \cdot 3 \cdot 7} = \frac{\cancel{2} \cdot \cancel{3} \cdot 3 \cdot 5}{\cancel{2} \cdot \cancel{3} \cdot 7} = \frac{3 \cdot 5}{7} = \frac{15}{7}$$

Addition of Fractions

Geometrically, we add fractions on the number line by the same process that we use to add whole numbers. To add

$$\frac{5}{7} \qquad \text{and} \qquad \frac{2}{3}$$

for example, we first measure a positive distance of $\frac{5}{7}$ and then a further distance of $\frac{2}{3}$. (See Fig. 3-4.)

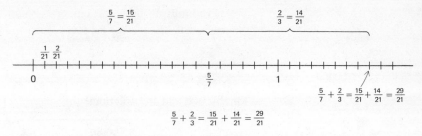

$$\frac{5}{7} + \frac{2}{3} = \frac{15}{21} + \frac{14}{21} = \frac{29}{21}$$

Figure 3-4 Addition of fractions (geometrical interpretation).

It is easy to add fractions that have the same denominator. We add their numerators and keep the original denominator. For example,

$$\frac{5}{13} + \frac{3}{13} = \frac{5+3}{13} = \frac{8}{13}$$

FLOW CHART LISTING THE BASIC STEPS IN THE ADDITION OF TWO FRACTIONS

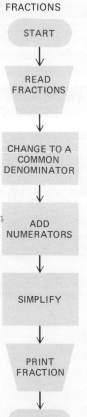

(See Fig. 3-5.) This follows from the fact that 5 "small" units added to 3 "small" units of the same size yields 8 "small" units. In general, if a/n and b/n are any two fractions with the same denominator, then

$$\frac{a}{n} + \frac{b}{n} = \frac{a+b}{n}$$

How do we add fractions that have different denominators, say $\frac{3}{7}$ and $\frac{5}{12}$? The first step is to "build up" these fractions into new fractions that have the same denominator. We can do this by using the rule

$$\frac{a}{b} = \frac{an}{bn}$$

that we discussed above. We normally use the product of the two original denominators as the common denominator.

For example, to add $\frac{3}{7}$ and $\frac{5}{12}$ we first multiply both the numerator and the denominator of $\frac{3}{7}$ by 12 (the denominator of the other fraction), obtaining

$$\frac{3}{7} = \frac{3 \cdot 12}{7 \cdot 12} = \frac{36}{84}$$

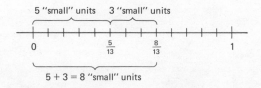

Figure 3-5 Addition of fractions with the same denominator: $\frac{5}{13} + \frac{3}{13} = \frac{8}{13}; a/n + b/n = (a+b)/n.$

We next multiply both the numerator and the denominator of $\frac{5}{12}$ by 7, obtaining

$$\frac{5}{12} = \frac{5 \cdot 7}{12 \cdot 7} = \frac{35}{84}$$

Finally, we add the fractions:

$$\frac{3}{7} + \frac{5}{12} = \frac{36}{84} + \frac{35}{84} = \frac{36 + 35}{84} = \frac{71}{84}$$

Example 1. (a) To add $\frac{1}{2}$ and $\frac{1}{3}$ we first write

$$\frac{1}{2} = \frac{1 \cdot 3}{2 \cdot 3} = \frac{3}{6} \quad \text{and} \quad \frac{1}{3} = \frac{1 \cdot 2}{3 \cdot 2} = \frac{2}{6}$$

Then

$$\frac{1}{2} + \frac{1}{3} = \frac{3}{6} + \frac{2}{6} = \frac{5}{6}$$

(b) $\frac{4}{9} + \frac{5}{12} = \frac{4 \cdot 12}{9 \cdot 12} + \frac{5 \cdot 9}{12 \cdot 9} = \frac{48}{108} + \frac{45}{108} = \frac{93}{108} = \frac{31 \cdot \cancel{3}}{36 \cdot \cancel{3}} = \frac{31}{36}$

Subtraction

Subtraction reverses the operation of addition. (See Fig. 3-6.) To subtract $\frac{3}{7}$ from $\frac{5}{9}$, for example, we first rewrite the fractions with the common denominator 63, then subtract. Observe that

$$\frac{5}{9} = \frac{5 \cdot 7}{9 \cdot 7} = \frac{35}{63} \quad \text{and} \quad \frac{3}{7} = \frac{3 \cdot 9}{7 \cdot 9} = \frac{27}{63}$$

so that

$$\frac{5}{9} - \frac{3}{7} = \frac{35}{63} - \frac{27}{63} = \frac{8}{63}$$

Example 2. (a) $\frac{1}{2} - \frac{1}{3} = \frac{1 \cdot 3}{2 \cdot 3} - \frac{1 \cdot 2}{3 \cdot 2} = \frac{3}{6} - \frac{2}{6} = \frac{1}{6}$

(b) $\frac{4}{9} - \frac{5}{12} = \frac{4 \cdot 12}{9 \cdot 12} - \frac{5 \cdot 9}{12 \cdot 9} = \frac{48}{108} - \frac{45}{108} = \frac{3}{108} = \frac{1 \cdot \cancel{3}}{36 \cdot \cancel{3}} = \frac{1}{36}$

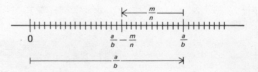

Figure 3-6 Subtraction of fractions (geometrical interpretation).

The operations of addition and subtraction of fractions obey the same general arithmetic laws as with the integers. Thus, for example, the order of adding two fractions is unimportant—we get the same answer if we add $\frac{2}{9}$ to $\frac{4}{13}$ as if we add $\frac{4}{13}$ to $\frac{2}{9}$. Similarly, if we add three fractions, such as $\frac{1}{5}$, $\frac{2}{7}$, and $\frac{4}{13}$, the grouping of terms is not important either. We get the same answer if we group these terms as

$$\left(\frac{1}{5} + \frac{2}{7}\right) + \frac{4}{13}$$

or as

$$\frac{1}{5} + \left(\frac{2}{7} + \frac{4}{13}\right)$$

The Babylonian Number System

The ancient Sumerian and Babylonian scholars wrote by pressing a wedge-shaped stylus into a mud tablet about the size of a postcard. Thousands of these hardened tablets, which are as impervious as bricks, have been unearthed in recent years and are stored in museums awaiting translation. The mathematical tablets that have been translated reveal a sophisticated system of writing integers and fractions that was much more advanced than the system used by the Egyptians.

The Babylonians used two basic symbols to represent all numbers, the symbols

$$\mathbf{Y} = 1$$

and

$$\mathbf{\langle} = 10$$

Numbers from 1 to 59 were written

by repeating these symbols within a fixed block of digits much as in the Egyptian system. For example, the following blocks of digits have the indicated values:

$$\text{YY} = 4 \qquad (4 \ ones)$$

$$\text{<<YY} = 24 \qquad (2 \ tens \text{ and } 4 \ ones)$$

and

$$\text{<{YY} = 35} \qquad (3 \ tens \text{ and } 5 \ ones)$$

Numbers greater than 59 were written by using several blocks of digits. The number 60 itself was written as the single block

$$\text{Y} \ = 60 \qquad (1 \ sixty)$$

followed by a space ("empty block"). The number 61 was written as the two blocks of digits

$$\text{Y} \quad \text{Y} = 61 \qquad (1 \ sixty \text{ and } 1 \ one)$$

Thus, for example,

$$\text{YYY} \quad \text{<{Y} = 332}$$

(5 *sixties*, 3 *tens*, and 2 *ones*)

$$\text{<{{YYY} \quad <{{YY} = 3294}}$$

(54 *sixties* and 54 *ones*)

Observe that the different blocks of digits are analogous to our digits when we write numbers in the standard decimal form. In our system the digit to the right indicates the number of *units*, the next digit the number of *tens*, the next the number of *hundreds*, and so on. In the Babylonian system, the right-hand block indicated the number of *units*, the next block indicated the number of *sixties*, the next block the number of 3600s (60 *sixties*), and so on. A block was left empty if there was no number of that type. Thus, 60 was written as

$$\text{Y}$$

which meant one *sixty* and no *units*.

As a further example, the symbols

$$\text{<<YY} \quad \text{{{YYYY} \quad YY}}$$

represent the number 81,963 (22 *3600s*, 46 *sixties*, and 3 *units*).

This system worked so well that it was taken one step further and used to represent fractions. By inserting an imaginary "decimal point," the Babylonians were able to let the blocks of digits also represent multiples of $1/60$, $1/60^2$, $1/60^3$, and so on. Thus, the single symbol Y could represent any of the numbers 1, 60, 60^2, 60^3, . . ., or the numbers $1/60$, $1/60^2$, $1/60^3$, . . ., depending on the context.

Similarly, the symbols

$$\text{<Y} \quad \text{<{Y}} \qquad \text{(literally 12 "large units" and 32 "small units")}$$

could represent

$$12 \cdot 60 + 32 = 752$$

or

$$12 \cdot 60^2 + 32 \cdot 60 = 45{,}120$$

or

$$12 \cdot 60^3 + 32 \cdot 60^2 = 2{,}707{,}200$$

or

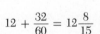

$$12 + \frac{32}{60} = 12\frac{8}{15}$$

or

$$\frac{12}{60} + \frac{32}{60^2} = \frac{752}{3600} = \frac{47}{225}$$

or an infinite number of other numbers.

EXERCISES

1. Simplify the following fractions by factoring the numerators and the denominators into prime factors.

   (a) $\frac{21}{36}$ (b) $\frac{108}{12}$ (c) $\frac{42}{70}$ (d) $\frac{24}{56}$

2. Calculate the following sums and differences. Simplify the final answer.

   (a) $\frac{20}{3} + \frac{19}{3}$ (d) $\frac{1}{6} + \frac{2}{6} + \frac{3}{6} - \frac{4}{6}$

   (b) $\frac{7}{15} + \frac{3}{15}$ (e) $\frac{3}{5} + \frac{12}{5} - \frac{7}{5}$

   (c) $\frac{9}{16} - \frac{2}{16} + \frac{3}{16}$ (f) $\frac{4}{7} + \frac{5}{7} - \frac{6}{7}$

3. Rewrite the fractions with common denominators; then add or subtract as indicated. Simplify the final answer. [*Hint:* In (c) write 2 as $\frac{2}{1}$.]

   (a) $\frac{5}{7} + \frac{8}{15}$ (d) $\frac{1}{2} + \frac{1}{3} + \frac{1}{4} + \frac{1}{5}$

   (b) $\frac{3}{8} + \frac{2}{13}$ (e) $\frac{8}{9} - \frac{2}{7} + \frac{1}{63}$

   (c) $2 - \frac{18}{17}$ (f) $\frac{4}{21} - \frac{2}{7} + \frac{8}{3}$

4. Construct diagrams similar to the indicated figures to illustrate the following facts:

   (a) $\frac{4}{4} = 1$ (Fig. 3-2)

   (b) $\frac{6}{4} = \frac{3}{2} = 1\frac{1}{2}$ (Fig. 3-3)

   (c) $\frac{1}{2} + \frac{1}{3} = \frac{3}{6} + \frac{2}{6} = \frac{5}{6}$ (Figs. 3-4, 3-5)

5. Construct a flow chart that can be used to test any two fractions a/b and c/d for *commutativity of addition*, that is, test if

$$\frac{a}{b} + \frac{c}{d} = \frac{c}{d} + \frac{a}{b}$$

6. Work through the steps of the flow chart in Exercise 5 with the following examples:

   (a) $\frac{a}{b} = \frac{12}{17}, \quad \frac{c}{d} = \frac{1}{3}$

   (b) $\frac{a}{b} = \frac{1}{2}, \quad \frac{c}{d} = \frac{1}{2}$

7. By example show that subtraction of fractions is not associative. That is, find particular fractions

$$\frac{a}{b}, \quad \frac{c}{d}, \quad \frac{e}{f}$$

such that

$$\left(\frac{a}{b} - \frac{c}{d}\right) - \frac{e}{f}$$

is not equal to

$$\frac{a}{b} - \left(\frac{c}{d} - \frac{e}{f}\right)$$

This shows that the order of grouping terms is important in subtraction even though it is not important in addition.

8. What numbers do the following symbols represent in the Babylonian (Sumerian) system? (Assume that the numbers are whole numbers—no fractions involved.)

   (a) ⟨⟨ ▼▼▼▼   (c) ⟨⟨▼▼ ⟨▼ ▼▼

   (b) ▼ ⟨▼   (d) ⟨▼ ▼▼ ⟨⟨▼▼

9. Write the following fractions in the Babylonian (Sumerian) system. List at least one other number that each set of Babylonian symbols could represent.

   (a) $\dfrac{7}{60}$ (c) $4\dfrac{7}{60}$

   (b) $\dfrac{3}{60} + \dfrac{2}{60^2}$ (d) $\dfrac{4}{60^2} + \dfrac{17}{60^3}$

10. Write each of the following fractions in the Egyptian system.
    (a) $\frac{1}{3}$ (c) $\frac{1}{271}$
    (b) $\frac{1}{7}$ (d) $\frac{2}{21}$

3-2 MULTIPLICATION AND DIVISION OF FRACTIONS

We can think of a fraction such as $\frac{8}{35}$ in terms of *area* as well as length. Recall that a square which has each side of length 1 has an area of *1 square unit*. Suppose we subdivide the base of the unit square into 5 equal parts and the altitude into 7 equal parts, and then partition the square into rectangles as in Figure 3-7. This process subdivides the original square into

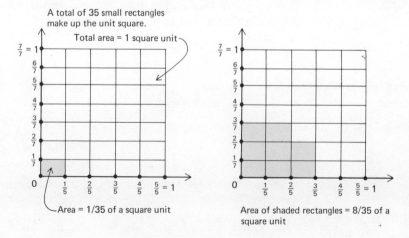

Figure 3-7 Fractions represented as areas—special case.

35 rectangles, all of the same size. The area of each of these rectangles is $\frac{1}{35}$ of a square unit. The combined area of eight of them is $\frac{8}{35}$. (See Fig. 3-7.)

In general, if we split the base of the "unit square" into m equal parts and the altitude into n equal parts, we partition the square into mn equal rectangles, each having area $1/mn$. A set of k of these small rectangles has area equal to k/mn. (See Fig. 3-8.)

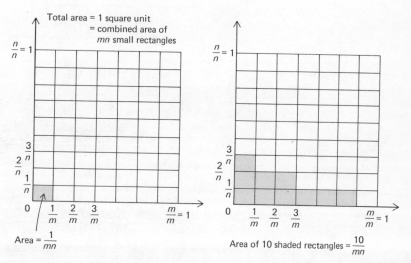

Figure 3-8 Fractions represented as areas—general case.

Multiplication

When we multiply fractions such as $\frac{3}{7}$ and $\frac{4}{9}$, we are actually calculating the area of a rectangle, one side of length $\frac{3}{7}$, the other of length $\frac{4}{9}$. This is represented geometrically in Figure 3-9. Observe that the shaded rectangle consists of $3 \cdot 4 = 12$ small rectangles, each of area $\frac{1}{63}$. Thus,

$$\frac{3}{7} \cdot \frac{4}{9} = \text{area of shaded rectangle}$$
$$= \text{area of 12 small rectangles} = \tfrac{12}{63}.$$

In general, we can represent two arbitrary fractions as a/m and b/n. When we multiply a/m by b/n, we are calculating the area of a large rectangle which contains ab small rectangles, each of area $1/mn$. Thus,

$$\frac{a}{m} \cdot \frac{b}{n} = \text{area of } ab \text{ small rectangles} = \frac{ab}{mn}$$

(See Fig. 3-10.) Thus, we are led to the following rule for multiplication of fractions:

$$\frac{a}{m} \cdot \frac{b}{n} = \frac{ab}{mn}$$

FLOW CHART TO VERIFY
THE COMMUTATIVE LAW OF
MULTIPLICTION OF
FRACTIONS IN PARTICULAR
CASES

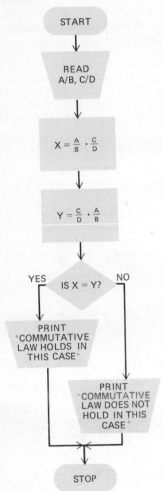

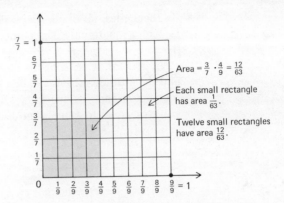

Figure 3-9 $\frac{3}{7} \cdot \frac{4}{9}$ = area of shaded rectangles = $\frac{12}{63}$.

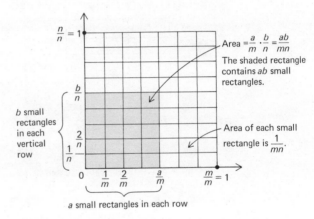

a small rectangles in each row

Figure 3-10 Multiplication of fractions—general case: $a/m \cdot b/n = ab/mn$.

The product of two fractions is a fraction with the numerator equal to the product of the two original numerators and the denominator equal to the product of the two original denominators.

Example 1. (a) $\dfrac{1}{3} \cdot \dfrac{1}{2} = \dfrac{1 \cdot 1}{3 \cdot 2} = \dfrac{1}{6}$

(b) $\dfrac{4}{9} \cdot \dfrac{3}{7} = \dfrac{4 \cdot 3}{9 \cdot 7} = \dfrac{12}{63} = \dfrac{4 \cdot 3}{21 \cdot 3} = \dfrac{4}{21}$

(c) $\dfrac{17}{5} \cdot \dfrac{5}{17} = \dfrac{17 \cdot 5}{5 \cdot 17} = \dfrac{85}{85} = 1$

Division

Division of fractions can be considered geometrically. When we divide $\frac{5}{3}$ by $\frac{8}{7}$, for example, we are calculating the height of a rectangle that has

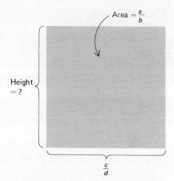

Area = $\frac{a}{b}$

Height = ?

$\frac{c}{d}$

Geometrically, the problem of dividing $\frac{a}{b}$ by $\frac{c}{d}$ is equivalent to finding the height of a rectangle that has $\frac{a}{b}$ for its area and $\frac{c}{d}$ for its base.

an area of $\frac{5}{3}$ square units and a base of $\frac{8}{7}$ units. Thus, when we perform this division, we are trying to find the number x which has the property that

$$\frac{8}{7} \cdot x = \frac{5}{3}$$

In other words,

$$x = \frac{5}{3} \div \frac{8}{7}$$

only in case

$$\frac{8}{7} \cdot x = \frac{5}{3}$$

It is easy to show that one such number is

$$x = \frac{5}{3} \cdot \frac{7}{8} = \frac{35}{24}$$

To show this, we observe that

$$\frac{8}{7} \cdot x = \frac{8}{7} \cdot \frac{35}{24} = \frac{\cancel{8}}{\cancel{7}} \cdot \frac{\cancel{7}}{\cancel{8}} \cdot \frac{5}{3} = \frac{5}{3}$$

Thus,

$$\frac{5}{3} \div \frac{8}{7} = \frac{5}{3} \cdot \frac{7}{8} = \frac{35}{24}$$

In the above example, to divide $\frac{5}{3}$ by $\frac{8}{7}$ we "inverted" the second fraction, obtaining $\frac{7}{8}$, and then multiplied. This process can be carried out in general. For example,

$$\frac{3}{8} \div \frac{2}{9} = \frac{3}{8} \cdot \frac{9}{2} = \frac{27}{16}$$

In general, if a/b and m/n are two fractions with $m \neq 0$, then

$$\frac{a}{b} \div \frac{m}{n} = \frac{a}{b} \cdot \frac{n}{m} = \frac{an}{bm}$$

We also indicate this division by writing

$$\frac{\dfrac{a}{b}}{\dfrac{m}{n}} = \frac{a}{b} \div \frac{m}{n} = \frac{a}{b} \cdot \frac{n}{m} = \frac{an}{bm}$$

The restriction $m \neq 0$ is necessary to avoid division by zero.

Example 2.   (a) $\frac{2}{3} \div \frac{1}{4} = \frac{2}{3} \cdot \frac{4}{1} = \frac{8}{3}$

(b) $\frac{\frac{2}{13}}{\frac{5}{12}} = \frac{2}{13} \div \frac{5}{12} = \frac{2}{13} \cdot \frac{12}{5} = \frac{24}{65}$

(c) $\frac{\frac{8}{5}}{\frac{7}{5}} = \frac{8}{5} \div \frac{7}{5} = \frac{8}{\cancel{5}} \cdot \frac{\cancel{5}}{7} = \frac{8}{7}$

(d) $\frac{6}{7} \div \frac{6}{9} = \frac{\cancel{6}}{7} \cdot \frac{9}{\cancel{6}} = \frac{9}{7}$

EXERCISES

1. Calculate the following numbers. Simplify the answers where it is convenient to do so.

   (a) $\frac{1}{3} \cdot \frac{3}{2}$ (f) $\frac{2}{3} \div \frac{4}{5}$

   (b) $\frac{2}{7} \cdot \frac{4}{5}$ (g) $\frac{13}{9} \div \frac{9}{13}$

   (c) $\frac{2}{7} \div \frac{4}{5}$ (h) $\frac{3}{8} \div \frac{2}{7}$

   (d) $\frac{9}{2} \cdot \frac{8}{5}$ (i) $\frac{12}{187} \cdot \frac{216}{41}$

   (e) $(\frac{1}{2} + \frac{1}{3}) \cdot \frac{7}{5}$ (j) $\frac{28}{43} \div \frac{28}{43}$

2. Construct a diagram similar to Figure 3-8 to illustrate the number $\frac{3}{20}$. (*Hint:* Partition the base of the unit square into 4 equal parts, partition the altitude into 5 equal parts.)

3. Construct a diagram similar to Figure 3-9 to illustrate the product $\frac{3}{4} \cdot \frac{2}{5}$. By counting the small rectangles, show that $\frac{3}{4} \cdot \frac{2}{5} = \frac{6}{20}$.

4. Construct a flow chart for the steps in forming the product of A/B and C/D where A, B, C, and D are given positive integers. (*Hint:* A similar flow chart is shown in the text for division of fractions.)

5. (a)   By example show that division of fractions is not commutative. (*Hint:* Find a particular pair of fractions a/b and c/d such that $a/b \div c/d \neq c/d \div a/b$.)

   (b)   What conditions are needed on the fractions a/b and c/d in order to have $a/b \div c/d = c/d \div a/b$. Find one such pair of fractions.

6. Make up three examples of your own to verify that multiplication of fractions is distributive over addition; that is, in each case show that

$$\frac{a}{b} \cdot \left(\frac{c}{d} + \frac{e}{f}\right) = \frac{a}{b} \cdot \frac{c}{d} + \frac{a}{b} \cdot \frac{e}{f}$$

3-3   DECIMAL FRACTIONS

Let us recall how to use the Hindu-Arabic notation for representing positive integers. When we write 3271, for example, we mean the number

$$3271 = 3 \cdot 10^3 + 2 \cdot 10^2 + 7 \cdot 10 + 1$$
$$= 3000 + 200 + 70 + 1$$

There is a simple extension of this notation that can be used to represent certain fractions. When we write 43.3271, we mean

$$43.3271 = 4 \cdot 10 + 3 + \frac{3}{10} + \frac{2}{10^2} + \frac{7}{10^3} + \frac{1}{10^4}$$

$$= 4 \cdot 10 + 3 + 3 \cdot \frac{1}{10} + 2 \cdot \left(\frac{1}{10}\right)^2 + 7 \cdot \left(\frac{1}{10}\right)^3 + \left(\frac{1}{10}\right)^4$$

That is, the digits to the *left* of the decimal point are multiplied by powers of 10, as in the standard Hindu-Arabic notation, while the digits to the *right* of the decimal point are multiplied by powers of $\frac{1}{10}$. In each case, the powers increase as we move away from the decimal point.

Example 1. (a) $0.3004 = \dfrac{3}{10} + \dfrac{0}{10^2} + \dfrac{0}{10^3} + \dfrac{4}{10^4}$

$$= \frac{3}{10} + \frac{4}{10,000} = \frac{3,000}{10,000} + \frac{4}{10,000}$$

$$= \frac{3,004}{10,000}$$

(b) $752.123 = 7 \cdot 10^2 + 5 \cdot 10 + 2 + \dfrac{1}{10} + \dfrac{2}{10^2} + \dfrac{3}{10^3}$

$$= 752\frac{123}{1000}$$

Fractions written in this extended Hindu-Arabic notation are called *decimal fractions*. The period is called the *decimal point*.

The beauty of using decimal fractions lies in the fact that all of the algorithms that we developed for use with integers can also be used with these numbers. It is only necessary to keep track of where the decimal point must be located.

For example, to add 59.84 and 31.7 we line up the digits in columns, making sure that the decimal points also are lined up, and add in columns,

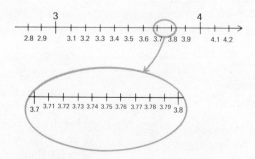

Figure 3-11 Decimals on the number line.

progressing from right to left, carrying wherever necessary. Subtraction is handled similarly.

Multiplication is a little different. To multiply 59.84 by 31.7, for example, we multiply the *integers* 5984 and 317 and then locate the decimal point so that the number of digits to the right of it is equal to the total number of digits to the right of the decimal point in the two original numbers combined. To see why this works in our example, note that

MULTIPLICATION WITH DECIMALS

```
 59.84  (two decimal places)
 31.7   (one decimal place)
 ─────
 41888
 5984
17952
 ──────────
1896.928  (three decimal places)
```

$$(59.84)(31.7) = \frac{5984}{100} \cdot \frac{317}{10}$$

$$= \frac{(5984)(317)}{1000} = \frac{1,896,928}{1000} = 1896.928$$

Division is best handled by modifying the problem so as to divide an integer into a decimal on an integer. For example, to divide 4.15 into 98.312, we observe that

$$\frac{98.312}{4.15} = \frac{(98.312)(100)}{(4.15)(100)} = \frac{9831.2}{415}$$

In other words, we obtain the same result by dividing 415 into 9831.2 as by dividing 4.15 into 98.312.

To carry out the process of dividing 415 into 9831.2, we proceed as in long division, except that we put a decimal point in the quotient directly above the one in the original number. We also add as many zeros as we wish to the right of the decimal point in the original number. We then carry out the division algorithm, not stopping until we get a zero remainder or until we have as many places to the right of the decimal point in the quotient as we need to give us a good approximation to the actual quotient. For example, if we wish five-place accuracy, we work out the quotient until we have six places to the right of the decimal point and then "round off" the answer at five decimal places. In this example, we get 23.689638 for the first six decimal places. Since the last digit is large, when we drop it we change the last digit retained by adding 1 to it, giving us 23.68964. Now this number is not the exact value of 9831.2/415. Instead, it is an approximation correct to five decimal places.

$$21.3214 \approx 21.321$$
$$481.7005 \approx 481.701$$
$$0.0398 \approx 0.040$$

When we divide decimals into decimals, we usually round-off the quotient to a fixed number of decimal places in order to get an approximate answer. We can make this approximation as close to the "true" quotient as we wish by calculating enough decimal places.

$$\frac{98.312}{4.15} = \frac{9831.2}{415} \approx 23.68964$$

When we work through the division algorithm with decimal fractions, we normally stop the process after a predetermined number of decimal places has been reached. We then drop the last digit. If this digit was 0, 1, 2, 3, or 4, we leave the other digits alone. If the dropped digit was 5, 6, 7, 8, or 9, we raise the last digit retained by 1. This process is known as "rounding-off."

Example 2. If we wish three-place accuracy, we calculate four places to the right of the decimal point, drop the fourth digit and "round-off."

(a) If the quotient has been calculated to be 21.3214, we round-off to 21.321. (The dropped digit was 4, which is less than 5.)

(b) If the quotient has been calculated to be 481.7005, we round-off to 481.701. (The dropped digit was 5; so we round up.)

(c) If the quotient has been calculated to be 0.0398, we round-off to 0.040. We keep the "0" in the third place to show that we have three-place accuracy rather than two-place accuracy as 0.04 would indicate.

The "rounding-off" process is necessary because most fractions cannot be expressed exactly as decimals. If we go back to our earlier example of 98.312/4.15, we find that the long-division algorithm never stops; that is, we never get a remainder of zero. Mathematicians make theoretical studies of how this "infinite-decimal" quotient behaves and even find ways to express all of the digits, but for most practical purposes they use two- or three-place accuracy.

A small table of *reciprocals* is in the margin. This table lists seven-place approximations to the numbers $\frac{1}{1}, \frac{1}{2}, \frac{1}{3}, \ldots, \frac{1}{20}$. Only eight of the numbers in the table are the exact values of the fractions. The other 12 are approximations. We can use this table to avoid long division when the denominator is one of the numbers 1, 2, 3, ..., 20. More extensive tables give the reciprocals of all numbers between 1 and 1000.

Example 3. Use the table of reciprocals to calculate $\frac{17}{20}$ and $\frac{13}{19}$.

Solution. (a) $\frac{17}{20} = 17 \cdot \frac{1}{20} = 17(0.0500000) = 0.85$
(b) $\frac{13}{19} = 13 \cdot \frac{1}{19} \approx 13(0.0526316) \approx 0.684211$

Because of their convenience, decimal fractions are used extensively in the business and scientific worlds. Consequently, nearly all of the fractions used in business and scientific calculations are approximations to the true values that are supposed to be used. This means that almost all calculations involving interest, service charges, rocket launchings, chemical reactions, and so forth, have built-in errors caused by the use of decimal fractions rather than common fractions.

BRIEF TABLE OF RECIPROCALS

$\frac{1}{1} = 1.0000000$
$\frac{1}{2} = 0.5000000$
$\frac{1}{3} \approx 0.3333333$
$\frac{1}{4} = 0.2500000$
$\frac{1}{5} = 0.2000000$
$\frac{1}{6} \approx 0.1666667$
$\frac{1}{7} \approx 0.1428571$
$\frac{1}{8} = 0.1250000$
$\frac{1}{9} \approx 0.1111111$
$\frac{1}{10} = 0.1000000$
$\frac{1}{11} \approx 0.0909091$
$\frac{1}{12} \approx 0.0833333$
$\frac{1}{13} \approx 0.0769231$
$\frac{1}{14} \approx 0.0714286$
$\frac{1}{15} \approx 0.0666667$
$\frac{1}{16} = 0.0625000$
$\frac{1}{17} \approx 0.0588235$
$\frac{1}{18} \approx 0.0555556$
$\frac{1}{19} \approx 0.0526316$
$\frac{1}{20} = 0.0500000$

The Invention of Decimal Fractions

The numeration system used by the Babylonians was a positional system similar to the decimal system that we use today. The Babylonians used their system to represent fractions as well as integers by varying the locations of the blocks of digits. Unfortunately, the principles developed by the Babylonians were not incorporated into the number systems that came later. Even the Hindus, who invented our number system, only used it to represent whole numbers and common fractions.

In medieval Europe a few tentative starts were made toward a positional system for fractions, but the changes were never accepted by the general public. Common fractions continued to be used for practical work, and Babylonian fractions (base 60) to be used for much of the scientific

STANDARD UNITS OF
LENGTH IN
THE METRIC SYSTEM

10 millimeters = 1 centimeter
100 centimeters = 1 meter
1000 meters = 1 kilometer

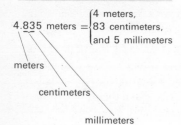

4.835 meters = { 4 meters,
                 83 centimeters,
                 and 5 millimeters

meters

centimeters

millimeters

work. Not until the seventeenth century was the decimal system used for fractions in Europe.

Simon Stevin (1548–1620), a Flemish engineer with broad interests, was the moving force behind the adoption of the decimal system for fractions. In his book *La disme* ("the tenth"), published in 1585, he argued the advantages of decimals over common fractions and urged that they be used for all monetary systems and systems of weights and measures.

Stevin in no sense invented decimal fractions. Various other systems, including that of the ancient Chinese, had included some of the basic principles, and the mathematicians of Stevin's day were all aware of the advantages of the system. Stevin's contribution was cultural and political—he made the common people aware of the advantages of decimals.

Stevin's original system did not include a decimal point. Digits were written with a small circled number above or to the right to indicate the decimal place. For example, 21.317 was written as

⓪①②③
2 1 3 1 7 or as 2 1 ⓪ 3 ① 1 ② 7 ③

The first use of the decimal point was made by *John Napier* (1550–1617), the inventor of logarithms. Napier referred to Stevin's work and suggested the use of a comma or dot to separate the integer part from the fractional part. In most European countries "decimal commas" are used today.

As Stevin noted in *La disme*, decimals offer many practical advantages over common fractions. (This is apparent to anyone who has ever tried to compute interest using the old British monetary system.) Unfortunately, to a large extent, custom dictates the system of weights and measures used in a country. As a consequence, the people of the United States have continued to use the archaic system of weights and measures that they inherited from the British colonies rather than the more efficient metric system.

The advantages of working with the metric system can be illustrated by the problem of calculating one-half of a distance of 5 meters, 83 centimeters, and 6 millimeters. We write the distance as 5.836 meters and divide by 2

$$\frac{5.836}{2} = 2.918$$

Thus, one-half of the total distance is 2 meters, 91 centimeters, and 8 millimeters. The reader should contrast the ease with which the above problem can be worked with the difficulty of working the equivalent problem of finding one-half of a distance of 13 feet, $7\frac{8}{64}$ inches.

Example   from *La disme*

⓪	①	②	③		
	2	7	8	4	7
	3	7	6	7	5
8	7	5	7	8	2
9	4	1	3	0	4

EXERCISES

Perform the calculations for Exercises 1 to 3. Use decimal fractions in all steps. Do not convert to common fractions.

1. (a) 5.873 + 13.2
   (b) 12.001 + 101.519
   (c) 87.1 − 0.0005
   (d) 0.0012 + 0.0715 + 0.9876
   (e) 10.314 − 0.189 + 12.76
   (f) 7.16 + 104.3 − 0.0972

2. (a) $(2.3) \cdot (5.1)$ (d) $(141.7) \cdot (13.001)$
    (b) $(7.01) \cdot (13.2)$ (e) $(87.16) \cdot (5.8)$
    (c) $(9.023) \cdot (0.057)$ (f) $(12.82) \cdot (17.99)$

3. (a) $\dfrac{19.72}{2}$ (d) $\dfrac{12.3}{4.4}$

    (b) $\dfrac{27.16}{4}$ (e) $\dfrac{62}{1.09}$

    (c) $\dfrac{82.13}{42}$ (f) $\dfrac{18.12}{0.0054}$

4. One inch is approximately equal to 2.54 centimeters. Convert the following from feet and inches to meters and centimeters.
    (a) 8 inches   (c) 4 feet, 8.5 inches
    (b) 2 feet, 3 inches (d) 20 feet, 9 inches

5. Convert from centimeters to feet and inches. (1 inch $\approx$ 2.54 centimeters.)
    (a) 23 centimeters   (c) 82 centimeters
    (b) 143 centimeters   (d) 19.2 centimeters

6. Draw a number line on a large sheet of graph paper. Use a scale in which one unit measures between 1 and 3 inches.
    (a) Locate the number 3.12 on the number line.
    (b) Describe a procedure that could be used to locate the number 1.270513 on the number line.

7. Use the table of reciprocals given near the end of this section to convert the following problems into problems involving multiplication. Calculate the decimal values of the fractions, rounding-off the answers to three decimal places.
    (a) $\frac{3}{14}$ (c) $\frac{41}{7}$
    (b) $\frac{7}{19}$ (d) $\frac{18}{5}$

8. Round-off the following numbers to three decimal places.
    (a) 41.824499   (d) 8.7166
    (b) 12.51479   (e) 0.0007
    (c) 4.0199   (f) 0.0004

9. Explain why $98.312/4.15 = 9831.2/415$. What rule of fractions did we use when we multiplied numerator and denominator by 100?

10. John needs to drill a hole that is as close to 0.19 of an inch as possible. The bits in his tool kit are graduated in increments of $\frac{1}{64}$ of an inch. (He has bits of $\frac{1}{64}$ inch, $\frac{2}{64}$ inch, $\frac{3}{64}$ inch, and so on.) Which bit should he use? (*Hint:* Multiply 0.19 by $\frac{64}{64}$.)

3-4 RATIO, PROPORTION, PERCENT

Many quantities are proportional to other quantities. For example, if the interest on a loan is 9 percent, then 9 cents in interest must be paid for every dollar borrowed. *The interest is proportional to the principal.* If we know either the interest or the principal, we can calculate the other quantity. If a man's take-home pay (after deductions) is $\frac{3}{5}$ of his gross pay, then he keeps $3 out of every $5 earned. *The take-home pay is proportional to the gross pay.* If we know either the take-home pay or the gross pay, then we can calculate the other quantity.

In each of these examples, we are concerned with *ratio* and *proportion.* For each part of one quantity there are a fixed number of parts of the other quantity. The ratio of interest to principal is 9 to 100 (9 parts interest to each 100 parts of principal); the ratio of take-home pay to gross pay is 3 to 5 (3 parts take-home pay to each 5 parts of gross pay).

As an example, suppose the amount borrowed on the loan (the principal) is $3000. Since the interest rate is 9 percent, then the amount of interest is

$$\text{interest} = \tfrac{9}{100} \cdot 3000 = \$270$$

Similarly, if the gross pay is $800 per month, then the net take-home pay is

$$\text{net pay} = \tfrac{3}{5} \cdot (\text{gross pay})$$
$$= \tfrac{3}{5} \cdot 800 = \$480$$

In the above examples, we were concerned with the fractions $\frac{9}{100}$ and $\frac{3}{5}$. These fractions are the *ratios.* The ratio of interest to principal is

$$\frac{\text{interest}}{\text{principal}} = \frac{9}{100} \qquad (\text{ratio of interest to principal})$$

The ratio of take-home pay to gross pay is

$$\frac{\text{take-home pay}}{\text{gross pay}} = \frac{3}{5} \qquad (\text{ratio of take-home pay to gross pay})$$

More generally, if r is the fraction x/y, that is,

$$r = \frac{x}{y}$$

we say that r is the *ratio of x to y.* A statement that two ratios are equal is called a *proportion.* In most cases we are given the ratio and one of the quantities (x or y) and must calculate the other quantity.

Example 1. A certain architect designs homes so that the number of square feet of usable space is approximately $\frac{4}{5}$ of the total number of square feet. (The remaining $\frac{1}{5}$ of the total number of square feet is taken up by walls, closets, utilities, and so forth.) His most recent two houses have quoted total areas of 6740 square feet and 4300 square feet, respectively. Estimate the amount of usable space in the two houses.

The first exhaustive study of ratio and proportion was made by the ancient Greeks. The Greeks used geometry as the basis for all mathematics and made an extensive study of such geometrical concepts as length, area, and volume. They found it necessary to develop an elaborate system of ratios and proportions in order to convert properties of one figure into equivalent properties of other figures.

If two quantities are proportional, then when one quantity doubles, so does the other. When one quantity triples, so does the other.

The ratio of solid circles to open circles is 2 to 1.

Solution. In each case

$$\text{usable space} \approx \tfrac{4}{5} \cdot \text{total space}$$

When the total space is 6740 square feet, then

$$\text{usable space} \approx \tfrac{4}{5} \cdot 6740 \approx 5400$$

When the total space is 4300 square feet, then

$$\text{usable space} \approx \tfrac{4}{5} \cdot 4300 \approx 3440$$

The notation

$$a:b = c:d$$

is frequently used for ratio and proportion. This means the same thing as the equality

$$\frac{a}{b} = \frac{c}{d}$$

Percents

A *percent* is the numerator of a fraction which has 100 as its denominator. Thus, 6 percent represents the fraction $\frac{6}{100} = 0.06$; $7\frac{1}{2}$ percent represents the fraction $7\frac{1}{2}/100 = 0.075$, and so on.

Although they are closely related, there is a basic difference between a percent and a fraction. A fraction such as $\frac{6}{100}$ is a number. It can be represented as a point on the number line; it can be used as an independent entity in addition, subtraction, multiplication, and division, and as a ratio for comparing two related quantities. A percent, on the other hand, has only one of these uses. It can only be used as a ratio. Thus, the expression 6 percent is meaningless usless it is used as a ratio in some type of comparison. It makes sense to speak of a growth rate of 6 percent or an interest rate of 6 percent, but 6 percent, by itself, is essentially meaningless.

When we use a percent in a calculation, it is always as a ratio involving the equivalent fraction. For example, if the interest rate is 6 percent, then the ratio of interest to principal is

$$\frac{\text{interest}}{\text{principal}} = \frac{6}{100}$$

If the growth rate of the gross national product (GNP) is 3 percent in a year, then the ratio of the *increase* in the GNP to the size of the GNP at the beginning of the year is

$$\frac{\text{increase in GNP}}{\text{GNP at beginning of year}} = \frac{3}{100}$$

Example 2. The population of a small city is approximately 150 percent of the population 10 years ago. If the present population is 90,000, estimate the size of the population 10 years ago.

Solution. We write the given information in terms of a ratio:

$$\frac{\text{present population}}{\text{old population}} = \frac{150}{100}$$

Then

$$\text{old population} = \frac{\text{present population}}{\dfrac{150}{100}} = \frac{90,000}{\dfrac{150}{100}}$$

$$= 90,000 \cdot \frac{100}{150} = 60,000$$

We now turn our attention to some examples that illustrate how ratios (sometimes involving percentages) are used and misused in our society.

Example 3.   Thomas Schnook recently borrowed $200 from a small loan company at a quoted annual interest rate of 12 percent. The loan is to be repaid in one month. Schnook finds that the interest plus unexplained "service charges" add to $6.00. Compute the true annual interest rate on the assumption that the service charges actually are a form of interest.

Solution.   We first calculate the monthly interest rate. When this rate is multiplied by 12, it yields the annual interest rate. Now

$$\text{monthly interest rate} = \frac{\text{interest}}{\text{principal}} = \frac{6}{200} = \frac{3}{100}$$

Thus, the annual interest rate is

$$12 \cdot \frac{3}{100} = \frac{36}{100}$$

In terms of percentages, the annual interest rate is 36 percent, not 12 percent as quoted by the loan company.

Example 4.   George bet $20 on Fleabag to win a horse race at odds of 8 to 5. How much money does George win if Fleabag wins the race?

Solution.   George will get his $20 back plus his winnings. (He had to leave the $20 with the bookie.) The ratio of winnings to the amount bet is $\frac{8}{5}$; that is, for each $5 bet, he wins $8. Thus,

$$\frac{\text{winnings}}{\text{amount bet}} = \frac{8}{5}$$

$$\text{winnings} = \frac{8}{5} \cdot (\text{amount bet})$$

$$= \frac{8}{5} \cdot 20 = \frac{160}{5} = 32$$

Example 5.   After three months of bargaining, the union local won a 20 percent across-the-board raise for the employees of the Honestman Dice Works. One month after the raise went into effect, President John N. Honestman announced a 20 percent reduction in all salaries (management excepted). The announcement explained that the reduction was necessary because of a decrease in sales and an increase in competition. It went on to note that the 20 percent reduction in salaries merely canceled out the 20 percent raise. The enraged union officials claim

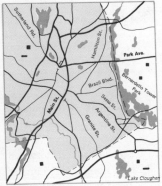

The *scale* of a map is a ratio. When we say, "One inch equals 15 miles," we mean

$$\frac{\text{map distance}}{\text{actual distance}} = \frac{1 \text{ inch}}{15 \text{ miles}}$$

that Honestman has actually lowered salaries below the preraise level. Honestman has just given a final statement to the press—"Everyone knows that a 20 percent increase followed by a 20 percent decrease leaves all employees with the same net salary!" Who is right? Honestman or the union?

Solution. Let us compare the salary of a typical employee before the raise, after the raise, and after the decrease. For simplicity, assume that before the raise he earned $100 for a certain number of hours' work. After the raise he earned $120 for the same amount of work. When Honestman ordered the decrease, it was based on the new salary, not the old. Thus, the salary was decreased by 20 percent of $120 = (0.20)(120) = \$24$, resulting in a net salary of

$$\$120 - \$24 = \$96$$

Since this line of reasoning applies to the salaries of all of the workers, then they all suffered a 4 percent net loss after the 20 percent increase and 20 percent decrease.

Example 5 illustrates a common way that numbers can be used to cheat people. When we compare percentages, we need to have a fixed base amount. Then we can express other quantities as percentages of the base amount. In the above example, Honestman deliberately changed the base amount (the salary used to determine the percentages) and then claimed that the 20 percent increase and 20 percent decrease are equivalent.

EXERCISES

1. In the following three situations, we are given two of the three quantities: *principal, interest,* and *interest rate.* Calculate the third quantity.
   (a) principal = $2000, interest rate = $\frac{1}{2}$ percent
   (b) principal = $4000, interest = $140
   (c) interest = $8, interest rate = 6 percent
2. An executive's take-home pay is approximately one and one-half times the size of his payroll deductions. Estimate the size of his take-home pay and his deductions if his gross salary is $2000 per month.
3. The Quik-Buildum Construction Company has standard floor plans available for many different houses. In all cases the usable floor space is between 80 and 85 percent of the total floor space.
   (a) Construct a flow chart showing the major steps in calculating upper and lower estimates for the amount of usable space in a home that has a given total floor space.
   (b) Use the flow chart from (a) to estimate the amount of usable floor space in the following homes:
      (1) Plan A: total floor space 4700 square feet
      (2) Plan B: total floor space 2800 square feet
4. The *true interest rate* over a period of time is computed by dividing the amount of interest paid at the end of the period by the amount of money borrowed. For example, if $100 is borrowed for one year

and $106 is paid back, then the true interest rate is 6% per year.

(a) A certain bank charges 7 percent interest, but *discounts* the interest from the loan. In other words, the interest is deducted from the principal in advance. Thus, a man who borrows $1000 for one year receives $930 and pays back $1000 at the end of the year. Calculate the true annual interest rate.

(b) Same problem as (a) except that the loan is for a period of three years. (*Hint:* First calculate the true interest rate for a three-year period, then divide this by 3 to obtain the true annual interest rate.)

(c) Bankers frequently claim that discounting is done "for the convenience of the borrower" or "to make payments easier." What is your opinion of these claims?

5. What percent decrease in salaries would exactly cancel out the 20-percent increase? (See Example 5.)

6. Which is better? A 20-percent raise in salary followed by a 20-percent decrease or a 20-percent decrease followed by a 20-percent raise?

7. A driver increases his average speed by $\frac{1}{5}$ from 50 miles per hour to 60 miles per hour. By what ratio does he decrease the time required for a 300-mile trip?

8. Show that if a driver increases his average speed by $\frac{1}{6}$, then he decreases the time required for the trip by $\frac{1}{7}$. (For example, at 60 miles per hour, a 14-mile trip takes 14 minutes. At 70 miles per hour—an increase of 10 miles per hour, which is $\frac{1}{6}$ of the original speed—the trip takes only 12 minutes.)

9. The Town of Bleaksburg charges a 10-percent tax on utility bills. This tax is not listed separately, but is included in the total tax bill. A note on the bill states, "If you wish to determine the amount of tax, divide the total bill by 11." Explain why this procedure works.

OLD BRITISH
MONETARY SYSTEM

2 halfpence = 1 penny
12 pence = 1 shilling
20 shillings = 1 pound

10. The old British monetary system (used until Britain applied for entry into the European Common Market in the early 1970s) is explained in the table in the margin.

(a) Calculate the interest due on 13 pounds (in terms of halfpence, pence, and shillings) at an interest rate of 7 percent.

(b) Calculate the interest due on $175 at an interest rate of 7 percent. How does this calculation compare in difficulty with the calculation in (a)?

3-5 NEGATIVE NUMBERS

Men have used positive integers and fractions for thousands of years. It is difficult to carry out even the simple acts of commerce without them. Negative numbers, on the other hand, were late in both development and acceptance.

The Hindus of the sixth to eighth centuries were the first people to have a clear understanding of negative numbers. They expressed them as "debts" rather than "assets" and allowed them as solutions to problems. It made sense to the Hindus that when a man traveled 10 steps in the positive direction and then backed up 15 steps, that he had traveled a net distance of "debt-five" steps, indicating that his final location was 5 steps in the negative direction.

We do not know why the Hindus decided to adopt these negative numbers—perhaps it was due to influence from the north. In a limited sense, the more sophisticated Chinese had used negative numbers for centuries before the Hindus used them. During those centuries the Chinese had used "counting rods" (small sticks placed on a "counting frame" marked off into squares) as a type of abacus. Two sets of rods were used—red rods for positive numbers and black rods for negative numbers. Unfortunately, the Chinese seem to have thought of negative numbers only in terms of subtraction of positive numbers. (Five red rods plus three black rods equals two red rods.) They never took the major step of considering negative numbers to be entities independent of their role in subtraction.

When the mathematical knowledge of the Hindus was acquired by the Arabs, negative numbers were largely ignored. The Arabs were very suspicious of these numbers and avoided their use, although their leading mathematicians were aware of the basic rules for adding, subtracting, multiplying, and dividing them. Later, when the Europeans adopted the mathematics of the Arabs, they almost completely ignored negative numbers.

Use of counting rods by the Japanese from a Japanese book published in 1795.

Positive divided by positive or negative by negative is affirmative. Cipher divided by cipher is naught. Positive divided by negative is negative. Negative divided by affirmative is negative. Positive or negative divided by cipher is a fraction with that for denominator.

Brahmagupta

Quoted from H. T. Colebrooke, *Algebra with Arithmetic and Mensuration, from the Sanscrit of Brahmagupta and Bhaskara* (1817).

Hindu mathematics was a blend of the very good and the very bad. Brahmagupta (seventh century) gave the above rules for division. The division of nonzero numbers was described correctly, but he went on to give the incorrect rule $0 \div 0 = 0$ and hedged on $a \div 0$ by saying that it was a fraction with zero for its denominator.

This avoidance of negative numbers caused no real difficulty at the time. During the Medieval and Renaissance European periods, practical mathematics was primarily a blend of common sense and cumbersome algorithms for arithmetic. It was almost always possible to restate a problem in such a way that a negative solution could be expressed as a positive solution. If this could not be done, then as far as the mathematicians were concerned the problem could not be solved.

Apparently the Hindu mathematicians made a careful study of algebraic rules such as

$$(a - b)(c - d) = ac - ad - bc + bd$$

which had been known since antiquity, and came to the conclusion that it was necessary to have

$$(-b)(-d) = bd$$

The algebraic rule mentioned above, which was known for the special case $a > b$, $c > d$, can be established by a geometrical argument. Observe that the *total area* of the large rectangle is equal to

$$ac = (a - b)(c - d) + ad + bc - bd$$

(The area represented by bd has been added in twice, once with ad and once with bc.) Thus,

$$(a - b)(c - d) = ac - ad - bc + bd$$

If we now expand $(a - b)(a - c)$ by use of the distributive law we have

$$(a - b)(c - d) = a(c - d) - b(c - d)$$
$$= ac + a(-d) + (-b)(c) + (-b)(-d)$$

On comparing this form with the one above we see that it is reasonable to suppose that

$$a(-d) = -ad$$
$$(-b)c = -bc$$

and

$$(-b)(-d) = bd$$

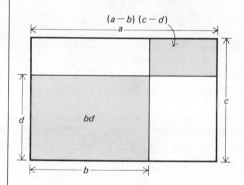

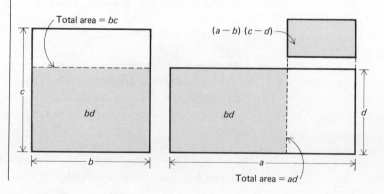

The invention of the printing press in the fifteenth century greatly influenced the development of mathematics. New notations as well as new advances in theory became known quickly over wide regions. In particular, the symbols $+$, $-$, $=$, $\sqrt{}$ became standard notations. The free use of these symbols had a major influence on the development of the laws of signed numbers by the eighteenth-century mathematicians.

The necessity for negative numbers became apparent when algebra was developed. In algebra we are concerned with solving problems by stating them as equations. These equations usually involve some unknown, which is represented by a symbol, such as x. At the time the equation is written down, there may be no way of knowing if x represents a positive quantity or a negative one. If, however, we are to use the rules of algebra to solve the problem, these rules must hold for negative quantities as well as positive ones.

Not until late in the eighteenth century did negative numbers become firmly established. By that time the number line had been invented, and it was easy to represent negative numbers geometrically as points. They quickly proved themselves so useful in mathematics that they became as indispensable as the positive integers and the fractions that had been invented earlier.

Graphically the negative numbers are represented by points on the left-hand side of the number line. "-2" is the point two units to the left of zero; "-3" is the point three units to the left of zero, and so forth. For every positive number "a" there is a corresponding negative number "$-a$" located on the number line a units to the left of zero.

Example 1. The point "3" is three units to the right of 0; the point "-3" is three units to the left of 0; the point "$-2\frac{1}{2}$" is two and one-half units to the left of zero.

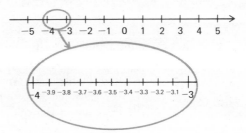

Figure 3-12   The number line.

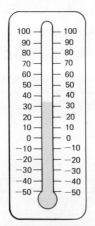

The thermometer scale furnishes a perfect model of negative numbers. A point is chosen to represent zero. Numbers are located at their distances from zero, the negative numbers below zero and the positive numbers above.

Before the invention of the thermometer, there was no generally accepted way to represent negative numbers geometrically by points.

Addition and Subtraction

Geometrically, addition is handled by a technique similar to the one that we used with positive numbers. To add two numbers, say 3 and -5, we start with the point representing 3 and then shift five units *to the left* giving us the number -2. Thus,

$$3 + (-5) = -2$$

The only significant change in the method is that we shifted to the left rather than to the right as we did with addition of positive numbers. (See Fig. 3-13.)

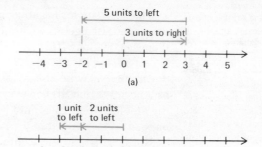

Figure 3-13 Addition on the number line. (a) $3 + (-5) = -2$. A shift of 3 units to the right of 0 followed by a shift of 5 units to the left results in a net shift of 2 units to the left of 0; (b) $(-2) + (-1) = -3$. A shift of 2 units to the left of 0 followed by a shift of 1 unit to the left results in a net of shift of 3 units to the left of 0.

Similarly, $(-2) + (-1) = -3$. A shift of two units to the left followed by a shift of one unit to the left results in a shift of three units to the left.

It is easy to see that this last process can be carried out for any two negative numbers, giving us the rule $(-a) + (-b) = -(a + b)$. *The sum of two negative numbers is the negative of the sum of the corresponding two positive numbers.*

Subtraction of positive numbers is an ancient operation. Traditionally, only small numbers could be subtracted from large ones. The invention of negative numbers allowed large numbers to be subtracted from small ones. In a sense negative numbers allow us to replace subtraction with addition by use of the rule

$$a - b = a + (-b)$$

in which subtraction of positive numbers is expressed in terms of addition of a positive number and a negative number.

Example 1. (a) $12 + (-11) = 12 - 11 = 1$
(b) $3 + (-5) = 3 - 5 = -2$
(c) $\dfrac{4}{7} + \left(-\dfrac{1}{3}\right) = \dfrac{4}{7} - \dfrac{1}{3} = \dfrac{4 \cdot 3}{7 \cdot 3} - \dfrac{7 \cdot 1}{7 \cdot 3} = \dfrac{12 - 7}{21} = \dfrac{5}{21}$
(d) $\left(-\dfrac{1}{2}\right) + \left(-\dfrac{1}{5}\right) = -\left(\dfrac{1}{2} + \dfrac{1}{5}\right) = -\left(\dfrac{5}{10} + \dfrac{2}{10}\right) = -\dfrac{7}{10}$
(e) $5 + (-5) = 5 - 5 = 0$

Multiplication

Let us look at negative numbers in a slightly different way: $-a$ *is the number we add to* a *in order to get zero*. As we can see, $2 + (-2) = 0$, $4 + (-4) = 0$, and so forth. We now ask, "What is $-(-a)$?" If we are to be consistent, then $-(-a)$ must be the number we add to $-a$ in order to get zero. Thus, for example, $-(-2)$ represents the number we add to -2 in order to get 0. Since $(-2) + 2 = 0$, this number is 2; so

$$-(-2) = 2$$

In general, since

$$(-a) + a = 0$$

then a is the number we add to $(-a)$ in order to get zero. Therefore,

$$-(-a) = a$$

This argument helps explain one of the rules for negative numbers that has mystified students for many years—why the negative of a negative number is positive.

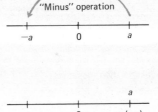

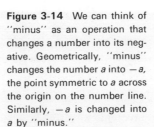

Figure 3-14 We can think of "minus" as an operation that changes a number into its negative. Geometrically, "minus" changes the number *a* into −*a*, the point symmetric to *a* across the origin on the number line. Similarly, −*a* is changed into *a* by "minus."

It can be shown that the associative, commutative, and distributive laws hold for addition and multiplication of negative, as well as positive, numbers. With this in mind it is easy to show that

$$(-1) \cdot a = -a \quad \text{and} \quad (-1) \cdot (-1) = 1.$$

First, observe that

$$(-1) \cdot a + a = [(-1) + 1] \cdot a = 0 \cdot a = 0$$

so that $(-1) \cdot a$ is the number we add to a in order to get zero. Therefore,

$$(-1) \cdot a = -a.$$

If we apply the rule $(-1) \cdot a = -a$ with $a = -1$ we obtain

$$(-1) \cdot (-1) = -(-1)$$

which is known to be equal to 1. Thus,

$$(-1) \cdot (-1) = -(-1) = 1.$$

With these two relationships established, we can handle all problems involving multiplication of negative numbers. We simply write each number $-a$ as $(-1) \cdot a$, regroup the terms and simplify, using the fact that $(-1)(-1) = 1$. The following example illustrates the technique.

Example 2.   (a) $(-3)(-4) = (-1) \cdot 3 \cdot (-1) \cdot 4 = (-1)(-1) \cdot 3 \cdot 4 = 12$
(b) $5(-7) = 5 \cdot (-1) \cdot 7 = (-1) \cdot 5 \cdot 7 = -35$
(c) $(-\frac{2}{3})(\frac{4}{5}) = (-1) \cdot \frac{2}{3} \cdot \frac{4}{5} = (-1) \cdot \frac{8}{15} = -\frac{8}{15}$
(d) $(-\frac{3}{2})(-\frac{2}{3}) = (-1) \cdot \frac{3}{2} \cdot (-1) \cdot \frac{2}{3} = (-1)(-1) \cdot \frac{3}{2} \cdot \frac{2}{3} = \frac{6}{6} = 1.$

Several rules of multiplication can be established without much trouble. These enable us to skip some of the steps illustrated in Example 2:
(1) $(-a) \cdot b = -ab$
This holds because

$$(-a) \cdot b = (-1) \cdot a \cdot b = -ab$$

The product of a *negative* number and a *positive* number is *negative*.

(2) $a \cdot (-b) = -ab$ [Exercise 8(a)]

The product of *two negative* numbers is *positive*.

(3) $(-a)(-b) = ab$ [Exercise 8(b)]

Fractions

Fractions involving negative numbers can be changed easily into fractions with positive numbers in the numerator and the denominator.
Observe first that

$$\frac{-1}{-1} = +1 \quad \left(\text{since } \frac{a}{a} = 1 \text{ if } a \text{ is any number different from zero}\right)$$

$$\frac{-1}{1} = -1 \quad \left(\text{since } \frac{a}{1} = a \text{ if } a \text{ is any number}\right)$$

and

$$\frac{1}{-1} = -1$$

This last fact follows from the fact that

$$\frac{1}{-1} = \frac{1}{-1} \cdot \frac{-1}{-1} = \frac{1 \cdot (-1)}{(-1)(-1)} = \frac{-1}{1} = -1$$

We now can use the same technique to simplify fractions that we used in Example 2: *Write all numbers of form* $-a$ *in the form* $(-1) \cdot a$.

Example 3. (a) $\dfrac{-2}{-5} = \dfrac{(-1) \cdot 2}{(-1) \cdot 5} = \dfrac{(\cancel{-1}) \cdot 2}{(\cancel{-1}) \cdot 5} = \dfrac{2}{5}$

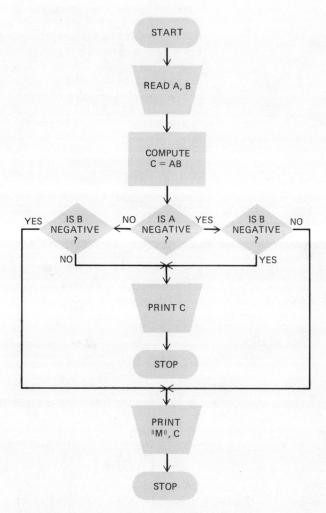

Figure 3-15 Multiplication on the STUPIAC computer (Example 4).

(b) $\dfrac{3}{-7} = \dfrac{1 \cdot 3}{(-1) \cdot 7} = \dfrac{1}{-1} \cdot \dfrac{3}{7} = (-1) \cdot \dfrac{3}{7} = -\dfrac{3}{7}$

(c) $\dfrac{-4}{-3} \cdot \dfrac{2}{-7} = 1 \cdot \dfrac{4}{3} \cdot (-1) \cdot \dfrac{2}{7} = 1 \cdot (-1) \dfrac{4}{3} \cdot \dfrac{2}{7} = -\dfrac{8}{21}$

(d) $\dfrac{-6}{-13} + \dfrac{1}{-2} = \dfrac{6}{13} - \dfrac{1}{2}$

$$= \dfrac{6 \cdot 2}{13 \cdot 2} - \dfrac{1 \cdot 13}{2 \cdot 13} = \dfrac{12}{26} - \dfrac{13}{26} = \dfrac{12 - 13}{26} = -\dfrac{1}{26}$$

Example 4.   The STUPIAC is a mythical simple computer that can only be used to add, multiply, subtract, and divide positive numbers and zero. The computer is programmed to recognize a negative number if an "M" is written in front of a positive number, but ignores the "M" in computation. Thus, if it is instructed to add M2 to 3, it will get 5 for the answer rather than 1, the correct answer. If it is instructed to subtract M7 from M4, it tries to subtract 7 from 4, sees that this is impossible in the system of nonnegative numbers, and prints an "error message" to the user. The flow chart of Figure 3-15 shows how the STUPIAC can be programmed to multiply any two numbers together regardless of whether they are positive, negative, or zero.

EXERCISES

1. Calculate
   (a) $12 + (-3)$
   (b) $12 + (-15) - 7$
   (c) $(-14) + (-2) - (-1)$
   (d) $-(-1) + 2(-3) - 4(5)$
   (e) $(-2.7) + 3.5$
   (f) $12.3 - 15.6$
   (g) $-(3.1)(-5.2)$
   (h) $(-2.1)(-3.2) + (7.5)(-0.01)$

2. Simplify the following fractions:
   (a) $\dfrac{-2}{-4}$
   (b) $\dfrac{3}{-8}$
   (c) $\dfrac{-21}{7}$
   (d) $\dfrac{(-12)(-7)}{(-4)(3)}$
   (e) $\dfrac{-(-3) + 7}{5 \cdot (-2)}$
   (f) $\dfrac{-12 + (-3)}{-17 - (-2)}$

3. Perform the following computations. Simplify the answers.
   (a) $\dfrac{3}{7} - \dfrac{8}{3}$
   (b) $\left(-\dfrac{1}{3}\right) + \left(-\dfrac{1}{4}\right) + \left(-\dfrac{1}{5}\right)$
   (c) $\left(-\dfrac{1}{2}\right) \cdot \dfrac{2}{3} - \dfrac{1}{5} \cdot \left(-\dfrac{1}{6}\right)$
   (d) $\left(-\dfrac{2}{-3}\right) \div \dfrac{1}{-5}$
   (e) $\dfrac{1}{12}\left(-\dfrac{2}{13}\right) + \left(\dfrac{-1}{5}\right) \cdot \dfrac{15}{26}$
   (f) $\dfrac{2}{3}\left(-\dfrac{1}{3}\right)$
   (g) $\left(\dfrac{-9}{7}\right) \cdot \left(\dfrac{-7}{9}\right)$
   (h) $\left(\dfrac{1}{2}\right)\left(\dfrac{-2}{3}\right) \cdot \left(\dfrac{3}{4}\right)\left(-\dfrac{4}{5}\right)$

4. Draw diagrams similar to Figure 3-13 illustrating the following additions and subtractions.

(a) $\frac{1}{2} + (-\frac{2}{3})$ (c) $(-\frac{1}{3}) + (-\frac{1}{4})$

(b) $(-2) + 3$ (d) $(-5) + 4$

5. Use a geometrical argument based on the number line to show that

$$-(7 + 3) = (-7) + (-3)$$

6. (a) Construct a flow chart listing the main steps in calculating a person's net worth without the use of negative numbers. (*Hint:* The basic problem is to compare the sizes of the total assets and the total liabilities.)

(b) Construct a flow chart listing the main steps in calculating a person's net worth utilizing negative numbers as fully as possible.

7. Obtain a copy of the Financial Statement Form used at a local bank by individuals who apply for loans. Does this form utilize negative numbers? (That is, are liabilities expressed as "negative assets"?) If not, write a brief paragraph explaining why you think the bankers choose to avoid negative numbers on the form.

8. Modify the argument given in Example 2 to establish the following rules of multiplication.

(a) $a(-b) = -ab$ (b) $(-a)(-b) = ab$

9. (a) Show that the following statements are true.

(1) The product of *two* negative numbers is *positive*.

(2) The product of *three* negative numbers is *negative*.

(3) The product of *four* negative numbers is *positive*.

(4) The product of *five* negative numbers is *negative*.

(b) What general rule would you conjecture after examining (a)?

10. The primitive STUPIAC computer is explained in Example 4. Make a flow chart for the addition of any two numbers A and B on the STUPIAC.

3-6 RATIONAL AND IRRATIONAL NUMBERS

The nonnegative and negative integers and fractions that we have discussed are known as rational numbers. To be more precise, a *rational number* is any number that can be written as a ratio of form a/b where a and b are integers and $b \neq 0$. For example, $5/7$, $(-3)/2$ and $12/(-2)$ are all rational numbers because they are ratios of two integers. The following example shows that certain numbers that may not appear to be rational can be rewritten in such a way that it becomes apparent that they are rational.

Example 1. (a) 5.3 is a rational number because it can be written as $53/10$.

(b) $-\frac{23}{5}$ is a rational number because it can be written as $(-23)/5$.

(c) 5 is a rational number because it can be written as $5/1$.

(d) The fraction $\frac{17}{23}/\frac{4}{7}$ is a rational number because it can be written as $\frac{119}{92}$:

$$\frac{17}{23} \div \frac{4}{7} = \frac{17}{23} \cdot \frac{7}{4} = \frac{17 \cdot 7}{23 \cdot 4} = \frac{119}{92}$$

It can be seen from Example 1 that the key words in the definition of a rational number are "... *can be written in the form* ...." To show that a number is rational, it is only necessary to rewrite it in such a way that it is a ratio of two integers.

The question naturally arises as to whether all numbers are rational. The answer is no; there do exist numbers that cannot be written as ratios of integers. To see what is involved, let us recall what we mean by a number. We have come a long way from the primitive ideas of our ancestors regarding numbers. Rather than require that a number count the objects in a collection or represent a ratio, we consider a number to be a location on the number line. Now this is equivalent to defining a number in terms of a *distance* and a *direction*. Each positive number is located a certain distance to the right of the origin, each negative number a certain distance to the left.

Thus, our original question becomes, "Does every possible distance that can be measured on the number line represent a rational number?"

As we shall see later, it is easy to construct certain distances (points on the number line) that cannot be represented by rational numbers. One such distance represents the ratio of the circumference of a circle to its diameter. It is customary to call the associated number by the Greek letter π (pi). As we see in Figure 3-16, it is easy to construct the number π by rolling a circle along the number line. The number we obtain on the line, however, is not a rational number. There is no pair of integers a, b with π as their ratio. Thus, π is an example of an *irrational* number.

Every irrational number can be approximated to any desired degree of accuracy by decimal fractions. In other words, if we start with an irrational

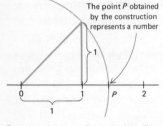

The point *P* obtained by the construction represents a number

Every point on the number line represents a number, either rational or irrational. This holds regardless of whether the point was obtained by measurement or by some type of geometrical construction.

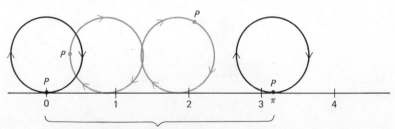

π = the distance the circle rolls in exactly one revolution

Figure 3-16 The geometrical construction of the number π. Place a circle with diameter equal to 1 unit on the number line with the point *P* at the origin. Roll the circle along the number line until the point *P* touches the number line again. The distance from the origin to the new location is exactly equal to π, the circumference of the circle.

number, such as π, and a certain number of decimal places, we can find a decimal that is approximately equal to the irrational number, the two numbers agreeing for the stated number of decimal places.

As an example,

$$\pi \approx 3.14 \qquad \text{(correct to two decimal places)}$$
$$\pi \approx 3.142 \qquad \text{(correct to three decimal places)}$$
$$\pi \approx 3.1416 \qquad \text{(correct to four decimal places)}$$
$$\pi \approx 3.14159 \qquad \text{(correct to five decimal places)}$$

and so on. This process can be continued for any number of decimal places, but the computations involved become more difficult. For example, correct to 11 decimal places, $\pi \approx 3.14159265359$.

At this point the reader is apt to think that most of the numbers on the number line are rational and that the few irrational numbers are oddities. Actually, the opposite is true. Every interval on the number line, regardless of how small, contains an infinite number of irrational numbers as well as an infinite number of rational numbers. In fact, in a certain technical sense, it has been shown that there are "more" irrational numbers than there are rational numbers.

The Square Root of 2

The very early mathematicians measured lengths with rather crude instruments. As a consequence they measured all lengths as rational numbers and never suspected that irrational numbers might exist. It came as quite a shock to the ancient Greeks to discover that there are lengths that cannot represent rational numbers. In particular, they discovered that a certain length represents a "square root" of 2. (To be more exact, a square with each side that length has two square units for its area. See Exercise 7.) It was only natural for them to assume that this length is a ratio of two integers and to search for the particular ratio. Much to their surprise, they eventually discovered that there is no way to represent this length as a ratio of two integers. In other words, there is no rational number a/b such that $(a/b)^2 = 2$.

To the Greeks the discovery of the irrationality of the square root of 2 meant that lengths were somehow different from "true" numbers. To get around this problem, they invented a cumbersome system involving positive integers, lengths of line segments, and ratios of lengths of line segments. Since ratios were the key to the entire system, they developed an elaborate set of rules for ratio and proportion that remained the bane of schoolboys until the nineteenth century.

We take a different point of view. Rather than considering lengths and numbers to be distinct entities, we assume that the points on the number line (and the associated distances) represent numbers, some of which are rational, some irrational. Thus, the square root of 2 is irrational, as is π.

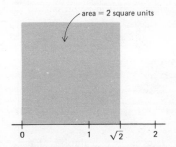
area = 2 square units

A square with each side equal to $\sqrt{2}$ (the square root of 2) has an area of exactly 2 square units.

The Irrationality of the
Square Root of 2

It is possible that the ancient mathematicians became aware of the existence of the square root of 2 by studying patterns of floor tiles. The figure shows a pattern in which four triangles make one standard square. The large square with each side equal to the diagonal of a small square is composed of eight triangles; so it has an area of two standard squares. It follows that the length of the diagonal of a unit square is equal to the square root of 2.

The ancient Pythagoreans used an ingenious argument to show that the square root of 2 is irrational. Essentially, they showed that if the square root of 2 could be written as a common fraction, then both the numerator and the denominator would have an infinite number of factors. Since no such integers exist, then the square root of 2 is not a ratio of two integers.

At one step in the argument, we need the following fact about integers:

If the square of an integer is even, then the integer itself is even.

This follows from the fact that every integer is either odd or even and that the square of an odd integer is odd.

The proof of the irrationality of the square root of 2 is as follows:

Assume that the square root of 2 is a rational number, say a/b, where a and b are integers. Then

$$\frac{a^2}{b^2} = 2$$

so that

$$a^2 = 2b^2$$

It follows from this last equation that a^2 is an even integer, so that a must be even. Thus, $a/2$ is an integer, say

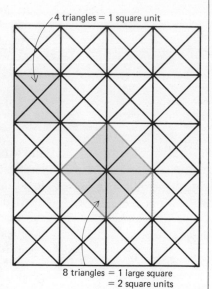

4 triangles = 1 square unit

8 triangles = 1 large square
           = 2 square units

$$\frac{a}{2} = a'$$

If we multiply both sides of this equation by 2, obtaining

$$a = 2a'$$

and substitute into the equation

$$a^2 = 2b^2$$

we obtain

$$(2a')^2 = 2b^2$$
$$4a'^2 = 2b^2$$
$$2a'^2 = b^2$$

Thus, b^2 is even, which implies that b must be even, say

$$b = 2b'$$

If we substitute this into the last of the above equations, we obtain

$$2a'^2 = b^2$$
$$2a'^2 = (2b')^2$$
$$2a'^2 = 4b'^2$$
$$a'^2 = 2b'^2$$

This last equation is identical in form to the equation

$$a^2 = 2b^2$$

that we started with. Thus, if we repeat the entire argument that we used above, we find that a' and b' are both even, say

$$a' = 2a'' \quad \text{and} \quad b' = 2b''$$

where

$$a''^2 = 2b''^2$$

We then can repeat the same argument again, and again, and again, *ad infinitum*. It follows that a and b both have 2 as a factor an infinite number of times. Since this cannot hold for any integers, then no such integers a and b can exist. Thus, the square root of 2 cannot be written as the ratio of two integers.

The Pythagoreans believed that all things can be explained from the properties of whole numbers. Their discovery of the irrationality of the square root of 2 destroyed this fundamental tenet of faith, because it showed that not even all distances can be expressed in terms of integers.

The discovery caused consternation in the secret society of the Pythagoreans. At first the society tried to suppress the news. (According to one legend, the member who first told the world of the discovery was drowned by his fellow members.) After a period of time, the Pythagoreans turned even more toward rigor in mathematical proofs and attempted to establish all results by arguments that could not be questioned. Eventually the society lost much of its secret nature and became, in essence, a professional society of mathematicians and mathematics teachers.

The numbers on the number line, rational and irrational, are known as *real numbers*.

Addition and subtraction are easy to consider geometrically with real numbers. To calculate $a + b$, for example, we first go to the point representing a and then shift b units to the point $a + b$. If $b > 0$, we shift to the right; if $b < 0$, we shift to the left. (See Fig. 3-17.) Subtraction can be handled by writing

$$a - b = a + (-b)$$

and then adding as above.

From a purely practical point of view, we normally use decimal approximations to real numbers in our work rather than the exact numbers. As an example, if $\sqrt{2}$ represents the square root of 2, then

$$\sqrt{2} \approx 1.414$$

We also know that

$$\pi \approx 3.142$$

The problems of dealing with irrational numbers were solved by the Greek mathematician *Eudoxus* (about 420 B.C.), who invented a sophisticated theory of ratio and proportion which allowed any two magnitudes of one type (whole numbers, fractions, lengths, distances, areas, and so forth) to be compared with any two magnitudes of any other type. The basic techniques developed by Eudoxus were used in the late nineteenth century to develop the properties of the real number system in a rigorous manner.

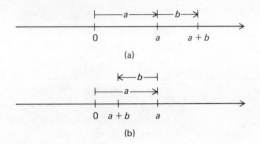

Figure 3-17 Addition on the number line. (a) $a + b$ $(b > 0)$. Shift to the right; (b) $a + b$ $(b < 0)$. Shift to the left.

Therefore,

$$\sqrt{2} + \pi \approx 1.414 + 3.142 \approx 4.556$$
$$\sqrt{2} - \pi \approx 1.414 - 3.142 \approx -1.728$$
$$\sqrt{2} \cdot \pi \approx (1.414)(3.142) \approx 4.443$$

and

$$\frac{\pi}{\sqrt{2}} \approx \frac{3.142}{1.414} \approx 2.222$$

Example 2. John works for a company that installs various types of insulating materials in public buildings. Today he must cut rectangles of foam installation to wrap around steam pipes that are 8 inches in diameter. How wide must he cut the insulation if he cuts it to the next largest $\frac{1}{16}$ of an inch?

Solution. The width of the insulation must equal the circumference of the pipe. Since the diameter is 8 inches, the circumference is

$$C = 8\pi \approx 8 \cdot (3.142) \approx 25.136 \text{ inches}$$

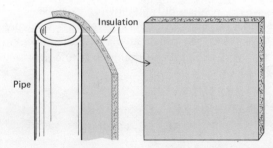

Figure 3-18 Example 2.

Now

$$\tfrac{1}{16} = 0.0625$$

$$\tfrac{2}{16} = 0.1250$$

and

$$\tfrac{3}{16} = 0.1875$$

Since 0.136 is between 0.1250 and 0.1875 and he cuts the insulation to the next largest $\tfrac{1}{16}$ of an inch, he must cut it $25\tfrac{3}{16}$ inches wide.

EXERCISES

1. Show that the following numbers are rational by writing them in the form a/b where a and b are integers.

   (a) 0.012   (c) $\dfrac{2\pi}{3\pi}$

   (b) 3.14159   (d) $\dfrac{21\pi}{\pi \cdot 16} - \dfrac{4\pi}{3\pi}$

2. Show that if π is divided by a rational number, the result is irrational. (*Hint:* If

   $$\frac{\pi}{a/b}$$

   were equal to the rational number c/d, show that π would be rational, contrary to the known fact that π is irrational.)

3. Without actually computing the numbers, decide which are rational and which are irrational. Use Exercise 2 and the results of this section.

   (a) $\dfrac{21 \cdot 3\pi}{(-5) \cdot 2}$ (c) $\dfrac{\pi}{7}$

   (b) $\dfrac{7\pi - 8\pi/3}{4\pi}$ (d) 5π

4. Show that there must be an infinite number of rational numbers between 0 and 1 by showing how an infinite set of these numbers can be constructed starting with 1/10, 1/100, 1/1,000, 1/10,000, and so on.

5. Show that there must be an infinite number of irrational numbers between 0 and 1 by showing how an infinite set of these numbers can be constructed starting with $\pi/10$, $\pi/100$, $\pi/1,000$, $\pi/10,000$, and so on. (*Hint:* Use Exercise 2.)

6. (a) Work Example 2 on the assumption that the pipes are 10 inches in diameter. Round off to the next largest $\tfrac{1}{16}$ of an inch.

   (b) Work Example 2 on the assumption that the pipes are 22 centimeters in diameter. Round off the result to the next largest millimeter.

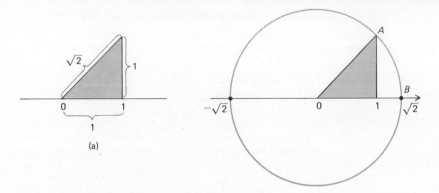

Figure 3-19   Geometrical construction of the square root of 2 (Exercise 7).

7. The length $\sqrt{2}$ (the square root of 2) can be constructed with a ruler and compass as follows (Fig. 3-19):

(1) Construct a right triangle with base and height equal to 1 unit as in the diagram in the margin.

(2) Place a compass with one point at 0 and the other at A. Construct a circle with center at 0. Let B be the point where the number line intersects the circle.

(a) If you recall the *Pythagorean Theorem,* use it to show that the distance from 0 to B, which is equal to the distance from 0 to A, is equal to $\sqrt{2}$.

(b) Make an accurate drawing on graph paper as described above and measure the distance from 0 to B. Compare your answer to the standard approximation $\sqrt{2} \approx 1.414$.

8. Make a drawing of the number line locating the irrational numbers π and $\sqrt{2}$ by the constructions described in this book. (Construct π by rolling a circle along the line; construct $\sqrt{2}$ as explained in Exercise 7.) Locate the following numbers on the number line by shifting distances geometrically:

(a) $\pi + \sqrt{2}$ (b) $\pi - \sqrt{2}$ (c) $\sqrt{2} - \pi$

9. Several times in this book we have mentioned that division by zero is not allowed and that fractions such as 6/0 are not defined. To see why this is the case, consider that the fraction a/b must have the property that when it is multiplied by b the product is a. For example,

$$\frac{2}{7} \cdot 7 = 2$$

$$\frac{(-12)}{5} \cdot 5 = -12$$

$$\frac{\pi}{7} \cdot 7 = \pi$$

and so on.

Suppose 6/0 were allowed as a number. Multiply this number by 0 and show that this leads to the conclusion that $6 = 0$, which is obviously not true. This illustrates why division by 0 is ruled out in mathematics.

3-7 NEGATIVE EXPONENTS; APPROXIMATIONS

In Chapter 2 we used positive exponents in certain expressions. For each number a we define

$$a^1 = a$$
$$a^2 = a \cdot a$$
$$a^3 = a \cdot a \cdot a$$
$$a^4 = a \cdot a \cdot a \cdot a$$
$$\cdots$$

These positive exponents obey four basic laws, which we illustrate with examples:

(1) *To multiply two powers of the same number we add the exponents.* For example,

$$4^2 \cdot 4^3 = (4 \cdot 4) \cdot (4 \cdot 4 \cdot 4) = 4 \cdot 4 \cdot 4 \cdot 4 \cdot 4 = 4^5$$

Thus,

$$4^2 \cdot 4^3 = 4^{2+3} = 4^5$$

(2) *To divide two powers of the same number we subtract the exponents.* For example,

$$\frac{3^4}{3^1} = \frac{3 \cdot 3 \cdot 3 \cdot \cancel{3}}{\cancel{3}} = 3 \cdot 3 \cdot 3 = 3^3$$

Thus,

$$\frac{3^4}{3^1} = 3^{4-1} = 3^3$$

Similarly,

$$\frac{2^2}{2^6} = \frac{\cancel{2} \cdot \cancel{2}}{2 \cdot 2 \cdot 2 \cdot 2 \cdot \cancel{2} \cdot \cancel{2}} = \frac{1}{2 \cdot 2 \cdot 2 \cdot 2} = \frac{1}{2^4}$$

so that

$$\frac{2^2}{2^6} = \frac{1}{2^{6-2}} = \frac{1}{2^4}$$

(3) *To raise a product to a power we raise each of the original factors to the power.* For example,

$$(2 \cdot 3)^4 = (2 \cdot 3) \cdot (2 \cdot 3) \cdot (2 \cdot 3) \cdot (2 \cdot 3)$$
$$= 2 \cdot 2 \cdot 2 \cdot 2 \cdot 3 \cdot 3 \cdot 3 \cdot 3 = 2^4 \cdot 3^4$$

Thus,

$$(2 \cdot 3)^4 = 2^4 \cdot 3^4$$

(4) *To raise a power to a new power we multiply the exponents.* For example,

$$(3^2)^3 = (3 \cdot 3)^3 = (3 \cdot 3)(3 \cdot 3)(3 \cdot 3) = 3 \cdot 3 \cdot 3 \cdot 3 \cdot 3 \cdot 3 = 3^6$$

Thus,

$$(3^2)^3 = 3^{2 \cdot 3} = 3^6$$

These laws can be stated for arbitrary numbers a and b and arbitrary exponents m and n as follows:

(1) $a^m \cdot a^n = a^{m+n}$ (Add exponents)

(2) $\begin{cases} \dfrac{a^m}{a^n} = a^{m-n} & \text{if } m > n \quad \text{(Subtract exponents)} \\[2mm] \dfrac{a^m}{a^n} = \dfrac{1}{a^{n-m}} & \text{if } n > m \quad \text{(Subtract exponents)} \end{cases}$

(3) $(ab)^n = a^n \cdot b^n$ (Raise each number in the product to the exponent)

(4) $(a^m)^n = a^{mn}$ (Multiply exponents)

RULES OF EXPONENTS

$a^m \cdot a^n = a^{m+n}$

$\dfrac{a^m}{a^n} = \begin{cases} a^{m-n} & \text{if } m > n \\ 1/a^{n-m} & \text{if } m < n \\ 1 & \text{if } m = n \end{cases}$

$(ab)^n = a^n b^n$
$(a^m)^n = a^{mn}$

Let us see what happens if we try to apply the second rule (subtracting exponents when we divide) in the special case where we have $2^3/2^3$. In that case the fraction is

$$\frac{2^3}{2^3} = \frac{8}{8} = 1$$

If the rule for subtracting exponents is to hold, we must have

$$\frac{2^3}{2^3} = 2^{3-3} = 2^0$$

Thus, 2^0 should be equal to 1 provided the rule for subtracting exponents

FLOW CHART TO VERIFY
THAT $(AB)^N = A^N B^N$ IN
PARTICULAR EXAMPLES

START

READ
A, B, N

COMPUTE
C = AB

COMPUTE
X = C^N

COMPUTE
D = A^N

COMPUTE
E = B^N

COMPUTE
Y = DE

IS
X = Y
?

NO   YES

PRINT "RULE
DOES NOT
HOLD IN
THIS
CASE"

PRINT "RULE
DOES HOLD
IN THIS
CASE"

STOP

is to hold when the exponents in the numerator and denominator are equal.

Similarly, we should have

$$\frac{5^2}{5^2} = 1$$

and

$$\frac{5^2}{5^2} = 5^{2-2} = 5^0$$

so that

$$5^0 = 1$$

Let us assume that a^0 is known to be equal to 1 when a is a nonzero number:

$$a^0 = 1 \quad \text{if} \quad a \neq 0$$

We can use this fact to show that a^{-m} should be equal to $1/a^m$.

Consider the example $4^2/4^5$. If the rule for subtracting exponents is to hold, we must have

$$\frac{4^2}{4^5} = 4^{2-5} = 4^{-3}$$

But we know that

$$\frac{4^2}{4^5} = \frac{\cancel{4} \cdot \cancel{4}}{4 \cdot 4 \cdot 4 \cdot \cancel{4} \cdot \cancel{4}} = \frac{1}{4^3}$$

Thus,

$$4^{-3} = \frac{1}{4^3}$$

Similarly,

$$\frac{7^8}{7^{13}} = 7^{8-13} = 7^{-5}$$

and

$$\frac{7^8}{7^{13}} = \frac{1}{7^{13-8}} = \frac{1}{7^5}$$

so that

$$7^{-5} = \frac{1}{7^5}$$

In general, if the law for subtracting exponents is to hold for all exponents, we must have

$$a^{-m} = \frac{1}{a^m} \quad \text{provided} \quad a \neq 0$$

The above arguments cannot be used to actually prove that

$$a^0 = 1 \quad \text{and} \quad a^{-m} = \frac{1}{a^m} \quad \text{if} \quad a \neq 0$$

because they were based on the underlying assumption that the law

$$\frac{a^m}{a^n} = a^{m-n}$$

holds in every case. If we have not properly defined negative exponents, then we cannot know that this law of exponents still holds. What the examples do show, however, is that we must define $a^0 = 1$ and $a^{-m} = 1/a^m$ if we wish to have $a^m/a^n = a^{m-n}$ in every case.

Mathematicians always try to develop mathematics in a consistent, systematic fashion. For this reason they define zero and negative exponents by the rules

$$a^0 = 1 \quad \text{and} \quad a^{-m} = \frac{1}{a^m}$$

provided $a \neq 0$.

One advantage of this definition is that we can combine the various parts of the law

$$\frac{a^m}{a^n} = a^{m-n} \quad \text{if} \quad m > n$$

$$\frac{a^m}{a^n} = \frac{1}{a^{n-m}} \quad \text{if} \quad m < n$$

$$\frac{a^m}{a^n} = 1 \quad \text{if} \quad m = n$$

into the one simple law

$$\frac{a^m}{a^n} = a^{m-n}$$

Example 1. (a) $\left(\dfrac{17}{13}\right)^{-1} = \dfrac{1}{\dfrac{17}{13}} = 1 \div \dfrac{17}{13} = 1 \cdot \dfrac{13}{17} = \dfrac{13}{17}$

(b) $\dfrac{6^2}{4^{-3}} = \dfrac{6^2}{\dfrac{1}{4^3}} = 6^2 \cdot \dfrac{4^3}{1} = 6^2 \cdot 4^3 = 36 \cdot 64 = 2304$

(c) 0^{-1} is not defined

(d) $(-13)^0 = 1$

(e) $\left(\dfrac{2}{5} - \dfrac{2}{5}\right)^0$ is not defined.

Example 2. Simplify the expression

$$\frac{2^{-7} + 2^{-10}}{2^{-11} + 2^{-9}}$$

Solution. Observe that if we multiply each of the individual terms by 2^{11}, we obtain a term with a positive exponent. In order to accomplish this, we multiply numerator and denominator by 2^{11}:

$$\frac{2^{-7} + 2^{-10}}{2^{-11} + 2^{-9}} = \frac{2^{-7} + 2^{-10}}{2^{-11} + 2^{-9}} \cdot \frac{2^{11}}{2^{11}} = \frac{2^{-7} \cdot 2^{11} + 2^{-10} \cdot 2^{11}}{2^{-11} \cdot 2^{11} + 2^{-9} \cdot 2^{11}}$$

$$= \frac{2^4 + 2^1}{2^0 + 2^2} = \frac{16 + 2}{1 + 4} = \frac{18}{5}$$

It can be shown that the four laws stated for positive exponents still hold if some or all of the exponents are negative or zero; that is,

(1) $a^m \cdot a^n = a^{m+n}$

(2) $\dfrac{a^m}{a^n} = a^{m-n}$

(3) $(ab)^m = a^m b^m$

(4) $(a^m)^n = a^{mn}$

Example 3. (a) $3^2 \cdot 3^{-3} = 3^{2+(-3)} = 3^{-1} = \frac{1}{3}$

(b) $(5/5^{-1})^{-1} = (5^1/5^{-1})^{-1} = (5^{1-(-1)})^{-1}$
  $= (5^{1+1})^{-1} = (5^2)^{-1} = 5^{2(-1)}$
  $= 5^{-2} = 1/5^2 = 1/25$

(c) $(5 \cdot 7^{-1})^{-1} = 5^{-1} \cdot 7^{(-1)(-1)}$
  $= 5^{-1} \cdot 7^1 = \frac{7}{5}$

RULES OF EXPONENTS (POSITIVE, NEGATIVE, ZERO)

$a^m \cdot a^n = a^{m+n}$

$\dfrac{a^m}{a^n} = a^{m-n}$

$(ab)^n = a^n b^n$

$(a^m)^n = a^{mn}$

These general laws of exponents can be established by rewriting all of the numbers as products or quotients involving positive exponents and then applying the rules stated previously for positive exponents. In establishing these laws, we must be sure that we have worked through all possible cases that could occur. As an example we shall verify that $(ab)^n = a^n b^n$ for the special cases where $a = 2$, $b = 7$, $n = -3$, and $n = 0$.

Example 4. (a) Show that

$$(2 \cdot 7)^{-3} = 2^{-3} \cdot 7^{-3}$$

(b) Show that

$$(2 \cdot 7)^0 = 2^0 \cdot 7^0$$

Solution. (a) $(2 \cdot 7)^{-3} = \dfrac{1}{(2 \cdot 7)^3} = \dfrac{1}{2^3 \cdot 7^3} = \dfrac{1}{2^3} \cdot \dfrac{1}{7^3} = 2^{-3} \cdot 7^{-3}$

(b) $(2 \cdot 7)^0 = 1 = 1 \cdot 1 = 2^0 \cdot 7^0$

Approximations Using Negative Exponents

Negative exponents can be used to simplify problems involving approximations to ratios. For example, we can approximate the ratio 0.00178/31 by first writing

$$0.00178 = 17.8 \cdot 10^{-4} \approx 18 \cdot 10^{-4}$$

and

$$31 = 3.1 \cdot 10 \approx 3 \cdot 10$$

Then

$$\frac{0.00178}{31} \approx \frac{18 \cdot 10^{-4}}{3 \cdot 10} \approx \frac{18}{3} \cdot 10^{-4-1} \approx 6 \cdot 10^{-5} \approx 0.00006$$

Scientific Notation

Scientists have been faced with the necessity of making approximations to products and quotients for years. In most cases the numbers involved have been obtained by measurement and are only approximate themselves. The problem has been to use these approximations in calculations in such a way that the computations can be carried out easily and the final answer does not lead one to think that the answer is too exact. From necessity they have developed the following method.

1. Write every nonzero number as a single nonzero digit followed by a decimal point and other digits, all multiplied by a power of 10. This scheme of writing the numbers is known as "scientific notation." For example, in scientific notation we write

SCIENTIFIC NOTATION

$271 = 2.71 \cdot 10^2$
$0.0051 = 5.1 \cdot 10^{-3}$
$20\pi \approx 6.28 \cdot 10^1$

$$271 \quad \text{as} \quad 2.71 \cdot 10^2 \quad \text{and} \quad 0.0051 \quad \text{as} \quad 5.1 \cdot 10^{-3}$$

2. Perform the indicated multiplications and divisions using the laws of exponents wherever possible.

3. Write the final answer in scientific notation. At this step we shorten the final answer to the *least number of decimal places* found among the original numbers.

Example 5. Use scientific notation to approximate ab/c where $a \approx 51$, $b \approx 0.00312$, and $c \approx 2134$.

Solution. We follow the steps listed above.

(1) $a \approx 5.1 \cdot 10^1$ (one decimal place)
    $b \approx 3.12 \cdot 10^{-3}$ (two decimal places)
    $c \approx 2.134 \cdot 10^3$ (three decimal places)

(2) $\dfrac{ab}{c} \approx \dfrac{(5.1 \cdot 10^1)(3.12 \cdot 10^{-3})}{2.134 \cdot 10^3}$

$\approx \dfrac{(5.1)(3.12)}{2.134} \cdot \dfrac{10^1 \cdot 10^{-3}}{10^3} \approx \dfrac{15.912}{2.134} \cdot 10^{1-3-3} \approx 7.456 \cdot 10^{-5}$

(3) We must keep only the least number of decimal places found among the original numbers when they were written in scientific notation. Since the least number of decimal places was in $a \approx 5.1 \cdot 10^1$ which had one decimal place, we keep only one decimal place in the final answer. This means that we must drop the quantity 0.056 from the number 7.456. Since the first digit we dropped is a "5," we "round up" the last digit we retain giving us $7.456 \approx 7.5$. The final answer is

$$\frac{ab}{c} \approx 7.5 \cdot 10^{-5}$$

or, in more conventional notation,

$$\frac{ab}{c} \approx 0.000075$$

The accuracy of the original measurements is shown by the number of decimal places when the numbers are written in scientific notation. In Example 5 the least accurate number had one decimal place. We round-off the final answer to the least number of decimal places in the original numbers in order to show that the answer is no more accurate than the least accurate of the original measurements.

EXERCISES

1. Use the laws of exponents to simplify the following expressions. The final answers should be free of negative and zero exponents.

(a) $\dfrac{3^7}{3^{-1}}$

(b) $(2^{-1} \cdot 3)^{-2}$

(c) $\dfrac{2^0 \cdot 5^3}{2^{-1} \cdot 5^2}$

(d) $\dfrac{7^3 \cdot 7^{-3}}{2^2}$

(e) $\dfrac{3^2 \cdot (21)^{-2}}{2^7 \cdot 3^4}$

(f) $\dfrac{17^{12} \cdot 5^3}{(17 \cdot 5)^2}$

(g) $\dfrac{(3 \cdot 4)^3}{(2 \cdot 6)^4}$

(h) $\dfrac{7^2 \cdot 0^{-1}}{7^1}$

2. Simplify the following expressions. The final answers should be free of negative and zero exponents.

(a) $\left(\dfrac{ab}{c}\right)^{-2}$

(d) $(a^2 \cdot b^3)^{-2}$

(b) $(a + b)^{-1}$

(e) $(a + b)^0$

(c) $(ab)^3 \cdot a^2 \cdot b^{-1}$

(f) $(a^3b)^2/(ab^2)^3$

3. Simplify

(a) $\dfrac{4^{-3} + 4^{-4}}{4^{-3} - 4^{-4}}$

(b) $\dfrac{3^7 + 3^9}{3^6 + 3^{10}}$

4. Write out all of the steps needed to give a direct argument that
(a) $7^5 \cdot 7^4 = 7^9$
(b) $3^{12} \cdot 3^4 = 3^{16}$
(c) $a^3 \cdot a^4 = a^7$

5. Write out all of the steps needed to give a direct argument that
(a) $(5^2)^3 = 5^6$ (b) $(a^3)^5 = a^{15}$
(c) $(a^m)^n = a^{mn}$ for any two *positive integers* m and n. [*Hint:* $(a^m)^n = a^m \cdot a^m \cdot \cdots \cdot a^m$ (n factors) and each factor a^m is $a^m = a \cdot a \cdot \cdots \cdot a$ (m factors).]

6. Construct a flow chart for the steps that would be needed to verify that

$$(a^m)^n = a^{mn}$$

for particular integers a, m, n.

7. Construct a flow chart that could be used to verify that $a^m/a^n = a^{m-n}$ for particular integers a, m, n where $a \neq 0$.

8. Write the following numbers in scientific notation:
(a) 7581.23   (c) 0.00078
(b) 4.12   (d) 0.00000000913

9. Use scientific notation to approximate the following numbers (where $a \approx 17.32$, $b \approx 0.00156$, $c \approx 183.2$):
(a) ab/c (b) abc (c) b/ac

10. The fact that 0^0 is not defined can be explained by the rule $a^m/a^n = a^{m-n}$. Interpret this rule for the special case $a \doteq 0$, $m = 5$, $n = 5$. Show that if the rule held in that case, we would have a fraction with zero in the denominator.

3-8 DECIMAL EXPANSIONS OF RATIONAL AND IRRATIONAL NUMBERS

Although decimals are the most convenient fractions to use, they do have two inherent limitations. First, they only can be used to represent rational numbers. If we want to express an irrational number, such as π, by decimals, we are forced to use an approximation, such as 3.14 or 3.14159.

A BRIEF TABLE OF RECIPROCALS
Only $\frac{1}{2}, \frac{1}{4}, \frac{1}{5}, \frac{1}{8},$ and $\frac{1}{10}$ have the exact
values shown.

$\frac{1}{2} = 0.5000000$
$\frac{1}{3} \approx 0.3333333$
$\frac{1}{4} = 0.2500000$
$\frac{1}{5} = 0.2000000$
$\frac{1}{6} \approx 0.1666667$
$\frac{1}{7} \approx 0.1428571$
$\frac{1}{8} = 0.1250000$
$\frac{1}{9} \approx 0.1111111$
$\frac{1}{10} = 0.1000000$

The second problem with decimals is that not even all of the rational numbers can be represented exactly by decimals. As a matter of fact, the rational numbers that can be expressed exactly as decimals are rather scarce in the entire set of rational numbers. The most familiar rational number that cannot be written exactly as a decimal is $\frac{1}{3}$. We can write

$$\frac{1}{3} \approx 0.3 \qquad \text{(one-decimal-place accuracy)}$$

$$\frac{1}{3} \approx 0.33 \qquad \text{(two-decimal-place accuracy)}$$

$$\frac{1}{3} \approx 0.333 \qquad \text{(three-decimal-place accuracy)}$$

and so on, but none of these approximations is exactly equal to $\frac{1}{3}$.

Every real number can be expressed exactly as a decimal if we allow an infinite number of decimal places. For example,

$$\frac{1}{3} = 0.33333\ldots \qquad \text{and} \qquad \frac{1}{6} = 0.16666666\ldots$$

the dots indicating that the digits repeat forever.

The basic difference between rational and irrational numbers becomes apparent when we consider their infinite decimal expansions. *Every rational number has an infinite decimal expansion in which, from some point on, a single block of digits repeats over and over.* Irrational numbers do not have these repeating blocks of digits.

For example,

$$\frac{1}{7} = 0.142857 \ \ 142857 \ \ 142857\ldots$$
$$\frac{5}{22} = 0.22 \ \ 72 \ \ 72 \ \ 72 \ \ 72 \ \ 72 \ \ 72\ldots$$

and

$$\frac{11}{26} = 0.423076 \ \ 923076 \ \ 923076\ldots$$

where the spaces separate the repeating blocks of digits. The number

$$0.010203040506070809010011012\ldots$$

in which every integer is listed with a "zero prefix" is irrational. This last number has no repeating block of digits.

If we are given a common fraction, such as $\frac{7}{22}$, we can find the repeating block of digits from the long-division algorithm. We simply continue the algorithm until the same remainder is obtained twice. From that point on the same block of digits will be repeated in the quotient.

An infinite decimal which has a repeating block can be rewritten as a common fraction by the technique illustrated in the following example.

Example 1. Write the following infinite decimals as common fractions:

(a) $a = 0.23\ 23\ 23\ 23\ldots$
(b) $b = 0.12\ 537\ 537\ 537\ldots$

Solution. We first rewrite each infinite decimal in blocks of digits equal in length to the repeating blocks. The number a already is in this form. We rewrite b as

$$b = 0.125\ 375\ 375\ 375\ldots$$

(a) The repeating block in the decimal expansion of

$$a = 0.23\ 23\ 23\ 23\ldots$$

has two digits. If we multiply a by 10^2, we shall simply move the decimal point two places to the right, obtaining another decimal with the same repeating block:

$$100a = 10^2 a = 23.23\ 23\ 23\ 23\ldots$$

We now subtract a from $100a$:

$$
\begin{aligned}
100a &= 23.23\ 23\ 23\ 23\ldots \\
-a &= -0.23\ 23\ 23\ 23\ldots \\
\hline
99a &= 23.00\ 00\ 00\ 00\ldots \\
a &= \frac{23}{99}
\end{aligned}
$$

Thus, $0.23\ 23\ 23\ 23\ldots = \frac{23}{99}$.
(b) The repeating block in the decimal expansion of

$$b = 0.125\ 375\ 375\ 375\ldots$$

has three digits. We multiply b by 10^3 and subtract b:

$$
\begin{aligned}
10^3 b = 1000b &= 125.375\ 375\ 375\ldots \\
-b &= -0.125\ 375\ 375\ldots \\
\hline
999b &= 125.250\ 000\ 000\ldots \\
b &= \frac{125.25}{999} = \frac{12{,}525}{99{,}900} \\
&= \frac{167 \cdot 25 \cdot 3}{1332 \cdot 25 \cdot 3} = \frac{167}{1332}
\end{aligned}
$$

The ancient Babylonians used tables of reciprocals for division. In order to divide 5 by 3, for example, they multiplied 5 by $\frac{1}{3}$.

Most of their tables did not have the reciprocals of 7 and 11 because these fractions could not be expressed as finite fractions in the Babylonian number system.

Irrational numbers such as π and $\sqrt{2}$ have decimal expansions that do not repeat in blocks. Regardless of the number of decimal places that are calculated, no apparent patterns of digits appear with these numbers.

With the limitations of decimals in mind, let us examine some of their advantages. First, they are easy to use. As long as we remember where to put the decimal point, we can use them in computations as easily as we use integers. Second, the use of decimals enables us to compare the sizes of two numbers with ease. It may be quite difficult to decide whether $\frac{22}{7}$ or π is larger, but the problem becomes simple if we write out the decimal approximations. We see that

$$\frac{22}{7} \approx 3.14286$$

and

$$\pi \approx 3.14159$$

so that $\frac{22}{7} > \pi$.

The third useful fact about decimals, the one that accounts for much of their convenience, is that they can be used to approximate both rational and irrational numbers to any desired degree of accuracy.

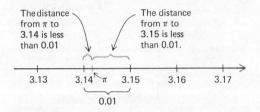

Figure 3-20 If π is between two numbers that are 0.01 unit apart, then the distance from π to either of the numbers is less than 0.01. Since π is between 3.14 and 3.15, then $\pi \approx 3.14$ and $\pi \approx 3.15$, both approximations correct to within 0.01 unit.

To explain this last statement in more detail, we consider the decimal approximations to π. First, since π is between 3.14 and 3.15 (see Fig. 3-20), we have

$$\pi \approx 3.14 \qquad \text{(small estimate, error less than 0.01)}$$

and

$$\pi \approx 3.15 \qquad \text{(large estimate, error less than 0.01)}$$

If we carry this one decimal place further, we have π between 3.141 and 3.142 (see Fig. 3-21). Thus,

$$\pi \approx 3.141 \qquad \text{(small estimate, error less than 0.001)}$$

and

$$\pi \approx 3.142 \qquad \text{(large estimate, error less than 0.001)}$$

$\pi \approx 3.14159265358979323846$
26433832795028841971

π has been approximated to over one-half million decimal places. There is no practical need for this type of approximation since six-decimal-place accuracy enables us to calculate the circumference of circle with a radius of 10 miles correct to one-third of an inch.

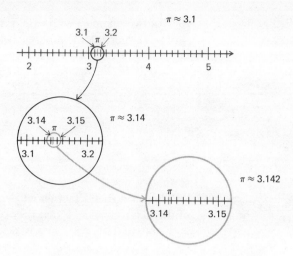

Figure 3-21 $\pi \approx 3.1$ (correct to one decimal place); π ≈ 3.14 (correct to two decimal places); $\pi \approx 3.142$ (correct to three decimal places).

When four decimal places are considered, we have π between 3.1415 and 3.1416 so that

$$\pi \approx 3.1415 \qquad \text{(small estimate, error less than 0.0001)}$$

and

$$\pi \approx 3.1416 \qquad \text{(large estimate, error less than 0.0001)}$$

This process can be carried out as far as we wish although it becomes more and more difficult to calculate the digits.

Round-off and Truncation Errors

We usually terminate the decimal approximation to a number after a certain number of decimal places have been calculated. There are two standard ways to do this: *round-off* and *truncation*.

The more accurate of the two methods is *round-off*. To apply it we locate the number between two decimal approximations and then choose the one that is closest to the original number. For example, if we wish to approximate π to four decimal places, we first locate it between 3.1415 and 3.1416. We then choose the one of these approximations that is closest to π. Correct to five decimal places, we have $\pi \approx 3.14159$. Since this number is closer to 3.1416 than it is to 3.1415, we have

$$\pi \approx 3.1416 \qquad \text{(correct to four decimal places)}$$

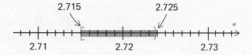

Figure 3-22 Any number that is greater than or equal to 2.715 and is less than 2.725 rounds off to 2.72. Consequently, when we see a number written as 2.72, we only know that it is within the above range of values.

In general, when we round-off a decimal expansion at a certain number of decimal places, we calculate one place more than we need, then drop this last digit. If the dropped digit is 0, 1, 2, 3, or 4, we leave the other digits as they were. If the dropped digit is 5, 6, 7, 8, or 9, we add *one* to the last digit retained. Thus, 2.718 would be rounded-off as 2.72, while 51.321 would be rounded-off as 51.32. (See Fig. 3-22.)

Example 1. The following numbers are rounded-off to four decimal places.
(a) $2.78539 \approx 2.7854$
(b) $0.00192 \approx 0.0019$
(c) $3.00285 \approx 3.0029$
(d) $21.53996 \approx 21.5400$

It should be clear from a study of Example 1 how the number of decimal places retained tells us the accuracy. In (a), for example, the approximation 2.7854 indicates a number between 2.78535 and 2.78545. (Any number in this range rounds-off to 2.7854.) In (d) we write the approximation as 21.5400, indicating a number between 21.53995 and 21.54005. If we had written the approximation as 21.54, only numbers between 21.535 and 21.545 would have been indicated.

Truncation is the second method that can be used to terminate the decimal expansion of a number. This method simply involves dropping all of the digits after the desired number of places. For example, if we have $\pi \approx 3.14159$ and we use truncation to approximate π to four decimal places, we simply drop the last digit obtaining $\pi \approx 3.1415$.

Both round-off and truncation have built-in errors, and since truncation obviously has the larger error, then why do we use it? The answer lies with the widespread use of the electronic computer. For certain technical reasons it is difficult to round-off with a computer; so it is programmed to truncate after a certain number of decimal places.

In our normal work with arithmetic, we rarely need to do more than three or four computations; so round-off and truncation errors do not cause us too much trouble. When we work with an electronic computer, however, we may have a problem that involves a thousand or more computations. (Such problems can be remarkably easy to program by the use of loops.) It is obvious that a thousand truncations can lead to a considerable error in the final answer.

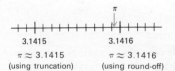

$\pi \approx 3.1415$ (using truncation) $\pi \approx 3.1416$ (using round-off)

Truncation is much less accurate than round-off. With truncation we get 3.1415 as a four-place approximation to π, even though π is much closer to 3.1416.

As an example, suppose the computer is programmed to calculate and add together 1000 numbers. All numbers are to be truncated after two decimal places. How much error can we expect in the final answer? Observe that the amount dropped from a number by truncation is roughly 0.000, 0.001, 0.002, 0.003, 0.004, 0.005, 0.006, 0.007, 0.008, or 0.009. On the average we would expect the error to be about 0.005 for each number that we truncate. When we add the 1000 numbers that have been truncated, we would expect the total error to be approximately $1000 \cdot (0.005) = 5$. If we were to add 1 million such numbers, the expected error would be approximately $1,000,000 \cdot (0.005) = 5000$.

Since computers are used for both speed and accuracy, an error of 5000 units in the final answer cannot be tolerated. It is the duty of the computer programmer to set up a "tolerance allowance" for the total error that can be allowed in the solution of a problem. He then must decide how many times the arithmetic operations can be carried out and how much accuracy is needed in all approximations in order to stay within the allowable error. For the problem involving 1 million additions, he might decide that four-place accuracy must be used rather than two-place accuracy in all approximations, or he might decide on a different method of solution that required fewer calculations.

EXERCISES

1. Calculate the decimal expansions of the following numbers correct to six decimal places using both round-off and truncation. When do the two approximations agree?

   (a) $\frac{2}{9}$ (c) $\frac{7}{9}$ (e) $\frac{4}{70}$
   (b) $\frac{5}{11}$ (d) $\frac{23}{140}$ (f) $\frac{19}{83}$

2. Use decimal approximations to decide which number in each pair is the larger.

   (a) $\frac{1}{6}, 0.16666666$ (b) $\pi, 3.1416$ (c) $\frac{18}{61}, \frac{5}{17}$

3. If round-off is used to approximate a number to two decimal places, show that the error can be approximately -0.005, -0.004, -0.003, -0.002, -0.001, 0.000, 0.001, 0.002, 0.003, 0.004, or 0.005.

   How large would you expect the total error to be when 1 million numbers are added, each being rounded-off to two decimal places? How does this compare with the expected error due to truncation as explained at the end of this section?

4. Every rational number either is an exact decimal or has an "infinite" decimal expansion that "repeats in blocks." For example, the decimal expansion of $\frac{1}{11}$ is

$$\tfrac{1}{11} = 0.09 \ 09 \ 09 \ 09 \ 09 \ldots$$

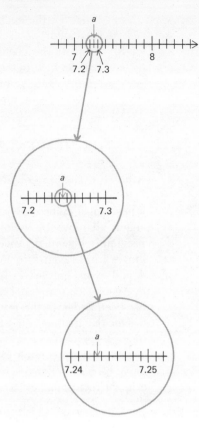

Figure 3-23 Exercise 6.

the block of two digits "09" repeating forever. Determine the repeating blocks for the decimal expansions of the following numbers.

(a) $\frac{1}{3}$ (b) $\frac{4}{9}$ (c) $\frac{1}{9}$ (d) $\frac{7}{11}$ (e) $\frac{2}{13}$

5. Express the following repeating decimals as common fractions.

    (a) 0.27 27 27 27...   (c) 0.314 412 412 412...

    (b) 0.103 103 103 103...   (d) 0.27 3015 3015 3015...

6. Every number on the number line either is an exact decimal or can be approximated by decimals to any desired degree of accuracy. As an example, consider the number a shown in Figure 3-23. We first locate the number between two consecutive integers. (a is between 7 and 8.) This gives us the first digit in our decimal expansion. Next we split the interval between 7 and 8 into ten equal subintervals and locate the number in one of these subintervals. (a is between 7.2 and 7.3.) Next partition the small interval containing the number into ten equal subintervals of length 0.01 and locate the number in one of these smaller subintervals.

(a) What are the endpoints of subintervals of length 0.01 in which *a* is located?

(b) Estimate the decimal expansion of *a* to three decimal places.

(c) Explain how in theory the process described above can be carried out indefinitely. What practical problems arise when you actually try to carry out this process to estimate *a* to several decimal places?

3-9   THE HEAVENLY HOTEL

A hotel in Heaven contains an infinite number of rooms numbered 1, 2, 3, 4, 5, and so on. Each of these rooms is occupied by one angel. When a new angel comes to visit, there is no shortage of rooms. The original angels all move to rooms numbered one greater than the rooms they previously occupied. Angel 1 moves from Room 1 to Room 2; Angel 2 moves from Room 2 to Room 3; Angel 3 moves from Room 3 to Room 4, and so on. This leaves Room 1 vacant for the new angel, who occupies it. Again each room is occupied by one angel even though one new angel has been added to the original collection. (See Fig. 3-24.)

One day the innkeeper finds it necessary to make drastic changes in the room assignments in order to accommodate an infinite number of new angels who are coming to the hotel for a convention. His solution is to move each of the angels into the room with a number double that of the original room number. At 10 o'clock in the morning, he rings a bell. At the sound of the bell, Angel 1 moves to Room 2; Angel 2 moves to Room 4; Angel 3 moves to Room 6; Angel 4 moves to Room 8; Angel 5 moves to Room 10, and so on. (See Fig. 3-25.) By 10:01 A.M. all of the

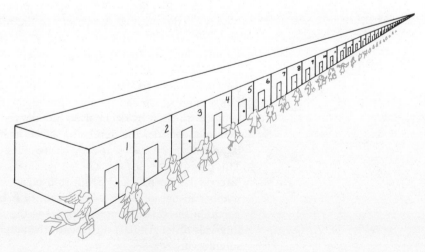

Figure 3-24   The heavenly hotel.

Angel number	Old room number	New room number		Angel number	Old room number	New room number
1	1	$\longrightarrow$ 2		1	1	$\longrightarrow$ 2
2	2	$\longrightarrow$ 3		2	2	$\longrightarrow$ 4
3	3	$\longrightarrow$ 4		3	3	$\longrightarrow$ 6
4	4	$\longrightarrow$ 5		4	4	$\longrightarrow$ 8
5	5	$\longrightarrow$ 6		5	5	$\longrightarrow$ 10
$\vdots$	$\vdots$	$\vdots$		6	6	$\longrightarrow$ 12
n	n	$\longrightarrow$ $n+1$		$\vdots$	$\vdots$	$\vdots$
$\vdots$	$\vdots$	$\vdots$		n	n	$\longrightarrow$ $2n$
				$\vdots$	$\vdots$	$\vdots$

(a)   (b)

Figure 3-25 (a) Old and new room assignments (first example); (b) old and new room assignments (second example).

angels are in their new rooms, leaving Rooms 1, 3, 5, 7, 9, 11, 13, and so on, empty. At 10:30 A.M. the new angels arrive and occupy the vacant rooms. Once again there is one angel to one room even though the original set of angels has been doubled.

These examples were devised by *Georg Cantor*, the founder of the branch of mathematics known as set theory. The first example shows that if one element is adjoined to an infinite set, then in a certain sense the number of elements is unchanged. The second example shows that even if an infinite number of elements are adjoined to an infinite set, the number of elements may be unchanged.

Georg Cantor (1845–1918), the developer of set theory, was one of the most creative and original mathematicians who ever lived. His radical ideas about set theory were so disturbing to some of his fellow mathematicians that they conspired to keep his works from being published. *Leopold Kronecker*, one of the prominent mathematicians of his day, led the fight against Cantor and personally intervened to keep Cantor from being appointed as a professor at the University of Berlin. Cantor's fragile grip on reality was weakened by the controversy and by the personal attacks by Kronecker. Eventually he had to be committed to a mental asylum, where he remained until his death.

The key to Cantor's work was the pairing of elements of two sets, such as the pairing of angels and hotel rooms. If such a pairing exists between two sets, then they have the same *rank* (or same *cardinality*). Cantor went on to rank all sets according to size in such a way that a larger set is always of rank greater than or equal to a smaller set. Turning his attention to number systems, Cantor was able to show that no pairing exists between the rational numbers and the irrational numbers (so they have different ranks), but that a pairing does exist between the rational numbers and a subset of the irrational numbers. This establishes that the rank of the rational numbers must be less than the rank of the irrational numbers. In other words, in the sense of Cantor's theory, there are more irrational numbers than there are rational numbers.

The fact that there are more irrational numbers than rational numbers has some surprising results. One result is that if we make a "blind stab" at the number line and pick a real number at random, it is almost certain to be irrational rather than rational. Thus, in a sense, the familiar rational numbers are oddities, while the irrational numbers are quite common.

Cantor's work has an interesting interpretation in terms of accuracy of measurement. Suppose that a carpenter cuts a board to a random length before he measures it. According to Cantor's theory the true length is almost certain to be an irrational number. Because his scale is marked only with rational numbers, however, the carpenter always measures the length as a rational number. This same reasoning applies to all lengths. Thus, almost all lengths are irrational, but almost all measured lengths are rational. It follows that almost all measured lengths are, of necessity, inaccurate approximations to the true lengths that they are supposed to measure.

A ONE-TO-ONE PAIRING BETWEEN THE POSITIVE INTEGERS AND THE POSITIVE EVEN INTEGERS
According to Cantor's theory, this pairing shows that there is the "same number" of positive integers as there are positive even integers. This is the case even though only every other positive integer is even.

Positive integers	Even positive integers
1	$\longleftrightarrow$ 2
2	$\longleftrightarrow$ 4
3	$\longleftrightarrow$ 6
4	$\longleftrightarrow$ 8
5	$\longleftrightarrow$ 10
$\vdots$	
n	$\longleftrightarrow$ $2n$
$\vdots$	

SUGGESTIONS FOR FURTHER READING

1. Ball, W. W. Rouse. *A Short Account of the History of Mathematics*, Chaps. 1, 7, 10, 11. New York: Dover, 1960.
2. Cooley, Hollis R., and Wahlert, Howard E. *Introduction to Mathematics*, Chaps. 2, 3, 4. Boston: Houghton-Mifflin, 1968.
3. Dantzig, Tobias. *Number, the Language of Science*, 4th ed. New York: Macmillan, 1954.
4. Kline, Morris. *Mathematics for Liberal Arts*, Chap. 4. Reading, Mass.: Addison-Wesley, 1967.

   The following selections can be found in *World of Mathematics*, James R. Newman, ed. New York: Simon & Schuster, 1956.
5. Ball, W. W. Rouse. "Calculating Prodigies." *World of Mathematics*, Vol. 1, Part III, Chap. 4.
6. Conant, Levi Leonard. "Counting." *World of Mathematics*, Vol. 1, Part III, Chap. 2.
7. Smith, David Eugene, and Ginsburg, Jekuthiel. "From Numbers to Numerals and from Numerals to Computation." *World of Mathematics*, Vol. 1, Part III, Chap. 3.

4

ELEMENTS OF GEOMETRY

4-1 INTRODUCTION TO GEOMETRICAL CONCEPTS

The study of geometry has its roots with the ancient Egyptians and Babylonians. Their geometry was an intensely practical kind, concerned with lengths of sides of figures, areas of triangles, and so on. In short, to them geometry was, for all practical purposes, the same as surveying. This is reflected in the term "rope stretcher," the Egyptian name for a surveyor.

Figure 4-1 The upper portion of the ancient Egyptian mural shows surveyors carrying a knotted rope to be used in measurements. The lower portion shows scribes recording the shipments of grain being placed in a storehouse. Surveying was highly developed by the ancient Egyptians. The annual flooding of the Nile River wiped out many of the boundary markers between farms and made it necessary to reestablish boundary lines. Egyptian surveyors used tightly stretched ropes, knotted at regular intervals, to measure distances. Accurate descriptions (and possibly even scale drawings) were then made of the boundary lines. Photo courtesy of The Metropolitan Museum of Art.

When the ancient Greeks took up the study of geometry, they transformed it from a descriptive science into a branch of theoretical mathematics. *Thales of Miletus* (about 624–548 B.C.) is usually given the credit for this change. Apparently he had traveled widely in the Egyptian and Babylonian worlds and had studied the known mathematical knowledge. He found a hodgepodge of approximation formulas and results that were true in special cases. He also found a general lack of understanding of when to apply the formulas and of when the special results were valid. When he returned to the Greek world, he attempted to develop a coherent theory of geometry that would remove all ambiguities and would state clearly what was exact and what was approximate. According to the ancient historians, Thales was the first man to give a deductive argument in mathematics.

Thales of Miletus

Thales of Miletus was one of the seven sages of the early Grecian world. These men were revered as law givers and philosophers. Thales also was noted as being a scholar, a mathematician, and a businessman. Thales remains a semimythical figure to us. Almost nothing is known about him other than stories written down centuries after his death.

According to the stories, Thales correctly predicted a solar eclipse (evidencing a knowledge of Babylonian astronomy). He was able to use geometry to calculate the distance between ships at sea and the height of the Great Pyramid. The year that a large olive crop was harvested, he bought up all of the available olive presses and rented them out to the farmers at exorbitant rates.

Thales was the typical absent-minded scholar. He once fell into a well as he walked along studying the stars. This prompted a nearby slave girl to say, "Thales is so interested in the heavens that he doesn't pay any attention to what happens under his feet here on earth."

In mathematics Thales is remembered for the "Theorem of Thales," which states that *a triangle inscribed in a semicircle must be a right triangle.* We believe that this was one of several theorems in which Thales attempted to study the foundations of mathematics and to develop the subject in a systematic manner.

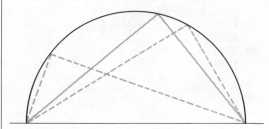

The Theorem of Thales: A triangle inscribed in a semi-circle is a right triangle.

The ancient Greek philosophers were concerned largely with abstractions. To them the abstract concept of *man* was more important than any single man who lived; the abstract concept of *line* was more important than any line that could be drawn, and so forth. They influenced the development of geometry along abstract lines, in the process divorcing it almost completely from practical considerations. They used mathematical techniques—deductive arguments based on abstractions—in much of their philosophy.

By 300 B.C. the Greeks had developed geometry into an intellectual discipline almost devoid of practical applications. They were so successful, in fact, that they eventually reduced all of mathematics to geometrical concepts. Unfortunately, this put a severe restriction on their work, because many topics in mathematics do not lend themselves to geometrical arguments. After a long period in which mathematics (meaning geometry) flourished under the Greeks, there was a progressive decline that lasted from the time of Caesar until the fifteenth century. When mathematics eventually began to flourish again, geometry was only a single branch of the subject.

In this chapter, we shall discuss several elementary concepts of geometry. The emphasis will be on understanding these concepts and applying them to the solution of practical problems. We then turn our attention to *graphing*—the modern use of geometry.

The remainder of this section is devoted to a brief review of three of the simplest geometrical concepts—*points, lines,* and *angles.*

Points

Any mathematical theory must begin with the use of primitive terms that are understood, but not actually defined. A precise definition of the word "line," for example, would require the use of related words, such as "straightness," "linearity," and so on. Since these words also would require definitions, we would be starting on an unending task if we attempted to define all of the words that we use.

Essentially, the word *point* means "location" in mathematics. Since a location has no size and covers no area, then a point has no size and covers no area—it simply marks a location on a plane or in three-dimensional space.

A great deal of confusion has arisen in the past about points having no size. When we indicate a point on paper, we put a dot or some other mark over the point and then speak of the dot as if it were the point. Because marks have size (as can be seen by examining them under a microscope), we are led to the mistaken impression that points have size. If we keep in mind that a point is a location while the dot is a blob of ink that covers the point, much of the confusion can be dispelled.

Lines

The word *line* means "straight line" in mathematics. A line is infinite in extent in both directions—it has no end. Since a line is composed of points, it has no width.

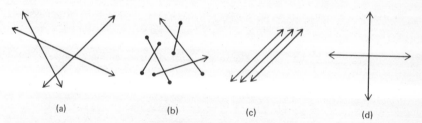

(a) (b) (c) (d)

Figure 4-2 (a) Lines. The arrowheads indicate that the lines continue on forever; (b) line segments and rays; (c) parallel lines; (d) perpendicular lines.

The portion of a line between two fixed points is called a *line segment*. A line segment has length, but no width. The line segment that connects the two points A and B is denoted by the symbol AB. The symbol $|AB|$ stands for the length of the line segment AB; it represents the distance between A and B.

In a plane, a line segment represents the shortest distance between two points. In other words, if a fly crawls from A to B, it travels the shortest distance if it moves along the line segment AB. This shortest distance is equal to $|AB|$.

A *ray* is a half-line. It can be drawn from a fixed point by extending a line out in one direction from the point [Fig. 4-2(b)].

Two lines in the plane that remain a constant distance apart regardless of how far they are extended are said to be *parallel*. Parallel lines never cross. Line segments and rays are said to be parallel if they are parts of parallel lines.

Angles

An *angle* is a geometrical figure that consists of two rays that emanate from a single point O. Several angles are shown in Figures 4-3 and 4-4.

One of the most important angles shown in Figure 4-4 is the *right angle*. If two intersecting lines cross in such a way that four equal angles are

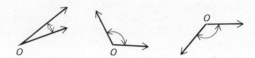

Figure 4-3 Angles. The vertex of each angle is located at the point O.

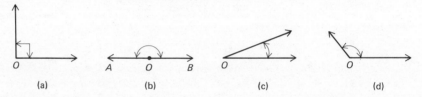

| (a) | (b) | (c) | (d) |

Figure 4-4 Special angles. (a) Right angle; (b) straight angle; (c) acute angle; (d) obtuse angle.

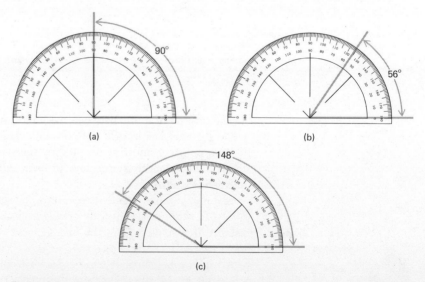

Figure 4-5 (a) A right angle has a protractor measure of 90°; (b) an acute angle has a protractor measure less then 90°; (c) an obtuse angle has a protractor measure greater then 90°.

formed then each of these angles is a right angle. Two lines (or rays or line segments) are said to be *perpendicular* if they form a right angle at their point of intersection.

There are several ways of assigning a measure to an angle. The most familiar involves the use of a *protractor*. A protractor scale is marked in degrees (°). A *degree* is defined by splitting a right angle into 90 equal small angles. Each of these small angles has a measure of one degree (1°). It follows that a right angle has a measure of 90°; half of a right angle has a measure of 45°, and so on. (See Fig. 4-5.)

If *A*, *O*, and *B* lie on a straight line with *O* between *A* and *B*, then the resulting angle *AOB* is called a straight angle. Since a straight angle can be decomposed into two right angles, then a straight angle has a protractor measure of 180°.

An angle is said to be *acute* if it has measure less than a right angle (less than 90°). It is said to be *obtuse* if it has measure greater than that of a right angle (greater than 90°). (See Fig. 4-5.)

EXERCISES

1. On a large sheet of good-quality paper draw the indicated figure, then fold the paper in such a way that the desired figure is constructed by the original lines and the creases in the paper (Fig. 4-6).
   (a) Construct a line and a point not on the line. Fold the paper so as to construct a line through the point that is perpendicular to the original line.
   (b) Construct an angle. Fold the paper so as to construct both an angle one-half the size of the original angle and an angle double the size of the original one.
   (c) Construct a line and a point not on the line. Fold the paper so as to construct a line through the point that is parallel to the original line.
2. Construct a horizontal line (the base line) and a line that intersects the base line so as to form a 60° angle. (Use a protractor.) Plot points that are located the proper distances away from these two lines. Decide if these instructions completely determine the locations of the points.

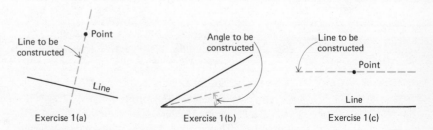

Figure 4-6

(a)   a line 3 inches above the base line and 2 inches to the left of the 60° angle;

(b)   a line 1 inch below the base line and $1\frac{1}{2}$ inches to the right of the 60° line.

3. Use a compass to construct a circle with radius equal to 2 inches. Show that any two points on the circle can be no farther apart than 4 inches.

4. Construct a line and a point P that is 3 units above the constructed line. Construct all points having the following properties.

(a)   The distance from a point to P is 3 units; the distance to the line is 1 unit.

(b)   The distance from a point to P is 2 units; the distance to the line is 2 units.

(c)   The distance from a point to P is 1 unit; the distance to the line is 1 unit.

(d)   The distance from a point to P is 1 unit; the distance to the line is 2 units.

5. Construct as straight a line as you possibly can using a good ruler and a sharp pencil. Examine the line under a strong magnifying glass. Does it still appear to be perfectly straight? Someone once said that it is not possible to construct a line that is truly straight. What did he mean?

6. Try to define the word "point" without actually using the word or any of its synonyms in the definition. Consult a good dictionary and see how this problem is resolved. (Or is it?)

4-2 SIMPLE CLOSED FIGURES

Circles

A circle is defined by choosing a point C in the plane (the *center*) and a distance r (the *radius*). The circle consists of all points that are exactly r units from C (Fig. 4-7).

Figure 4-7 The circle consists of all points located exactly r units from the center C.

The Ptolemaic Theory of Planetary Motion

The ancient Greek philosophers believed in harmony and perfection. To them it was clear that the gods had formed the universe in a perfect manner. Consequently, they argued that the stars and planets must revolve about the earth in circular orbits (circles being perfect figures).

After several centuries of observation, it became obvious that the planets could not revolve about the earth in circular orbits. Eventually the astronomer *Ptolemy* modified the theory to the form illustrated in the figure. Each planet was believed to revolve in a circular orbit which had its center on the circumference of a large circle with the earth as its center. This theory, which seems ludicrous to us, allowed the accurate determination of the planets in the sky.

Because of the great respect rightly accorded the Greek philosophers and mathematicians by later scholars and theologians, this incorrect theory became an accepted part of the Christian dogma and one of the common beliefs of all educated Europeans. Not until the sixteenth century did astronomers establish that the planets revolve about the sun, not the earth, and that the orbits are elliptical, not circular.

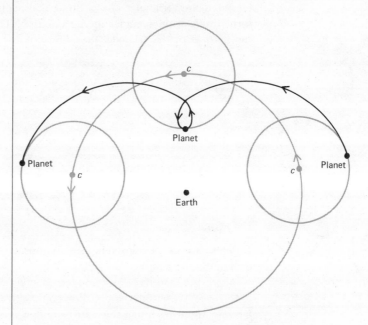

The Ptolemaic Theory: The planet moves in orbit about the small circle which, in turn, moves in a circular orbit about the earth.

Remark

A circle is an example of a simple closed figure. Roughly speaking, a geometrical figure is said to be a *simple closed figure* if it is possible to trace a pencil point around its boundary and eventually arrive back at the starting point without ever backing up or crossing the boundary. Figure 4-8(a) shows several simple closed figures. In Figure 4-8(b) several figures that are not simple closed figures are shown for contrast.

(a) (b)

Figure 4-8 (a) Simple closed figures; (b) figures that are not simple closed figures.

Triangle

A *triangle* is a simple closed figure composed of three straight line segments. We can construct a triangle by connecting any three points *A*, *B*, and *C* that do not lie in a straight line with straight line segments. The reason for the name "triangle" (*tri* = three) is obvious because the line segments form three distinct angles at *A*, *B*, and *C*. These three points are called the *vertices* of the triangle. (See Fig. 4-9.)

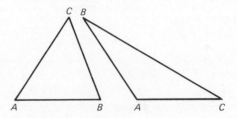

Figure 4-9 Triangles. The vertices of the triangles are at *A, B,* and *C.*

Polygons

Polygons are simple closed figures composed of straight line segments. The triangle is the simplest polygon.

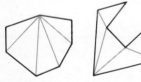

Any polygon with seven sides can be decomposed into five nonoverlapping triangles.

A polygon can have any number of sides. As we saw above, a polygon with three sides is a triangle. Polygons with four sides are called *quadrilaterals;* those with five sides are *pentagons;* those with six sides are *hexagons,* and so forth. (See Fig. 4-10.)

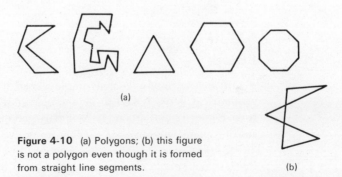

(a)

Figure 4-10 (a) Polygons; (b) this figure is not a polygon even though it is formed from straight line segments.

(b)

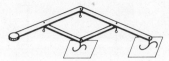

The pantograph is a drafting machine used for copying drawings on an enlarged or reduced scale. Its design is based on the geometrical properties of the parallelogram.

Rectangles, Squares, Parallelograms

The most important of the quadrilaterals (four-sided polygons) are those in which the pairs of opposite sides are parallel. These figures are called *parallelograms*. In a parallelogram the pairs of opposite sides are equal in length as well as parallel. (See Fig. 4-11.)

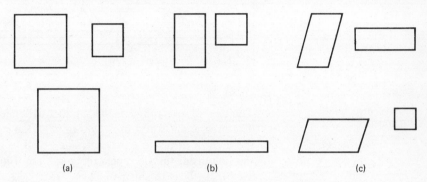

Figure 4-11 (a) Squares; (b) rectangles; (c) parallelograms.

A *rectangle* is a quadrilateral in which all of the angles are right angles. It is easy to see that the opposite sides are parallel; so a rectangle is a special kind of parallelogram. It follows that the opposite sides of a rectangle are equal.

A *square* is a rectangle in which all four sides are equal.

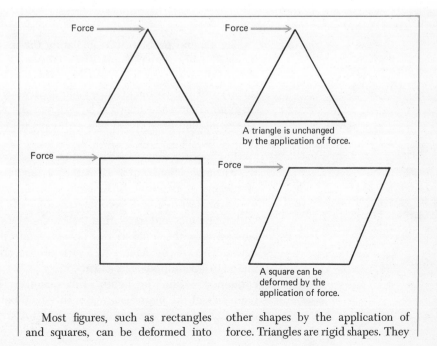

A triangle is unchanged by the application of force.

A square can be deformed by the application of force.

Most figures, such as rectangles and squares, can be deformed into other shapes by the application of force. Triangles are rigid shapes. They cannot be deformed into other shapes without bending or breaking their sides. For this reason, triangles are widely used in structures that must resist great stresses without being deformed.

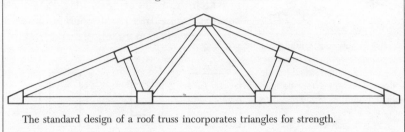

The standard design of a roof truss incorporates triangles for strength.

Area

HEIGHTS AND BASES OF TRIANGLES

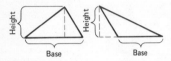

Any side can be chosen as the base. The height is the distance from the base to the opposite vertex. The height is measured along a line perpendicular to the base.

The *area of a rectangle* is the product of the lengths of two adjacent sides. The area is measured in *square units*. (One square unit is the measure of the area of a square with each side equal to one unit of length.) Thus, if the lengths of the sides of a rectangle are 3 inches and 5 inches, then the area is $3 \cdot 5 = 15$ square inches.

It can be shown that the area of a triangle with base B and height H is $BH/2$. To do this we make a duplicate triangle and cut it along the dotted line as shown in Figure 4-12. We then reassemble the three triangles to form a rectangle with base B and height H. Since the original triangle has one-half the area of the rectangle (because the rectangle was formed from two triangles), then

$$\text{area of triangle} = \tfrac{1}{2} \text{ area of rectangle}$$
$$= \tfrac{1}{2} \cdot BH$$

The general problem of calculating the area of a simple closed figure is considerably more difficult. Essentially, we try to "fill up" the figure with rectangles and measure their combined areas as in Figure 4-13. Obviously, since all of the rectangles are contained within the figure, then in every case the area of the figure is larger than the total area of the rectangles. The combined area of the rectangles is thus seen to be a lower approximation to the area of the figure.

By choosing enough rectangles and making them as small as we need, we can make the total area of the rectangles to be as close to the area of the figure as we wish. In other words, the area of the figure can be approximated to any desired degree of accuracy by "filling up" the figure with rectangles and measuring their area. Mathematicians use an idea that is essentially the same as this one to *define* the area of a simple closed figure. The area is defined to be the largest number that can be approximated to any desired degree of accuracy by the combined area of rectangles contained within the figure. This definition is adequate, provided the boundary of the figure is a reasonably well-behaved curve, and the result is consistent with our intuitive ideas about area.

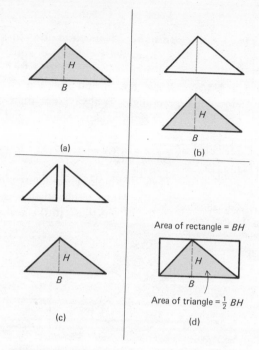

Figure 4-12 The area of a triangle is one-half the product of the base and the height. (a) Original triangle; (b) construct duplicate triangle; (c) cut duplicate triangle; (d) reassemble to form a rectangle.

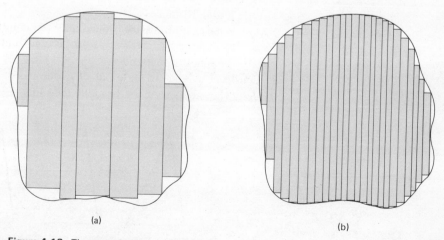

Figure 4-13 The area of a figure is the largest number that can be approximated by rectangles that lay completely within the figure. (a) The total area of the rectangles is a small approximation to the area of the figure; (b) by choosing very thin rectangles, we can make the total area of the rectangles as close to the area of the figure as we wish.

Area of a Circle

In more advanced courses in mathematics, it is proved that the area of a circle with radius r is equal to πr^2. (Recall that π is the ratio of the circumference of a circle to its diameter: $\pi \approx 3.14159$.) In other words, if we try to "fill up" the circle with rectangles, the total area of the rectangles is always less than πr^2, but can be made as close as we want to πr^2 by making the rectangles small enough.

The area of a circle with radius r is πr^2. By choosing the rectangles thin enough, we can make the total area of the rectangles as close to πr^2 as we wish.

EXERCISES

1. (a) Which of the figures in Figure 4-14 are simple closed figures?
   (b) Make freehand drawings of three simple closed figures different from those in the text.

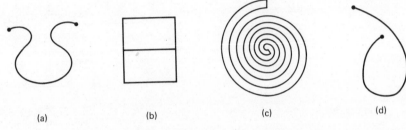

(a)   (b)   (c)   (d)

Figure 4-14 Which of these are simple closed figures (Exercise 1)?

2. (a) Construct a triangle with sides of length 3, 8, and 9 inches.
   (b) Show that no triangle can be constructed with sides of length 2, 7, and 10 inches.
   (c) What conditions are needed on three numbers a, b, and c in order to ensure that a triangle can be constructed with a, b, and c as the lengths of its sides?
3. Construct a large triangle out of good-quality paper. Fold the paper along the medians of the triangle. (See Fig. 4-15.) Note that the three

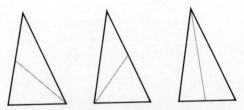

Figure 4-15 The medians of a triangle. A line drawn from a vertex of a triangle to the midpoint of the opposite side is called a *median* of the triangle (Exercise 3).

medians intersect in a single point. Show by measurement that this point is located $\frac{2}{3}$ of the distance from each vertex to the midpoint of the opposite side.

4. Show that each of the polygons with n sides in Figure 4-10 can be decomposed into $n - 2$ nonoverlapping triangles. (That is, show that the polygon with 4 sides can be decomposed into 2 triangles, those with 5 sides can be decomposed in 3 triangles, and so on.) See if you can construct a polygon with 10 sides that *cannot* be decomposed into 8 triangles. What general result would you conjecture from this exercise?

5. Construct a large parallelogram with base 5 units and height 3 units using Figure 4-16 as a model. Show that the parallelogram can be cut along a line and reassembled into a rectangle which has an area equal to 15 square units. Carry out the same construction with a parallelogram with base B and height H. (The *height* and *base* of a parallelogram are defined similarly to the way they are for a triangle.) In general, this method can be used to show that for a parallelogram

$$\text{area} = \text{base} \cdot \text{height}$$

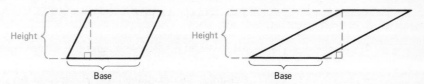

Figure 4-16 The area of a parallelogram is equal to the product of the base and the height; area = base·height (Exercise 5).

6. A *trapezoid* is a quadrilateral with two parallel sides. Construct a large trapezoid different from either of the two shown in Figure 4-17. Show that this trapezoid can be cut along a line and reassembled to form

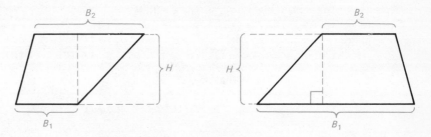

Figure 4-17 The area of a trapezoid with bases B_1 and B_2 and height H is

$$\text{area} = \frac{B_1 + B_2}{2} \cdot H$$

The *bases* of a trapezoid are the lengths of the two parallel sides. The *height* is the distance between the parallel sides (Exercise 6).

a parallelogram. Then use the result of Exercise 5 to cut the parallelogram and form a rectangle. If B_1 and B_2 are the two bases and H is the height of the trapezoid, it can be shown by this method that

$$\text{area} = (B_1 + B_2) \cdot H/2$$

7. Buy a large sheet of graph paper marked with small squares. Choose a scale of measure in which 1 unit is 3 to 4 inches long. Construct a circle with radius equal to 2 units. Measure the area of the circle in *square units* by counting the number of small squares inside the circle. Compare your answer with the value of the area obtained from the formula

$$A \overset{\bullet}{=} \pi r^2 \qquad (\pi \approx 3.14)$$

8. Explain why some of the parallelograms in Figure 4-11(c) are rectangles, but none of the rectangles in Figure 4-11(a) are parallelograms (*or are they?*).

4-3 SIMILARITY

More than 2000 years ago, the Greek philosopher Thales of Melitus astounded the Egyptian pharaoh by computing the height of the Great Pyramid. Thales placed a stick straight up in the sand and measured the length of its shadow (the distance $|AB|$ in Fig. 4-18). At the same instant his assistant measured the distance from a point directly under the vertex of the pyramid to the tip of its shadow (the distance $|A'B'|$ in Fig. 4-18). He then computed that

$$\frac{\text{height of pyramid}}{\text{length of pyramid's shadow}} = \frac{\text{height of stick}}{\text{length of stick's shadow}}$$

or, using our symbols,

$$\frac{|B'C'|}{|A'B'|} = \frac{|BC|}{|AB|}$$

so that

$$|B'C'| = |A'B'| \cdot \frac{|BC|}{|AB|}$$

Since the distances $|A'B'|$ (the length of the pyramid's shadow), $|BC|$ (the length of the stick), and $|AB|$ (the length of the stick's shadow) had all been measured, it was a simple matter to compute $|B'C'|$ (the height of the pyramid).

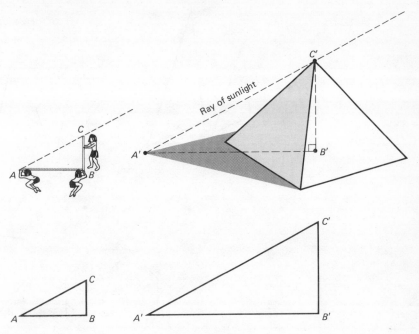

Figure 4-18 Thales' computation of the height of the Great Pyramid.

$$\frac{|BC|}{|AB|} = \frac{|B'C'|}{|A'B'|}$$

The key to an understanding of Thales' approach is pictured in the lower part of Figure 4-18. Essentially, the problem reduces to a consideration of the two triangles ABC and $A'B'C'$. Thales recognized that these triangles have exactly the same shape. Because of this he knew that *the ratios of the corresponding sides are equal.* Thus, the quotient obtained by dividing the height of one triangle by its base is equal to the height of the other triangle divided by its base.

Similarity

Two geometric figures are said to be *similar* if they have exactly the same shape, possibly differing only in size and position. For example, the two triangles in Figure 4-18 are similar. We also consider two figures to be similar if one figure is the "mirror image" of a figure having the same shape as the other figure. Similar figures may, or may not, have the same size. Several pairs of similar figures are shown in Figure 4-19.

There is a natural correspondence between the points, sides, and angles of one figure and the points, sides, and angles of a similar figure. This correspondence can best be described by imagining the larger of the two figures to be reduced to the size of the smaller one and placed directly over it so that the two figures coincide exactly. The correspondence is

The picture on a slide is similar to its projection on a screen.

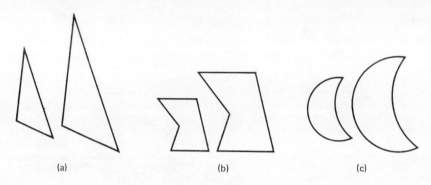

(a)   (b)   (c)

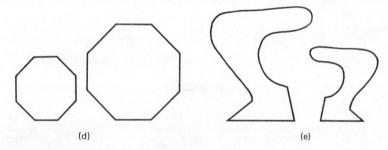

(d)   (e)

Figure 4-19 Pairs of similar figures. Two figures are similar if they have exactly the same shape, or one is the mirror image of a figure having the same shape as the other.

CORRESPONDING POINTS
ON SIMILAR FIGURES

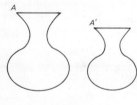

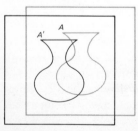

If a scale drawing is made of the larger figure the same size as the smaller one and is placed over the smaller figure, then A and A^1 coincide.

between the points (or sides or angles) on one figure and the points that then coincide with them on the other one.

Figure 4-20 shows several pairs of similar figures. In each pair two or more points, labeled A, B, and so on, are shown on one figure, and the corresponding points A', B', and so on, are shown on the other figure.

Two important properties of similar figures account for most of their applications. First, *corresponding angles on similar figures are equal.* In Figure 4-20, for example, each pair of similar figures has corresponding angles labeled A and A'. In every case these two angles are equal.

The second important property is that the ratios of corresponding distances on similar figures are equal. Each pair of figures in Figure 4-20 has points A, B, C and corresponding points A', B', C'. In each case the ratios of corresponding distances are equal. For example,

$$\frac{|AB|}{|BC|} = \frac{|A'B'|}{|B'C'|}$$

between each pair of figures.

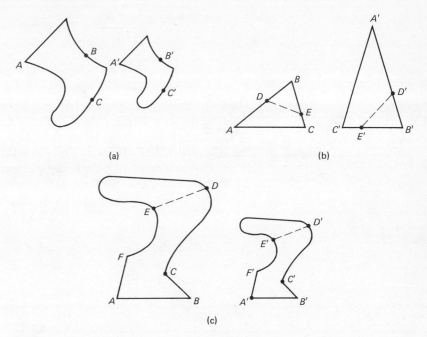

Figure 4-20 Corresponding points on pairs of similar figures. In each pair, angle A = angle A', and ratios of corresponding distances are equal.

Figure 4-21 shows two similar triangles. One has sides measuring 6, 8, and 10 units; the other has sides 9, 12, and 15 units. Because the triangles are similar, then each pair of corresponding distances has the same ratio as every other pair. We can exploit this fact to compute the distance $|A'B'|$. Since the distance $|AB| = 6$, then

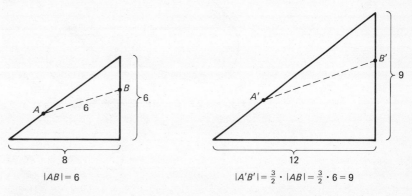

$$|AB| = 6$$

$$|A'B'| = \tfrac{3}{2} \cdot |AB| = \tfrac{3}{2} \cdot 6 = 9$$

Figure 4-21 Distances on one figure can be used to compute the corresponding distances on a similar figure.

$$\frac{|A'B'|}{|AB|} = \frac{12}{8} = \frac{3}{2}$$

$$|A'B'| = \frac{3}{2}|AB| = \frac{3}{2} \cdot 6 = 9$$

Thus, as we see, the distance $|A'B'|$ can be computed from the corresponding distance $|AB|$ without the need for direct measurement.

This type of analysis can be extended to any pair of similar figures. (See Fig. 4-22.) If A, B, C, D are any four points on one figure and A', B', C', D' are the corresponding points on the other figure, then we can measure $|A'B'|$ and $|AB|$ and compute their ratio. If this ratio is k, then

$$|A'B'| = k \cdot |AB|$$

Since all other distances on the figure $A'B'C'D'$ are in the same proportion to the distances on figure $ABCD$, then

$$|A'C'| = k \cdot |AC|$$
$$|B'C'| = k \cdot |BC|$$
$$|C'D'| = k \cdot |CD|$$

and so on.

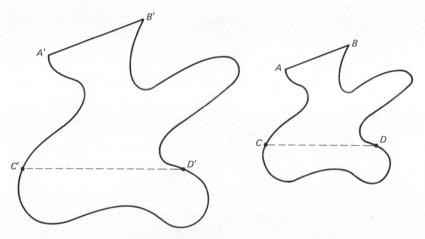

Figure 4-22 Corresponding distances on similar figures are proportional.

If $\dfrac{|A'B'|}{|AB|} = k$, then

$$|A'C'| = k|AC|,$$
$$|B'C'| = k\,|BC|$$
$$|C'D'| = k\,|CD|$$

The area of the large figure is k^2 times the area of the small one.

Properties of similar figures are used whenever measurements are taken from scale drawings.

The property of distances being proportional in similar figures is the basis for the use of scale drawings by architects and designers. A scale drawing can be made of a set of bookshelves having proportions pleasing to the designer, and then all measurements needed for cutting the lumber can be obtained by measuring the drawing.

How do the areas of similar figures compare? Suppose figure ABC has each dimension equal to k times the corresponding dimension of figure $A'B'C'$. What relationship holds between their areas? It can be shown that

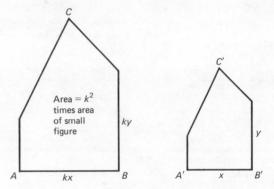

Figure 4-23 Each dimension of the large figure is k times the corresponding dimension of the small similar figure. The area of the large figure is k^2 times the area of the small figure.

Each dimension of the large figure is twice the corresponding dimension of the small figure. This proportion is preserved in all measurements along lines and curves. Thus, the distance from A to B is twice the distance from A' to B'; the length of the curve connecting D and E is twice the length of the curve connecting D' and E'.

The area of the large figure is four times the area of the small figure. This property is preserved in all corresponding parts. The area of triangle ABC is four times the area of triangle $A'B'C'$; the area of the curved portion at the bottom of the large figure is four times the area of the corresponding portion of the small figure.

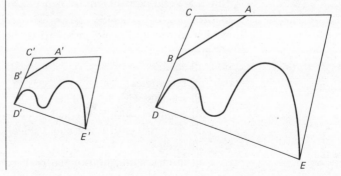

the area of figure ABC is k^2 times the area of $A'B'C'$ (Fig. 4-23). This relationship holds for any two similar figures. Essentially, it follows from the fact that area is a product of two dimensions—*length* and *width*. If each dimension of a figure is multiplied by k, then the area is multiplied by k^2. This relationship is easy to establish for rectangles. It can be proved for more complicated figures by using rectangles to approximate their areas as discussed at the end of Section 4-2.

EXERCISES

1. Two similar figures are shown in Figure 4-24.
    (a)   Calculate the distances $B'C'$ and AD and the measure of the angle at D'.
    (b)   Calculate the area of figure $A'B'C'D'$.

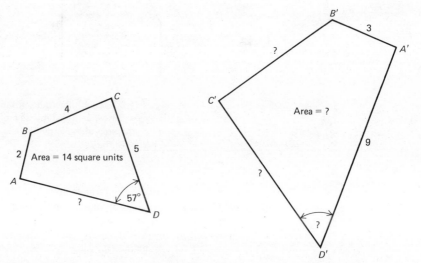

Figure 4-24   Exercise 1.

2. Two similar figures are *congruent,* provided corresponding distances are equal. Several pairs of similar figures are shown in Figure 4-25. Which of these are congruent?
3. Which of the following geometrical figures are similar? Which are congruent?
    (a)   all squares
    (b)   all rectangles
    (c)   all squares with sides of length 3
    (d)   all triangles with two equal sides
    (e)   all triangles with three equal sides
    (f)   all circles
    (g)   all circles with radius equal to 1

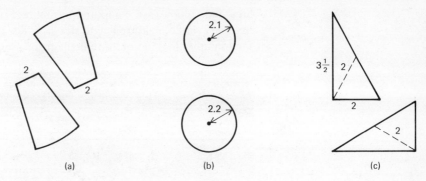

Figure 4-25 Exercise 2.

4. The top of a vertical post is 4 feet above the ground. (See Fig. 4-26.) At 3:00 P.M. it casts a shadow 3 feet long. At the same instant a nearby tree casts a shadow 43 feet long. How tall is the tree?

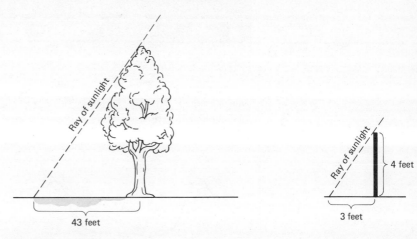

Figure 4-26 Exercise 4.

5. Let two rectangles be similar with k equal to the constant of proportionality. Show that the area of one rectangle is k^2 times the area of the other.

4-4 INDIRECT MEASUREMENT

A small child's height can be measured by placing the child against a yardstick. Measurements of this type, where a direct comparison can be made with a scale, are called *direct measurements*. Length is one of the few properties that can be measured directly. Almost all others, such as area and volume, are computed by indirect measurement. Often it is not feasible even to measure length directly. The height of a distant mountain, for example, may be quite difficult to measure directly.

Photo courtesy of NASA.

Almost all methods for indirect measurement involve geometrical concepts. The depths of craters on the moon can be computed from the lengths of shadows in photographs. The computation involves knowing the exact angle of the rays of sunlight that strike the craters.

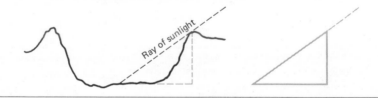

One of the best methods for the indirect measure of length involves the use of similar right triangles. Recall that if two figures are similar, then the ratios of corresponding distances are equal. For example, the diagram in Figure 4-27 shows two similar right triangles. The small triangle measures 5 units, 12 units, and 13 units on the three sides. We then know immediately that the ratios of the three sides of the large triangle are

$$\frac{|AB|}{|AC|} = \frac{|A'B'|}{|A'C'|} = \frac{12}{13}$$

$$\frac{|AB|}{|BC|} = \frac{|A'B'|}{|B'C'|} = \frac{12}{5}$$

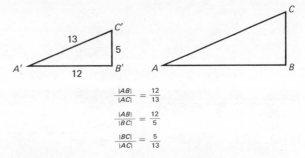

$$\frac{|AB|}{|AC|} = \frac{12}{13}$$

$$\frac{|AB|}{|BC|} = \frac{12}{5}$$

$$\frac{|BC|}{|AC|} = \frac{5}{13}$$

Figure 4-27 The sides of a known triangle can be used to calculate the ratios of the sides of an unknown, but similar, triangle.

and

$$\frac{|BC|}{|AC|} = \frac{|B'C'|}{|A'C'|} = \frac{5}{13}$$

Thus, as we see above, a known right triangle can be used to compute the ratios of the sides of an unknown but similar triangle.

In general, it may be difficult to decide when two geometric figures are similar. In the case of triangles, however, it is quite easy to decide. Two triangles are similar if, and only if, the angles of one are equal to the angles of the other. Because two right triangles automatically have one angle equal, it follows that two right triangles are similar provided one of the acute angles of one right triangle is equal to one of the acute angles of the other right triangle.

The concepts mentioned above are the basis for a powerful method of indirect measurement known as *trigonometry*. The following example illustrates a primitive form of trigonometry.

Example 1. A farmboy wishes to measure the distance across a river on his farm. He knows that a tree on the water's edge is exactly 37 feet tall. At 9:23 one morning he notices that the shadow of the tree falls directly across the river with the tip of the shadow just touching the opposite bank. He immediately measures the length of the shadow of a nearby fence post and finds that it is 11 feet long. He knows that the fence post is 5 feet high. What is the width of the river?

Solution. The two similar triangles involved are shown in Figure 4-28. If x is the distance across the river, then

$$\frac{x}{37} = \frac{|AB|}{|BC|} = \frac{|A'B'|}{|B'C'|} = \frac{11}{5}$$

Thus,

$$x = \frac{11}{5} \cdot 37 = \frac{407}{5} \approx 81 \text{ feet}$$

The river is approximately 81 feet wide at that point.

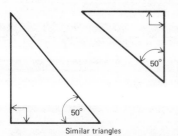

Similar triangles

If two right triangles have an acute angle equal, then they are similar.

The technique of using ratios of right triangles for indirect measurement was used by the ancient Hindus, who had "instant recall" of the values of these ratios. The Hindus used various words as synonyms for the digits. *Two* was remembered as "bird" (*two* wings), or "arms" (*two* arms), and so on. *Three* was remembered as "brothers" (the *three* brothers of the god Rama), and so on. The trigonometric tables were written in the form of an epic poem, using the "code words" for the digits, which was memorized by the students and mathematicians.

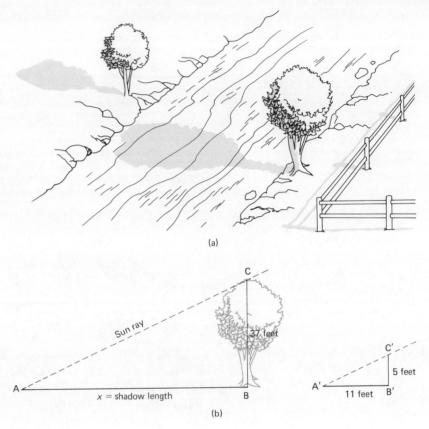

Figure 4-28   Example 1.

In most of our work we shall deal with right triangles. Centuries ago it was realized by the ancient Babylonians that it is not actually necessary to construct a similar triangle in order to compute the lengths of a given right triangle. Since we only need to know the *ratios* of sides of triangles, we can make a table of these ratios and use the table over and over again.

There is a standard nomenclature for the sides of a right triangle. The longest side is called the *hypotenuse*. The side that intersects with the hypotenuse to form the angle α is called the *side adjacent to angle α*. The third side is called the *side opposite angle α*.

STANDARD NOMENCLATURE
FOR RIGHT TRIANGLES

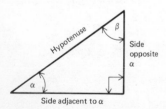

The Plimpton Mathematical Tablet
(*About* 1900–1600 B.C.)

The Plimpton mathematical tablet is one of the most important of the ancient documents in the history of mathematics and science. The num-bers on the tablet are the lengths of the sides and one of the trigonometric ratios for angles between 31° and 45°, in increments of almost exactly one

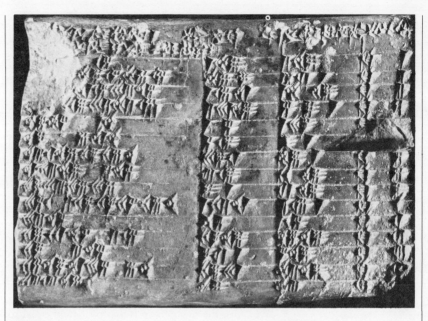

Photo courtesy of Plimpton Collection, Columbia University Libraries.

degree. The lengths of the sides of the triangles are in whole numbers selected so that the trigonometric ratio, expressed in the base-60 notation used by the Babylonians, is finite, rather than infinite, in each case. Because such numbers are almost impossible to obtain by direct measurement, it seems likely that at least a few of the Babylonian mathematicians had developed an understanding of some rather deep mathematical relationships.

THE TRIGONOMETRIC RATIOS

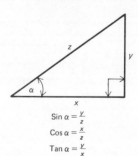

For each angle α between $0°$ and $90°$, we define the following trigonometric ratios:

$$\sin \alpha = \text{sine of } \alpha = \frac{\text{opposite side}}{\text{hypotenuse}}$$

$$\cos \alpha = \text{cosine of } \alpha = \frac{\text{adjacent side}}{\text{hypotenuse}}$$

$$\tan \alpha = \text{tangent of } \alpha = \frac{\text{opposite side}}{\text{adjacent side}}$$

Table I at the end of the book lists four-place decimal approximations to the trigonometric ratios of angles between $0°$ and $90°$. For example,

$$\cos 23° \approx 0.9205$$
$$\sin 57° \approx 0.8387$$
$$\tan 83° \approx 8.1443$$

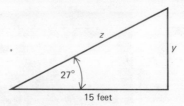

Figure 4-29 Example 2.

Example 2.   The right triangle shown in Figure 4-29 has the side adjacent to the 27° angle equal to 15 feet. Calculate the lengths of the other two sides.

Solution.   We use Table I.

$$\cos 27° = \frac{15}{z} \approx 0.8910$$

$$z \approx \frac{15}{0.8910} \approx 16.84$$

Thus, the hypotenuse is approximately equal to 16.84 feet.

$$\tan 27° = \frac{y}{15} \approx 0.5095$$

$$y \approx 15(0.5095) \approx 7.64 \text{ feet}$$

The side opposite the 27° angle is approximately equal to 7.64 feet.

Surveyors use an instrument called a transit to measure angles. The following example shows how the heights can be measured by calculations based on the use of a transit.

Example 3.   A surveyor sets up a transit at the bottom of a canyon 500 feet from a canyon wall (Fig. 4-30). He measures the angle of elevation to the top of the canyon wall as 63°. Assuming that the transit is 5 feet above the floor of the canyon, calculate the height of the canyon wall.

Solution.   Let x represent the distance from the level of the transit to the top of the canyon wall. Then

$$\tan 63° = \frac{x}{500} \approx 1.9626$$

$$x \approx 500(1.9626) \approx 981 \text{ feet}$$

Since the transit is 5 feet above the floor of the canyon, then the height of the canyon wall is

$$\text{height} = x + 5 \approx 981 + 5 \approx 986 \text{ feet}$$

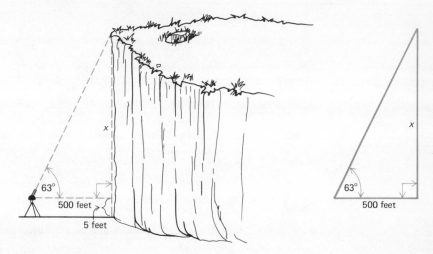

Figure 4-30 Example 3.

Example 4. A parallelogram has two sides of length 7 inches and two of length 10 inches. The angles formed by the four sides are 27°, 153°, 27°, and 153°. (See Fig. 4-31.) Calculate the area of the parallelogram.

Solution. The area (see Exercise 5, Section 4-2) is the product of the base and the height, where the height is equal to the distance between two horizontal bases. (See Fig. 4-31.) We use the sine function to compute the height:

$$\sin 27° = \frac{\text{height}}{7} \approx 0.4540$$

$$\text{height} \approx 7(0.4540) \approx 3.178$$

The area is

$$\text{area} = \text{base} \cdot \text{height} \approx 10 \cdot 3.178$$
$$\approx 31.8 \text{ square inches}$$

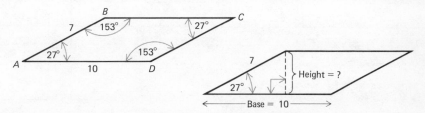

Figure 4-31 The area of a parallelogram is: area = base · height (Example 4).

EXERCISES

1. (a)   Use Table *I* (at the end of the book) to find the following trigonometric ratios:
    (1) cos 59°   (2) sin 12°   (3) tan 89°

   (b)   Use Table I to find the angles with the following trigonometric ratios:
    (1) cos $\alpha \approx 0.9986$ (2) sin $\beta \approx 0.8988$ (3) tan $A \approx 0.6494$

2. In all parts of this problem angle $C = 90°$, a is the side opposite angle A, b is the side opposite angle B, c is the side opposite angle C (Fig. 4-32).

   (a)   If angle $A = 17°$ and $b = 15$ inches, calculate a and c.

   (b)   If $a = 27$ feet, $c = 40$ feet, calculate angle A, angle B, and the length of side b.

   (c)   If angle $B = 59°$ and $c = 17$ millimeters, calculate the area of the triangle.

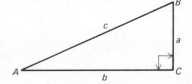

Figure 4-32 Standard labeling of the parts of a right triangle. *A* and *B* are the acute angles; *C* is the right angle; *a, b,* and *c* are the sides opposite angles *A, B,* and *C,* respectively (Exercise 2).

3. A small city park is shaped like a trapezoid. The park, bounded by four streets, two of which are parallel, is shown in the scale drawing in Figure 4-33. Calculate the area of the park in square feet. (See Exercise 6, Section 4-2, for the formula for the area of a trapezoid.)

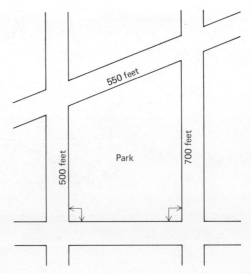

Figure 4-33 Exercise 3.

4. Construct a flow chart that can be used to calculate the area of a trapezoid $ABCD$ if the lengths of the four sides and the measure of the acute angle A are given as in Figure 4-34. Test your flow chart with the data given for the trapezoid in Exercise 3. (See Exercise 6, Section 4-2, for the formula for the area of a trapezoid.)

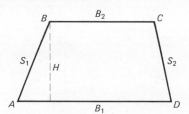

Figure 4-34 Exercise 4.

5. A surveyor is located at point A due south of a mountain peak (point C). A second surveyor, located at point B one mile due east of A, measures angle B [Fig. 4-35(a)] to be 87°. Calculate the distance from A to C.

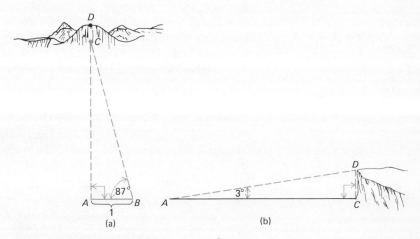

Figure 4-35 Exercises 5 and 6.

6. The surveyor at point A [see Exercise 5 and Fig. 4-35(a)] measures the angle of elevation to the top of the peak to be 3° [Fig. 4-35(b)]. Use the distance $|AC|$ calculated in Exercise 5 to calculate the height of the mountain peak.

4-5 THE COORDINATE PLANE; GRAPHS

Streets on a city map are located from the index by means of coordinates. Most maps have a reference scheme similar to the one shown in Figure

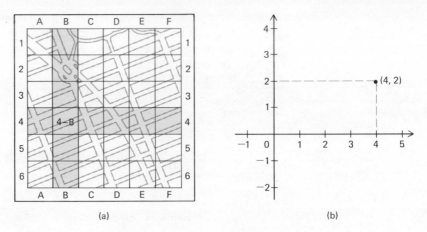

(a)   (b)

Figure 4-36 (a) Coordinates on a map. The coordinates of the square 4-B are "4" and "B"; (b) coordinates in the number plane. The coordinates of the point (4, 2) are "4" and "2."

4-36. The map is divided into squares by horizontal and vertical lines. The letters A, B, C, and so on, are written across the top, and the numbers 1, 2, 3, and so on, are written down the side. Each square is assigned the letter above it and the number to the side of it. The square 4-B, for example, is located to the right of the number 4 and below the letter B [Fig. 4-36(a)]. The number "4" and the letter "B" are called the *coordinates* of the square 4-B.

Points in the plane are described by coordinates in a manner similar to the way the squares are described by coordinates on the map. We draw two number lines, one horizontal and one vertical, that cross at their origins. The horizontal number line, called the *x-axis*, has its positive direction to the right of the origin. The vertical line, the *y-axis*, has its positive direction above the origin.

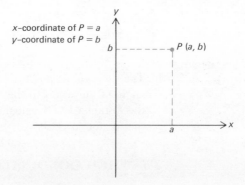

Figure 4-37 Coordinates of the point $P(a, b)$ in the number plane.

We can now locate points by their coordinates on the two lines. The point with coordinates (4, 2), for example, is directly above the point "4" on the x-axis and directly to the right of the point "2" on the y-axis [Fig. 4-36(b)].

More generally, to locate the point (a, b), we draw a vertical line through the point "a" on the x-axis and a horizontal line through the point "b" on the y-axis. The point (a, b) is at the intersection of those two lines (Fig. 4-37).

We write $P = (a, b)$ or $P(a, b)$ to indicate that P is the point with (a, b) as coordinates. The number a is called the *x-coordinate* of P, and b is called the *y-coordinate* of P. Every point in the plane can be assigned a pair of coordinates by the scheme described above.

Example 1. Several points, including (2, 3), (4, −3), (−2, 7), (0, 0), (0, −1), and (3, 1), are pictured in Figure 4-38. The reader should label each point with its coordinates.

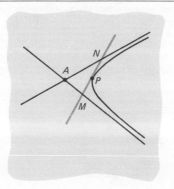

Coordinates have been used extensively to represent points for over 2000 years. One of the first persons to utilize them was the great Greek geometer *Apollonius* (third century B.C.), who made an exhaustive study of various geometrical curves. After drawing a curve, he constructed lines that related naturally to it, then proved that the important properties of the curve could be expressed in terms of the distances measured along these or other lines. Essentially, this involved the use of coordinate axes that were not at right angles to each other.

The diagram shows a typical result of Apollonius. The curve is part of a *hyperbola* (a special curve considered in mathematics). The two intersecting lines that cross at A are called *asymptotes*. These asymptotes approximate the curve as we trace along it. Apollonius showed that a line that touches the curve at any point P without crossing it must intersect the two asymptotes at points M and N equidistant from P.

Graphs

The representation of points by pairs of numbers is one of the most valuable tools ever invented for business and science because it enables us to draw figures, called *graphs*, that picture various types of relationships in visual form.

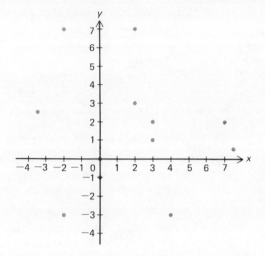

Figure 4-38 Label each point with its coordinates (Example 1).

Example 2. Figure 4-39 is a graph showing the speed of an automobile at each second as it accelerated from 0 to 60 miles per hour. At the end of 1 second, the speed was 10 miles per hour; at the end of 2 seconds, it was 17 miles per hour, and so on. Observe that the speed increased much more rapidly between 0 and 4 seconds than it did between 12 and 16 seconds.

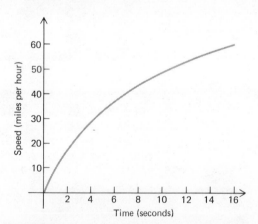

Figure 4-39 Velocity graph of an automobile (Example 2).

One of the most useful types of graph is one that shows how some quantity (such as temperature, profit, speed, stock market average) changes with time. The graph discussed in Example 2 is such a graph—it shows how speed changes with time.

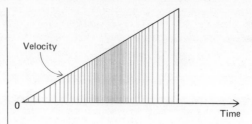

Velocity graph of an object that starts from rest and increases its velocity with a constant acceleration.

Although coordinates have been used for well over 2000 years to describe properties of curves, the modern type of graph is a comparatively late invention. To our knowledge, the first person to use a graph to represent quantities that varied jointly was *Nicole Oresme* (1323?–1382), a French scholar and churchman, who eventually became Bishop of Lisieux. Oresme studied the problem of calculating the distance traveled by an object that started from rest and increased its velocity with a constant acceleration. Somehow he came up with the idea of drawing a picture that represented the velocity at any given instant. The resulting graph was a straight line through the origin as seen in the drawing. Oresme went on to argue that the area of the triangle formed by the *x*-axis and the line is proportional to the distance traveled. Apparently he considered the triangle to be decomposed into an infinite number of vertical lines, each of which represented the distance traveled over an infinitely short period of time. By adding the "areas" of the line segments, he found that the total distance was equal to the area of the triangle.

Example 3. Figure 4-40 is a hygrothermograph chart. The graphs were drawn by a recording device in which pen arms move up and down recording the temperature and relative humidity while touching a revolving drum. This machine makes a continuous record of the temperature and relative humidity over a seven-day period.

Example 4. Figure 4-41 shows the marriage rate per 10,000 persons in the United States over a three-year period. This graph was prepared from data obtained by the National Center for Health Statistics, Public Health Service, U.S. Department of Health, Education, and Welfare. The graph in Figure 4-41(a) was constructed by plotting the number of marriages per 10,000 persons at the point representing the month and then connecting these points by straight line segments. Observe that there is a seasonal peak in marriages during the summer months and a seasonal low during the winter months.

The graph in Figure 4-41(b) shows the same data in a different form. Bars are drawn from the *x*-axis up to the points representing the number of marriages per 10,000 persons each month.

Example 5. Figure 4-42(a) shows the closing stock market quotations for Amalgamated Conglomerates, Ltd., over an eight-month period. As is customary in such graphs, the points on the *x*-axis give dates of the year. The stock closed at

Broken-line or "curved-line" graphs are best for representing quantities that vary continuously, such as the speed of an automobile at each instant of time. Bar graphs are best for representing quantities that take values at discrete intervals of time, such as the number of births or deaths reported each month during the year.

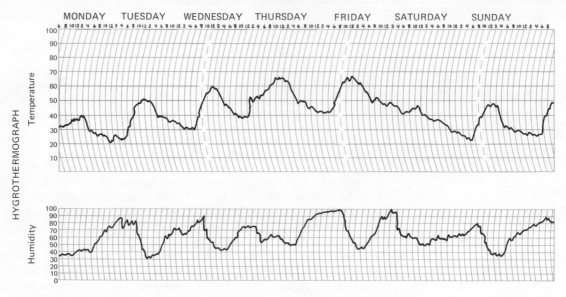

Figure 4-40 Hygrothermograph chart. The "vertical" lines are curved to match the radius of the pen arms. On Tuesday the temperature varied from 22° at 2:00 A.M. to 50° at 4:00 P.M. The relative humidity varied from 88 per cent at 4:00 A.M. to 31 per cent at 2:00 P.M. (Example 3).

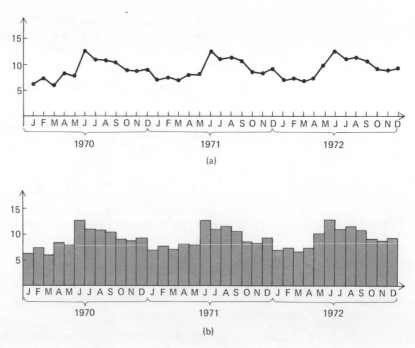

Figure 4-41 The monthly marriage rate per 10,000 persons in the United States. (a) Broken-line graph; (b) bar graph.

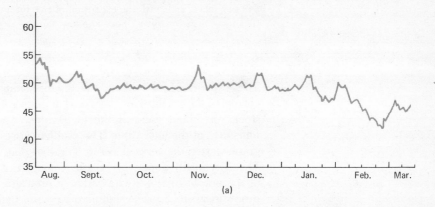

(a)

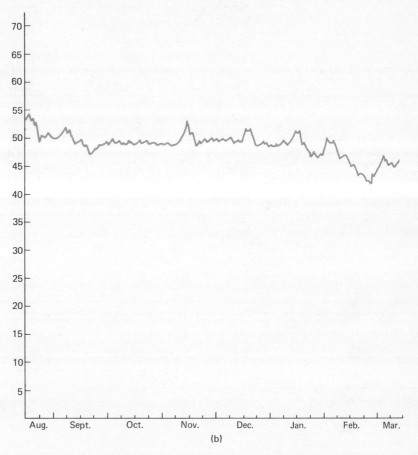

(b)

Figure 4-42 Closing stock market quotations for Amalgamated Conglomerates, Ltd. (Example 5).

a high of just over $54 per share in August; dropped almost immediately to about $49 per share, and, with the exception of two brief surges, has remained at about that level or below ever since. The low value was $42.50 per share in February.

Remark

The graph in Figure 4-42(a) is visually misleading even though it is factually correct. The graph gives the visual impression that the low value of the stock was about one-half the high value, when, in fact, it was almost four-fifths of the high value. The problem arises because the makers of the graph started the vertical axis at 35 rather than at 0 in order to save space. This is typically done when the bottom part of the scale contains no part of the graph. Figure 4-42(a) is based on a graph prepared for the use of professional stock brokers and would cause them no trouble—they watch out for such misleading impressions. The entire graph is shown in Figure 4-42(b). Observe that this graph gives the correct visual impression as well as the correct information.

Many people change scales on graphs, leave out parts of graphs, and so on, in a deliberate attempt to mislead the casual user. The graphs in Figure 4-43 give a good illustration of how this can be done. The graph in Figure 4-43(a) is based on one that was printed in a number of newspapers. It apparently shows an increase in the cost of living of about 400 percent in one year. In fact, since no scale is given on the y-axis, this is the only interpretation that can be made. Figure 4-43(b) shows the correct graph for the same information. Note that it shows that the cost of living actually increased by about 6 percent. The misleading graph in (a) was obtained by leaving off the bottom part of the vertical scale, exaggerating the remaining part of the vertical scale and removing the reference numbers.

Any graph on which reference numbers are not shown, scales are exaggerated, or portions of axes are left out should be treated with great suspicion. It is possible that the graph was modified in order to save space or to make the information clearer, but it also is possible that the graph represents a deliberate attempt to mislead the casual user.

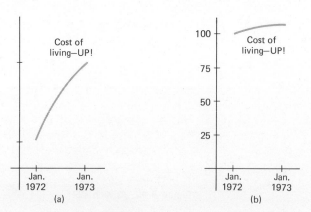

Figure 4-43   (a) Increase in cost of living—newspaper graph;
(b) increase in cost of living—correct graph.

EXERCISES

1. Plot the following points in the coordinate plane: $(1, 2)$, $(-2, 3)$, $(4, -5)$, $(-6, -7)$, $(0, 3)$, $(-1, 0)$.

2. Use Figure 4-40 to determine the temperatures at 2:00 and 3:30 for each day. When was the maximum temperature for each day recorded? The minimum temperature? What do you consider to be the "average" temperature? Explain why a bar graph would not be very satisfactory for a temperature graph.

3. The population of the United States in 1972 was approximately 205 million. Use this figure and the graph in Figure 4-41 to estimate the number of marriages in the United States during July of 1972.

4. The following data give the approximate number of deaths per 10,000 persons in the United States for the 12 months of 1971. Construct both a broken-line graph and a bar graph using the data. Which type of graph is superior for this type of data? Why?
   Jan.: 8.2; Feb.: 8.3; March: 8.1; April: 7.9; May: 7.5; June: 7.8; July: 7.5; Aug.: 7.1; Sept.: 7.7; Oct.: 7.4; Nov.: 7.9; Dec.: 7.9.

5. Copy the graph for Amalgamated Conglomerate stock on graph paper (Fig. 4-42). Study the trends of the ups and downs of the stock. Continue the graph for the next two months showing how *you think* the stock will behave, based on these trends. Do you think this would be a good time to buy additional shares of the stock or to sell stock that you already own?

6. Check your newspapers and news magazines for examples of graphs with exaggerated scales on the axes, missing reference numbers, axes that do not cross at zero, and so forth. In your opinion, which of these graphs were deliberately designed to mislead the casual reader?

7. The following statistics list the number of robberies in Bleaksburg over a three-year period: three years ago, 761; two years ago, 830; last year, 841.

   (a) You are working part time for the campaign to reelect the mayor of Bleaksburg who must answer questions about the rising number of robberies during his administration. Prepare a graph that is factually correct, but misleading, to convince the voters that the increase in the number of robberies has been minimal and that they are secure in their homes.

   (b) You also are working part time for the candidate for mayor on the reform ticket. Prepare a graph that is factually correct, but misleading, to convince people that the rise in the number of robberies has been very great and that they are in danger of being robbed if the present mayor continues in office.

8. The population of Bleaksburg has increased over the past three years from 17,000 to 18,500 (two years ago) to 21,000 (last year). Prepare a graph showing the number of robberies per 100 persons over the three-year period. (See Exercise 7.) Who would you expect to use this

graph—the mayor or the reform candidate? Which graph gives the most useful information, the one that shows the rate of robberies or the one that shows the total number of robberies?

4-6 GRAPHS DEFINED BY MATHEMATICAL RELATIONSHIPS

The branch of mathematics concerned with mathematical graphs is known as *analytic geometry*. The basic principle of analytic geometry is that every curve in the plane can be described by an equation (which may, of course, be very complicated), and that the points satisfying any equation form a geometrical figure, usually a curve.

Many graphs can be constructed from well-defined mathematical laws. In most cases, these laws are expressed by equations relating x and y, such as $y = 3x$ or $y = x^2$. The graph of such an equation consists of all points $P(x, y)$ that have coordinates satisfying the equation.

The graph of $y = 3x$, for example, consists of all points $P(x, y)$ where the y-coordinate is three times the x-coordinate. Thus, the points (5, 15), (0, 0), (−1, −3), and (2, 6) are on the graph of $y = 3x$, but (1, 4) is not (Fig. 4-44).

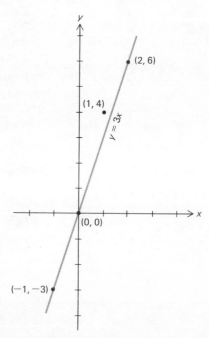

Figure 4-44 The graph of $y = 3x$ consists of all points $P(x,y)$ that have the y-coordinate equal to three times the x-coordinate.

The most direct way to graph an equation involving x and y is to calculate a very large number of points that lie on the graph and then connect them with a smooth curve (or a straight line). This method can lead to an incorrect graph in some cases, however, because we may not actually plot enough points to accurately determine the shape of the graph.

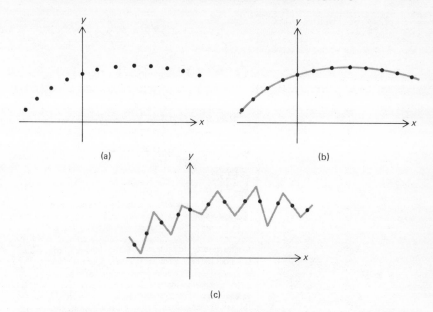

Figure 4-45 (a) Points plotted in the plane; (b) is this the graph? (c) is this the graph?

The diagram in Figure 4-45 illustrates the problem. The plotted points seem to indicate a fairly smooth graph such as in the second figure. The rather jagged broken-line graph in the third figure also goes through these points, however, and may indeed be the correct graph.

Example 1. Sketch the graph of $y = 2x$.

Solution. We plot several points on the graph and examine them to see if any pattern is apparent.

$$
\begin{array}{llll}
\text{If} & x = -2, & y = 2 \cdot (-2) = -4; & \text{point } (-2, -4) \\
\text{If} & x = -1, & y = 2 \cdot (-1) = -2; & \text{point } (-1, -2) \\
\text{If} & x = 0, & y = 2 \cdot (0) = 0; & \text{point } (0, 0) \\
\text{If} & x = 1, & y = 2 \cdot (1) = 2; & \text{point } (1, 2) \\
\text{If} & x = 2, & y = 2 \cdot (2) = 4; & \text{point } (2, 4) \\
\text{If} & x = 3, & y = 2 \cdot (3) = 6; & \text{point } (3, 6)
\end{array}
$$

The six points calculated above are plotted in Figure 4-46(a). Observe that these points appear to lie on a straight line. If we put a ruler over them, we see that this is indeed the case. The graph is a line through the origin.

The first major use of lines and curves to represent equations was made by *René Descartes* (1596–1650), the founder of analytic geometry. Descartes, who was primarily a philosopher of science rather than a mathematician, made major discoveries in mathematics while a young man

Des minimes de Paris

LA

GEOMETRIE.

LIVRE PREMIER.

Des problefmes qu'on peut conftruire fans
y employer que des cercles & des
lignes droites.

TOu s les Problefmes de Geometrie fe
peuuent facilement reduire a tels termes,
qu'il n'eft befoin par aprés que de connoi-
ftre la longeur de quelques lignes droites,
pour les conftruire.

Et comme toute l'Arithmetique n'eft compofée, que
de quatre ou cinq operations, qui font l'Addition, la
Souftraction, la Multiplication, la Diuifion, & l'Extra-
ction des racines, qu'on peut prendre pour vne efpece
de Diuifion : Ainfi n'at'on autre chofe a faire en Geometrie touchant les lignes qu'on cherche, pour les pre-
parer a eftre connuës, que leur en adioufter d'autres, ou
en ofter, Oubien en ayant vne, que ie nommeray l'vnité
pour la rapporter d'autant mieux aux nombres, & qui
peut ordinairement eftre prife a difcretion, puis en ayant
encore deux autres, en trouuer vne quatriefme, qui foit
à l'vne de ces deux, comme l'autre eft a l'vnité, ce qui eft
e mefme que la Multiplication ; oubien en trouuer vne
quatriefme, qui foit a l'vne de ces deux, comme l'vnité

(margin: Commēt le calcul d'Arithmeti- que fe rapporte aux ope- rations de Geome- trie.)

P p   eft

Title page of "Discours."

in the army. During that time he took frequent leaves of absence for travel in order to discuss mathematics and philosophy with the leading European scholars.

We would consider Descartes to be something of a hypochondriac—he constantly worried about his health and took great pains to provide comfortable surroundings for his life of contemplation. During one winter campaign with the army, he remained in bed in a heated room until 10:00 every morning, thinking about mathematics and establishing the foundations for his system of philosophy. These results were published almost 20 years later in the monumental work on philosophy, *Discours de la méthode pour bien conduire sa raison et chercher la vérité dans les sciences* (*Discourse on the Method of Reasoning Well and Seeking Truth in the Sciences*). The "method" was illus-

trated in three appendixes to the book. One of these, *La géométrie*, explained the development of analytic geometry. This one appendix provided much of the impetus for the further development of coordinate geometry and the eventual discovery of the calculus. Ironically, the development of the calculus, based on Descartes' geometry, undermined much of the science to which Descartes had devoted a large part of his life.

After a lifetime of pampered leisure, Descartes accepted a charge by the queen of Sweden to found an academy of science. Unfortunately, his duties also included the private tutoring of the queen at five o'clock every morning in an unheated room. After only 11 weeks of this austere life, Descartes contracted influenza and died at the age of 54.

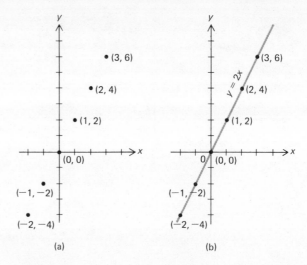

Figure 4-46 The graph of $y = 2x$ (Example 1). (a) Points on the graph; (b) the graph of $y = 2x$.

The Graph of $y = mx$

As we saw in Section 3-4, two quantities x and y are proportional if their ratio is a constant, say m. In other words, x and y are proportional if there is a constant m such that

$$y = mx$$

The graph of $y = mx$ is one of the simplest of all of the mathematical graphs. If we plot the points on the graph as we did in Example 1, we shall find that the graph is a straight line through the origin. (It is understood, of course, that the number m is a constant and that x and y are the proportional quantities.)

Since the graph of $y = mx$ is a straight line through the origin, we can construct the graph by calculating the coordinates of any one point different from the origin and then drawing a line through the plotted point and the origin. Thus, we do not need to plot a large number of points to determine the graph—one point different from the origin is sufficient.

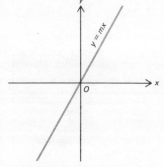

The graph of $y = mx$ is a line through the origin.

Example 2. Sketch the following graphs:
(a) $y = 3x$
(b) $y = 4x/3$
(c) $3y = -4x$

Solution. All of the graphs are straight lines through the origin because all of the equations can be written in the form $y = mx$. In each case only one additional point must be found.

(a) $y = 3x$. If $x = 1$, then $y = 3$. Thus, $(0, 0)$ and $(1, 3)$ are two points on the graph [Fig. 4-47(a)].

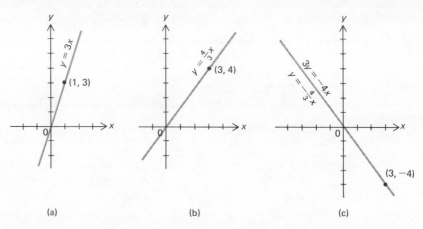

Figure 4-47   (a) $y = 3x$; (b) $y = \frac{4}{3}x$; (c) $3y = -4x$, $y = -\frac{4}{3}x$ (Example 2).

(b) $y = 4x/3$. One point is $(1, \frac{4}{3})$ obtained by choosing $x = 1$, but this point may be difficult to plot exactly. Note that if $x = 3$, then $y = (\frac{4}{3}) \cdot 3 = 4$; so $(3, 4)$ is on the graph. The graph is the line through $(0, 0)$ and $(3, 4)$ [Fig. 4-47(b)].

(c) $3y = -4x$. We rewrite the equation as

$$y = -\tfrac{4}{3}x$$

If $x = 3$, then $y = -4$, so that $(0, 0)$ and $(3, -4)$ are on the line. The graph is similar to the one in (b) except that it is oriented downward to the right rather than upward to the right [Fig. 4-47(c)].

The number m is called the *slope* of the line $y = mx$. The slope completely determines the orientation of the line. If $m > 0$, the line is oriented *upward to the right*. If $m = 0$, the line is the *x-axis*. If $m < 0$, the line is oriented *downward to the right*. The larger the value of m, the more the line is inclined toward the vertical (Fig. 4-48).

A natural question arises at this point. If we are given the equation $y = mx$, why do we need the graph? After all, any value of y that we need can be computed from the equation more accurately than from the graph, so why bother with the graph at all? One answer to this question can be illustrated with the graph in the following example. The equation relates lengths measured in feet and inches to lengths measured in centimeters. Almost every calculation from the formula is cumbersome and time-consuming. The graph, however, can be used quite easily to get the approximate number of centimeters equal to a given number of inches.

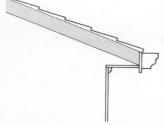

The slope of a line is comparable to the pitch of a roof. Each measures the amount of change in the vertical direction that corresponds to a given change in the horizontal direction.

Example 3.   The number of inches in a given length is proportional to the number of centimeters.

(a) Use the relationship

$$100 \text{ centimeters} \approx 39.37 \text{ inches}$$

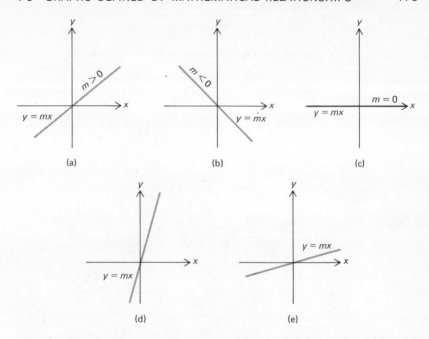

Figure 4-48 Properties of slopes of lines. (a) Positive slope; (b) negative slope; (c) zero slope; (d) large slope; (e) small slope.

to sketch a graph relating centimeters (y) and inches (x).

(b) Write the equation for the graph in the form $y = mx$.

(c) Use the graph to estimate the number of centimeters in 17 inches.

Solution. (a) Because x and y are proportional, the graph is a straight line through the origin. Thus, the equation for the graph is of form

$$y = mx$$

When $y = 100$, $x \approx 39.37$ so that the graph is a line through the origin and the point (39.37, 100). (See Fig. 4-49.)

(b) To write the equation in the form $y = mx$, we substitute the values $x \approx 39.37$, $y = 100$ into the equation and solve for m.

$$y = mx$$
$$100 = mx \approx m \cdot (39.37)$$
$$m \approx \frac{100}{39.37} \approx 2.54$$

The equation can be written as

$$y \approx 2.54x$$

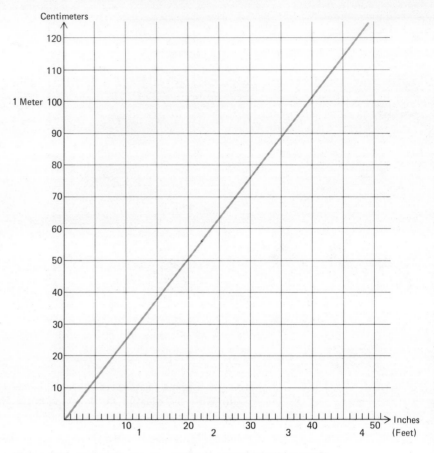

Figure 4-49   Conversion graph—inches to centimeters (Example 3).

or, more accurately, as

$$y = mx \quad \text{where} \quad m \approx \frac{100}{39.37} \approx 2.54$$

(c) When $x = 17$, we see from the graph that $y \approx 43$ centimeters.

Straight lines are the simplest of the mathematical graphs. If an equation is *not* of form $y = mx$, then its graph is not a line through the origin (although it may yet be a line), and other techniques must be used to draw its graph. Usually we must plot a large number of points and then connect them with a smooth curve. The next example illustrates the procedure.

Example 4.   Sketch the graph of $y = x^2$.

Solution.   We first calculate the values of y that correspond to several values of x:

If $x = -3,$ $y = (-3)^2 = 9;$ point $(-3, 9)$
If $x = -2,$ $y = (-2)^2 = 4;$ point $(-2, 4)$
If $x = -1,$ $y = (-1)^2 = 1;$ point $(-1, 1)$
If $x = 0,$ $y = 0^2 = 0;$ point $(0, 0)$
If $x = 1,$ $y = 1^2 = 1;$ point $(1, 1)$
If $x = 2,$ $y = 2^2 = 4;$ point $(2, 4)$
If $x = 3,$ $y = 3^2 = 9;$ point $(3, 9)$

When we plot these points, we obtain the pattern shown in Figure 4-50(a). Observe that the points descend as we move toward the right until we reach the origin and then start to ascend. There also is a form of symmetry about the y-axis. It would appear that the graph should descend as x increases until $x = 0$, and then should ascend as x increases through positive values. We connect up the points with a smooth curve as shown in Figure 4-50(b).

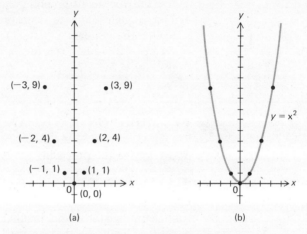

Figure 4-50   The graph of $y = x^2$ (Example 4). (a) Selected points on the graph of $y = x^2$; (b) the graph of $y = x^2$.

EXERCISES

1. Sketch the graphs of the following lines through the origin. What are the slopes of the lines?
   (a) $y = -x/2$ (c) $x - 5y = 0$
   (b) $3y = 2x$ (d) $x = y/2$
2. Copy the graph in Figure 4-49 on a large sheet of graph paper. [Use the point (39.37, 100) to determine the graph.] Use the graph to convert from feet and inches to centimeters or from centimeters to feet and inches.
   (a) 2 feet 6 inches   (d) 23 centimeters
   (b) 14 inches   (e) 87 centimeters
   (c) 4 feet   (f) 127 centimeters
3. Copy the diagram of Figure 4-51 on a sheet of graph paper. Draw two

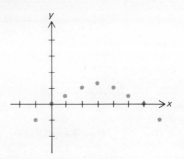

Figure 4-51 Exercise 3.

completely dissimilar graphs that pass through the points. What does this tell you about graphing equations by plotting points?

4. Plot enough points to determine the shapes of the following graphs. Sketch the graphs.

(a) $y = 2x - 1$ (c) $y = x^2 + 1$

(b) $3y = 5x - 7$ (d) $y = -x^2 + 2$

4-7   MORE ON RATIO AND PROPORTION

As we saw in the preceding section, the graph of

$$y = mx$$

is a line through the origin. In other words, if x and y represent proportional quantities, the graph that we get by plotting y against x is a straight line through the origin. Because ratio and proportion problems occur with great frequency in our society, we consider several of these problems in this section.

Example 1.   A loan shark charges Joe Schnook 8 percent interest per week on the $1000 that Joe borrowed two years ago to finance a 10-day vacation in Las Vegas. (Joe had a good time, but had to leave when he went broke after only four days.) Under the terms of their agreement, Schnook must pay the interest out of his weekly paycheck. Unfortunately, after paying the interest he never has any money left over to pay on the principal.

(a) Make a graph to show the total amount of money that Schnook had paid after x weeks.

(b) Use the graph to calculate how much money he had paid after 6 months, 18 months, and 2 years.

Solution.   (a) Each week Schnook must pay 8 percent of $1000, which amounts to

$$(0.08) \cdot \$1000 = \$80$$

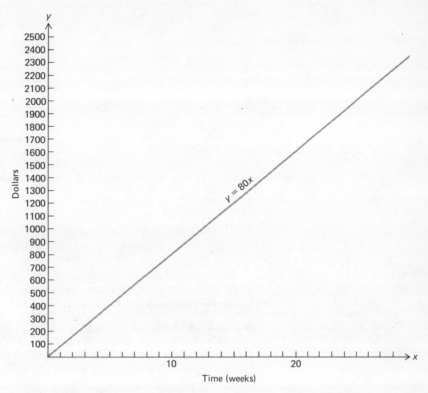

Figure 4-52 Graph of $y = 80x$ (Example 1). The vertical scale has been compressed to save space. Technically, since the interest is paid at the end of the week, only the points representing the ends of the week on the graph give the exact interest.

in interest. Thus, the total amount of interest y paid after x weeks is

$$y = 80x$$

The graph is shown in Figure 4-52.

(b) At the end of 6 months (26 weeks), Joe had paid approximately $2100 in interest. At the end of 18 months (78 weeks), he had paid approximately $6200 in interest. At the end of 2 years (104 weeks), he had paid approximately $8300 in interest.

Example 2. Tele-Ratings Co. conducts surveys to determine the popularity of television shows. In a recent survey of 200 families made between 10:00 and 10:30 P.M., the following results were obtained:

> 67 families were either not at home or not watching television;
> 43 families were watching "Fat Detective";
> 58 were watching "Redheaded Screwball";
> 30 were watching "Down Home with the Folks";
> 2 were watching a mathematics lecture on public television.

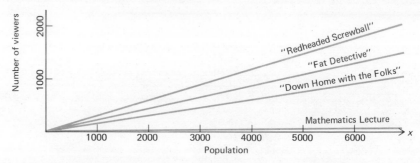

Figure 4-53   Number of viewers watching selected television programs as determined by a sample of 200 families.

Assume that the results of the survey yield proportions that hold nationwide.

(a) Construct graphs on one set of axes which can be used to estimate the number of individuals watching each program.

(b) Assume that approximately 120 million persons are in the potential 10:00 P.M. nationwide television audience. Approximately how many persons watched each show?

Solution.   (a) A total of 43 out of 200 families watched "Fat Detective." Thus, that show has a proportion of $43/200 = 21.5$ percent of the potential market. If y represents the number of viewers watching "Fat Detective" out of x potential viewers, then

$$y = 0.215x$$

The mathematical formulas for the other shows can be calculated similarly. The graphs are shown in Figure 4-53.

(b) The total number of viewers watching "Fat Detective" is estimated to be approximately

$$(0.215)(120,000,000) \approx 26,000,000$$

The total number of viewers for the other programs can be estimated similarly.

Statistical Graphs

The type of estimation made in Example 2 is very risky. It is dangerous (mathematically) to predict what millions of people will do based on a sample of a few hundred. It is just as risky to predict what 200 persons will do based on a sample of 200 million.

Two quantities may be proportional, but the exact relationship may be impossible to calculate. Although this seems like a contradiction in terms, it actually is not. Television ratings furnish a good example. The total number of persons watching a program is a certain proportion of the total population. The only way the exact proportion can be determined is to ask all of the persons in the United States and Canada if they watched the program. Since this is not feasible, other methods are used to estimate the total number of viewers.

The simplest way to estimate the total number of viewers of a television program is to survey a small number of viewers (as in Example 2), find

The Nielson Ratings are the most widely used television ratings. These ratings are based on what a selected group of people actually watch on television rather than on what they say they watch. The Nielson Ratings are based on a sample of 1200 households.

how many of them watch the program and then assume that the same proportion holds nationwide. Obviously, this approach can lead to gross inaccuracies unless great care is taken to poll a representative sample.

A variation of this method is to make several surveys and then determine the proportionality relationship that "best fits" the collected data. One method for doing this is to plot the points (x, y) representing the data and then determine the straight line through the origin that "best fits" these points. The following two examples show how this can be done and how the results, can be used.

Example 3.   Calculate the line through the origin that "best fits" the points (1, 3), (5, 13), (1, 1), (3, 4), and (5, 11). What is the slope of this line? What mathematical relationship between x and y does this line determine?

Solution.   We first plot the points [Fig. 4-54(a)]. Next we place a clear ruler on the graph paper with the edge of the ruler passing through the origin. We rotate the ruler, keeping it against the origin, until the plotted points appear to be distributed as equally as possible on each side of the ruler and the sum of the distances of the points from the ruler on one side appears to be equal to the sum of the distances on the other side. We then draw the line at that position [Fig. 4-54(b)]. In our example the line passes through the point (5, 10.7)—approximately.

Because the line passes through the origin, its equation is of form

$$y = mx$$

A large number of techniques exist for "fitting" data to graphs. Most of these methods are based on simple geometrical concepts, but involve complicated calculations. The examples in the text were chosen for ease in calculation.

where the slope m must be determined. When $x = 5$, $y \approx 10.7$; so

$$10.7 \approx m \cdot 5$$

$$m \approx \frac{10.7}{5} \approx 2.1$$

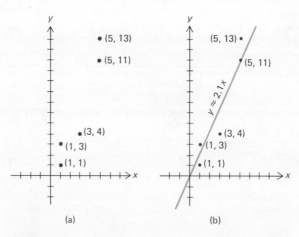

(a)   (b)

Figure 4-54   The line of "best fit" through the origin (Example 3). (a) The plotted points; (b) the line of "best fit."

The slope of the line is $m \approx 2.1$. The relationship between x and y determined by the line is

$$y \approx 2.1x$$

The following example shows how the line of "best fit" can be used to solve a meaningful problem.

Example 4.   The minister of human resources of the underdeveloped country of Bonga-Bong-Bong is concerned about the high birthrate of the country. He hopes to eventually introduce birth-control measures, but in the meantime he needs to determine the birthrate in order to predict the number of schools that should be built during the next five years. Unfortunately, vital statistics are not uniformly collected; so no one knows the exact number of births. The available data from the eight cities in which good records were kept are summarized in Table 4-1.

(a) Find the line of "best fit" through the origin based on the data in Table 4-1.

(b) Use the line of "best fit" to estimate the birthrate for the country.

(c) The total population of Bonga-Bong-Bong is approximately 30 million. How many births can be expected during the next year?

Solution.   (a) The points obtained from the data are plotted in Figure 4-55(a). (Each unit on the x-axis represents 10,000 persons; each unit on the y-axis represents 1,000 births.) The line of "best fit" through the origin is shown in Figure 4-55(b).

(b) The line in Figure 4-55(b) passes through the point (10, 8.3)—approximately. Thus, when the population is 100,000, the number of births is approximately 8,300. (Recall that one unit on the x-axis represents 10,000 persons and one unit on the y-axis represent 1,000 births.) Thus, the birthrate is approximately

$$\frac{8,300}{100,000} = \frac{83}{1,000} = 83 \text{ per thousand}$$

(c) The total number of births that can be expected in Bonga-Bong-Bong in one year is approximately

$$\frac{83}{1,000} \cdot 30,000,000 = \frac{2,490,000,000}{1,000}$$

$$\approx 2\tfrac{1}{2} \text{ million}$$

UNITED STATES BIRTHRATE
PER 1000 PERSONS OVER
A 30-YEAR PERIOD

1940	17.9
1945	19.5
1950	23.6
1955	24.6
1960	23.7
1965	19.4
1970	18.2

Table 4-1

City	Approximate Population	Number of Births
A	123,000	9,461
B	117,000	10,942
C	187,000	14,619
D	95,000	8,000 (est.)
E	85,000	9,000 (est.)
F	224,000	18,063
G	140,000	10,861
H	74,000	6,210

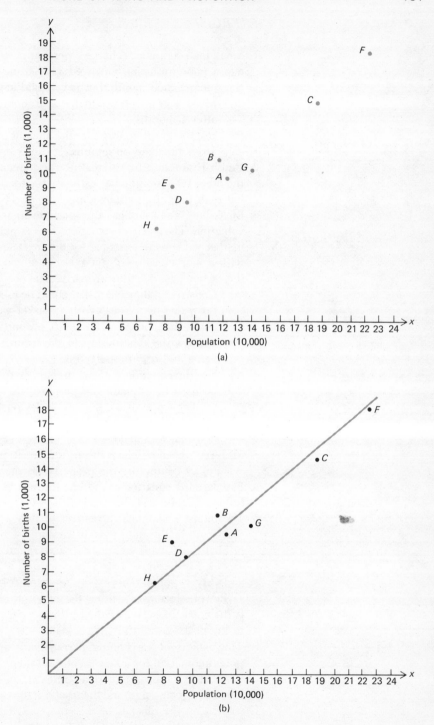

Figure 4-55 (a) Points obtained from the data; (b) the line of "best fit" (Example 4).

EXERCISES

1. Small loan companies charge 3 percent interest per month. Three years ago a poor freshman borrowed $100 from such a company and paid the interest each month, but never paid anything on the principal.
   (a)   Make a graph showing the total amount of interest he had paid after x months ($0 \leq x \leq 36$).
   (b)   Use the graph to estimate the amount of interest he had paid after 15 months and 36 months.

2. Finish Example 2 by calculating the percentages of viewers watching the other television shows. Estimate the total number of viewers watching each program nationwide.

3. Bleaksburg Township has a property tax that is 4 percent of the assessed valuation. The assessed valuation (for tax purposes) is 20 percent of the market value as determined by the tax assessor. For example, a house with a market value of $40,000 has an assessed value of $8,000, and the owner pays $320 in property taxes.
   (a)   Construct a flow chart that could be used to calculate the property tax y due on a house with a market value of $x.
   (b)   Show that the property tax is proportional to the market value. Express the relationship in the form $y = mx$, where $x =$ market value and $y =$ property tax.
   (c)   Use the equation obtained in part (b) to construct a graph that can be used to estimate the property tax y due on property with a market value of $x. Choose scales on the two axes that make it convenient to estimate the tax when the market value is less than or equal to $80,000.
   (d)   Test the flow chart in (a), the equation in (b), and the graph in (c) by computing the property tax on a house worth $50,000.

4. Estimate the equation of the line through the origin that "best fits" the following points: $(1, 1)$, $(1, 2)$, $(2, 4)$, $(2.5, 4)$, $(3, 5)$, $(4, 5.5)$, $(3.5, 7)$.

5. The minister of human resources of Bonga-Bong-Bong has discovered that some of the data used in Example 4 were in error. The number of births in City E was actually 5,641, and the number in City F was 14,691. Furthermore, City I, population 120,000, reported 7,691 births, and City J, population 87,000, reported 6,427 births.

   Rework Example 4 using the new information given above. What is the corrected value of the birthrate? How many births are expected in the next 12 months?

6. The makers of La Phuma perfume are attempting to estimate the potential market for a new scent, La Phuma Number Eight. The inhabitants of five typical American cities are being subjected to an intense advertising campaign, and the results of the campaign are being carefully studied by the officials of the company. The following table lists the number of sales in the cities over a 30-day period:

City	Adult Population	Sales (No. of Bottles)
A	250,000	4,100
B	160,000	2,500
C	180,000	2,300
D	210,000	2,800
E	180,000	3,400

(a) Find the line through the origin that "best fits" the data. Express the equation in the form $y = mx$, where y is the number of bottles of perfume and x is the population.

(b) Estimate the number of sales nationwide that could be expected out of an adult population of 125 million.

SUGGESTIONS FOR FURTHER READING

1. Ball, W. W. Rouse. *A Short Account of the History of Mathematics,* Chap. 4. New York: Dover, 1960.
2. Cooley, Hollis R., and Wahlert, Howard E. *Introduction to Mathematics,* Chaps. 8, 10, 11. Boston: Houghton-Mifflin, 1968.
3. Kline, Morris. *Mathematics for Liberal Arts,* Chaps. 6, 7, 8, 12. Reading, Mass.: Addison-Wesley, 1967.
4. Waerden, B. L. van der. *Science Awakening.* New York: Oxford University Press, 1961.

5

LINEAR EQUATIONS

5-1 INTRODUCTION TO ALGEBRAIC METHODS

In this chapter we consider techniques for solving problems by the use of elementary algebra. To a large extent, algebra is the study of equations and methods of solving problems involving equations.

As an introductory example, suppose that a certain car averages 12 miles per gallon of gasoline. If the driver is planning an 840-mile trip, he can use algebra to estimate the amount of gas he will need.

Let

$$x = \text{the number of gallons of gas needed for the trip}$$

Then $12x$ is equal to the total distance he can travel, so that

$$12x = 840$$

This last equation completely describes the situation and can be used to find the value of the unknown quantity x. If we divide both sides of the equation by 12, we get

$$12x = 840$$
$$\frac{12x}{12} = \frac{840}{12} = 70$$
$$x = 70$$

for the number of gallons.

The procedures used in the above example are typical of those we shall use throughout this chapter. We start with a problem involving an unknown quantity that we wish to calculate; let x (or some other symbol) represent the unknown quantity; write an equation involving x that describes the situation; modify the form of the equation until we can find the value of x.

The key to success in algebra is an understanding of the rules that can be used to change an equation such as

$$12x = 840$$

into a simpler equation such as

$$x = 70$$

The rules of algebra, which are really quite simple, were worked out by the Arabs of the eighth and ninth centuries. Apparently the Arabs, who lacked established scholarly traditions of their own, absorbed as much mathematics as they could from the contemporary Persians and Hindus

$$K^\gamma \beta \Delta^\gamma \iota \beta s \gamma$$

The Greek mathematician *Diophantus* (150 A.D.?) was one of the first persons outside of the Babylonian world to consider algebraic methods of solving problems. His work represents a complete break with the classical Greek tradition of basing all mathematics on geometry.

To facilitate his work, Diophantus invented an advanced form of algebraic symbolism. The above expression represents

$$2x^3 + 12x^2 + 3$$

as Diophantus would have written it:

K^γ = cube of the unknown
$\beta = 2$
Δ^γ = square of the unknown
$\iota \beta = 12$
s = the unknown
$\gamma = 3$

Although his work involved algebraic methods, the problems solved by Diophantus were quite difficult, and his results were not of the type encountered in an elementary course of algebra. Diophantus was primarily concerned with finding special solutions to indeterminate problems (problems that have many possible solutions). In most cases, he was interested in solutions that were either integers or rational numbers. A typical problem was to *find two rational numbers such that if either is added to the square of the other, the sum is a square*. His rather special method of solution leads to the numbers 3/13 and 19/13, ignoring the other solutions to the problem.

The work of Diophantus involved so many special techniques that were based on the properties of the particular numbers used in the problems that it is difficult to discover his underlying methods. One scholar stated that one could study 100 examples of how Diophantus had solved a particular type of problem and have no idea of how to solve the 101st example.

and from the writings of the ancient Greeks. They found rudimentary ideas about equations in all of these sources. Up to that time equations had been used to keep track of computational steps much as we write $\frac{1}{2} + \frac{3}{7} = \frac{7}{14} + \frac{6}{14} = \frac{13}{14}$. The genius of the Arabs was in the realization that equations could be dealt with as entities in themselves. The Arabs wrote equations involving unknown quantities and then manipulated them until they could find the values of the unknowns.

During the ninth century, the Arab mathematician *al-Khowarizmi* wrote the second most influential book on mathematics that the world has known. (The most influential was the 13-volume series by Euclid.) al-Khowarizmi wrote *Al-jabr wa'l muquābalah*, a textbook on elementary algebra that explained the rules for working with equations to the common people. The *Al-jabr* and other books by al-Khowarizmi were eventually translated into the European languages where the title *Al-jabr* was changed to *algebra*. The name of the author was corrupted into the word *algorithm*.

In this section we shall concentrate on the first step in the solution of a problem by algebra—the writing of an equation that describes the problem.

"A man dies leaving two sons behind him and bequeathing one-third of his property to an outsider. He leaves 10 dirhems of property and a claim of 10 dirhems upon one of the sons. How much does each person receive?"

Many of the illustrative examples used by the ancient Arab algebraists involved inheritance problems. The inheritance laws at that time were so complicated that it was difficult to settle some estates without mathematical analysis.

"One-third of a herd of elephants and three times the square root of the remaining part of the herd were seen on a mountain slope. In a lake was seen a male elephant along with three female elephants, constituting the ultimate remainder of the herd. How many elephants were there?"

Mahaviracarya (about 850 A.D.)

This ancient problem is typical of the exercises posed by the Hindu mathematicians. The Hindus used square roots freely in their work, considering them to be proper numbers.

"Three men with denarii found a purse containing denarii. The first man told the second, 'If I take this purse, I will have twice as much as you.' The second man told the third, 'If I take the purse, I will have three times as much as you.' The third man told the first, 'If I take the purse, I will have four times as much as you.' How much was in the purse and how much did each man have?"

The above problem is from the *Liber abaci* (1202) by Leonardo of Pisa. It probably is a translation of an older Arab problem.

Algebra in the Ancient World

The ancient Babylonians used a form of algebra to solve many problems. In their system, words were used to represent both unknowns and operations. For example, a product might be represented by the word "area," and the individual factors by the words "length" and "width." Such terms were added and subtracted without regard to whether or not the operations made sense in the physical world. For example, one ancient problem required the determination of "length" and "width" if "area" plus "length" less "width" is equal to 183 and "length" plus "width" is equal to 27. (In our notation, find x and y if $xy + x - y = 183$ and $x + y = 27$.)

Over the ages most of the Babylonian algebra was lost to subsequent civilizations. The ancient Greeks ignored it almost completely, basing their mathematics on geometry. During the second to third centuries A.D., Diophantus worked with algebraic methods, but not with the content of elementary algebra.

During the sixth to eighth centuries, the Hindus developed certain algebraic topics to a high art. It is believed that the Hindu mathematicians were influenced by the works of Diophantus and by the vestigial remains of Babylonian algebra in the Persian world. Unfortunately, the Hindu mathematicians were members of the leisure class and saw no reason to solve the problems that faced the common man. Much of their algebra was of the "ivory tower" variety—beautiful to behold, but with no practical application. Most of their mathematical problems were written up as puzzles in a flowery poetic style.

When the Arabs absorbed the Hindu knowledge, they completely changed the content of algebra. These practical people were concerned with applications that could be appreciated by the average person. To accomplish this aim, they made a thorough study of equations, concentrating on the rules that could be used to change equations into simpler equations with the same solutions. Once the scholars had developed these methods, they were written up in books that could be understood by the average person in the population.

When we first encounter a problem that involves the calculation of an unknown quantity, we may be bewildered about how to proceed. Unfortunately, the examples that we give cannot show the most important step—the work done in puzzling over the problem trying to figure out what to do. The reader should know that most of our examples require a considerable amount of time to be spent in studying the problem before the first equation can be written down. There is no simple way to read a problem and decide automatically how to proceed. It must be carefully thought out.

The following steps may help in setting up a problem as an equation.

1. Read the problem carefully. Study it, think about it, puzzle over it until all parts of the problem are understood. The reader should not attempt to solve the problem until he clearly understands what is given and what must be calculated. He should completely understand all of the relationships that exist among the various quantities.

Eleventh-century Arab astronomy book based on an earlier Hindu source. Photo courtesy of The British Library.

From 750 until 1150 Baghdad was the intellectual center of the Western world. During that period almost all of the surviving manuscripts from the ancient Greek and Roman Empires were collected in the great library of that city along with the more current texts from the Hindu and Persian worlds. Much of our knowledge of the ancient civilizations is based on the Arabic translations of these books.

2. Assign symbols to represent the unknown quantities. In most of our work, we use the symbol x to represent an unknown. This is a matter of custom, not necessity. Any symbol can be used, such as α, β, a, b, $\cap$, and so on. Once a symbol is chosen to represent a quantity in a problem, however, it cannot be used to represent any other quantity—different quantities must be represented by different symbols.

3. Write down the equations (using the symbols for the unknowns) that describe the relationships in the problem.

As we mentioned above, the illustrative examples give a misleading impression of the work done in writing down an equation to solve a problem. Step 1 is entirely omitted, and usually only equations that actually have some bearing on the problem are given in Step 3.

The reader must be warned not to expect mathematical solutions to all problems. Many practical problems do not lead to equations. Furthermore, some problems do not provide enough information for a complete mathematical solution, while others provide superfluous information.

We now turn our attention to some specific examples.

Example 1.   Thomas has scored 61, 47, 58, and 80 on his mathematics tests. The final examination counts as two tests. Write an equation that can be used to calculate the grade that he must make on the final examination in order to have a final average of 70.

Solution.   Let x denote the final exam grade. Tom must average the six grades

$$61, 47, 58, 80, x, \text{ and } x$$

(recall that the final examination counts as two grades). If the final average is to be 70, then

$$\text{the average of the six grades} = 70$$
$$\frac{61 + 47 + 58 + 80 + 2x}{6} = 70$$

This gives the desired equation.

Example 2.   The U-Fill-Um Auto Rental Company rents old taxicabs at bargain rates: $5 per day plus 3 cents per mile, you furnish the gas. Joe has allotted $150 for automobile expenses for a four-day trip. Write an equation that describes how many miles he can travel on the $150 if gas costs him 5 cents per mile.

Solution.   Let x represent the number of miles he can travel. The costs for the trip are

$0.03x$ for the mileage rental of the car
$0.05x$ for gas
20 for the rental of the car for four days

The total cost must equal $150; so

$$0.03x + 0.05x + 20 = 150$$

Example 3. The produce manager for a grocery store bought a shipment of 1000 heads of lettuce for $150. He finds that one-fourth of the lettuce is bad and must be discarded. At what price per head should he sell the remaining lettuce in order to make a total profit equal to 5 cents per head on each head sold?

Solution. Let x denote the selling price per head of lettuce. Note that 250 heads were discarded and 750 remain to be sold. The total income that will be brought in by selling the 750 heads of lettuce is

$$\text{income} = 750x$$

The total cost of the lettuce was $150. Thus, the profit is

$$\text{profit} = \text{income} - \text{cost} \overset{\cdot}{=} 750x - 150$$

Since the profit is to be equal to 5 cents per head sold, we also have

$$\text{profit} = 0.05(750)$$

By setting the two expressions for the profit equal to each other, we get the equation

$$750x - 150 = 0.05(750)$$

EXERCISES

1. Rewrite the equation for Example 1 on the assumption that the test grades were 85, 72, 90, 59 and the final average is to be 80.
2. Rewrite the equation for Example 2 on the assumption that the car costs $11 per day and 11 cents per mile, gas included, and that Joe has $200 for automobile expenses for the trip.
3. Rewrite the equation for Example 3 on the assumption that the bad lettuce can be sold at half price and that the total profit is to be 10 cents per head sold.
4. A student on academic probation has taken 87 credit hours of work before this semester and has earned 157 quality points. [A = 4, B = 3, C = 2, D = 1, F = 0 quality points.] If he has not earned at least a 2.00 cummulative quality point average by the end of this semester, he will be required to leave school. He is currently taking 15 hours of work. Write an equation that can be used to calculate the minimum number of quality points that he must earn this semester in order to stay in school. (Let x be the number of quality points he must earn.)
5. The manager of a wine and cheese shop has a 50-percent markup on all items. (If he buys a bottle of wine for $1.00, he sells it for $1.50.) The operating costs of the store are $1600 per month. Write an equation that describes the total monthly sales necessary to make a net profit of $500. (Let x be the total sales for the month.)
6. Almost nothing is known about the life of Diophantus except for the following puzzle written after his death:

Diophantus' youth lasted $\frac{1}{6}$ of his life; he grew a beard after $\frac{1}{12}$ more. After $\frac{1}{7}$ of his life he married; five years later he had a son. The son lived exactly $\frac{1}{2}$ as long as his father, and Diophantus died just four years after his son. All of this adds up to the number of years that Diophantus lived.

Write an equation that describes the information given in this 1500-year-old puzzle. (Let x be the number of years that Diophantus lived.) How old was Diophantus when he died?

5-2   ALGEBRAIC EXPRESSIONS

In our work with algebra later in this section, we shall find it necessary to work with algebraic expressions such as

$$(3x + 5)(x + 1) \quad \text{and} \quad \frac{5x - 2}{x + 7}$$

Although such work will not be extensive, we do need a brief introduction to the basic techniques.

Sums and Differences

To add or subtract algebraic expressions, we add or subtract like terms.

Example 1.
(a) $(3x + 2) + (5x + 3) = (3x + 5x) + (2 + 3) = 8x + 5$
(b) $(5x + 3) + (8x - 7) = (5x + 8x) + (3 - 7) = 13x - 4$
(c) $(x - 18) - (2x - 3) = x - 18 - 2x + 3 = -x - 15.$
(d) $(x^2 - 2x + 3) + (x + 2) = x^2 + (-2x + x) + (3 + 2) = x^2 - x + 5$

Much of the confusion that surrounds work with more complicated algebraic expressions can be dispelled if we remember that the symbol x (or y or z) represents a number. If we remember the rules for working with numbers, we should be able to deal with these expressions without too much difficulty.

As a simple example, recall that, by the use of the distributive law,

$$3(6 + 5) = 3 \cdot 6 + 3 \cdot 5 = 18 + 15 = 33$$

In a similar way, we obtain

$$x(2x + 5) = x \cdot 2x + x \cdot 5$$
$$= 2x^2 + 5x$$

There is a useful technique that may help the reader when he is faced with an algebraic expression that must be multiplied or divided by another expression and he does not know how to proceed. *Substitute a number*

*for x in the expression and carry out the indicated operations using that
number for x.* This experience should help him understand the equivalent
steps that are required to work out the original problem. For example,
if he needs to multiply

$$x(2x + 5)$$

as above, he might first substitute 3 for x in the expression. The new
expression is then

$$3(6 + 5)$$

which he can work out as

$$3(6 + 5) = 18 + 15 = 33$$

This method of solution should help him realize that

$$x(2x + 5) = x \cdot 2x + x \cdot 5 = 2x^2 + 5x$$

We now consider a slightly more complicated problem.

Example 2. Multiply

$$(x + 2)(x + 5)$$

Solution. We must multiply the number $x + 2$ by the number $x + 5$. Observe
that we can multiply a number by $x + 5$ if we use the distributive law:

$$(x + 2)(x + 5) = (x + 2) \cdot x + (x + 2) \cdot 5$$

We now can use the distributive law on each of these terms, obtaining

$$\begin{aligned}
(x + 2)(x + 5) &= (x + 2) \cdot x + (x + 2) \cdot 5 \\
&= x^2 + 2x + 5x + 10 \\
&= x^2 + 7x + 10
\end{aligned}$$

There is an alternate way of computing the product that we considered
in Example 2. We write the product as

$$\begin{array}{r} x + 2 \\ times \quad \underline{x + 5} \end{array}$$

We multiply each term in the top expression by x (the first term in the
bottom expression), obtaining

$$x \cdot (x + 2) = x^2 + 2x$$

then multiply each term in the top expression by 5, obtaining

$$5 \cdot (x + 2) = 5x + 10$$

We write these two expressions, lining up similar terms in columns, then add the columns.

$$
\begin{array}{r}
x + 2 \\
times \quad x + 5 \\
\hline
x^2 + 2x \\
5x + 10 \\
\hline
x^2 + 7x + 10
\end{array}
$$

When we use this method of multiplication, it is convenient to line up the terms in columns according to the powers of x. The first column usually contains the highest power of x, the second column the next highest power, and so on. This method can be used for more complicated products, such as

$$(x^2 - x + 2)(3x^2 + x - 7)$$

Example 3.   Calculate the following products.
(a) $(3x + 5)(2x + 7)$
(b) $(x - 8)(7x - 1)$
(c) $(x^2 - x + 2)(3x^2 + x - 7)$

Solution.

(a)
$$
\begin{array}{r}
3x + 5 \\
2x + 7 \\
\hline
6x^2 + 10x \\
21x + 35 \\
\hline
6x^2 + 31x + 35
\end{array}
$$

(b)
$$
\begin{array}{r}
x - 8 \\
7x - 1 \\
\hline
7x^2 - 56x \\
- x + 8 \\
\hline
7x^2 - 57x + 8
\end{array}
$$

(c)
$$
\begin{array}{r}
x^2 - x + 2 \\
3x^2 + x - 7 \\
\hline
3x^4 - 3x^3 + 6x^2 \\
x^3 - x^2 + 2x \\
- 7x^2 + 7x - 14 \\
\hline
3x^4 - 2x^3 - 2x^2 + 9x - 14
\end{array}
$$

Fractions

We shall be concerned with several algebraic expressions that involve fractions later in this chapter. The rules for working with these expressions are similar to those for fractions involving numbers.

(1) *Simplification of Fractions.* Recall that

$$\frac{5 \cdot 3}{7 \cdot 3} = \frac{5 \cdot \cancel{3}}{7 \cdot \cancel{3}} = \frac{5}{7}, \qquad \frac{9 \cdot 8}{13 \cdot 8} = \frac{9 \cdot \cancel{8}}{13 \cdot \cancel{8}} = \frac{9}{13}$$

and so on. In general,

$$\frac{a \cdot m}{b \cdot m} = \frac{a \cdot \cancel{m}}{b \cdot \cancel{m}} = \frac{a}{b}$$

if m is a nonzero number. Similarly,

$$\frac{(3x-2)(4x+7)}{(x+3)(4x+7)} = \frac{(3x-2)\cancel{(4x+7)}}{(x+3)\cancel{(4x+7)}} = \frac{3x-2}{x+3}$$

$$\frac{(5x+1)(x+12)}{(x-6)(x+12)} = \frac{(5x+1)\cancel{(x+12)}}{(x-6)\cancel{(x+12)}} = \frac{5x+1}{x-6}$$

and so on.

Any nonzero common factor of the numerator and the denominator can be canceled using the same rule as for fractions involving numbers. Similarly, fractions can be "built up" by using this rule. For example,

$$\frac{x-1}{x+1} = \frac{(x-1)(x+2)}{(x+1)(x+2)} = \frac{x^2+x-2}{x^2+3x+2}$$

(2) *Addition.* To add fractions with the same denominator, we add the numerators. The sum has the same denominator as the original fractions.

Example 4.

(a) $\dfrac{3}{7} + \dfrac{2}{7} = \dfrac{5}{7}$

(b) $\dfrac{x+2}{3x^2-5} + \dfrac{2x-1}{3x^2-5} = \dfrac{3x+1}{3x^2-5}$

(c) $\dfrac{x^2+9}{x-1} + \dfrac{2x^2+3x-7}{x-1} = \dfrac{3x^2+3x+2}{x-1}$

(d) $\dfrac{x+1}{x^2-5x+2} - \dfrac{2x-1}{x^2-5x+2} = \dfrac{(x+1)-(2x-1)}{x^2-5x+2}$

$$= \frac{x+1-2x+1}{x^2-5x+2} = \frac{-x+2}{x^2-5x+2}$$

To add fractions with different denominators, we first "build up" the fractions so that they have a common denominator using the rule

$$\frac{a}{b} = \frac{am}{bm}$$

Example 5.

(a) $\dfrac{5}{9} + \dfrac{3}{7} = \dfrac{5\cdot7}{9\cdot7} + \dfrac{3\cdot9}{7\cdot9} = \dfrac{35}{63} + \dfrac{27}{63} = \dfrac{62}{63}$

(b) $\dfrac{2}{x-1} + \dfrac{3}{x} = \dfrac{2\cdot x}{(x-1)x} + \dfrac{3(x-1)}{(x-1)\cdot x} = \dfrac{2x}{x^2-x} + \dfrac{3x-3}{x^2-x} = \dfrac{5x-3}{x^2-x}$

(3) *Multiplication.* To multiply fractions, we multiply their numerators and their denominators.

Example 6.

(a) $\dfrac{2}{3} \cdot \dfrac{2}{5} = \dfrac{4}{15}$

(b) $\dfrac{x-1}{x+1} \cdot \dfrac{2x-1}{x+3} = \dfrac{(x-1)(2x-1)}{(x+1)(x+3)} = \dfrac{2x^2 - 3x + 1}{x^2 + 4x + 3}$

(c) $\dfrac{x+2}{x^2+3} \cdot \dfrac{x-6}{2x+1} = \dfrac{(x+2)(x-6)}{(x^2+3)(2x+1)} = \dfrac{x^2 - 4x - 12}{2x^3 + x^2 + 6x + 3}$

(4) *Division.* To divide fractions, we "invert" the divisor and multiply the resulting fractions.

Example 7.

(a) $\dfrac{3}{8} \div \dfrac{2}{5} = \dfrac{3}{8} \cdot \dfrac{5}{2} = \dfrac{15}{16}$

(b) $\dfrac{x-1}{x+1} \div \dfrac{2x}{x+7} = \dfrac{x-1}{x+1} \cdot \dfrac{x+7}{2x} = \dfrac{(x-1)(x+7)}{(x+1) \cdot 2x} = \dfrac{x^2 + 6x - 7}{2x^2 + 2x}$

(c) $\dfrac{\dfrac{4}{7}}{\dfrac{5}{13}} = \dfrac{4}{7} \div \dfrac{5}{13} = \dfrac{4}{7} \cdot \dfrac{13}{5} = \dfrac{52}{35}$

(d) $\dfrac{\dfrac{5x-2}{x+3}}{\dfrac{x+7}{x-1}} = \dfrac{5x-2}{x+3} \div \dfrac{x+7}{x-1} = \dfrac{5x-2}{x+3} \cdot \dfrac{x-1}{x+7}$

$= \dfrac{(5x-2)(x-1)}{(x+3)(x+7)} = \dfrac{5x^2 - 7x + 2}{x^2 + 10x + 21}$

EXERCISES

1. Calculate the following sums and differences.
   (a) $(15x + 8) + (9x - 6)$
   (b) $(3x^2 + 2) + (x^2 - x + 4)$
   (c) $(x^2 + 5x - 2) - (2x^2 - 3x + 8)$
   (d) $(2x - 5) - (-x^2 - 3x + 2)$
   (e) $(x^3 + x^2 + 1) + (3x + 7)$
   (f) $(3x^4 + x^3 + 2x - 1)$
   $\quad - (x^4 - x^2 - 1)$

2. Calculate the following products.
   (a) $(3x + 9)(x - 4)$
   (b) $(8x - 1)(3x - 2)$
   (c) $(x + 4)(x - 2)(x + 3)$
   (d) $x(x - 1)(x + 1)$
   (e) $(x^2 + x + 5)(3x - 2)$
   (f) $(-x - 1)(-2x^2 + 3x + 2)$
   (g) $(x^2 - 1)(x^2 + x + 1)$
   (h) $(2x^2 + 3x + 1)(x^2 + x - 2)$

3. Simplify the following fractions.
   (a) $\dfrac{(x+2)(3x+1)}{(4x-1)(3x+1)}$
   (b) $\dfrac{(2x-3)(7x-2)}{(7x-2)(x+4)}$
   (c) $\dfrac{(x+8)(x-2)}{(2-x)(x+1)}$
   (d) $\dfrac{x(x+1)(2x-1)}{(x-1) \cdot x \cdot (x+1)}$

4. Calculate the following products and quotients.
   (a) $\dfrac{x+2}{x-1} \cdot \dfrac{x+3}{x-2}$
   (b) $\dfrac{x+1}{2x-5} \cdot \dfrac{x^2+7}{x+2}$

(c) $\dfrac{x-1}{x+1} \cdot \dfrac{x+1}{x+5}$

(f) $\dfrac{x+1}{2x-5} \div \dfrac{x^2+7}{x+2}$

(d) $\dfrac{4x-2}{x^2+3} \cdot \dfrac{2x^2-8}{x+1}$

(g) $\dfrac{\dfrac{x-1}{x+1}}{\dfrac{x+1}{x+5}}$

(e) $\dfrac{x+2}{x-1} \div \dfrac{x+3}{x-2}$

(h) $\dfrac{4x-2}{x^2+3} \div \dfrac{2x^2-8}{x+1}$

5-3 SOLUTIONS OF EQUATIONS

The solution of problems by algebra involves two basic steps:

(1) Write an equation that describes the problem. This equation involves one or more symbols, such as x, y, z, α, β, and so on, that represent the unknown quantities. (This step was discussed in Section 5-1.)

(2) Find all numbers that make the equation a true statement when they are substituted for the symbols into the equation. The numbers found in this step are called *solutions* of the equation.

Example 1. Show that -5 is a solution and that 3 is not a solution of the equation

$$3x - 7 = 5x + 3$$

Solution. We substitute each number into both sides of the equation. If the two sides are equal to the same number after a substitution, then that value of x is a solution of the equation. If the two sides are not equal, then that value of x is not a solution.

If $x = -5$:

left-hand side is $3x - 7 = 3(-5) - 7 = -15 - 7 = -22$
right-hand side is $5x + 3 = 5(-5) + 3 = -25 + 3 = -22$

Since the two sides of the equation are equal to the same number when -5 is substituted for x, then -5 is a solution.

If $x = 3$:

left-hand side is $3x - 7 = 3 \cdot 3 - 7 = 9 - 7 = 2$
right-hand side is $5x + 3 = 5 \cdot 3 + 3 = 15 + 3 = 18$

Since the two sides are not equal when 3 is substituted for x, then 3 is not a solution.

The method illustrated in Example 1 can always be used to test individual numbers to find whether or not they are solutions of an equation. Unfortunately, substituting values for x in an equation is usually not enough to find the solutions. It is only a matter of luck if we actually find a solution by that method.

The basic methods for solving equations were worked out by the Arabs

$$3x - 7 = 5x + 3$$

$$-2x = 10$$

$$x = -5$$

We solve an equation by constructing a chain of progressively simpler equations that have the same solution.

of the ninth century. These methods are quite simple and are easy to apply to the equations that we shall encounter in this chapter. We shall start with an equation and successively modify its form, obtaining a chain of equations which have the same solution as the original one. We shall construct these new equations in such a way that they become progressively simpler. The process is stopped when we get an equation that is so simple that its solution is obvious.

There are two operations that we can use to modify the form of an equation so that the new equation has the same solution:

(1) Add (or subtract) the same quantity to both sides of the equation. This quantity may be a number or an expression involving x.

(2) Multiply (or divide) both sides of the equation by the same nonzero number.

Example 2. Solve the equation

$$3x - 3 = x + 6$$

Solution. Our plan of attack is as follows: First, we rewrite the equation with all of the terms involving x on one side and all other terms on the other side of the equality sign. Then we divide both sides of the equation by the coefficient of x.

To write the equation with all of the x's on one side, we need to get rid of the x term on the right. We can do this by subtracting x from both sides of the equation.

$$3x - 3 = x + 6$$
$$3x - 3 - x = 6 \qquad \text{(subtracting } x \text{ from both sides)}$$
$$2x - 3 = 6$$

Next, we add 3 to both sides of the equation in order to get rid of the "-3" on the left:

$$2x - 3 = 6$$
$$2x = 6 + 3 = 9 \qquad \text{(adding 3 to both sides)}$$

At this step we have an equation with all of the x terms on one side and all constant terms on the other side. Now we divide both sides by 2, the coefficient of x:

$$2x = 9$$
$$x = \tfrac{9}{2} = 4\tfrac{1}{2} \qquad \text{(dividing both sides by 2)}$$

We have successively modified our original equation as follows:

$$3x - 3 = x + 6$$
$$2x - 3 = 6$$
$$2x = 9$$
$$x = \tfrac{9}{2}$$

This last equation has the obvious solution $x = \frac{9}{2}$. What we have done is to show that if x satisfies the original equation, then x must be equal to $\frac{9}{2}$. In other words, we have restricted the possible solutions of the original equation to the one number $x = \frac{9}{2}$. We now check our work by substituting $x = \frac{9}{2}$ into the original equation $3x - 3 = x + 6$.

$x = \frac{9}{2}$:

$$\text{left-hand side is}\qquad 3x - 3 = 3 \cdot \tfrac{9}{2} - 3 = \tfrac{27}{2} - 3 = 10\tfrac{1}{2}$$
$$\text{right-hand side is}\qquad x + 6 = \tfrac{9}{2} + 6 = 4\tfrac{1}{2} + 6 = 10\tfrac{1}{2}$$

Thus, $x = \frac{9}{2}$ is the solution of the equation

$$3x - 3 = x + 6$$

Example 2 was analyzed in considerable detail in order to give the reasons behind the steps used in reducing the equation

$$3x - 3 = x + 6$$

to the new equation

$$x = \frac{9}{2}$$

The two essential steps involved (1) adding quantities to both sides of the equation and (2) dividing both sides by a nonzero constant. These two steps can be carried out freely whenever we deal with equations. The reason is as follows: Suppose the original equation represents a true statement for some value of x, say $x = x_0$. If we add a quantity to both sides of the equation, then we are preserving the equality when $x = x_0$. This is roughly equivalent to adding equal weights to both sides of a balance scale. If the scales were originally in balance, then they will remain in balance when the equal weights are added. Similarly, multiplying both sides of the equation by the same number preserves the equality when $x = x_0$.

Example 3. Solve the equation

$$13x - 5 + 2x = 7 + 4x - 3$$

Solution. We first simplify the equation by collecting the like terms on both sides of the equation:

$$13x - 5 + 2x = 7 + 4x - 3$$
$$15x - 5 = 4x + 4$$

We now need to get all of the terms involving x to the left-hand side of the equation and all of the other terms to the right-hand side. We subtract $4x$ from both sides and add 5 to both sides. Finally we divide both sides by the coefficient of x.

$$5x - 7 = x + 2$$

$$5x - x = 7 + 2$$

The Arabs considered *transposition* to be a basic operation of algebra. It made sense to them to move terms across the equality sign, provided the signs of the terms were changed. In the above example, the -7 is transposed to the right-hand side of the equation, being changed to a 7 in the process, and the x is transposed to the left-hand side.

Rules of transposition were taught (without proof) as part of the standard algebra course until a very few years ago.

$$15x - 5 = 4x + 4$$
$$15x - 4x - 5 = 4 \qquad \text{(subtracting } 4x \text{ from both sides)}$$
$$11x - 5 = 4$$
$$11x = 4 + 5 \qquad \text{(adding 5 to both sides)}$$
$$11x = 9$$
$$x = \tfrac{9}{11} \qquad \text{(dividing both sides by 11)}$$

We now substitute the value $x = \tfrac{9}{11}$ back into the original equation as a check on our work.

$$x = \frac{9}{11}:$$

left-hand side is $\quad 13x - 5 + 2x = 13 \cdot \dfrac{9}{11} - 5 + 2 \cdot \dfrac{9}{11}$

$$= \frac{13 \cdot 9 - 5 \cdot 11 + 2 \cdot 9}{11} = \frac{117 - 55 + 18}{11} = \frac{80}{11}$$

right-hand side is $\quad 7 + 4x - 3 = 7 + 4 \cdot \dfrac{9}{11} - 3$

$$= \frac{7 \cdot 11 + 4 \cdot 9 - 3 \cdot 11}{11} = \frac{77 + 36 - 33}{11} = \frac{80}{11}$$

Since the value $x = \tfrac{9}{11}$ checks out in the equation, it is a solution.

An equation can be pictured as a balance scale. If we start with the scale in balance and perform the same operations to both sides, then the scale remains in balance.

2x + 7 = 23

(a)

2x = 16

(b)

x = 8

(c)

(a) Original equation; (b) remove 7 units from both sides; (c) remove half the units from both sides.

EXERCISES

1. Decide which, if any, of the given numbers are solutions of the equations by substituting the numbers into the equations.

   (a) $3x + 5 = 7x + 8$; $x_1 = \frac{7}{3}$, $x_2 = -\frac{3}{4}$

   (b) $21x - 7 + 2x = 13x + 24 + x - 5$; $x_1 = 5$, $x_2 = \frac{26}{9}$, $x_3 = \frac{7}{38}$

2. Use the methods of this section to solve the following equations. Check your answer by substituting it back into the equation.

   (a) $x + 12 = 3x - 4$ (d) $-x + 12 + 2x = 35$

   (b) $13x - 2 = 6 - 3x$ (e) $25x - 11 + 2x + 8 = 0$

   (c) $15x + 7 = 2x + 2$ (f) $2x - 5 + x = 3 + 2x - 8$

3. Solve the equation in Exercise 3, Section 5-1.

4. Solve the equation in Exercise 4, Section 5-1.

5. Solve the equation in Exercise 5, Section 5-1.

6. Solve the equation in Exercise 6, Section 5-1.

5-4 THE LINEAR EQUATION $ax = b$

As we saw in the examples of the preceding section, equations that can be reduced to the form

$$ax = b$$

where $a \neq 0$, are particularly easy to solve. We divide both sides of the equation by a, obtaining the following chain of equations

$$ax = b$$

$$\frac{ax}{a} = \frac{b}{a} \qquad \text{(divide both sides by } a\text{)}$$

$$x = \frac{b}{a}$$

This shows that the only possible solution is $x = b/a$. If we substitute this value back in the original equation, the left-hand side is

$$ax = a\frac{b}{a} = \frac{ab}{a} = b$$

so that $x = b/a$ is indeed a solution of the equation.

The above argument shows that the only solution of the equation

$$ax = b$$

where $a \neq 0$, is $x = b/a$.

Example 1.

(a) The only solution of $7x = 13$ is $x = \frac{13}{7}$.

(b) The only solution of $-\frac{5}{3}x = \frac{4}{7}$ is $x = \frac{4}{7}/(-\frac{5}{3}) = -\frac{12}{35}$.

(c) The only solution of $-x = 1$ is $x = 1/(-1) = -1$.

The invention of the printing press in the fifteenth century had a major impact on the development of algebraic symbolism. Until that time the algebraic symbols had, for the most part, been abbreviations of the names of the operations. The widespread disseminations of the printed German mathematics books helped to establish + and − notation in the rest of Europe.

Equations that involve x raised to only a first power are called *linear equations*, because they can be interpreted geometrically in terms of graphs of straight lines. (This will be discussed in more detail later in this section.) For example, the equations

$$3x - 2 = x + 5$$

and

$$4x + 7 = 3x - 32$$

are linear equations.

The steps involved in solving a linear equation are outlined in Figure 5-1. These are the steps that we considered in the preceding section with particular examples. The main steps involve adding quantities to both sides of the equation (in order to get all of the terms involving x on the left-hand side of the equation and all other terms on the right-hand side) and dividing both sides of the equation by the coefficient of x.

Many equations can be reduced to the form $ax = b$ even when the original form is quite different. This is illustrated in the following two examples. Whenever an equation of some other form is modified to the form $ax = b$, it is very important that the final answer be checked by substituting it back into the original equation. Aside from the danger of a computational error, there is the very real possibility that we may have changed the solution when we changed the form of the equation. This sometimes occurs when we multiply or divide both sides of the equation by an expression involving x rather than a constant.

Example 2.   Solve $2x^2 + 11x + 12 = 2x^2 - 27x + 13$.

Solution.

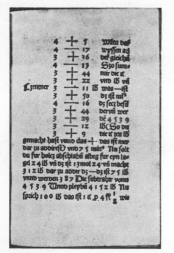

Photo courtesy of The British Library.

The + and − signs were first used in print in *Behende und hubsche Rechnung auf allen Kauffmanschafft* (1489) by Johann Widman.

$$
\begin{aligned}
2x^2 + 11x + 12 &= 2x^2 - 27x + 13 \\
11x + 12 &= -27x + 13 \quad \text{(subtracting } 2x^2 \text{ from both sides)} \\
11x + 12 + 27x &= 13 \quad \text{(adding } 27x \text{ to both sides)} \\
38x + 12 &= 13 \\
38x &= 13 - 12 \quad \text{(subtracting 12 from both sides)} \\
38x &= 1 \\
x &= \tfrac{1}{38} \quad \text{(dividing both sides by 38)}
\end{aligned}
$$

We now must check our answer by substituting it back into the original equation $2x^2 + 11x + 12 = 2x^2 - 27x + 13$.

$$x = \frac{1}{38}:$$

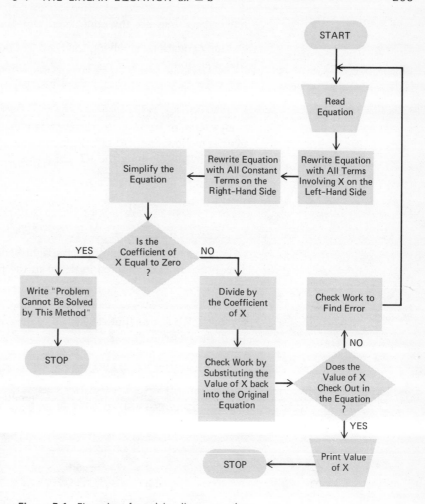

Figure 5-1 Flow chart for solving linear equations.

The $=$ sign was first used in print in *The Whetstone of Witte* (1557) by Robert Recorde.

left-hand side is
$$2x^2 + 11x + 12 = 2 \cdot \left(\frac{1}{38}\right)^2 + 11 \cdot \frac{1}{38} + 12$$

$$= \frac{2}{38^2} + \frac{11 \cdot 38}{38^2} + \frac{12 \cdot 38^2}{38^2}$$

$$= \frac{2 + 418 + 17{,}328}{38^2} = \frac{17{,}748}{38^2}$$

right-hand side is
$$2x^2 - 27x + 13 = 2 \cdot \left(\frac{1}{38}\right)^2 - 27 \cdot \frac{1}{38} + 13$$

$$= \frac{2}{38^2} - \frac{27 \cdot 38}{38^2} + \frac{13 \cdot 38^2}{38^2}$$

$$= \frac{2 - 1{,}026 + 18{,}772}{38^2} = \frac{17{,}748}{38^2}$$

Since the values are the same, then $x = \frac{1}{38}$ is a solution of the equation.

Many equations involving fractions can be reduced to linear equations. The equation

$$\frac{x^2 - 1}{x + 2} = x + 3$$

is an example. In order to reduce this equation to a linear equation, we must somehow "get rid of" the denominator $x + 2$. This can be accomplished by multiplying the left-hand side of the equation by $x + 2$. The left-hand side then becomes

$$\frac{x^2 - 1}{x + 2}(x + 2) = \frac{(x^2 - 1)(x + 2)}{x + 2} = x^2 - 1$$

Of course, we must also multiply the right-hand side by the same expression. This problem is considered in detail in the following example.

Example 3. Solve $(x^2 - 1)/(x + 2) = x + 3$.

Solution. We first multiply both sides of the equation by $x + 2$ in order to get rid of the fraction:

$$\frac{x^2 - 1}{x + 2} = x + 3$$

$$\frac{x^2 - 1}{x + 2}(x + 2) = (x + 3)(x + 2) \quad \text{(multiplying both sides by } x + 2)$$

$$x^2 - 1 = (x + 3)(x + 2) = x(x + 2) + 3(x + 2)$$

$$x^2 - 1 = x^2 + 2x + 3x + 6$$

$$x^2 - 1 = x^2 + 5x + 6$$

$$-1 = 5x + 6 \quad \text{(subtracting } x^2 \text{ from both sides)}$$

$$-5x = 1 + 6 \quad \text{(subtracting } 5x, \text{ adding 1 to both sides)}$$

$$-5x = 7$$

$$x = \frac{7}{(-5)} = -\frac{7}{5}$$

We now must check this value in the original equation.

$$x = -\frac{7}{5}:$$

left-hand side is
$$\frac{x^2 - 1}{x + 2} = \frac{(-\frac{7}{5})^2 - 1}{-\frac{7}{5} + 2} = \frac{\frac{49}{25} - 1}{-\frac{7}{5} + 2}$$

$$= \frac{\frac{49}{25} - \frac{25}{25}}{-\frac{7}{5} + \frac{10}{5}} = \frac{\frac{24}{25}}{\frac{3}{5}} = \frac{\overset{8}{\cancel{24}}}{\underset{5}{\cancel{25}}} \cdot \frac{\cancel{5}}{\cancel{3}} = \frac{8}{5}$$

right-hand side is
$$x + 3 = -\frac{7}{5} + 3 = -\frac{7}{5} + \frac{15}{5} = \frac{8}{5}$$

Since the value $x = -\frac{7}{5}$ checks out in the equation, it is a solution.

The following example shows how the solution of an equation may be changed when we reduce it to the form $ax = b$.

Example 4.   Show that the equation

$$\frac{2}{x} + 3x - 5 = \frac{1}{x} - (2x + 5) + \frac{1}{x}$$

has no solution.

Solution.   We collect like terms on both sides of the equation:

$$\frac{2}{x} + 3x - 5 = \frac{1}{x} - (2x + 5) + \frac{1}{x}$$

$$\frac{2}{x} + 3x - 5 = \frac{2}{x} - 2x - 5$$

$$3x - 5 = -2x - 5 \qquad \text{(subtracting } 2/x \text{ from both sides)}$$

$$3x = -2x \qquad \text{(adding 5 to both sides)}$$

$$3x + 2x = 0 \qquad \text{(adding } 2x \text{ to both sides)}$$

$$5x = 0$$

$$x = 0$$

The only possible number that can be a solution is $x = 0$. When we try to substitute this value into the original equation, however, we see that it cannot be a solution because the expressions $2/x$ and $1/x$ are meaningless when $x = 0$. Thus, the original equation has no solution even though the final equation has the solution $x = 0$.

Solution by Graphs

The equation $ax = b$ can be solved by using the graph of $y = ax$. Most of us have done this at one time or another as in the following example.

Example 5.   The local tax office distributes small graphs to taxpayers showing the property tax for assessed valuation [Fig. 5-2(a)]. The tax can be calculated from the equation

$$y = 0.043x$$

where x is the assessed value of the property or from the graph.

The graph also can be used to figure the assessed value from the property tax. Figure 5-2(b) shows one of these graphs after someone used it to calculate the assessed value of his property:

(1) He drew a horizontal line through the number 344 on the y-axis, the value of the property tax.

(2) He located the point where this horizontal line crosses the tax line.

(3) He drew a vertical line from the point of intersection down to the x-axis in order to find that the assessed value of his property is approximately equal to $8000.

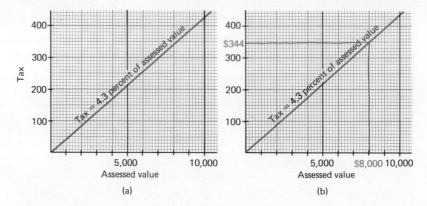

Figure 5-2   Tax graph furnished by local tax office (Example 5). (a) The tax graph; (b) the tax graph used to compute the assessed value from the tax.

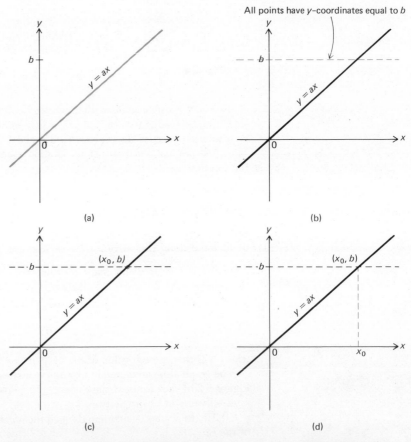

Figure 5-3   Graphical method of solving the equation $ax = b$. (a) Draw the line $y = ax$; (b) draw a horizontal line through the point b on the y-axis; (c) locate the point of intersection of the two lines; (d) draw a vertical line from the point of intersection down to the x-axis. Locate x_0, the solution of $ax = b$.

The procedure of Example 5 can always be used to solve the equation $ax = b$ graphically (Fig. 5-3):

1. Draw the line $y = ax$.

2. Draw a line parallel to the x-axis through the point $y = b$ on the y-axis. (All of the points on this line have y-coordinates equal to b.)

3. Locate the point of intersection of the two lines. If x_0 is the x-coordinate of this point, then the y-coordinate is $ax_0 = b$. (It follows that x_0 is the solution of the equation $ax = b$.)

4. To obtain the value of x_0, draw a line parallel to the y-axis through the point (x_0, b), the point of intersection. This line crosses the x-axis at the point x_0, the value of which can be read from the number scale.

The usefulness of the above method becomes apparent when we use previously prepared graphs to approximate the solution of $ax = b$. It is easier to use a graph to find the approximate number of inches equal to 71 centimeters, for example, than to solve the equation $2.54x = 71$ algebraically.

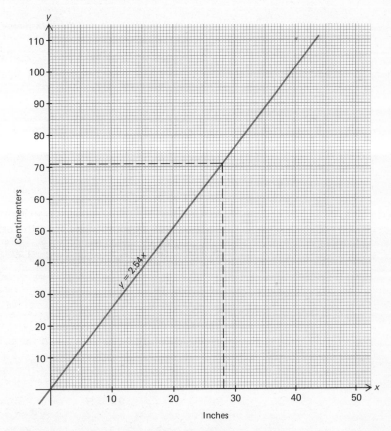

Figure 5-4 Graph for converting inches to centimeters; 71 cm $\approx$ 28 in. (Example 6).

Example 6. The equation for converting inches (x) to centimeters (y) is

$$y = 2.54x$$

(a) Use the equation $y = 2.54x$ to approximate the number of inches equal to 71 centimeters.

(b) Use the graph in Figure 5-4 to estimate the number of inches equal to 71 centimeters.

Solution. (a) We must solve the equation

$$2.54x = 71$$

If we divide both sides of the equation by 2.54, we obtain

$$x = \frac{71}{2.54} \approx 27.95 \text{ inches}$$

(b) If we use the graph in Figure 5-4, we draw a horizontal line through the point 71 on the y-axis, locate the point of intersection of the two lines, then draw a vertical line down to the x-axis. After carrying out this procedure, we obtain

$$x \approx 28 \text{ inches}$$

EXERCISES

1. Find the solutions of the following equations.
   (a) $2x = 8$ (e) $x - 15 = 0$
   (b) $3x = -6$ (f) $3x + 7 = 0$
   (c) $4x = -13$ (g) $14 = -8x$
   (d) $12x = -73$ (h) $287x = -591$
2. Work through the steps of Figure 5-1 with the following examples.
   (a) $2x = 9$ (c) $x - 8 + 2x = 14 + 3x - 22$
   (b) $5x - 13 = 2x + 7$ (d) $-7x + 5 + x = 8 - x + 3$
3. Reduce the following equations to the form $ax = b$ and solve. Check each answer by substituting it in the original equation.
   (a) $x^2 + 2x + 3 = x^2 - x + 12$ (d) $(1 - x)(5 + x) + x^2 - 2$
   $$= 3$$
   (b) $2x^2 + x = 3x^2 - 2 - x - x^2$ (e) $x(x + 1) = (x - 1)(x + 2)$
   (c) $(x - 1)(x + 1) = x^2 + x$ (f) $(x - 4)(x + 1)$
   $$= (x + 4)(x - 1)$$
4. Reduce the following equations to the form $ax = b$ and solve. Check each answer by substituting it in the original equation.

   (a) $\dfrac{x - 1}{x} = 2$ (d) $\dfrac{x + 2}{x + 3} = \dfrac{x + 1}{x + 4}$

   (b) $\dfrac{2x^2 + 10}{x} = 2x + 5$ (e) $\dfrac{x - 1}{x + 1} = \dfrac{-x}{1 - x}$

   (c) $\dfrac{x^2 + 5}{x + 2} = x - 3$ (f) $\dfrac{x(x - 1)}{x - 1} = 2x + 3$

5. Use the graph in Figure 5-4 to estimate the number of inches equal to
   (a) 27 centimeters   (c) 86 centimeters   (e) 200 centimeters
   (b) 49 centimeters   (d) 98 centimeters   (f) 387 centimeters
6. The county property tax is calculated as follows: The assessed value of property is 15 percent of the market value. The property tax per year is 3 percent of the assessed value.
   (a) Draw a graph for the tax y due on property with a market value of x dollars.
   (b) Use the graph in (a) to calculate the market values (as listed in the county tax office) for properties on which the following taxes are charged:
   (1) $270
   (2) $810.
7. Jim plans to buy a new car and pay for it over a three-year period. The interest and finance charges amount to 6 percent of the original loan per year. These charges are calculated for the full three years regardless of when the loan is repaid. The largest monthly payment that Jim can afford to pay is $150.
   (a) Write an equation for the largest purchase price that Jim can consider.
   (b) Solve the equation.
8. The business manager for a movie star charges a 15-percent commission on the money he earns for the star. How much of the star's money should he invest at 5.5 percent interest (a safe investment) in order to earn a net return for himself of approximately $1000?

5-5 CONDITIONAL EQUATIONS, INCONSISTENT EQUATIONS, AND IDENTITIES

The mere writing of an equation does not guarantee that it is meaningful. There are, in fact, three types of equations that we can write down. Only one of these types is meaningful for the solution of problems by algebra.
   (1) Equations such as

$$5 = 7$$

or

$$x - 1 = x$$

Inconsistent equations have no solution.

$x - 1 = x$

are, essentially, gibberish written with mathematical symbols. There is no way that either of these equations could have a solution because the left-hand and right-hand sides are incompatible with each other. These equations are called *inconsistent equations*.

(2) Equations such as

$$3 = 3$$

or

$$0 = 0$$

or

$$x - 2 = -(1 - x) - 1$$

$x - 2 = -(1 - x) - 1$

Identities are satisfied by all values of x.

are satisfied by all values of x. These equations are called *identities*.

(3) Equations such as

$$3x = 9$$

or

$$x^2 - 13 = 3$$

$x^2 - 13 = 3$

Conditional equations are satisfied by some (but not all) values of x.

If we obtain an inconsistent equation in our solution to a problem, it is either a sign that the problem has no solution or that we have made an error.

are satisfied by certain numbers, but not by all numbers. (The first equation has only the solution $x = 3$; the second has only the two solutions $x = -4$ and $x = 4$.) These equations are called *conditional equations*.

Thus, an inconsistent equation has no solution at all; an identity has every number x as a solution; a conditional equation is solved by certain values of x, but not all values.

Conditional equations are important for solving problems by algebra. When we start with a problem to be solved, there are many possible equations involving the unknowns that we can write down. Most of these equations are identities. These are the equations that do not fully utilize all of the information. The major task is to write down a meaningful conditional equation that actually describes the relationships in the problem we are trying to solve.

Example 1. Dimmy works as a clerk in a candy store. This morning he took a telephone order for a mixture of two kinds of candy, Bridge Mix at $1.25 per pound and Cashew Delight at $2.25 per pound. When he took the order, he wrote down the total amount (3 pounds) and the total bill ($5.00), but forgot to write down the amount of each type of candy. He now is trying to use algebra to figure out the quantity of each type.

Dimmy started his computation by letting x represent the number of pounds of Bridge Mix. Since the total amount is 3 pounds, he correctly reasoned that the amount of Cashew Delight is $3 - x$. Now the total amount is equal to the amount of Bridge Mix *plus* the amount of Cashew Delight; so

total amount = (amount of Bridge Mix) + (amount of Cashew Delight) which gives the equation

$$3 = x + (3 - x)$$

When Dimmy tries to solve this equation, he reduces it to the equation

$$3 = 3$$

which gives no information. What is wrong?

Solution. Dimmy did not use all of the information. He only used the information about the weight and not that about the price. He wrote down a correct relationship involving the two weights (amount of Cashew Delight $= 3 - x$) and then used the same information to write down the final equation. This led to the identity

$$3 = x + (3 - x)$$

which is satisfied by every value of x.

The proper solution to the problem involves the use of the prices as well as the weights:

$$\text{cost of Bridge Mix} = 1.25x$$
$$\text{cost of Cashew Delight} = 2.25 \cdot (3 - x)$$
$$\text{total cost} = \$5.00$$

Thus,

$$\text{total cost} = (\text{cost of Bridge Mix}) + (\text{cost of Cashew Delight})$$
$$5.00 = 1.25x + 2.25 \cdot (3 - x)$$
$$5.00 = 1.25x + 6.75 - 2.25x$$
$$5.00 - 6.75 = -1.00x$$
$$-1.75 = -1.00x$$
$$1.00x = 1.75$$
$$x = 1.75 = 1\tfrac{3}{4} \text{ pounds}$$

Therefore,

$$\text{amount of Bridge Mix} = x = 1\tfrac{3}{4} \text{ pounds}$$
$$\text{amount of Cashew Delight} = 3 - x = 3 - 1\tfrac{3}{4} = 1\tfrac{1}{4} \text{ pounds}$$

Although we normally deal with conditional equations, we must be able to recognize identities and inconsistent equations when we encounter them. There is a simple test that can be used to identify the types of linear equations. We apply the methods developed in this chapter to rewrite each equation, putting all nonzero terms on the left-hand side of the equation and zero on the right-hand side. We then combine all like terms. After we simplify the equation, we apply the following test.

Test for Classifying Linear Equations

(1) If a linear equation reduces to the form

$$0 = 0$$

it is an *identity*.

(2) If a linear equation reduces to the form

$$\text{nonzero number} = 0$$

it is *inconsistent*.

(3) If a linear equation reduces to the form

$$ax - b = 0$$

where $a \neq 0$, it is a *conditional equation*.

Example 2.   Decide if the following equations are conditional, inconsistent, or are identities.

(a) $5 = (x - 1) - x + 6$
(b) $3x - 4 + x = 3x - 7$
(c) $2x + 5 - x = x + 14$

Solution.   We must rewrite the equations with all of the terms on the left, then combine terms and simplify. Our work will be easier if we combine like terms on each side of the original equation before we proceed further.

(a)
$$5 = (x - 1) - x + 6$$
$$5 = 5 \qquad \text{(combining terms on the right)}$$
$$0 = 0 \qquad \text{(subtracting 5 from both sides)}$$

The equation is an identity.

(b)
$$3x - 4 + x = 3x - 7$$
$$4x - 4 = 3x - 7 \qquad \text{(combining terms on the left)}$$
$$4x - 3x - 4 + 7 = 0 \qquad \text{(subtracting } 3x, \text{ adding 7 to both sides)}$$
$$x + 3 = 0 \qquad \text{(combining terms)}$$

The equation is conditional. (The only solution is $x = -3$.)

(c)
$$2x + 5 - x = x + 14$$
$$x + 5 = x + 14 \qquad \text{(combining terms on the left)}$$
$$x - x + 5 - 14 = 0 \qquad \text{(subtracting } x + 14 \text{ from both sides)}$$
$$-9 = 0$$

The equation is inconsistent.

EXERCISES

1. Use the test explained in this section to decide if the following equations are conditional, inconsistent, or are identities.

   (a) $6 = 6$

   (b) $3x + x - 5 = 2x + 7 - 2$

   (c) $x - 3 = -(x + 3) + 2x$

   (d) $x^2 - 4 + x = -(4 - x^2) - x$

   (e) $x + 7 - 2x = -(x - 6) + 1$

   (f) $x^3 - 2x + 5 = -(x - x^3 + 1) - x + 6$

   (g) $2x - x^2 = -x^2 + x$

   (h) $x - 1 = x + 5$

2. Construct a flow chart for the steps in the test for equations described in this section. Work through the flow chart with the following examples.

   (a) $2x + 3 - x = x + 3$

   (b) $5x + 1 = 6x + 2 - x$

3. A bootlegger has run off 172 gallons of 190-proof whiskey (190-proof = 95 percent alcohol). He needs to dilute it with branch water until he obtains an 86-proof blend (86-proof = 43 percent alcohol). He has written down the following equations and relationships as part of an attempt to calculate the amount of water he needs to add:

   (1) $x =$ amount of water to add

   (2) $172 =$ original amount of liquid

   (3) $172 + x =$ total amount of mixture

   (4) $172 = (172 + x) - x$

   (5) $x = (172 + x) - 172$

   (6) original amount of alcohol $= (0.95) \cdot (172) = 163.4$

   (7) amount of alcohol in mixture $= (0.43) \cdot (172 + x)$

   (a)   Which, if any, of the above relationships give meaningful information for solving the problem?

   (b)   Write an equation that can be solved for x. Calculate the amount of water that must be added.

4. A chemist has a beaker containing 45 cubic centimeters of a 20-percent alcohol—80 percent water solution. He needs to pour out a certain amount and replace it with a 50-percent alcohol solution in order to obtain 45 cubic centimeters of 30 percent solution.

   (a)   Which of the following equations give meaningful information for the problem of calculating the quantity of solution to pour out and replace?

   (1) $x =$ amount of solution to pour out and replace

   (2) original amount $= 45$

   (3) final amount $= 45$

   (4) final amount $=$ (amount poured out) $+$ (amount replaced)

   (5) $45 = (45 - x) + x$

   (6) amount of alcohol in original solution $= 0.20(45)$

   (7) amount of water in original solution $= 0.80(45)$

   (8) amount of alcohol to be added $= 0.50x$

(9) amount of alcohol in final solution $= (0.30)(45)$

(b)   Calculate the amount of 50 percent alcohol solution that must be added.

5-6   THE LINE $y = mx + b$

In Section 4-6 we studied the graph of $y = mx$. We saw that the graph is a line through the origin with slope m. In this section we consider the closely related equation $y = mx + b$, where m and b are given constants.

Example 1.   Sketch the graphs of $y = 2x$ and $y = 2x + 3$.

Solution.   The graph of $y = 2x$ is the line through the origin with slope $m = 2$. For any number x_0 on the x-axis, the point $(x_0, 2x_0)$ is on the line. Thus, for example, the line passes through the points $(0, 0)$ and $(1, 2)$ [Fig. 5-5(a)].

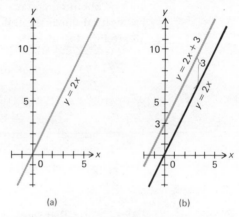

(a)   (b)

Figure 5-5   The graph of $y - 2x + 3$ is a line through $(0,3)$ parallel to the line $y = 2x$. (a) The graph of $y = 2x$; (b) the graph of $y = 2x + 3$.

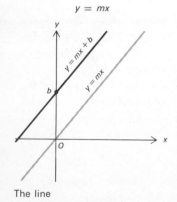

The line

$$y = mx + b$$

has *slope* equal to m. This line is parallel to the line

The coordinates of the points on the graph of $y = 2x + 3$ can be found in two steps. For any number $x = x_0$ on the x-axis, we first find $2x_0$, then $y_0 = 2x_0 + 3$. The first calculation gives us the point $(x_0, 2x_0)$ on the line $y = 2x$; the second calculation gives us the point $(x_0, 2x_0 + 3)$, which is three units above the first point that we found. This last point is on the graph of $y = 2x + 3$. Since we can do this for every number x_0 on the x-axis, we see that the graph of $y = 2x + 3$ is obtained by shifting the graph of $y = 2x$ three units in the vertical direction. It follows that the graph of $y = 2x + 3$ is a line through the point $(0, 3)$ that is parallel to the graph of $y = 2x$ [Fig. 5-5(b)].

In general, the graph of $y = mx + b$ is the line through the point $(0, b)$ that is parallel to the line $y = mx$.

The number m is called the *slope* of the line $y = mx + b$. The slope

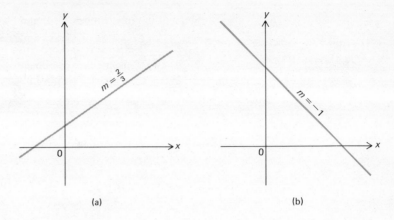

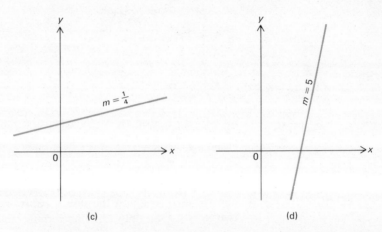

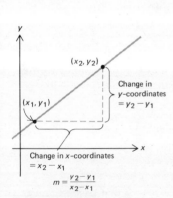

$$m = \frac{y_2 - y_1}{x_2 - x_1}$$

Figure 5-6 Properties of lines as determined by their slopes. (a) Lines with positive slope are oriented upward to the right; (b) lines with negative slope are oriented downward to the right; (c) lines with slopes close to zero are nearly horizontal; (d) lines with large slopes are nearly vertical.

The slope of a line can be calculated from any two points on the line. It is equal to the ratio of the change in the y-coordinates to the change in the x-coordinates:

$$m = \frac{\text{change in } y\text{-coordinates}}{\text{change in } x\text{-coordinates}}$$

$$= \frac{y_2 - y_1}{x_2 - x_1}$$

measures the steepness of the line with respect to the x-axis. More precisely, *the slope is the ratio of the change in the y-coordinates between two points to the corresponding change in the x-coordinates.* If the two points are $P_1(x_1, y_1)$ and $P_2(x_2, y_2)$, then

$$m = \frac{\text{change in } y\text{-coordinates}}{\text{change in } x\text{-coordinates}} = \frac{y_2 - y_1}{x_2 - x_1}$$

Example 2. (a) The slope of the line through $(2, 2)$ and $(7, 5)$ is

$$m = \frac{\text{change in } y\text{-coordinates}}{\text{change in } x\text{-coordinates}} = \frac{5 - 2}{7 - 2} = \frac{3}{5}$$

(b) The slope of the line through $(5, 4)$ and $(-1, -3)$ is

$$m = \frac{\text{change in } y\text{-coordinates}}{\text{change in } x\text{-coordinates}} = \frac{(-3) - 4}{(-1) - 5} = \frac{7}{6}$$

(See Fig. 5-7.)

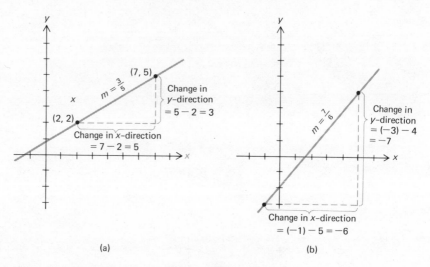

(a)   (b)

Figure 5-7   (a) $m = \frac{3}{5}$; (b) $m = \frac{7}{6}$ (Example 2).

The procedure for graphing the equation $y = mx + b$ is similar to the one we used to graph the line $y = mx$. We calculate the coordinates of two points, then draw the line through them.

Example 3.   Calculate the slopes and sketch the graphs of
(a) $y = 2x - 1$
(b) $3y + 5 = -2x - 7$

Solution.   (a) $y = 2x - 1$. This equation is in the form $y = mx + b$ with $m = 2$ and $b = -1$. Consequently, it is the line with slope $m = 2$ that passes through the point $(0, -1)$. To find another point on the line, we choose a convenient value of x, say $x = 1$, and calculate y.

$$x = 1: \qquad y = 2x - 1 = 2 \cdot 1 - 1 = 1; \qquad \text{point } (1, 1)$$

The graph is the line through $(0, -1)$ and $(1, 1)$, shown in Figure 5-8(a).

(b) $3y + 5 = -2x - 7$. We first change the form of the equation to $y = mx + b$:

$$\begin{aligned}
3y + 5 &= -2x - 7 \\
3y &= -2x - 7 - 5 \qquad \text{(subtracting 5 from both sides)} \\
3y &= -2x - 12
\end{aligned}$$

$$y = \frac{-2x - 12}{3} \qquad \text{(dividing both sides by 3)}$$

$$y = -\frac{2}{3}x - 4$$

The slope is $m = -\frac{2}{3}$.

We now calculate two points on the graph corresponding to convenient values of x:

$$x = 0: \qquad y = -\tfrac{2}{3} \cdot 0 - 4 = -4; \qquad \text{point } (0, -4)$$
$$x = 3: \qquad y = -\tfrac{2}{3} \cdot 3 - 4 = -2 - 4; \qquad \text{point } (3, -6)$$

The graph is shown in Figure 5-8(b).

Any pair of numbers x_0, y_0 that makes the equation $y = mx + b$ an identity is called a *solution* of that equation. The solution $x = x_0, y = y_0$ corresponds to the point (x_0, y_0) on the graph.

Example 4. Show that $x = 2, y = 5$ is a solution of the equation

$$y = 2x + 1$$

and that $x = 3, y = -1$ is not a solution.

Solution. If we substitute $x = 2, y = 5$ into the equation, the left-hand side is 5 and the right-hand side is $2 \cdot 2 + 1 = 5$. Since the two sides are equal, then $x = 2, y = 5$ is a solution.

If we substitute $x = 3, y = -1$ into the equation, the left-hand side is $y = -1$

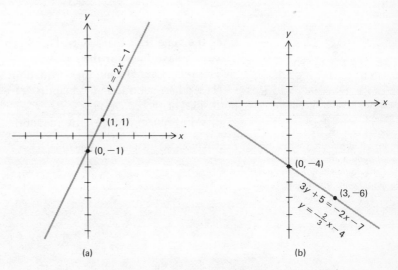

(a)   (b)

Figure 5-8 (a) Graph of $y = 2x - 1$; (b) graph of $3y + 5 = -2x - 7$, $y = -\frac{2}{3}x - 4$ (Example 3).

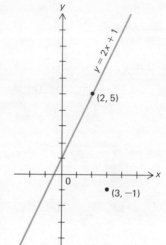

Figure 5-9 $y = 2x + 1$. The point (2,5) is on the graph. The point (3, −1) is not on the graph. $x = 2$, $y = 5$ is a solution of the equation; $x = 3$, $y = -1$ is not a solution (Example 4).

and the right-hand side is $2x + 1 = 2 \cdot 3 + 1 = 7$. Since the two sides are not equal, then $x = 3$, $y = -1$ is not a solution. See Figure 5-9.

Equations such as

$$3x + 2y - 5 = 2x + 8y + 9$$

can be reduced to the form

$$y = mx + b$$

Equations such as

$$5x + 3y = 37$$

are said to be *indeterminate* because they have an infinite number of solutions. One of the basic problems considered by the Hindus in the sixth century was that of finding integer solutions or rational solutions to indeterminate equations. For the above equations, they might have found the solutions $x = 2$, $y = 9$; $x = 5$, $y = 4$; or $x = -1$, $y = 14$.

using the same rules that we used for equations of one variable. Thus, such equations have lines for their graphs.

The rules for reducing these equations are:

(1) Add (or subtract) any quantity (or expression involving x and y) to both sides of the equation.

(2) Multiply (or divide) both sides of the equation by a nonzero number.

Example 5. (a) Reduce the equation

$$3x + 2y - 5 = 2x + 8y + 9$$

to the form

$$y = mx + b$$

(b) Sketch the graph of the equation.

(c) Find four solutions of the equation.

Solution. (a) We must get all of the terms involving y on the left-hand side of the equation and all other terms on the right-hand side, then divide both sides by the coefficient of y.

$$3x + 2y - 5 = 2x + 8y + 9$$

$3x - 6y - 5 = 2x + 9$ (subtracting $8y$ from both sides)

$-6y = -x + 14$ (subtracting $3x$, adding 5 to both sides)

$y = \dfrac{-x + 14}{-6} = \dfrac{x}{6} - \dfrac{14}{6} = \dfrac{x}{6} - \dfrac{7}{3}$ (dividing both sides by -6)

$$y = \frac{1}{6}x - \frac{7}{3}$$

(b) The graph is the line through $(0, -\frac{7}{3})$ with slope $m = \frac{1}{6}$. We must calculate two points on the graph. If we experiment with several values of x, we find that if $x = 2$ or $x = -4$ the corresponding points are easy to plot:

$x = 2$: $y = \dfrac{-x + 14}{-6} = \dfrac{-2 + 14}{-6} = \dfrac{12}{-6} = -2$;   point $(2, -2)$

$x = -4$: $y = \dfrac{-x + 14}{-6} = \dfrac{-(-4) + 14}{-6} = \dfrac{18}{-6} = -3$;   point $(-4, -3)$

The graph is the line through $(2, -2)$ and $(-4, -3)$. (See Fig. 5-10.) As a check observe that the line passes through $(0, -\frac{7}{3})$.

(c) Any point (x_0, y_0) on the graph determines a solution $x = x_0, y = y_0$. Because $(2, -2)$, $(-4, -3)$, and $(0, -\frac{7}{3})$ are known to be on the graph, then

$$x = 2, y = -2 \text{ is a solution;}$$
$$x = -4, y = -3 \text{ is a solution;}$$
$$x = 0, y = -\tfrac{7}{3} \text{ is a solution.}$$

Since the problem calls for four solutions, we only need to find one additional solution. We can choose any value of x and calculate y from the equation $y = x/6 - 7/3$. For example, if

$$x = 5, \quad \text{then} \quad y = \tfrac{5}{6} - \tfrac{7}{3} = \tfrac{5}{6} - \tfrac{14}{6} = -\tfrac{9}{6} = -\tfrac{3}{2}$$

so that $x = 5, y = -\frac{3}{2}$ is a fourth solution.

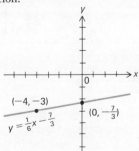

Figure 5-10 Example 5.

We close this section with two examples that involve equations in two variables.

Example 6.   The Bleaksburg Electric Company charges each customer a flat fee of $2.00 per month plus 3 cents per kilowatt-hour of electricity used.

(a) Write an equation for the total bill y when x kilowatt-hours of electricity are used.

(b) Sketch the graph of the equation. Use the graph to estimate the bill when 253 kilowatt-hours are used.

Solution.   (a) The total bill is

$$y = 2.00 + 0.03 \cdot \text{(number of kilowatt-hours)}$$
$$y = 2.00 + 0.03x$$

(b) The graph is shown in Figure 5-11. We see from the graph that when $x = 253$, $y \approx 9.60$. Thus, when 253 kilowatt-hours are used, the total bill is approximately $9.60.

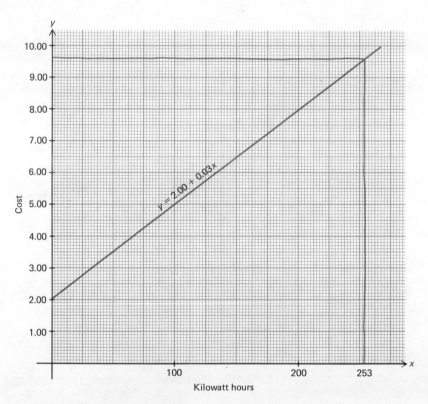

Figure 5-11 Rate graph for the Bleaksburg Electric Company. Note that different scales are used on the two axes (Example 6).

Example 7. The owner of Ye Olde Gifte Shoppe has found that he can make a profit and maintain reasonable sales if he has a 100-percent markup on all items. (If he buys an item for $2.75, he sells it for $5.50.)

The overhead costs and employee expenses amount to $1900 per month.

(a) Let y be the net profit and x the total sales for a month as read from the cash register tapes. Write an equation relating x and y. Sketch the graph of the equation.

(b) Interpret the meaning of a negative value of y. What minimum value of x is required in order not to lose money?

(c) Use the graph to calculate the approximate total sales necessary in order to make a net profit of $1000 for the month.

Solution. (a) If the sales are x dollars for the month, then the items that were sold cost $x/2$ dollars. (Recall that one-half of the selling price is markup, one half is cost.) The overhead and employee costs are $1900. Thus,

$$\text{net profit} = \text{amount received} - \text{expenses}$$

$$y = x - \left[\frac{x}{2} + 1900\right] = x - \frac{x}{2} - 1900$$

$$y = \frac{x}{2} - 1900$$

The graph is shown in Figure 5-12.

(b) Negative values of y indicate a "negative profit" or *loss*. The break-even point occurs when $y = 0$. We see from the graph that this happens when $x = 3800$. Thus, the owner loses money if the total sales are less than $3800 for the month.

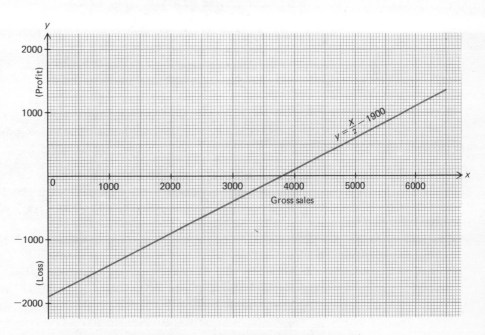

Figure 5-12 Profit (and loss) graph for Ye Olde Gifte Shoppe (Example 7).

(c) It appears from the graph that the total sales must be a little less than $6000 in order to have a net profit of $1000. We can calculate this value exactly from the equation

$$y = \frac{x}{2} - 1900$$

If $y = 1000$, then

$$1000 = \frac{x}{2} - 1900$$

$$2900 = \frac{x}{2}$$

$$\frac{x}{2} = 2900$$

$$x = 2 \cdot (2900) = 5800$$

Thus, the net profit is $1000 if sales are $5800 for the month.

EXERCISES

1. Sketch the graphs of the following pairs of equations on the same sheet of graph paper.
   (a) $y = 2x$, $y = 2x - 1$ (d) $y = -\frac{3}{4}x$, $y = -\frac{3}{4}x + 1$
   (b) $y = 3x$, $y = 3x + 2$ (e) $2y = x$, $2y = x + 1$
   (c) $y = \frac{2}{3}x$, $y = \frac{2}{3}x + \frac{1}{3}$ (f) $4y = -3x$, $4y = -3x + 6$

2. Decide if the given pair of numbers is a solution of the equation.
   (a) $y = \frac{2}{3}x + \frac{1}{3}$; $x = 3, y = 2$ (d) $8x + 3y = 5$; $x = -2, y = 7$
   (b) $y = \frac{1}{7}x - \frac{3}{7}$; $x = 3, y = 0$ (e) $2x - 5y = 3$; $x = 4, y = 1$
   (c) $y = 2x + \frac{3}{8}$; $x = -1, y = -\frac{13}{8}$ (f) $3x + 7y + 9 = 0$; $x = 2, y = -2$

3. Find the slope m of the line through the pair of points:
   (a) $(2, 7), (5, 4)$
   (b) $(-1, 3), (2, -4)$
   (c) $(6, 8), (-4, -5)$
   (d) $(25, 43), (-23, 74)$

4. Find the slope m; locate three points on the graph; draw the graph for each of the following equations.
   (a) $y = \frac{1}{7}x + \frac{2}{7}$ (e) $3y + 5 = 0$
   (b) $y + 2x = -\frac{1}{2}$ (f) $2y = 7$
   (c) $5x - y + 3 = 0$ (g) $12x + 13y = 8$
   (d) $3x - 4y + 1 = 0$ (h) $8x + 17y = -9$

5. The monthly local utility bill is based on the following formula: The water bill is $1.00 plus 10 cents per 100 gallons of water used; the sewer

service bill is 80 percent of the water bill; the garbage collection bill is $3.00.

(a)   Construct a flow chart for the total utility bill y when x hundred gallons of water are used.

(b)   Write an equation for the total utility bill y when x hundred gallons of water are used. Simplify the equation as much as possible.

(c)   Draw a graph for the total utility bill y when x hundred gallons of water are used ($0 \le x \le 500$).

(d)   Test your flow chart, equation, and graph by using them to calculate the utility bill when 5000 gallons of water are used.

(e)   Use the graph or the equation to calculate the amount of water that was used when the total bill is $15.06.

6. Due to increased competition, the gift shop owner has had to cut the markup to 75 percent. Furthermore, he has fired two clerks, lowering the operating costs to $1200 per month (Example 7).

(a)   Let x be the monthly total sales and y the net profit (or loss). Write an equation for y in terms of x. Sketch the graph of the equation.

(b)   What minimum value of x is required in order not to lose money?

(c)   Use either the graph or the equation to calculate the total sales necessary in order to make a net profit of $900 per month.

5-7   SYSTEMS OF TWO EQUATIONS IN TWO UNKNOWNS

Many problems involve two related quantities. Most of these can be solved provided we can obtain two different equations relating the quantities. The following example illustrates the situation.

Example 1.   The state highway department has just announced the statistics for road construction for the past two years.

Two Years Ago: 260 miles of primary roads (interstate system or equivalent) and 100 miles of secondary roads were constructed at a total cost of $360 million.

One Year Ago: 140 miles of primary roads and 550 miles of secondary roads were constructed at a total cost of $290 million.

Write two equations that describe the costs of road construction.

Solution.   If we let x be the average cost of constructing one mile of primary highway and y be the average cost of constructing one mile of secondary highway, we can write two equations in x and y that describe the total cost:

Two Years Ago: $\underbrace{260x}_{\substack{\text{total cost of} \\ \text{primary roads}}} + \underbrace{100y}_{\substack{\text{total cost of} \\ \text{secondary roads}}} = \underbrace{360 \text{ (million)}}_{\text{total cost}}$

One Year Ago: $\underbrace{140x}_{\substack{\text{total cost of} \\ \text{primary roads}}} + \underbrace{550y}_{\substack{\text{total cost of} \\ \text{secondary roads}}} = \underbrace{290 \text{ (million)}}_{\text{total cost}}$

Thus, we have two equations relating x and y:

$$\begin{cases} 260x + 100y = 360 \\ 140x + 550y = 290 \end{cases}$$

The values of x and y will be calculated in Example 4.

The basic problem that we consider in this section is that of finding a pair of numbers x_0, y_0 that simultaneously satisfy two linear equations in x and y, such as

$$\begin{cases} 260x + 100y = 360 \\ 140x + 550y = 290 \end{cases} \quad \text{or} \quad \begin{cases} 5x + 3y = 8 \\ 3x - y = 9 \end{cases}$$

We use the braces to indicate that we want a common solution of the two equations. We speak of the two equations, braced together, as a *system of equations*.

We have seen that any solution $x = x_0$, $y = y_0$ of a linear equation corresponds to a point (x_0, y_0) on the graph. Thus, a common solution of a system of equations corresponds to the common point on their graphs. In other words, $x = x_0$, $y = y_0$ is a solution to a system of equations if, and only if, (x_0, y_0) is a point of intersection of the two lines. (See Fig. 5-13.)

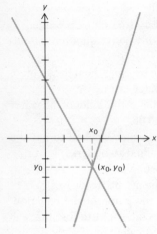

The pair of numbers $x = x_0$, $y = y_0$ is a solution of a system of equations if and only if the point (x_0, y_0) is on both of the graphs.

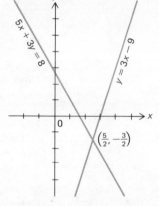

Figure 5-13 The solution of the system

$$5x + 3y = 8$$
$$3x - y = 9$$

is $x = \frac{5}{2}$, $y = -\frac{3}{2}$. The point $(\frac{5}{2}, -\frac{3}{2})$ is on both of the graphs (Example 2).

Example 2. Use graphs to solve the system of equations

$$\begin{cases} 5x + 3y = 8 \\ 3x - y = 9 \end{cases}$$

Solution. We rewrite both equations in the form $y = mx + b$. The first equation becomes

$$y = -\tfrac{5}{3}x + \tfrac{8}{3}$$

which has the line through $(0, \frac{8}{3})$ with slope $m = -\frac{5}{3}$ as its graph. The second equation becomes

$$y = 3x - 9$$

which has the line through $(0, -9)$ with slope $m = 3$ as its graph.

The graphs of the two equations are shown in Figure 5-13. We see that the lines cross at the point $(\frac{5}{2}, -\frac{3}{2})$. Thus, $x = \frac{5}{2}$, $y = -\frac{3}{2}$ is the common solution of the two equations.

There is a simple algebraic method for solving a system of two equations. We solve one of the equations for y in terms of x and substitute this expression for y into the other equation. This gives us a linear equation in x, which we solve. Once we get the value of x, we substitute it into either one of the original equations and calculate y. The following example illustrates the method.

Example 3.   Solve the system

$$\begin{cases} 6x + y = 4 \\ 2x - 3y = 3 \end{cases}$$

Solution.   We solve one of the equations for y and substitute into the other equation. (Since the first equation has 1 for the coefficient of y, we have fewer computational difficulties if we solve that equation for y.)

$$6x + y = 4$$
$$y = -6x + 4$$

We substitute this expression for y into the second equation and solve for x.

$$2x - 3y = 3$$
$$2x - 3(-6x + 4) = 3$$
$$2x + 18x - 12 = 3$$
$$20x = 3 + 12 = 15$$
$$x = \tfrac{15}{20} = \tfrac{3}{4}$$

We now substitute this value of x into the first equation and solve for y:

$$y = -6(\tfrac{3}{4}) + 4 = \tfrac{-18}{4} + \tfrac{16}{4} = -\tfrac{2}{4} = -\tfrac{1}{2}$$

The solution is $x = \frac{3}{4}$, $y = -\frac{1}{2}$. (See Fig. 5-14.)

Example 4 (Continuation of Example 1). Calculate the average costs of primary and secondary roads as described in Example 1.

Solution.   In Example 1 we obtained the system of equations

$$\begin{cases} 260x + 100y = 360 \\ 140x + 550y = 290 \end{cases}$$

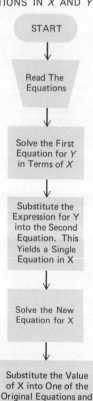

TECHNIQUE FOR SOLVING
A SYSTEM OF TWO LINEAR
EQUATIONS IN X AND Y

START

Read The
Equations

Solve the First
Equation for Y
in Terms of X

Substitute the
Expression for Y
into the Second
Equation. This
Yields a Single
Equation in X

Solve the New
Equation for X

Substitute the Value
of X into One of the
Original Equations and
Solve the Resulting
Equation for Y

The Values of
X and Y
Constitute the
Solution

STOP

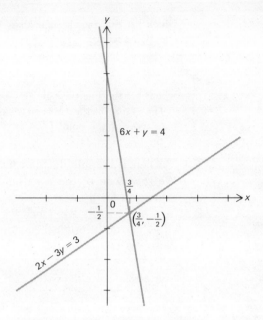

Figure 5-14 The solution of the system

$$\begin{cases} 6x + \ y = 4 \\ 2x - 3y = 3 \end{cases}$$

is $x = \frac{3}{4}$, $y = -\frac{1}{2}$. The point $(\frac{3}{4}, -\frac{1}{2})$ is on both of the graphs (Example 3).

where x and y represent the average costs (in millions of dollars) of constructing one mile of primary and one mile of secondary road, respectively.

Before proceeding further, we simplify the equations, dividing the first one by 20 and the second one by 10. The system reduces to

$$\begin{cases} 13x + \ 5y = 18 \\ 14x + 55y = 29 \end{cases}$$

We now solve the first equation for y,

$$y = \frac{-13x + 18}{5}$$

and substitute into the second equation:

$$14x + 55y = 29$$

$$14x + 55 \left(\frac{-13x + 18}{5} \right) = 29$$

$$14x + \overset{11}{\cancel{55}} \left(\frac{-13x + 18}{\cancel{5}} \right) = 29$$

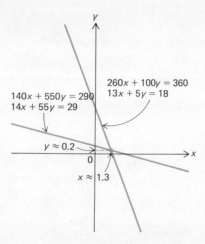

Figure 5-15 The two lines cross at the point with $x \approx 1.3$, $y \approx 0.2$. The cost of building primary road is approximately $\$1,300,000$ per mile, the cost of building secondary road is approximately $\$200,000$ per mile (Examples 1 and 4).

$$14x + 11(-13x + 18) = 29$$
$$14x - 11 \cdot 13x + 11 \cdot 18 = 29$$
$$14x - 143x + 198 = 29$$
$$-129x = 29 - 198 = -169$$
$$x = \frac{-169}{-129} \approx 1.3 \text{ (million)}$$

To find y we substitute $x \approx 1.3$ into the equation

$$y = \frac{18 - 13x}{5}$$

$$y \approx \frac{18 - 13(1.3)}{5}$$

$$\approx \frac{18 - 16.9}{5} \approx \frac{1.1}{5} \approx 0.2 \text{ (million)}$$

All methods of solving systems of equations algebraically involve the reduction of the several equations to a single equation which can be solved. The solution of the single equation usually is a part of the solution of the original system.

Thus, the average cost of one mile of primary road is approximately $\$1.3$ million $= \$1,300,000$, and the average cost of one mile of secondary road is approximately $\$0.2$ million $= \$200,000$. See Figure 5.15.

A number of problems can be reduced to solving systems of two equations in two unknowns. The following problem is typical.

Example 5. (a) Find the equation of the line through the points $(2, -3)$ and $(5, 7)$.
(b) Where does the line cross the y-axis? The x-axis?

Solution. (a) The line has an equation of form

$$y = mx + b,$$

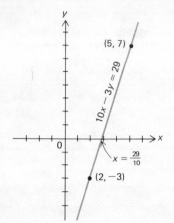

Figure 5-16 The line through (2, −3) and (5, 7) has the equation 10*x* − 3*y* = 29 (Example 5).

where m and b are to be determined. (See Fig. 5-16.) Since $(2, -3)$ is a point on the line, then $x = 2$, $y = -3$ is a solution of the equation. If we substitute these values into the equation $y = mx + b$, we obtain

$$-3 = m \cdot 2 + b$$

$$\boxed{2m + b = -3}$$

Similarly, since $(5, 7)$ is a point on the line, then $x = 5$, $y = 7$ is a solution. Therefore,

$$y = mx + b$$
$$7 = m \cdot 5 + b$$

$$\boxed{5m + b = 7}$$

Thus, we see that the numbers m and b satisfy the system of equations

$$\begin{cases} 2m + b = -3 \\ 5m + b = 7 \end{cases}$$

where m and b are the unknowns rather than x and y.

To solve the system, we solve one equation for b in terms of m and substitute into the other equation. If we solve the second equation for b, we have

$$b = 7 - 5m \qquad \text{(the second equation)}$$

We substitute this expression into the first equation, obtaining

$$
\begin{aligned}
2m + b &= -3 & \text{(the first equation)} \\
2m + (7 - 5m) &= -3 & \text{(substituting the value of } b) \\
2m + 7 - 5m &= -3 \\
-3m &= -3 - 7 = -10 \\
m &= \frac{(-10)}{(-3)} = \frac{10}{3}
\end{aligned}
$$

To find b we substitute this value of m into the second equation:

$$b = 7 - 5m$$
$$= 7 - 5 \cdot \tfrac{10}{3} = \tfrac{21}{3} - \tfrac{50}{3} = -\tfrac{29}{3}$$

The equation of the line is

$$y = \tfrac{10}{3}x - \tfrac{29}{3}$$

which also can be written as

$$3y = 10x - 29$$

or

$$10x - 3y = 29$$

(b) The line crosses the y-axis at the point corresponding to the solution which has $x = 0$. To find this point, we set $x = 0$ in the equation:

$$10x - 3y = 29$$
$$0 - 3y = 29$$
$$y = -\tfrac{29}{3}$$

The line crosses the y-axis at the point $(0, -\tfrac{29}{3})$.
To find where the line crosses the x-axis, we set $y = 0$ in the equation:

$$10x - 3y = 29$$
$$10x - 0 = 29$$
$$x = \tfrac{29}{10}$$

The line crosses the x-axis at the point $(\tfrac{29}{10}, 0)$.

EXERCISES

1. By substituting into the equations decide whether or not x, y is a solution of the system.

   (a) $\begin{cases} 2x - 3y = -8 \\ 8x + 2y = -4 \end{cases} \qquad x = -1, y = 2$

   (b) $\begin{cases} x - y = 1 \\ 14x - 7y = 29 \end{cases} \qquad x = 3, y = 2$

   (c) $\begin{cases} 4x + 3y = 17 \\ 5x - y = 5 \end{cases} \qquad x = 2, y = 3$

2. Draw accurate graphs of the equations in each system. Use the point of intersection of the two graphs to approximate the solution x_0, y_0 correct to one decimal place.

   (a) $\begin{cases} 4x - 2y = -13 \\ x + y = -2 \end{cases}$ (b) $\begin{cases} 7x + y = 13 \\ 2x - y = 5 \end{cases}$

(c) $\begin{cases} 12x + y = 2 \\ 5x - 3y = -5 \end{cases}$ (e) $\begin{cases} 2x + y = 11 \\ x - y = 1 \end{cases}$

(d) $\begin{cases} x + 2y = 2 \\ x - y = -6 \end{cases}$ (f) $\begin{cases} 5x - 3y = 1 \\ 4x + y = -1 \end{cases}$

3. Solve the following systems of equations.

(a) $\begin{cases} 3x + 17y = 0 \\ 8x - y = 0 \end{cases}$ (d) $\begin{cases} 3x + y = -2 \\ x + 5y = -11 \end{cases}$

(b) $\begin{cases} 7x - 8y = 4 \\ x + y = 7 \end{cases}$ (e) $\begin{cases} 7x - y = 20 \\ x + 5y = 3 \end{cases}$

(c) $\begin{cases} 2x - y = -11 \\ x + 2y = 2 \end{cases}$ (f) $\begin{cases} 11x + 2y = 9 \\ 5x - 8y = 12 \end{cases}$

4. Calculate the equation of the line through each pair of points. What is the slope of the line? Where does it cross the x-axis? The y-axis?

(a) $(2, -1), (5, 3)$ (d) $(4, 3), (-2, 7)$

(b) $(-1, 5), (0, 0)$ (e) $(-2, -4), (1, 9)$

(c) $(-8, -1), (2, -3)$ (f) $(1, 1), (7, 3)$

5. Let F and C denote the Fahrenheit and Centigrade measures of temperature, respectively. The equation relating the two measures is of form

$$C = mF + b$$

where m and b are constants.

(a) Use the fact that $C = 0$ when $F = 32$ (freezing point of water) and $C = 100$ when $F = 212$ (boiling point of water) to find the values of m and b.

(b) Draw a graph that can be used to convert from Fahrenheit to Centigrade.

(c) For what temperature is $F = C$?

5-8   SYSTEMS OF EQUATIONS

There is an alternate method for solving systems of equations that may be more convenient than the one we used in Section 5-7. The following example illustrates the method.

Example 1. Solve the system

$$\begin{cases} 3x + 5y = -2 \\ 7x + 2y = 5 \end{cases}$$

Solution. We need to reduce the system to a single equation in one of the unknowns, say x, solve that equation for x, then substitute into one of the original equations in order to solve for y.

To carry out the reduction to a single equation, we multiply the first equation by 2 (the coefficient of y in the second equation), obtaining

$$6x + 10y = -4$$

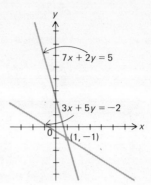

Figure 5-17 Example 1.

then multiply the second equation by -5 (the *negative* of the coefficient of y in the first equation), obtaining

$$-35x - 10y = -25$$

then add these two equations together:

$$
\begin{array}{r}
6x + 10y = -4 \\
add \quad -35x - 10y = -25 \\
\hline
-29x + 0 \cdot y = -29 \\
-29x = -29 \\
x = 1
\end{array}
$$

To calculate y we substitute the value $x = 1$ into one of the original equations and solve for y. If we use the first equation, we obtain

$$
\begin{aligned}
x = 1: \quad 3x + 5y &= -2 \\
3 \cdot 1 + 5y &= -2 \\
5y &= -2 - 3 = -5 \\
y &= -1
\end{aligned}
$$

The solution is $x = 1$, $y = -1$. See Figure 5-17.

The procedure illustrated in Example 1 can always be used to solve a system of equations if a unique solution exists. The procedure is as follows:

(1) Multiply the first equation by the coefficient of y from the second equation.

(2) Multiply the second equation by the negative of the coefficient of y from the first equation.

(3) Add the two equations obtained in steps (1) and (2). The resulting equation has no terms involving y.

(4) Solve the equation obtained in (3) for x.

(5) Substitute the value of x obtained in (4) into one of the original equations and solve for y.

Example 2. Solve the system

$$\begin{cases} 4x + 5y = 7 \\ 8x + 3y = -2 \end{cases}$$

Solution. Multiply the first equation by 3 and the second by -5, obtaining the system

$$\begin{cases} 12x + 15y = 21 \\ -40x - 15y = 10 \end{cases}$$

Add the two equations and simplify:

$$\begin{array}{r} 12x + \ \ 15y = 21 \\ -40x - \ \ 15y = 10 \\ \hline \end{array}$$
$$\begin{array}{rl} adding & -28x + 0 \cdot y = 31 \\ & -28x = 31 \\ & x = -\frac{31}{28} \end{array}$$

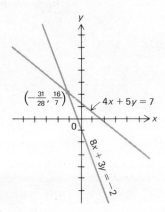

$$\left(-\frac{31}{28}, \frac{16}{7}\right)$$

$$4x + 5y = 7$$

$$8x + 3y = -2$$

Figure 5-18 Example 2.

To solve for y we substitute into the first equation. (We could just as well have substituted into the second equation.)

$$4x + 5y = 7$$
$$4(-\tfrac{31}{28}) + 5y = 7$$
$$-\tfrac{31}{7} + 5y = 7$$
$$5y = \tfrac{31}{7} + 7$$
$$5y = \tfrac{31}{7} + \tfrac{49}{7} = \tfrac{80}{7}$$
$$y = \frac{\tfrac{80}{7}}{5} = \tfrac{80}{7} \cdot \tfrac{1}{5} = \tfrac{16}{7}$$

The solution is $x = -\frac{31}{28}$, $y = \frac{16}{7}$. See Figure 5-18.

Inconsistent and Dependent Systems

The two equations in a system can be represented by lines in the xy-plane. Either they are represented by the same line or by different lines. In the latter case, either the lines are parallel or they cross at a single point. Thus, one of the following three situations must occur (Fig. 5-19):

> A system of two equations in x and y has a *unique solution* if the graphs cross at a single point.

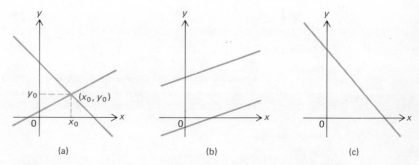

Figure 5-19 Graphs of systems of equations. (a) The graphs cross at the point (x_0, y_0). The system has the unique solution $x = x_0$, $y = y_0$; (b) the graphs are parallel lines. There is no common solution of the two equations. The system of equations is inconsistent; (c) the two equations have the same graph. Every point on the line determines a solution of the system of equations. The system is dependent.

(1) *The two lines may cross at a single point* (x_0, y_0). In this case, $x = x_0$, $y = y_0$ is the unique solution of the system.

> A system of two equations in x and y is *inconsistent* if it has no solution at all. The graphs are parallel lines.

(2) *The two lines may be parallel.* In this case the lines never cross, so the system has no solution. We say that the system of equations is *inconsistent*.

(3) *The two equations may have the same line for a graph.* In this case every point on the line determines a solution of the system, so the system has an infinite number of solutions. We say that the system is *dependent*.

> A system of two equations in x and y is *dependent* if it has an infinite number of solutions. The two graphs define the same straight line.

Example 3. Use graphs to decide if the following systems have unique solutions, are inconsistent, or are dependent.

(a) $\begin{cases} 3x - 2y = 5 \\ x + y = 7 \end{cases}$ (b) $\begin{cases} x - 2y = 3 \\ -2x + 4y = -6 \end{cases}$ (c) $\begin{cases} 2x - 4y = 8 \\ -x + 2y = 5 \end{cases}$

Solution. (a) The graphs of the two equations cross at a single point. The system has a unique solution $x \approx 3.8$, $y \approx 3.2$. [See Fig. 5-20(a).]

(b) The graph of each of the equations is the line through $(0, -\frac{3}{2})$ with slope $m = \frac{1}{2}$. Every point on the line corresponds to a solution of the system. The system has an infinite number of solutions. It is dependent. [See Fig. 5-20(b).]

(c) The graph of the first equation is the line through $(0, -2)$ with slope $m = \frac{1}{2}$. The graph of the second equation is the line through $(0, \frac{5}{2})$ with slope $m = \frac{1}{2}$. The two lines are parallel, so there is no common solution. The system is inconsistent. [See Fig. 5-20(c).]

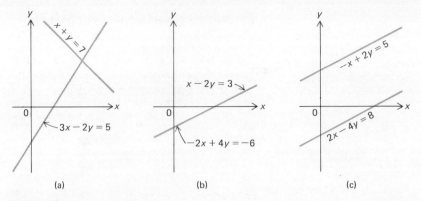

Figure 5-20 (a) $3x - 2y = 5$, $x + y = 7$. The system has a unique solution; (b) $x - 2y = 3$, $-2x + 4y = -6$. Every point on the line determines a solution of the system. The system is dependent; (c) $2x - 4y = 8$, $-x + 2y = 5$. The system is inconsistent (Example 3).

When we solve a system of equations algebraically, we can apply a simple test to decide if it has a unique solution, is inconsistent, or is dependent. We simply try to solve it by one of the methods that we have developed. Recall that the first major step in both of our methods is to reduce the system to a single equation in x.

(1) If the final equation in x has a unique solution, then the original system has a unique solution.

(2) If the final equation in x reduces to the identity

$$0 \cdot x = 0$$

then the original system is dependent.

(3) If the final equation in x reduces to an inconsistent equation of form

$$0 \cdot x = \text{nonzero number}$$

then the original system is inconsistent.

Example 4. Decide if each of the systems of equations in Example 3 has a unique solution, is inconsistent, or is dependent. Use algebraic methods.

Solution. (a) $\begin{cases} 3x - 2y = 5 \\ x + y = 7 \end{cases}$

We multiply the second equation by 2 and add:

$$\begin{array}{r} 3x - 2y = 5 \\ add \quad \underline{2x + 2y = 14} \\ 5x + 0 \cdot y = 19 \\ x = \tfrac{19}{5} \end{array}$$

The system has a unique solution.

We now substitute into the second equation of the original system and solve for y:

$$x + y = 7$$

$$\frac{19}{5} + y = 7$$

$$y = 7 - \frac{19}{5} = \frac{35 - 19}{5} = \frac{16}{5}$$

The unique solution is $x = \frac{19}{5}$, $y = \frac{16}{5}$.

(b) $\begin{cases} x - 2y = 3 \\ -2x + 4y = -6 \end{cases}$

We multiply the first equation by 4, the second equation by 2 and add:

$$\begin{array}{r} 4x - 8y = 12 \\ \text{add} \quad \underline{-4x + 8y = -12} \\ 0 \cdot x + 0 \cdot y = 0 \end{array}$$

This equation is an identity. It is satisfied by every value of x. Each solution of this equation corresponds to a solution of the original system. Consequently, the original system has an infinite number of solutions. It is dependent.

(c) $\begin{cases} 2x - 4y = 8 \\ -x + 2y = 5 \end{cases}$

We multiply the first equation by 2, the second equation by 4 and add:

$$\begin{array}{r} 4x - 8y = 16 \\ \text{add} \quad \underline{-4x + 8y = 20} \\ 0 \cdot x + 0 \cdot y = 36 \end{array}$$

This equation is inconsistent. It has no solution. Consequently, the original system is inconsistent. It also has no solution.

EXERCISES

1. Solve the systems of equations by the methods of this section.

(a) $\begin{cases} 3x + 4y = -2 \\ -2x + 5y = 9 \end{cases}$ (d) $\begin{cases} 7x - 8y = 15 \\ 2x + 9y = -7 \end{cases}$

(b) $\begin{cases} 5x + 7y = 10 \\ 3x - 13y = 6 \end{cases}$ (e) $\begin{cases} 3x + 11y = 9 \\ 5x - 6y = 8 \end{cases}$

(c) $\begin{cases} 2x - 5y = 11 \\ 3x - 4y = 6 \end{cases}$ (f) $\begin{cases} x - 4y = 3 \\ 9x + y = 16 \end{cases}$

2. Use graphs *and* the method illustrated in Example 4 to decide if the systems of equations have unique solutions, are dependent, or are inconsistent.

(a) $\begin{cases} 3x - 2y = 7 \\ 2x + 5y = 8 \end{cases}$ (c) $\begin{cases} x - 3y = 9 \\ -2x + 6y = -18 \end{cases}$

(b) $\begin{cases} -x + y = 5 \\ 2x + 2y = 7 \end{cases}$ (d) $\begin{cases} 2x + y = 5 \\ -4x - 2y = -6 \end{cases}$

3. (a) Make up an inconsistent system of equations different from those
in this section. Show that if you try to solve it by the methods of
this section, you obtain an inconsistent equation in x.

(b) Make up a dependent system of equations different from those in
this section. Show that if you try to solve it by the methods of this
section, you get a single equation in x that is an identity.

5-9 APPLICATIONS OF SYSTEMS OF EQUATIONS

We close this chapter with several applications that lead naturally to
systems of equations. In each case, the final determination of x and y is
left for the reader.

Example 1. The Electric Company charges one rate for the first 300 kilowatts
of power used each month and a different rate for all over 300 kilowatts. The
author received the following two bills.

April: 874 kilowatts, total bill $24.85

May: 798 kilowatts, total bill $22.95

What are the two rates?

Solution. Let x cents be the cost per kilowatt-hour for the first 300 kilowatt-hours
and y cents be the cost for each kilowatt-hour over 300.

The first bill is based on 300 kilowatt-hours at x cents and 574 at y cents.
Therefore,

$$300x + 574y = 2485$$

Similarly,

$$300x + 498y = 2295$$

This gives us the system

$$\begin{cases} 300x + 574y = 2485 \\ 300x + 498y = 2295 \end{cases}$$

to solve for x and y (Exercise 1).

Example 2. The Hunkburger Company sells quick service hamburgers. (The
specialty is the "Big Hunk" at $0.85.) The meat is bought daily in quantity at
discount rates, is ground on the premises into hamburger, and then is mixed with
a vegetable protein compound to bring the total cost down to $0.70 per pound.
The actual proportions of meat and vegetable protein vary each day according
to the price of the meat.

On a given day the meat costs $1.05 per pound, and the vegetable protein
costs $0.30 per pound. How much of each should be mixed in order to get 500
pounds of the mixture at a total cost of $0.70 per pound?

Solution. Let x be the quantity of meat and y the quantity of vegetable protein.

Then

$$\underbrace{x}_{\substack{\text{weight} \\ \text{of meat}}} + \underbrace{y}_{\substack{\text{weight of} \\ \text{vegetable} \\ \text{protein}}} = \underbrace{500}_{\substack{\text{total} \\ \text{weight}}}$$

If we consider the cost of each commodity, we have

$$\underbrace{1.05x}_{\substack{\text{cost of} \\ \text{meat}}} + \underbrace{0.30y}_{\substack{\text{cost of} \\ \text{vegetable} \\ \text{protein}}} = \underbrace{(0.70)(500) = 350}_{\substack{\text{total cost} \\ \text{of mixture}}}$$

This gives us the system

$$\begin{cases} x + y = 500 \\ 1.05x + 0.30y = 350 \end{cases}$$

to solve for x and y (Exercise 2).

Example 2 is a typical mixture problem. These problems can be solved using only one variable, but usually the work is easier if two variables are used, as in Example 2.

Example 3. The radiator on Joe's car contains 20 quarts of a 20-percent antifreeze solution. How much should he withdraw and replace with pure antifreeze in order to get a 60-percent antifreeze solution?

Solution. Let x be the amount that is withdrawn and replaced; let y be the amount of the original solution that is left in the radiator. Because the final amount is 20 quarts, we have

$$x + y = 20$$

After the antifreeze is replaced, the total amount of antifreeze in the tank is

$$\underbrace{1.00x}_{\substack{\text{amount of} \\ \text{pure antifreeze} \\ \text{added}}} + \underbrace{0.20y}_{\substack{\text{amount of} \\ \text{antifreeze left} \\ \text{in the tank in} \\ \text{the original} \\ \text{solution}}} = \underbrace{(0.60)(20) = 12}_{\substack{\text{total amount of} \\ \text{antifreeze in the} \\ \text{tank}}}$$

The two equations give us the system

$$\begin{cases} x + y = 20 \\ x + 0.20y = 12 \end{cases}$$

to solve for x and y (Exercise 3).

Example 4. The Lustre-Feel Carpet Company manufactures two grades of carpet: Perfection and Supreme. The productions of July and August were as follows.

July: 8,000 square yards of Perfection, 12,000 square yards of Supreme; total cost $92,000

August: 10,000 square yards of Perfection, 5,000 square yards of Supreme; total cost $65,000

What are the costs per square yard of producing the two grades of carpet?

Solution. Let x be the cost per square yard of Perfection and y the cost per square yard of Supreme.

The costs for the two months can be summarized in the following equations.

July: $8,000x + 12,000y = 92,000$

August: $10,000x + 5,000y = 65,000$

We can simplify these equations by dividing the first one by 4,000 and the second one by 5,000. We obtain the system

$$\begin{cases} 2x + 3y = 23 \\ 2x + \ y = 13 \end{cases}$$

(See Exercise 4.)

EXERCISES

1. Finish Example 1.
2. Finish Example 2.
3. Finish Example 3.
4. Finish Example 4.
5. The bank charges a flat monthly service charge on each checking account *plus* a small service charge for each check processed. In January, seventeen checks were processed in one account for a total service charge of $1.99. In February, twenty-six checks were processed for a total service charge of $2.17. What are the amounts of the two charges?
6. The Lustre-Feel Carpet Company has decided to separate its expenses into two parts—overhead and production. The overhead expenses amount to $17,000 each month. (This amount was included in the $92,000 and the $65,000 costs.) Rework Example 4 in order to calculate the *production costs* for each type of carpet.
7. A chemist has one supply of 15 percent alcohol solution and another supply of 60 percent alcohol. How much of each should he mix in order to get 36 cubic centimeters of 40 percent alcohol solution?

SUGGESTIONS FOR FURTHER READING

1. Cooley, Hollis R., and Wahlert, Howard E. *Introduction to Mathematics,* Chaps. 6 and 7. Boston: Houghton-Mifflin, 1968.
2. Eves, Howard. *An Introduction to the History of Mathematics,* 3rd ed., Chap. 7. New York: Holt, Rinehart & Winston, 1969.
3. Kline, Morris. *Mathematics for Liberal Arts,* Chap. 5. Reading, Mass.: Addison-Wesley, 1967.

6

THE COMPUTER

6-1 INTRODUCTION: A BRIEF HISTORY OF COMPUTING

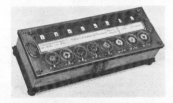

Pascal's adding machine. Photo courtesy of IBM Corp.

Blaise Pascal.
Brown Brothers

Pascal was a mathematical genius of the highest order. He independently discovered many of the theorems of plane geometry at age 12, participated in meetings with leading French mathematicians at 14, and proved important theorems in geometry at 16. When he was 27, he abandoned the study of mathematics and devoted most of the remainder of his short life to religious contemplation.

Man has been calculating with his brain (and his fingers) for thousands of years. In Chapter 2 we described the *counting board* or *abacus*, the first mechanical aid to calculation. Similar devices were used extensively in parts of Europe (and still are used in the Orient), but they did not become popular with the accountants and bookkeepers in the large European cities. By the seventeenth century, these men were adding columns containing hundreds (or even thousands) of numbers. The times required a mechanical calculator.

The first adding machine, the forerunner of today's desk calculator, was invented by *Blaise Pascal* (1623–1662), the French philosopher and mathematician, when he was 19 years old. Pascal's machine was similar in principle to the small hand adding devices that now can be bought in novelty stores for a few dollars. The user had to manually set each number, which slowed down the work considerably. For this reason, Pascal's machine and its immediate successors were never used extensively.

By the middle 1800s mechanical calculators had been improved greatly and were widely utilized for business and scientific work in the Western world. Asian bookkeepers continued to use the abacus, which was much faster than the best of the adding machines.

The first major advance in modern calculating came as a direct result of the 1890 United States census. Officials planning that census realized that it would be impossible to complete it in less than 10 years. The 1880 census was taking over nine years to tabulate, and it was obvious that the 1890 census would be both larger and more complicated. The sheer bulk of the data would make it difficult to analyze in any reasonable length of time.

The officials realized that the situation would continue to worsen with each succeeding census as the population increased and Congress requested more information. They decided that new methods were required and started an intensive search for ideas. *Herman Hollerith* (1860–1927), a young inventor working in the U.S. Patent Office, decided that a method for adding numbers automatically was needed. By some technique, numbers had to be processed without the necessity of humans entering each number in sequence. Hollerith found the method on a visit to a textile plant. Punched cards had been used to control automatic weaving machines for 150 years. Hollerith realized that similar punched cards could be used to enter numbers into adding machines. After several years of experimenting, Hollerith invented the first *automatic data-processing machine*. This machine could read millions of punched cards and tabulate the results of the census from the numbers punched on the cards.

One major problem remained. The millions of cards that were needed had to be cut to exact size. Hollerith found that no commercial plant in the world had the equipment to do the job. The United States Mint was the only facility that Hollerith could discover with the capacity to cut such

The Jacquard Loom and Jacquard Loom Card. Photos courtesy of IBM Corp.

A belt of punched cards automatically controlled the patterns woven by the Jacquard Loom. A similar method was used by Hollerith to code information for his calculator.

a large quantity of cards to the same size. Because all of its cutting machines were built to cut U.S. currency, the cards had to be cut the exact size of the 1890 dollar bill. The dollar bill has changed in size since then, but not the punched card. It remains the size that necessity forced on Hollerith in 1890.

Using Hollerith's data-processing machine, the 1890 census was processed in 3 years rather than the projected 10 years.

By the early 1930s, a number of data-processing machines were available to sort cards, count them, and to perform simple calculations on the numbers punched on the cards. These were special-purpose machines. Each could perform exactly the task that it was built for, and no others.

The next major step was the invention of the *programmable calculator*, a machine which could be used to solve a variety of different problems by changing the program. The first of these was the *Mark I*, a mechanical programmable calculator, started in 1937 and completed in 1944. The *ENIAC*, the first programmable calculator to use vacuum tubes instead of wheels and gears, was completed in 1946. Both the Mark I and the ENIAC received their instructions, one at a time, from a device external to the computer. The programs were punched on paper tapes and loaded in a machine which read them. During the execution of a program, an instruction was read into the calculator; it was executed; the next instruction was read, and so on.

About this time *John von Neumann,* one of the greatest mathematicians of the twentieth century, became interested in calculators. Von Neumann was one of the pioneers of modern computing. He, and other men, realized that the mechanical process of reading instructions one at a time slowed the programmable calculator to a fraction of its potential speed. He believed that it was necessary to store the program in the calculator's memory unit along with the numbers it was calculating.

In 1946 von Neumann and *H. H. Goldstine* (an electrical engineer) began directing the construction of the *EDVAC*, a true computer that stored its programs internally. A number of delays occurred which postponed completion until 1952. During that time the world's first modern computer, the *EDSAC*, was completed at Cambridge University under the direction of *M. V. Wilkes.*

Both the EDVAC and the EDSAC stored their programs internally. Each program was broken down into a series of simple steps that the computer could execute. Each step in a program was written as a numerical code word that the computer recognized as an instruction. Finally, the code words were punched on IBM cards (or paper tape) which were read into the computer's memory unit. For all practical purposes, the memory unit was split into two parts. One part stored the code words that made up the program; the other stored data. During execution of a program, the first instruction was read, it was executed; the second instruction was read, it was executed, and so on. These machines had the capacity to change their internally stored programs according to the values stored in their memory units.

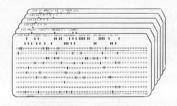

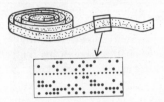

*Punched cards
and paper tape.*

Most programs are typed on punched cards or paper tape which are read into the computer. The punched card was invented by Hollerith. Paper tape was developed for use on commercial teletype networks.

Since the early 1950s, the world has experienced an explosion in computer technology. The basic principles of the computer remain the same as developed by von Neumann and the other pioneers, but the modern computer has many times the capabilities of the EDVAC in both speed and storage capacity. Actually, the modern computer is more properly called a *computer system*. It is a sophisticated system composed of a main computer, one or more secondary computers which organize the work of the main computer, many storage units, and a large number of input–output devices—card readers, teletype terminals, magnetic tape machines, high-speed printers, cathode ray tubes, and so on.

There are indications that the computer revolution may be just beginning. No single invention since the printing press has the potential to make such an impact on society. Many people think that the world may change as much in the next 50 years because of the computer as it has changed since the invention of printing.

*Mr. Babbage's Great
Calculating Machine*

Every science has its share of visionaries who are decades (or even centuries) ahead of their times. *Charles Babbage* (1791–1871), an English mathematician and inventor, was the actual inventor of the modern computer.

In 1823 Babbage obtained government support for the development of a *difference machine*, a limited-purpose automatic calculator that could calculate certain functions of numbers to 20 decimal places. As he continued work on this machine, he began to realize the potential of in-

ternally stored programs. This led to the *analytical engine*, a modern-type mechanical computer designed to read data from punched cards. In 1842 the British government turned down Babbage's request for additional funds, dooming the project to failure.

Babbage spent the rest of his life developing the principles of his analytical engine. He was aided by Lady Lovelace, the daughter of Lord

Charles Babbage and model of part of Babbage's difference machine. Photos courtesy of IBM Corp.

Byron, the poet, who worked out several programs for his computer.

Babbage's machine was properly designed, but the technology of the nineteenth century was not advanced enough to allow him to build a working computer. Because the vacuum tube had not been invented, he had to use gears and wheels instead of electrical relays as in modern computers. There was no way to make these parts perfect enough or strong enough to function in a large-scale mechanical computer. Years later a model of Babbage's analytical machine was built, using modern materials and milling techniques. It worked perfectly.

Babbage went to his grave convinced that his machine held the key to future scientific development. His principles were not rediscovered until the 1930s when the modern age of computers began.

TOPICS FOR FURTHER INVESTIGATION

1. Read the biography of Pascal in *Men of Mathematics,* by E. T. Bell. Pascal has been called "the greatest might-have-been in mathematics." What facets of his life give support to this description?
2. Read the biography of Charles Babbage in *Mathematics in the Modern World,* readings from *Scientific American.* Should Babbage be given credit for inventing the modern computer?
3. The book, *A Computer Perspective,* by Charles and Ray Eames, gives a pictorial history of the development of computing machinery. Write a short report on the history of the high-speed electronic computer based on information in this book. Include references to Howard Aiken, Alan Turing, Presper Eckert, John Mauchly, H. H. Goldstine, John von Neumann, and Maurice Wilkes.

6-2   APPLICATIONS OF THE COMPUTER

The electronic computer was specifically designed to solve scientific problems. In order to have flexibility in the programs, some *logical statements* which allow the computer to make simple decisions were included. By utilizing these logical facilities, many nonscientific applications which were not anticipated by its original developers have been found for the computer.

In this section we discuss a few typical problems that the computer can solve. All of these problems have one common feature. *They must be capable of solution in a fixed number of simple steps of the type that the computer can perform.*

Most of the problems that we encounter in society involve value judgments or subjective feelings. Problems such as "How can I make myself more popular?" and "What is a fair income tax law?" cannot be solved on the computer. There is no way that such problems can be solved in a fixed number of steps of the kind the computer can execute.

We now turn our attention to some problems that have been solved by the computer. In each case we shall discuss one typical problem.

Population Problems

Based on current trends, what will be the population of the United States in the year 2050?

There are a number of standard formulas, based on birth, death, and immigration rates, that can be used to calculate the size of the population at any time in the future. A program for the solution of this problem simply works through the steps of one of these formulas. The final answer is used by government workers, businessmen, and others to help plan the brave new world.

Unfortunately, there is no guarantee that the final answer obtained by the computer will accurately predict the population in the year 2050. The key words in the problem are "based on current trends." We can almost be certain that these trends will not remain unchanged until 2050. (The birthrate, for example, has changed dramatically over the past 20 years.) The method for solving this problem, however, requires the programmer to assume a knowledge of these rates. The solution to the problem can be calculated in a few seconds on the computer, but may be meaningless after it has been calculated.

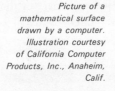

Picture of a mathematical surface drawn by a computer. Illustration courtesy of California Computer Products, Inc., Anaheim, Calif.

Computers can be used in conjunction with drafting machines to draw graphs of theoretical curves and surfaces.

Engineering Problems

What is the maximum load that a 100-foot interstate highway bridge of standard design can hold?

For this problem, as well, there is a standard formula that can be used. The computer program simply follows the steps involved in working out the formula. There is a major difference between this answer and the one obtained for the population problem, however. This answer can be expected to be accurate. The engineering formulas are based on sound principles and yield reasonably accurate answers. We can say "reasonably accurate" because no two bridges are exactly alike, even if their engineering specifications are identical. Minor variations in the materials or the workmanship may cause minor changes in the maximum load.

Medical Problems

How much radium is needed and where should it be placed in order to cure a cancer of the tongue?

One of the common methods for treating a cancer of the tongue is to partially fill several nail-like cylinders with radium and stick them through the tongue, surrounding the cancer. After three or four days the nails are removed, (hopefully) having cured the cancer. The original placement of the nails is in a circle around the cancer, and the same amount of radium is used in each nail. If the nails were to remain in their original positions, there would be no problem. The fact is, however, that the tongue is very responsive to being injured, and it both swells and contracts, drastically

A picture drawn by a computer. Illustration courtesy of California Computer Products, Inc., Anaheim, Calif.

changing the relative positions and orientations of the nails. The net result is that the cancer may not get enough radiation, and healthy tissue may receive too much. It is necessary to take frequent measurements of the locations of the nails (every six hours is standard) and to change the amounts of radium so that each part of the tongue receives the proper radiation.

The formula for calculating the radiation received from the nails was worked out more than 40 years ago. This formula is so complicated that it takes several days for an expert using a desk calculator to compute the amounts. The electronic computer makes the calculations in a few seconds, draws a graph showing the radiation levels for each part of the tongue, and tells how much to change the quantity of radium in each nail for proper treatment.

Hospital Diets

Which foods and how much of them should be eaten by a hospital patient suffering from one or more diseases which impose dietary restrictions?

George Octogenarian is a patient in Mercy Hospital suffering from high blood pressure. The treatment imposes certain restrictions on George's diet—very little meat, no salt, no butter or eggs or other foods high in saturated fats. The situation is complicated by the fact that he also has diabetes—a disease which imposes other severe restrictions on George's diet. The problem faced by the hospital dietitian is that of properly feeding George from the list of available foods.

The problem is easily solved on the computer. The list of foods available for the day is read into the computer along with the patient's dietary requirements (so many grams of carbohydrates, so much calcium, so much protein, and so on). The computer compares the foods with tables showing the composition of each food, performs some complicated calculations, and prints out a menu: 3 ounces stewed squash, $1\frac{1}{2}$ ounces baked bluefish, and so on. The entire process takes only a second or two on a fast computer.

Actually, the computer can do much more in planning menus. The majority of patients in the hospital eat the same meal. Only a minority have dietary problems. After the computer is given the wholesale costs of the available foods, it can plan all of the special meals (as indicated above), and then plan the standard meals for the rest of the patients so as to minimize the total cost. At the end it can print out a list of menus for the next day along with a shopping list telling how many crates of lettuce, how many sacks of potatoes, and so on, will be needed.

Search Problems

Find a name (or number) with certain properties from a list of 200,000 names.

The state police are looking for a red 1972 sedan that was used in a robbery. The license plate was remembered by a witness to be ——Q–78—, where the first two letters and the final digit are not remembered (all of the license plates in the state have three letters followed by three digits). Thus, the license plate number could be any one of AAQ-780, ABQ-783, ZZQ-789, and so on.

All information about automobiles registered in the state is stored on magnetic tapes, which are available to a computer. The computer searches through the entries on the tapes looking for license plates with Q as the third letter and 78 as the first two digits. When it finds one of these, it compares the number with a list of stolen license plates, then checks to see if the other requirements are met (1972, red, sedan). At the end of the search, it prints out four lists:

(1) A list of stolen license plates that could have been used in the robbery.

(2) A list of the license plate numbers (and the owners' names and addresses) that meet all of the additional requirements (1972, red, sedan).

(3) A list of the license plate numbers (and the owners' names and addresses) that meet at least one of the additional requirements (red, 1972, or sedan).

(4) A list of the license plate numbers (and the owners' names and addresses) that do not meet any of the additional requirements.

All of these lists could have been made by hand. The computer is not an essential factor at all. The only advantage of the computer is its speed. It can do all of the searching in a few minutes. The same task performed by humans could take weeks.

Sort Problems

Sort a list of 100,000 numbers into ascending or descending numerical order.

Each year the computer in the state tax office prints a list of all persons with large gross incomes. The incomes are listed, along with the names and addresses of the taxpayers, in descending numerical order—the person with the largest gross income listed first, the person with the next largest listed second, and so on. This list is used to establish a priority list for audits.

There are a number of different computer programs for sorting numbers into numerical order. They all follow standard procedures for sorting lists. For example, the programmer may decide to search the list, find the largest number, put it first, then find the next largest number, put it second, and so on. Obviously, some methods are more efficient than others. Only the efficient methods are extensively used.

Payroll Problems

Make a payroll for a large corporation. Calculate each worker's wages, his payroll deductions, print his check and check stub along with a copy of his check stub for the company. Print a final summary for the company and pertinent lists for the Internal Revenue Service, Social Security Administration, the union, the pension fund, and so on.

This was one of the earliest business problems solved by using the computer. Each employee's payroll record is stored on magnetic tape. This record contains his name, social security number, address, number of dependents, and hourly (or yearly) wage rate, along with a list of his total gross salary and total amounts withheld for taxes, social security, insurance, retirement, and pension fund for the elapsed part of the year. There is a master program that tells the computer the steps to follow in making up the payroll. At the beginning of the process, this master program is read into the computer, and the magnetic tapes containing the information about the workers are loaded on the tape machines. Punched cards containing each worker's social security number and the total number of hours that he worked are then read into the computer. When a punched card is processed, the computer finds the worker's record on magnetic tape, multiplies the number of hours that he worked by the hourly wage rate, and calculates the standard pay and the overtime pay. Then it calculates his social security payment, his withholding taxes, union dues, insurance and retirement payments. As each of these is calculated, it is added to the total amount listed on his magnetic tape record. Thus, the worker's record on tape is always up to date. These amounts also are added to running totals stored in the computer. It next adds the worker's name and other pertinent information to tapes containing current lists for the Internal Revenue Service, Social Security Administration, and so on. Finally, the computer either prints his paycheck or puts the information on another magnetic tape that can be run later on a different machine to print all of the paychecks at once. After all of the punched cards have been processed, the computer uses the running totals it has maintained to print lists and summaries for the company, the Internal Revenue Service, and so on.

Misuses of the Computer

The computer is a machine. It works through the steps of a program doing exactly what it was programmed to do—no more, no less. It can only be used to solve well-formulated problems that can be solved in a fixed number of simple steps. It cannot make value judgments. There are many misconceptions about what the computer can do, and in practice a great many abuses occur.

A typical misconception occurs on election night when the TV anchorman says, "With less than 1 percent of the votes in, the computer declares Houlihan the winner in New Jersey." Now a computer has no authority to declare anyone a winner in New Jersey. The computer has made a prediction based on returns from sample precincts, which have been closely analyzed for past voting performances. The statement that the anchorman should make is similar to the following: "If the voters in the sample precincts continue to be typical of the voters in the entire state, and if the computer program was prepared by experts and has no major errors, and if the computer is not malfunctioning, then it appears almost certain that Houlihan will carry New Jersey."

Many people think that the computer works by a process similar to black magic. Give it a problem, and it immediately gives the correct answer. These people put great faith in results announced on the basis of computer programs. They are quite willing to accept that Houlihan has carried New Jersey with only 1 percent of the votes tabulated. They feel cheated if they awake the next morning and find that Houlihan lost the election. These people should realize that the computer does exactly what it was programmed to do. If the program is defective or if incorrect data are given the computer, then the results are meaningless. As computer experts say, *GIGO—garbage in, garbage out.*

One of the major potential abuses of the computer comes from the willingness of some persons to let it make value judgments. As we pointed out previously, the computer cannot make such judgments, but it can print out statements that *appear* to be value judgments. Consider the following example.

Thomas Lawnorder, the head programmer for the Local Institute for Crime Analysis, is very rigid in his outlook, hates young people, and is paranoid in his fear of juvenile delinquency. Experts disagree about the root causes of juvenile delinquency, and how best to deal with it, but Lawnorder has solved the problem in his own way. He is the only person who thoroughly understands the complicated computer program that is used to analyze crime statistics. He has just added a few statements to the program that cause the computer to print the following message whenever juvenile crime is more than 20 percent of all reported crime:

JUVENILE CRIME OUT OF CONTROL. A NATIONAL CURFEW MUST BE IMPOSED AT ONCE ON ALL UNMARRIED PERSONS UNDER 25 YEARS OF AGE.

The first time this message appears, it will cause a stir if word is leaked to the newspapers. Many people will give the recommendation serious consideration because it was made by a computer. They should properly ask, "Why was this recommendation made by the computer? What steps in the program caused the message to be printed? How can a computer make the value judgments implicit in the recommendation?"

The most common abuses of the computer are caused by programming inadequacies. Many programs are not sophisticated enough to allow for all eventualities. This is most obvious when the computer is programmed to interact directly with the public.

An automobile-financing company programmed its computer to send dunning letters to customers whose accounts were two weeks in arrears and to order repossession of the automobiles if no payments were made for four months. There was no mechanism to control the computer in the event that exceptional circumstances kept individuals from making payments for a short period of time (serious illness, for example, or temporarily frozen bank accounts) or if errors were made in posting payment checks by employees of the financing company. Several persons wrote the company explaining that because of medical expenses they would be one month late in making a payment and would pay the required extra finance charge with the payment. They received permission from the company to make this adjustment and then later received threatening letters from the computer.

One man discovered that the final payment on his car had not been credited to his account when he received a dunning letter from the computer. He sent a photograph of the canceled check to the company only to receive a form letter (written by the computer) that threatened repossession of the car. Several weeks later, on orders sent directly to the repossession company by the computer, his car was legally repossessed. He had to sue the finance company for damages before he was able to get his car back.

TOPICS FOR FURTHER INVESTIGATION

The book *The Computer Revolution*, by Nigel Hawkes, contains hundreds of applications of the computer to the modern world. Write brief reports based on the following chapters.
1.   Chapter 2: "Computers in Business"
2.   Chapter 3: "Computers in Science"
3.   Chapter 5: "The Computer and the Arts"
4.   Chapter 8: "The Future"

6-3   INTRODUCTION TO BASIC

We can think of a computer as having three basic components:

(1) A *storage (memory) unit.* This unit contains thousands of storage locations in which numbers can be stored. The computer recognizes these locations by numbered addresses much like the numbers used to address post office boxes. Both the statements of the program and the numbers used in calculation are stored in these locations.

11704	15	32000	11705
11705	11	00000	32000
11706	35	00043	32000
11707	13	32000	32000
11710	35	00126	32000
11711	47	11712	11156
11712	21	00126	17466
11713	11	17466	00131
11714	11	00126	32000

Part of a program written in machine language for an early UNIVAC computer. The first programmers had to write all of the instructions to the computer as numerical code words.

(2) An *arithmetic-control unit.* This unit reads and executes the program step by step. It reads numbers from storage, performs calculations, makes simple decisions, stores numbers in assigned locations, and so forth.

(3) An *input–output unit.* This unit is used to communicate with the computer. It is used to put programs and data into the computer and to print output from the computer. It is wired to card readers, high-speed printers, magnetic tape machines, teletype terminals, and other devices.

A *program* is a detailed set of instructions for solving a specific problem. Each program must be stored in the computer in coded instructions. These instructions are stored as numerical "code words" that the computer recognizes. The "code words" form the *machine language* in which the computer actually does its work.

The early programmers had to write their programs in machine language. The process was tedious and time-consuming. Errors were easy to make and difficult to locate. One of the major advances in programming occurred in 1957 with the invention of *FORTRAN (FORmula TRANslation)*, the first of the translator languages.

In this book we shall use the translator language *BASIC*, which is similar to FORTRAN. BASIC was developed in 1965 by *John Kemeny* and *Thomas Kurtz* of Dartmouth College. It was specifically designed for teaching the art of programming and is one of the easiest of the translator languages to learn and understand.

BASIC = **B**eginner's **A**ll-Purpose **S**ymbolic **I**nstruction **C**ode

BASIC consists of a large number of statements similar to the formulas we use in algebra. After a program is written in BASIC, it is read into the computer. The computer has a permanently stored program, which is used to translate each BASIC statement into machine language. The final result is a program, written in machine language, that solves the same problem as the one written in BASIC. Once the program is translated, the computer executes it step by step. At the end of the program, the computer stops work and proceeds to the next program.

Before we begin our study of programming, we must make one point clear. The language BASIC has many dialects. Some of these are especially designed for use with teletype terminals, others for IBM card readers. In this book we shall attempt to use only statements that are common to all of the dialects. This means that we must restrict ourselves to the simplest statements. As a consequence, some of our programs will not be as efficient as they might be otherwise.

We now consider a sample program. This program contains examples of most of the statements that we shall study.

Example 1. The following program, written in BASIC, causes the computer to read five pairs of numbers and to print each pair along with the sum, product, and quotient. (See Fig. 6-1.)

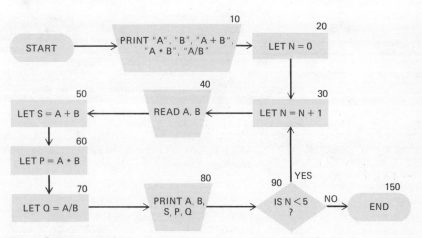

Figure 6-1 Example 1.

```
 5   REMARK - SAMPLE PROGRAM   (REMARK statement; not executed)
1Ø   PRINT ''A'', ''B'', ''A+B'', ''A*B'', ''A/B''     (PRINT column headings)
2Ø   LET N=Ø     (Assignment statement)
3Ø   LET N=N+1
4Ø   READ A,B     (READ statement)
5Ø   LET S=A+B
6Ø   LET P=A*B
7Ø   LET Q=A/B
8Ø   PRINT A,B,S,P,Q     (PRINT statement; print values of A,B,S,P,Q)
9Ø   IF N<5 THEN 3Ø     (Transfer statement)
1ØØ   DATA 2,5.1,3,−7,14,2     (DATA statement)
11Ø   DATA − 1,8,13,13
15Ø   END     (END statement; last statement of program)
```

Observe that all of the statements are numbered. These numbers must be in the proper numerical order. It is customary to skip numbers in going from one statement to the next so that other statements can be inserted later. For example, if we were to decide to add a statement between 50 and 60, we could number it 55.

When the program is first read into the computer, it translates all of the statements into its own language. At that time it also makes up a *data list*. This is a list of all of the numbers that are on DATA statements, the numbers being listed in the order they appear on the statements. In Example 1 the data list contains the numbers 2, 5.1, 3, −7, 14, 2, −1, 8, 13, 13, which are found on the DATA statements 100 and 110.

After the program is translated and the data list is compiled, the computer executes the instructions in the order they are listed until it receives an instruction to change the order (Statement 90). We now consider the statements of the program in order.

DATA LIST
2
5.1
3
−7
14
2
−1
8
13
13

The numbers on the DATA statements are read onto the data list when the program is read and translated into machine language.

5 REMARK-SAMPLE PROGRAM

This is a REMARK statement. It is not executed. It is a note written into the program to help the reader. Whenever we wish to insert a comment about the program, we simply write REM and then the comment. The computer will print the entire statement, but will ignore it during the execution of the program.

1Ø   PRINT ''A'', ''B'', ''A+B'', ''A*B'', A/B''

This is a PRINT statement. The computer separates each line into 5 zones of 15 spaces each. Each expression in quotation marks is printed, one expression to each zone. Statement 10 causes the computer to print the line of column headings

A   B   A+B   A*B   A/B

2Ø   LET N= Ø

ASSIGNMENT STATEMENTS

```
2Ø  LET N = Ø
3Ø  LET N = N + 1
```

Statement	Value of N
2Ø	Ø
3Ø	1

When the computer executes an assignment statement, it first calculates the expression on the right-hand side of the equality sign, then stores the result in the storage location named on the left-hand side.

READ STATEMENT

```
4Ø   READ A, B
```

Storage	Locations
A	B
2	5.1

A READ statement causes the computer to read the numbers from the data list and store them in the named locations.

This is an ASSIGNMENT statement. The computer stores the number zero in the storage location it knows as N. This instruction sets up the loop to be executed.

$$3Ø \text{ LET } N = N + 1$$

The computer works with the right-hand side of the expression, then stores the result as indicated on the left-hand side. It adds 1 to the number stored in location N ($0 + 1 = 1$) and stores the sum in location N. After this instruction, location N contains the number 1. This instruction starts the loop.

$$4Ø \text{ READ A, B}$$

This is a READ statement. The computer reads the first two numbers on the data list and stores them in locations A and B. After this instruction has been executed, location A contains 2 and B contains 5.1.

$$5Ø \text{ LET } S = A + B$$

The sum of the numbers stored in A and B is calculated, and the result, 7.1, is stored in location S.

$$6Ø \text{ LET } P = A*B$$

The symbol "*" indicates *multiplication* in BASIC. The product of the numbers stored in A and B is calculated, and the result, 10.2, is stored in location P.

$$7Ø \text{ LET } Q = A/B$$

The quotient $2/5.1 \approx .392157$ is calculated and stored in location Q.

$$8Ø \text{ PRINT A, B, S, P, Q}$$

The numbers stored in locations A, B, S, P, Q are printed on one line, one number to each zone. This instruction causes the line

2	5.1	7.1	1Ø.2	.392157

to be printed.

$$9Ø \text{ IF } N < 5 \text{ THEN } 3Ø$$

This is a CONDITIONAL TRANSFER statement. The computer checks to see if the number stored in location N is less than 5. Since it is less than 5 ($N = 1$), it transfers control to Statement 30 and works through the loop again.

3Ø to 9Ø. The second time through the loop the value of N is raised by 1 in Statement 30 (making $N = 2$); the numbers 3 and -7 are stored in locations A and B (Statement 40); the sum, -4, is stored in S (Statement 50); the product, -21, is stored in P (Statement 60); the quotient, $-.428571$, is stored in Q

(Statement 70); and the line

3	-7	-4	-21	$-.428571$

is printed (Statement 80). Since the value of N is less than 5, then Statement 90 transfers control to Statement 30 and the loop begins again. The computer works through this loop five times.

On the fifth time through the loop, the computer finds that N = 5 when it executes Statement 90. Since the condition N < 5 is not met, the computer ignores Statement 90 and proceeds on to Statements 100 and 110. These statements are ignored because the DATA statements were read when the program was translated. It then proceeds to Statement 150.

$$15\emptyset \quad \text{END}$$

This is an END statement. It must be the last statement in the program. When the computer executes this statement, it stops work on this program, calculates the bill, and goes on to the next program.

The total output for the program follows:

A	B	A+B	A*B	A/B
2	5.1	7.1	1$\emptyset$.2	.392157
3	-7	-4	-21	$-.428571$
14	2	16	28	7
-1	8	7	-8	$-.125$
13	13	26	169	1

Almost all of the statements in BASIC that we shall study are illustrated in the above program. The reader should work through the steps, comparing them with the steps in Figure 6-1, until he understands each statement.

EXERCISES

1. The following statements are legal in BASIC. Describe the effect of executing each statement in the program.

```
  20   READ A,B
  21   DATA 7,42,5,9,13
  24   LET C=A+B
  25   PRINT A,B,C,A*B
  26   PRINT "A =",A,"B =",B
  27   REMARK * TEST FOR THE END OF THE LOOP
  28   IF A=7 THEN 20
  30   END
```

2. (a) Work through the steps of the following program exactly as the computer would do. List the printout that would be obtained. (All statements are legal.)

```
10   REM — SAMPLE PROGRAM
20   READ A
30   READ B
40   DATA 1
50   DATA 2,3,4,5,6,7
60   LET C=A+B
70   LET C=A+C
80   PRINT A,B,"C=",C
90   IF C=4 THEN 20
100  END
```

(b) Run the program on the computer. See if you correctly predicted the output in (a).

3. The computer is programmed to check each program for possible errors. The following "program" contains several errors. Put it on the computer and study the "error messages" that you get back. Using these messages and the format of the statements in Exercises 1 and 2 as a guide, try to correct each illegal statement.

```
     REMARK — "PROGRAM" WITH SEVERAL ERRORS
20   LET A=10
30   LET A+B=A
40   LET B=7
25   LET N=N+1
60   LET N=−1
250  IF N<0 THEN 35
260  GO TO 40
800  PRINT N
900  STOP
```

6-4 READ AND INPUT STATEMENTS

The various instructions and operations in BASIC are discussed in more detail in the next few sections than in Section 6-3. You should study each topic as much as necessary in order to work the assigned problems. Do not try to memorize these sections. Refer to them when you need the information.

Storage Locations, Variables

Storage locations are named by single letters or by single-letter–single-digit combinations. Thus,

A, B, C7, A1, B5, Z and M3

are legal names for storage locations, but AB, M13, and 5T are not. This assignment makes 286 storage locations available to each program (26 single letters and 260 letter–digit combinations).

Storage locations are commonly known as *variables*. We frequently say, "The variable N has the value 5," or just, "N = 5," to indicate that the number 5 is stored in location N.

READ Statements

READ statements instruct the computer to read one or more numbers from the data list and store them in the named locations. These statements follow one of the formats

(statement no.) READ (variable)

or

(statement no.) READ (variable), (variable), . . .

In the second format, any number of variable names can be listed.

Examples of READ Statements

```
2Ø   READ A
3Ø   READ B1, C7, B, Z6, A3, M, K
```

Statement 20 causes the computer to read the first number from the data list and store it in location A. Statement 30 causes it to read the next seven numbers and to store them, in order, in locations B1, C7, B, Z6, A3, M, and K.

Example 1. The following two programs have the same effect. Each causes A to be assigned the value 7, B the value −3, C the value 2.5, and D the value 0.

```
21Ø   READ A,B,C,D          11Ø   READ A,B,C
22Ø   DATA 7                12Ø   READ D
23Ø   DATA −3,2.5           13Ø   DATA 7,−3,2.5,Ø
24Ø   DATA Ø                14Ø   END
25Ø   END
```

Example 2.   The statement

15 READ A+B

is *not legal* in BASIC. The expression A + B is not the name of a storage location.

INPUT Statements (*Teletype Terminals Only*)

INPUT statements allow you, the operator, to insert data during the execution of a program. When the computer executes an INPUT statement, it types a *question mark* as a signal that input is needed. The operator must then type in as many numbers as required in the program (separating the numbers by commas) and then press the *return* key. INPUT statements have the same format as READ statements.

Examples of INPUT Statements

```
275   INPUT A
593   INPUT B, C3, M, V
```

Suppose that we want to input the value A = 7, B = 5, C3 = 3.14, M = 500, V = 5. At Statement 275 the computer types ? and we type 7 and press the *return* key, resulting in the line
?7 *return*
The computer stores the number 7 in location A.
    At Statement 593 the computer types ? and we type in the other numbers, resulting in the line
?5, 3.14, 5ØØ, 5 *return*
The computer stores the number 5 in location B, 3.14 in C3, 500 in M, and 5 in V.

PRINT statements can be combined with *INPUT* statements to tell the operator the type of data that is needed.

Example 3.   The following program is used to convert inches to centimeters. At the question mark, one number must be typed as input. The program continues until the word STOP is typed at the request for data or until the *break* key is pressed.

```
 1Ø   REMARK * PROGRAM TO CONVERT INCHES TO CENTIMETERS
 2Ø   PRINT "NUMBER OF INCHES"
 3Ø   INPUT I
 4Ø   LET C=2.54*I
 5Ø   PRINT I, "INCHES EQUAL",C,"CENTIMETERS"
 6Ø   PRINT
 7Ø   GO TO 2Ø
 8Ø   END
```

A typical run produces the following output:

```
NUMBER OF INCHES
?17
 17              INCHES EQUAL     43.18              CENTIMETERS

NUMBER OF INCHES
?28.7
 28.7            INCHES EQUAL     72.898             CENTIMETERS
```

EXERCISES

1. Which of the following are legal names for variables (storage locations)?

   (a) M   (e) 5F   (i) E7
   (b) C3   (f) F5   (j) AB
   (c) I2   (g) E13   (k) A * B
   (d) Z9   (h) E − 1   (l) W

2. (a) Make a list showing the values of all of the variables after each step of the following program.

   (b) Run the program on the computer. Could you predict the output?

```
       1Ø   READ A,B
       2Ø   LET K=1
  →    3Ø   READ C
  ⌐    4Ø   LET K=K+2
  |    5Ø   LET D=A+B+C
  |    6Ø   PRINT A,B,C,D
  └    7Ø   IF K<6 THEN 3Ø
       8Ø   PRINT "THATS ALL FOLKS"
      1ØØ   DATA 1,2,3,4,5,6,7,8,9,1Ø,11,12
      11Ø   END
```

3. *(For terminal users only)* Modify the program in Exercise 2 so that INPUT statements are used rather than READ statements. Run the program using you own data. How many numbers must you type when the computer prints a question mark?

4. (a) Modify the program in Example 3 so that FEET and INCHES are required for data and METERS and CENTIMETERS are the output. Run the program using your own data.

   (b) Modify the program in Example 3 so that READ and DATA statements are used instead of INPUT statements. Run the program using your own data.

   (c) Same as (b) for the program in (a).

5. (a) Construct a program (patterned after Example 3) that converts

temperatures from Fahrenheit to Centigrade. [The conversion formula is $C = 5 * (F - 32)/9$.] Run the program using your own data.

(b) Modify the program so that READ and DATA statements are used instead of INPUT statements. Run the program using your own data.

6. *Insufficient data.* This exercise teaches you how the computer terminates a program if it does not receive sufficient data. This may be the most practical way to stop a program if different quantities of DATA are used each time the program is run. Run the following program on your computer.

```
→10    READ A,B
 20    PRINT A,B,A*B
└30    GO TO 10
 40    DATA 7,−3,41,8,9
 50    END
```

6-5 ASSIGNMENT STATEMENTS; OPERATIONS

Statements such as

```
20   LET A=5
30   LET M1=(A+5)*2
40   LET B=A
```

are called *assignment statements*. They assign the value on the right-hand side of the equality sign to the variable on the left-hand side. Assignment statements are of the form

(statement no.) LET (variable) = (expression)

The expression to the right of the equality sign can be a *number*, a *variable*, or any *arithmetic expression* involving numbers and variables.

Examples of Assignment Statements
___

```
15    LET B=0
25    LET M=M+1
37    LET A=B
1873  LET Z1=C8−(A*2+5)/3
```

When the computer executes an assignment statement, it first calculates the expression on the right-hand side of the equality sign (using the current values of the variables), then it stores the result in the location named on the left-hand side. Thus, statements such as

$$25 \quad \text{LET } M = M + 1$$

are perfectly correct in BASIC, even though equations such as

$$x = x + 1$$

are meaningless in algebra.

Operations

The five arithmetic operations that the computer can perform are listed below:

Operation	Symbol
Addition	$+$
Subtraction	$-$
Multiplication	$*$
Division	$/$
Exponentiation	$\uparrow$ (teletype terminals) $**$ (card records)

Examples of Statements Involving Operations

```
1Ø   LET C=3/2
2Ø   LET A=5+7
3Ø   LET B=(A*C)↑2
4Ø   LET M=((A+3)*9)−1
5Ø   PRINT A, B, M, M↑5, A*B, A↑(M*3)
```

Example 1. Write the following mathematical expressions in BASIC.
(a) $(15 + 2^2)/3$
(b) $a^3 - bc$
(c) $5a + 3b^2$
(d) $5a + (3b)^2$
(e) $5a + 3^{b^2}$

Solution. (a) $(15+(2\uparrow2))/3$
          (b) $(A\uparrow3)-(B*C)$
          (c) $(5*A)+(3*(B\uparrow2))$
          (d) $(5*A)+((3*B)\uparrow2)$
          (e) $(5*A)+(3\uparrow(B\uparrow2))$

What does the computer do with the expression

$$A + B/C \uparrow 2?$$

As far as the reader could know at this point, the computer could interpret this expression to be any of the following:

$$[(A + B)/C]^2, A + (B/C^2), A + (B/C)^2, (A + B)/C^2, \text{ or } [A + (B/C)]^2$$

The computer has its own order of executing operations within an expression. First, it performs all *exponentiation;* next, it calculates all *products* and *quotients* (proceeding from left to right); and finally, it calculates all *sums* and *differences* (proceeding from left to right). Thus, it interprets $A + B/C \uparrow 2$ as $A + (B/C^2)$.

If we remove all unnecessary parentheses from the expressions in Example 3, we can write them as

(a) (15＋2↑2)/3
(b) A↑3－B*C
(c) 5*A＋3*B↑2
(d) 5*A＋(3*B)↑2
(e) 5*A＋3↑(B↑2)

The beginning student is advised to use as many parentheses as needed to avoid ambiguity. It is better to use unnecessary parentheses than to have the computer interpret an expression incorrectly.

Example 2. The diagram next to the program shows the values assigned to the variables A, B, C, and D after each of the instructions is executed. At the beginning of the program, B, C, and D have not been assigned values. (This is indicated by ". . . ." in the boxes.) This does not mean that the values are zero. Whatever numbers were last assigned to those locations during previously executed programs are still assigned to them at the beginning of this program.

```
17   LET A＝5
2Ø   LET B＝A
3Ø   LET A＝A＋2
4Ø   LET D＝A↑2
5Ø   LET C＝B
6Ø   LET B＝A－C/3＋1
7Ø   END
```

Statement Number	A	B	C	D
17	5	. . . .	. . . .	. . . .
2Ø	5	5	. . . .	. . . .
3Ø	7	5	. . . .	. . . .
4Ø	7	5	. . . .	49
5Ø	7	5	5	49
6Ø	7	6.33333	5	49
7Ø	7	6.33333	5	49

EXERCISES

1. Write the following mathematical expressions in BASIC.
   (a) $(a^2 - b^2)/c$ (d) $x + y/2 - z$
   (b) $(3^4 - 5^a)/a$ (e) $(x + y)/2 - z$
   (c) $(3/a) + b^{5a}$ (f) $x + y/(2 - z)$

2. What is the effect of each of each of the following assignment statements? Assume that A = 5 and B = 2 when Statement 275 is executed.
   (a) 275 LET C=A*B
   (b) 275 LET A=A+1
   (c) 275 LET A=3
   (d) 275 LET B=((A−B)*(2+B))/(A−1)

3. (a) Work through all of the steps of the following program. Run the program on the computer. Did you predict the output correctly?

```
    11Ø   LET N=Ø
    12Ø   LET A=2
 ┌→ 13Ø   LET N=N+1
 │  14Ø   LET A=A↑2
 │  15Ø   PRINT A
 └─ 16Ø   IF N<6 THEN 13Ø
    17Ø   END
```

   (b) Change Statement 160 to read

   16Ø   IF N<6 THEN 14Ø

   How does this change the output? How does it affect the running of the program?

   (c) Change State 160 to read

   16Ø   IF N<6 THEN 12Ø

   How does this change the output? How does it affect the running of the program?

4. *Infinite looping.* Occasionally all programmers inadvertently instruct the computer to execute a loop an infinite number of times. This causes no major problems on the card reader—the computer automatically terminates the program after a certain number of seconds. An infinite looping process on the terminal, however, can be quite serious. It ties up part of the computer until the operator aborts the program. On most terminals this is accomplished by pressing the *break* key while the program is being executed.

   The following program contains infinite looping caused by an error in Statement 50.

```
 5   REMARK * PROGRAM CONTAINING AN ERROR - INFINITE LOOPING
1Ø   LET A = −1
2Ø   LET N = Ø
3Ø   LET N = N + A
4Ø   PRINT "VALUE OF N = ",N
5Ø   IF N < 5 THEN 3Ø
6Ø   END
```

(a) Work through the steps of the program until you can predict the output.

(b) Run the program on your computer. If you are on a terminal, use the *break* key (or other mechanism as told by your instructor) to terminate the program.

(c) Change Statement 50 so that the loop will be executed five times. Run the modified program on your computer.

5. Make a diagram similar to the one in Example 2, showing the values of the variables A, B, C, and N after each statement in the following program is executed.

```
 2Ø   LET N = Ø
 3Ø   LET N = N + 1
 4Ø   READ A,B
 5Ø   LET C = A/B
 6Ø   PRINT "A/B=",C
 7Ø   IF N < 3 THEN 3Ø
 8Ø   DATA 1,2,3,4,5,6
1ØØ   END
```

6-6   TRANSFER STATEMENTS; COMPARISONS; stop AND end STATEMENTS

A go to transfer statement such as

$$15 \quad \text{GO TO } 73$$

causes control to be transferred immediately to the named statement (Statement 73).

The early computers were called "giant brains" by newspaper writers who had the mistaken notion that these machines could think. The logical facilities of the computer enable it to change the sequence of steps in a program according to the size of numbers stored in certain locations (by means of conditional transfer statements). The computer is incapable of any other type of "thought."

Statements such as

$$83 \quad \text{IF } N=7 \text{ THEN } 18\emptyset$$
$$92 \quad \text{IF } X+Y<Z-5 \text{ THEN } 71$$

are called *conditional transfer statements*. If the stated condition is satisfied, then control is transferred to the named statement. If the condition is not satisfied, then control passes on to the next statement in sequence.

The format for a conditional transfer statement is

(statement no.) IF (comparison) THEN (statement no.)

Examples of Conditional Transfer Statements

$2\emptyset$ IF $X+Y<Z$ THEN 15
25   IF $N=3$ THEN 8
$3\emptyset$ IF $Y-A7 >= B+5$ THEN 42
$4\emptyset$ IF $M<>4$ THEN 9

The comparisons that the computer can make are shown in Table 6-1. Observe that different symbols are used on the card reader than on the teletype terminal. Furthermore, some terminals require the use of "$<>$" for "not equal to" and others require "$\#$." Your instructor will tell you which symbols are used with your computer.

Table 6-1

Comparison	Symbol	
	Teletype Terminal	*Card Reader*
Equals	$=$	$=$ or EQ
Greater than	$>$	GT
Less than	$<$	LT
Greater than or equal to	$> =$	GE
Less than or equal to	$< =$	LE
Not equal to	$< >$ or $\#$	NE

Conditional transfer statements are at the heart of most programs. They allow the computer to make simple decisions based on the numbers stored in various locations.

Example 1. Search Problem. The following program instructs the computer to read three numbers and to search through the numbers, as it reads them, to find the largest.

The steps of the program are shown in the flow chart of Figure 6-2. Each

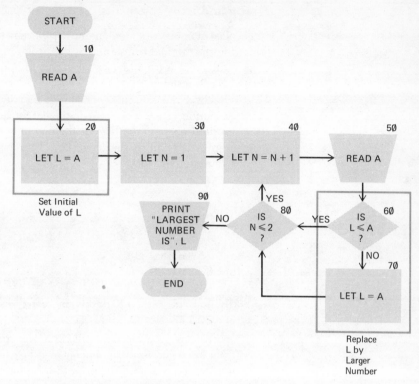

Figure 6-2 The search problem (Example 1).

number is stored in location A as it is read. The first number read also is stored in L. As each succeeding number is read, it is compared with L. If it is larger than L, it replaces the old value of L. Thus, L is always equal to the largest of the numbers that have been read.

```
5   REMARK * PROGRAM TO FIND THE LARGEST OF 3 NUMBERS
1Ø   READ A
2Ø   LET L=A
3Ø   LET N=1
4Ø   LET N=N+1
5Ø   READ A
6Ø   IF L >= A THEN 8Ø
7Ø   LET L=A
8Ø   IF N <= 2 THEN 4Ø
9Ø   PRINT "LARGEST NUMBER IS ",L
11Ø   DATA −2Ø,17,6
12Ø   END
```

STOP and END Statements

The last numbered statement in a program must be an END statement. During the original translation of the program into machine language, the END statement is used as a signal to the computer that the entire program has been translated.

Whenever the computer encounters a STOP or an END statement during the execution of a program, it immediately stops work on that program.

Terminal Users Only. One of the common methods to stop a program in progress is to press the *break* key. Your instructor will inform you of how your system works.

EXERCISES

1. Which of the following are legal transfer statements?
   - (a)   75 GO TO 9
   - (b)   8Ø GO TO 8Ø
   - (c)   10Ø IF X+Y≤Z THEN 12Ø
   - (d)   175 IF X<3+K THEN 5
   - (e)   20Ø IF N>X THEN STOP

2. Construct a program that will cause the computer to read 20 numbers (from READ and DATA statements or INPUT statements according to your preference), compute and print the absolute value of each number, with appropriate labels.

   Recall that the *absolute value of a positive number* is the number itself. The *absolute value of a negative number* is the *negative* of the number. For example,

$$|3| = 3$$
$$|(-7.2)| = -(-7.2) = 7.2$$

   Run the program using your own data.

3. (a)   Run the program in Example 1 on the computer using your own data. You may change the READ statements to INPUT statements if you use a terminal.
   - (b)   Make the following modifications in the program in Example 1. (First construct a new flow chart.) Run the modified program with your own data.
     - (1)   Read M numbers rather than 3 numbers, where M is read at the beginning of the program.
     - (2)   Find the *largest* and the *smallest* of the numbers. Print both numbers at the end of the program with appropriate labels.

   (Terminal users may use INPUT statements rather than READ statements if they so desire.)

6-7 PRINT **STATEMENTS**

A simple PRINT statement, such as

$$375 \quad \text{PRINT A, B, C}$$

is quite adequate for getting information from the computer. If a great many of these simple PRINT statements are executed, however, the persons reading the output may need detailed sets of instructions in order to decipher it. For this reason, rather sophisticated PRINT statements are available that cause the computer to print column headings, explanations of its output, titles, and so forth.

The general format for a PRINT statement is

(statement no.) PRINT (text or expression), (text or expression), . . .

The *text* can be any set of symbols enclosed in quotation marks (excepting quotation marks themselves). The computer will print exactly what is written between the quotation marks, including typographical errors. The *expression* can be a *number*, the name of a *variable*, or any *arithmetic expression* involving numbers and names of variables.

Examples of PRINT Statements

```
4Ø   PRINT " ", "A=", A
5Ø   PRINT A,A+B,"A−B*C↑2=",A−B*C↑2
6Ø   PRINT 1,2,3,4,5,6,7,8,9
7Ø   PRINT
8Ø   PRINT "A", "B", "C", "D", "E"
9Ø   PRINT "AA", "BB", "CC", "DD", "EE", "FF"
```

Assuming that A = 2, B = 3, and C = 1 at the time the above statements are executed, we get the following output (the dashed lines separate the five zones):

```
|            |A=         |           |           |
| 2          | 5         | 2         |           |
| 1          | 2         |A−B*C↑2=   |−1         |
| 6          | 7         | 3         | 4         | 5
|            |           | 8         | 9         |
|A           |B          |C          |D          |E
|AA          |BB         |CC         |DD         |EE
|FF          |           |           |           |
```

The computer separates each line into 5 zones of 15 spaces each. Each PRINT statement starts a new line. Each item to be printed begins in the first available zone. The computer uses as many zones (and lines) as it needs to print each statement.

20 PRINT A; B; C

Students writing BASIC on teletype terminals can use semicolons in PRINT statements rather than commas. These cause the computer to partition each line into 12 zones rather than 5, allowing much more information to be printed on each line.

Dangling commas

A comma at the end of a PRINT statement causes the next PRINT statement to be treated as a continuation of the one with the dangling comma. The output of the next PRINT statement will start in the next available zone of the same line.

Example 1. The following program causes the computer to print powers of 2 as shown. Observe the effect of the dangling comma in Statement 70. If the comma had been deleted, the 25 numbers would have been printed one to a line.

```
10   REMARK * PROGRAM TO PRINT FIRST 25 POWERS OF 2
20   PRINT "POWERS OF 2"
25   PRINT
30   LET A=1
40   LET N=0
50   LET N=N+1
60   LET A=A*2          Note dangling comma
70   PRINT A,
80   IF N<25 THEN 50
90   END

POWERS OF 2
```

2	4	8	16	32
64	128	256	512	1024
2048	4096	8192	16384	32768.
65536.	131072.	262144.	524288.	1.04858E+06
2.09715E+06	4.19430E+06	8.38861E+06	1.67772E+07	3.35544E+07

Scientific Notation (E-Notation)

The computer prints most numbers just as we would write them. Very large or very small numbers, however, are printed in *scientific notation*. An "E" is written before the power of 10 to indicate that scientific notation is being used.

Example 2. The computer writes 3.54366E + 21 for the number

$$3.54366 * 10^{21} = 3,543,660,000,000,000,000,000$$

It writes 5.71326E − 12 for the number

$$5.71326 * 10^{-12} = \frac{5.71326}{10^{12}} = 0.00000000000571326$$

EXERCISES

1. Convert the following numbers from "E-notation" to standard decimal notation.
   - (a) 5.31264E17
   - (c) 9.78462E–8
   - (b) 1.814268E9
   - (d) 4.300612E–26

2. Write PRINT statements that cause the computer to perform the following operations.
   - (a) Skip a line.
   - (b) Print "A = 5" in the *second* zone.
   - (c) Print "A computer is a girl's best friend," starting in the *third space* of the *first* zone.
   - (d) Print the two expressions "THE QUICK BROWN FOX" and "JUMPED OVER THE LAZY DOG" so that only a single space separates the "X" and the "J." (*Hint:* Put several spaces at the beginning of the first expression.)

3. (*Terminal users only!*) Rewrite the program in Example 1 with a *semicolon* at the end of Statement 70 instead of the comma. Run the program. What effect does the change have on the output?

4. The following program can be used to show how the computer prints output in the five zones.
   - (a) Run the program exactly as written. (Do not forget the dangling comma at the end of Statement 150.) Draw vertical boundary lines to outline the five zones on the output.
   - (b) (*Terminal users only!*) Change all of the commas to semicolons in Statements 130, 150, and 160. Run the modified program. What effect does this change have on the output?

```
1ØØ   REMARK * PROGRAM ILLUSTRATING OUTPUT OF PRINT STATEMENTS
11Ø   LET A=5
12Ø   LET B=15
13Ø   PRINT " ","A=",A,"B=",B
14Ø   PRINT
15Ø   PRINT "FOURTEEN SPACE","#FIFTEEN#SPACES",
16Ø   PRINT "$SIXTEEN$SPACES$","=SEVENTEEN=SPACES"
17Ø   END
```

6-8 FUNCTIONS

BASIC has a number of *mathematical functions* built into its framework. Each of these functions is a small program for the automatic calculation of certain numbers. The four functions that we shall consider are the *absolute value function* ABS( ), the *integer function* INT( ), the *square root function* SQR( ), and the *random number function* RND( ). These

sin 27° = 0.4540

|−7.2| = 7.2

A mathematical function is a law for assigning values to given quantities. For example, the *sine function* assigns the value

sin 27° = 0.4540

to the quantity 27°. The *absolute value function* assigns the value

|−7.2| = 7.2

to the number −7.2. A large number of other mathematical functions have been devised. Several of these are built in to the BASIC language.

functions can be used in any arithmetic expression. We simply write a number in the parentheses, and the computer calculates the functional value.

ABS( )

The *absolute value* function causes the computer to calculate the absolute value of the number in parentheses. Recall that $|a| = a$ if $a \geq 0$ and $|a| = -a$ if $a < 0$.

Example 1.   (a) ABS(3) = 3
  (b) ABS(−7) = 7
  (c) ABS(0) = 0
  (d) ABS(−9.327) = 9.327

Example 2.   If a and b are points on the number line, then $|a - b|$ is the distance between them (Fig. 6-3). The following program instructs the computer to read five pairs of numbers (A and B), print the numbers and the distance between the corresponding points. The output follows the program.

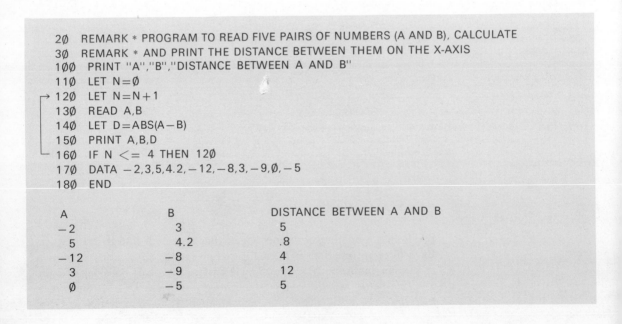

Figure 6-3   The distance between a and b on the number line is $|a - b|$. (a) Case 1: $a < b$; (b) Case 2: $a > b$.

```
 20   REMARK * PROGRAM TO READ FIVE PAIRS OF NUMBERS (A AND B), CALCULATE
 30   REMARK * AND PRINT THE DISTANCE BETWEEN THEM ON THE X-AXIS
100   PRINT "A","B","DISTANCE BETWEEN A AND B"
110   LET N=Ø
120   LET N=N+1
130   READ A,B
140   LET D=ABS(A−B)
150   PRINT A,B,D
160   IF N <= 4 THEN 12Ø
170   DATA −2,3,5,4.2,−12,−8,3,−9,Ø,−5
180   END
```

A	B	DISTANCE BETWEEN A AND B
−2	3	5
5	4.2	.8
−12	−8	4
3	−9	12
Ø	−5	5

SQR( )

The square root of a positive number a is the positive number we square in order to get a. For example, the square root of 9 is 3 ($3^2 = 9$); the square root of 16 is 4 ($4^2 = 16$); and the square root of 2 is approximately equal to 1.414 ($1.414^2 \approx 2$). In mathematics we use the symbol $\sqrt{}$ to indicate the square root. Thus,

$$\sqrt{9} = 3$$
$$\sqrt{16} = 4$$
$$\sqrt{2} \approx 1.414$$

In BASIC we use SQR( ) to indicate the square root. Thus,

$$SQR(9) = 3$$
$$SQR(16) = 4$$
$$SQR(2) \approx 1.414$$

Example 3. The following program causes the computer to print a table of square roots of the numbers 0.1, 0.2, 0.3, 0.4, . . . , 0.9. The output follows the program.

```
 10   REMARK * PROGRAM TO PRINT TABLE OF SQUARE ROOTS
 20   PRINT "N","SQUARE ROOT OF N"
100   LET K=0
110   LET K=K+1
120   PRINT K/10,SQR(K/10)
130   IF K<9 THEN 110
140   END
```

N	SQUARE ROOT OF N
.1	.316228
.2	.447214
.3	.547723
.4	.632456
.5	.707107
.6	.774597
.7	.83666
.8	.894427
.9	.9486683

INT( )

The *integer function* causes the computer to calculate the largest integer that is less than or equal to the number in parentheses. If X is between two integers, then INT(X) is the integer to the left of X on the number line. If X is an integer, then INT(X) = X.

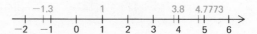

Figure 6-4   (a) INT(3.8) = 3; (b) INT(4.7773) = 4; (c) INT(1) = 1; (d) INT(−1.3) = −2 (Example 4).

Example 4.   (a) INT(3.8)=3
        (b) INT(4.7773)=4
        (c) INT(1)=1
        (d) INT(−1.3)=−2
(See Fig. 6-4.)

The INT function is very useful in problems involving dollars and cents. As a simple example, consider the problem of calculating compound interest at 0.625 percent per month (7.5 percent yearly interest). If the principal is $376.43, the interest for the month is

$$(0.00625) \cdot (376.43) = 2.3526875$$

which we (and the bank) round-off to $2.35. Unfortunately, the computer does not round-off at two decimal places. It registers the interest as 2.35269 and uses this figure in future calculations. If a very large number of such calculations are involved, the error may be considerable.

The INT function can be used in a simple algorithm to round-off monetary calculations to the nearest cent. The first step is to use *cents* instead of dollars and cents for all amounts. The problem then reduces to rounding-off numbers to the nearest integer.

The correct expression for rounding-off a number X to the nearest integer is

$$INT(X+.5)$$

To see why this expression rounds-off X to the nearest integer, suppose that X is between the integers N and N + 1. First, if

$$N \leq X < N + 0.5$$

then

$$N < X + 0.5 < N + 1$$

so that

$$INT(X + 0.5) = N$$

the integer nearest to X. If

$$X \geq N + 0.5$$

then

$$N + 1 \leq X + 0.5 < N + 2$$

so that

$$INT(X + 0.5) = N + 1$$

the integer nearest to X.

In either case $INT(X + 0.5)$ is equal to the integer that is nearest to X. (See Fig. 6-5.)

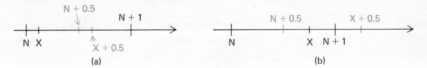

Figure 6-5 $INT(X + 0.5)$ is the integer closest to X. (a) If $N \leq X < N + 0.5$, then $INT(X + 0.5) = N$; (b) if $N + 0.5 \leq X < N + 1$, then $INT(X + 0.5) = N + 1$.

As an example, we use this round-off method to calculate interest at 0.625 percent per month. If the principal is 37,643 cents ($376.43), the rounded-off interest is

INT(.00625*37643+.5)
    =INT(235.26875+.5)=INT(235.76875)
    = 235 cents = $2.35

If the principal is 86,045 cents ($860.45), the interest is

INT(.00625*86045+.5)
    =INT(537.78125+.5)=INT(538.28125)
    = 538 cents = $5.38

In each case the computer rounds-off to the nearest cent.

Example 5. This program is designed to calculate the total interest due on a loan of $1000 after N months at an interest rate of 0.625 percent per month. The program is based on the flow chart of Figure 6-6.

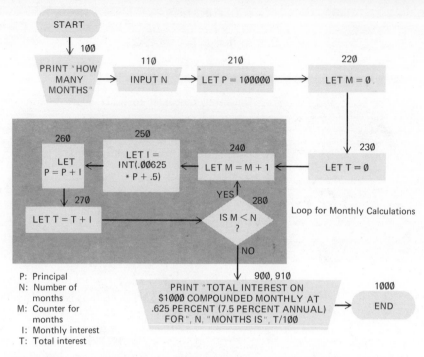

P: Principal
N: Number of months
M: Counter for months
I: Monthly interest
T: Total interest

Figure 6-6 Compound interest at .625 percent per month (Example 5).

```
1Ø    REMARK * PROGRAM TO CALCULATE THE TOTAL INTEREST DUE ON A LOAN OF
2Ø    REM *      $1ØØØ AFTER N MONTHS AT AN INTEREST RATE OF .625% PER MONTH
1ØØ   PRINT "HOW MANY MONTHS",
11Ø   INPUT N
21Ø   LET P=1ØØØØØ.
22Ø   LET M=Ø
23Ø   LET T=Ø
24Ø   LET M=M+1
25Ø   LET I=INT(.ØØ625*P+.5)
26Ø   LET P=P+I
27Ø   LET T=T+I
28Ø   IF M<N THEN 24Ø
9ØØ   PRINT "TOTAL INTEREST ON $1ØØØ COMPOUNDED MONTHLY AT .625% FOR"
91Ø   PRINT N, "MONTHS IS",T/1ØØ
1ØØØ  END
```

If we run the program twice, first with N = 8, then with N = 24, we get the following output:

```
HOW MANY MONTHS?8
TOTAL INTEREST ON $1ØØØ COMPOUNDED MONTHLY AT .625% FOR
8               MONTHS IS          51.12

HOW MANY MONTHS?24
TOTAL INTEREST ON $1ØØØ COMPOUNDED MONTHLY AT .625% FOR
24              MONTHS IS          161.31
```

RND( )

The *random number function* generates numbers that are randomly distributed on the interval between zero and one. We write RND(X) where X is any positive number. Each time RND(X) is encountered in the program, the computer calculates a new random number.

Example 6. This program causes the computer to calculate and print 25 random numbers as shown in the output. Figure 6-7 shows these numbers plotted on the interval $0 \le x \le 1$. Observe how the dangling comma is used to conserve space on the printout.

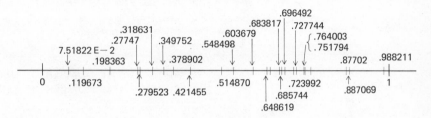

Figure 6-7 Twenty-five random numbers (Example 6).

```
1Ø   REMARK * PROGRAM TO CALCULATE 25 RANDOM NUMBERS
2Ø   LET N=Ø
3Ø   LET N=N+1          Note dangling comma
4Ø   PRINT RND(1),
5Ø   IF N<25 THEN 3Ø
6Ø   END
```

.349752	.723992	7.51822E−Ø2	.378902	.877Ø2
.988211	.656807	.751794	.683817	.27747
.887Ø69	.198363	.421455	.774579	.548498
.685744	.727744	.279523	.764ØØ3	.318631
.648619	.514876	.119673	.6Ø3679	.696492

The random number function can be used to simulate many games of chance. The following program simulates the tossing of a coin 20 times, prints the result of each toss and a final summary. Since random numbers are evenly distributed on the interval $0 \le x \le 1$, approximately half of them are less than 0.5, and the other half are greater than or equal to 0.5. In this program the computer recognizes a random number less than 0.5 as a toss of HEADS and a random number greater than or equal to 0.5 as TAILS.

Example 7. This program simulates the tossing of a coin 20 times. The flow chart for the program is Figure 6-8. The output for one run follows the program.

```
100   REMARK * BASIC COIN TOSS GAME
110   PRINT "TOSS","RESULT"
120   LET H=Ø
130   LET T=Ø
140   LET N=Ø
150   LET N=N+1
160   LET X=RND(1)
170   IF X<.5 THEN 21Ø
180   LET T=T+1
190   PRINT N, "TAILS"
200   GO TO 23Ø
210   LET H=H+1
220   PRINT N,"HEADS"
230   IF N <= 19 THEN 15Ø
240   PRINT
250   PRINT "TOTAL NUMBER OF HEADS =",H
260   PRINT "TOTAL NUMBER OF TAILS =",T
270   END
```

TOSS	RESULT
1	TAILS
2	TAILS
3	TAILS
4	TAILS
5	HEADS
6	TAILS
7	TAILS
8	HEADS
9	TAILS
1Ø	TAILS
11	TAILS
12	HEADS
13	TAILS
14	HEADS
15	TAILS
16	TAILS
17	HEADS
18	HEADS
19	HEADS
2Ø	HEADS

TOTAL NUMBER OF HEADS =	8
TOTAL NUMBER OF TAILS =	12

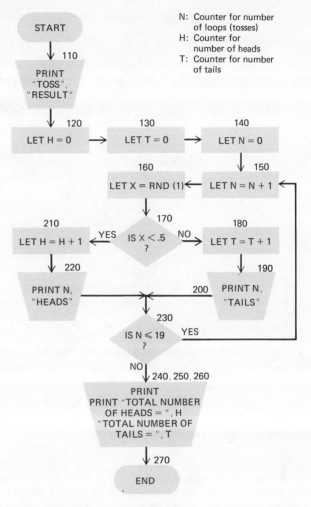

Figure 6-8 Coin-toss game. A random number less than 0.5 counts as *heads;* a random number greater than or equal to 0.5 counts as *tails* (Example 7).

EXERCISES

1. Calculate the values of the following functions. (Let the computer do your work for you, if you wish.)

   (a) SQR(25)   (e) INT(4)   (i) ABS((−2) ∗ (−3))
   (b) SQR(1)   (f) INT(−8.9)   (j) INT(2.531 + 0.5)
   (c) SQR(49)   (g) ABS(9.2 − 5.7)   (k) INT(5.496 + 0.5)
   (d) INT(5.372)   (h) ABS(−8.36)   (l) INT(−3.72 + 0.5)

2. (a) Run the program in Example 2 using your own data. Plot the numbers on a graph of the *x*-axis and verify that in each case the computer calculates the distance between the points.

(b)   (*For terminal users.*) Modify the program in Example 2 by making the following changes.
(1) Delete Statements 110, 120, 160, 170.
(2) Change 130 to an INPUT statement.
(3) Change 160 to
    160 GO TO 130

Run the modified program. Input your own data until you get tired of the program. To stop the program, type STOP *return* when the computer requests input or use the special instructions for your computer.

3. (a)   Work through the steps of the program in Example 3 several times as the computer would do. What is the relationship between the "N" of the output and the "K" of the program?
   (b)   Modify the program so that the computer will print a table of the square roots of 5.0, 5.1, 5.2, ..., 7.0. Run the modified program on the computer.
   (c)   What effect on output would result from a comma at the end of Statement 120?

4. In the text we claim that the expression

$$INT(X + 0.5)$$

rounds X off to the nearest integer. The following program can be used to test this claim in particular examples. To stop the program, type STOP at a request for input. (If you wish, you may change the INPUT statement to a READ statement and add a DATA statement.)

```
1Ø   REMARK * PROGRAM TO ROUND OFF TO THE NEAREST INTEGER
2Ø   PRINT
3Ø   PRINT "WHAT IS THE NUMBER",
4Ø   INPUT X
5Ø   PRINT "THE NEAREST INTEGER IS ",INT(X+.5)
6Ø   GO TO 2Ø
7Ø   END
```

5. (a)   Run the program in Example 5 on the computer using your own data.
   (b)   Modify the program by adding statement
   92Ø GO TO 1ØØ
   How does this change the execution of the program? Run the modified program.
   (c)   Modify the program so that the interest rate and the number of months are read from a DATA statement. Run the modified program with your own data.

6. Run the program in Example 7 several times on the computer. Do you get the same output each time?

7. Occasionally we may wish to duplicate a set of random numbers. This can be done by starting each list with RND(-1). The following program lists the same set of random numbers twice. Run the program on the computer.

```
 5Ø   REMARK * PROGRAM FOR DUPLICATING LISTS OF RANDOM NUMBERS
1ØØ   LET K=Ø
11Ø   LET K=K+1
12Ø   PRINT
13Ø   PRINT RND(−1),
14Ø   LET N=Ø
15Ø   LET N=N+1
16Ø   PRINT RND(1),
17Ø   IF N <= 23 THEN 15Ø
18Ø   IF K <= 1 THEN 11Ø
19Ø   END
```

8. Modify the program in Example 7 as follows:
   (1) You start with $10.
   (2) If the result is HEADS, you win $1. If the result is TAILS, you lose $1.
   (3) The program terminates after 20 tosses *or* when you go broke.
   (4) The final summary tells you how much you won or lost, congratulates you if you won, orders you to leave if you are broke.
   Run the modified program on the computer.

6-9 FOR . . . NEXT LOOPS

Loops are sets of instructions that are executed over and over. The loops that we have considered thus far are similar in form to

```
1ØØØ   LET N=Ø
1ØØ1   LET N=N+1
   ⋮
1999   IF N<5Ø THEN 1ØØ1
```

Statement 1000 sets up the loop; 1001 starts it; 1999 terminates it. The computer works through the steps of the loop 50 times, then goes to the next statement.

When a problem requires several loops, it is easy to make costly errors in the program—errors such as working through a loop an infinite number of times, or the wrong number of times, or mixing up the counters for two loops.

To simplify the looping process, a single instruction has been included in BASIC that allows us to specify at the beginning of a loop the number of times it is to be executed. The format of the statement is

(statement no.) FOR N=(number or variable) TO (number or variable)

The end of the loop is indicated by the statement

(statement no.) NEXT N

We are not restricted to the variable N in setting up a FOR . . . NEXT loop. Any variable that does not occur in other parts of the program can be used.

Examples of FOR . . . NEXT Loops

```
 ┌→ 100   FOR  N=1 TO 20 ⎫    "N"-loop is
 │          ⋮             ⎬    executed 20
 └─ 175   NEXT N          ⎭    times.

 ┌→ 200   FOR  K=50 TO 61 ⎫    "K"-loop is
 │          ⋮             ⎬    executed 12
 └─ 225   NEXT K          ⎭    times.

 ┌→ 300   FOR  M=L TO Z   ⎫    "M"-loop is
 │          ⋮             ⎬    executed
 └─ 500   NEXT M          ⎭    Z − L + 1 times.
```

FOR . . . NEXT Loops can be nested as the following diagram.

```
 ┌──→ 100   FOR  N=1 TO 100
 │ ┌→ 110   FOR  K=1 TO N
 │ │ ┌→120   FOR  L=1 TO K
 │ │ │          ⋮
 │ │ └─ 130   NEXT L
 │ │ ┌→140   FOR  J=1 TO 10
 │ │ │          ⋮
 │ │ └─ 150   NEXT J
 │ │            ⋮
 │ └─── 160   NEXT K
 │              ⋮
 └───── 170   NEXT N
```

FOR . . . NEXT loops cannot overlap. Loops such as

```
   100   FOR N=1 TO 10
   110   FOR K=1 TO 7
   130   NEXT N
   140   NEXT K
```

are strictly illegal.

Example 1. The following program causes the computer to print 21 random numbers, three to a line. Technically, the program calls for the computer to print seven lines (the "K-loop") with three random numbers to each line (the "M-loop").

```
   100   REMARK * PROGRAM FOR PRINTING 7 LINES OF RANDOM
         NUMBERS 3 TO A LINE
   120   FOR K=1 TO 7
   130   FOR M=1 TO 3
   140   PRINT RND(M),
   150   NEXT M
   160   PRINT
   170   NEXT K
   180   END
```

Infinite Looping

FOR . . . NEXT loops help the programmer to avoid infinite looping—one of the most serious programming errors. The following program contains an error which causes infinite looping.

```
   10   REMARK * PROGRAM WITH INFINITE LOOPING
   20   LET N=0
   30   LET N=N+1
   40   PRINT N,
   50   IF N<5 THEN 20
   60   END
```

Example 2. This program causes the computer to continually print the number 1. (If you do not see why this occurs, work through the steps of the program exactly as the computer would do.) The error is in Statement 50. The programmer should have written

$$50 \quad \text{IF } N<5 \text{ THEN } 30$$

This would have caused the computer to print

1   2   3   4   5

and then to stop. This error also could have been avoided with a FOR . . . NEXT loop. The program would have been similar to the following:

```
      1Ø   REM * CORRECTED PROGRAM
  ┌→  3Ø   FOR N=1 TO 5
  │   4Ø   PRINT N,
  └─  5Ø   NEXT N
      6Ø   END
```

EXERCISES

1. The following program demonstrates how the ordinary rules of algebra may fail on the computer. We add 1/1000 to itself 1000 times. The correct value of the sum is 1, which we do not get. This is a consequence of the way numbers are stored in the computer's memory. Run the program on your computer.

```
  1ØØ   REMARK * PROGRAM TO ADD 1/1ØØØ TO ITSELF 1ØØØ TIMES
  2ØØ   LET A=1/1ØØØ
  3ØØ   LET S=Ø
→ 4ØØ   FOR N=1 TO 1ØØØ
┌ 5ØØ   LET S=S+A
└ 6ØØ   NEXT N
  7ØØ   PRINT "1/1ØØØ ADDED TO ITSELF 1ØØØ TIMES ON THE COMPUTER"
  8ØØ   PRINT "RESULTS IN THE ANSWER",S
  9ØØ   END
```

2. Make a program containing a FOR . . . NEXT loop to print a list of 40 random numbers, 4 to each line. Run your program on the computer.
3. Make a program to print a table of *squares* and *square roots* of the numbers 0.1, 0.2, 0.3, . . . , 1.9, 2.0. Use a FOR . . . NEXT loop to control each line of output. Run your program on the computer.
4. Rewrite the coin-toss program in Example 7 of Section 6-8, using a FOR . . . NEXT loop instead of the IF . . . THEN loop. Run your program on the computer.

Project 1 TABLE OF SQUARES, CUBES, AND SQUARE ROOTS

Construct a computer program that instructs the computer to print a table of squares, cubes, and square roots for the 100 numbers 0.1, 0.2, 0.3, . . . , 9.9, 10.0. Print column headings and an appropriate title.

The first part of the output should be similar to the following sample:

TABLE OF SQUARES, CUBES, AND SQUARE ROOTS
OF NUMBERS BETWEEN .1 AND 10.0

NUMBER	SQUARE	CUBE	SQUARE ROOT
.1	.Ø1	.ØØ1	.316228
.2	.Ø4	.ØØ8	.447214
.3	.Ø9	.Ø27	.547723

Project 2   PAINTING A ROOM

What does it cost to paint a room? Obviously, we must calculate N, the *number of gallons* of paint needed, and multiply N by P, the *price per gallon*. A typical paint covers 500 square feet per gallon. Because too much paint must be purchased in preference to too little, we must always "round up" to the next largest number of gallons if the exact number is not an integer.

For example, if the area to be painted is 1200 square feet, we need approximately

$$\frac{1200}{500} = 2\frac{2}{5} \text{ gallons}$$

and we must buy 3 gallons.

If we know the length (L), width (W), and height (H), we can calculate the area of the four walls and ceiling (2LH + 2WH + LW). If we subtract from this number the area of the windows and doors that remain unpainted (U), we get the total area (A) that is to be painted.

Project

Write a program with the following steps.

(1) The computer should ask, with appropriate questions, for the following items as input:

    (a)   price of paint (P),

    (b)   length of room (L),

    (c)   width of room (W),

    (d)   height of room (H),

    (e)   total area to remain unpainted (U).

(2) The computer should calculate the following quantities:

    (a)   total area to be painted (A),

    (b)   total number of gallons of paint (P) to buy, "rounding up" to the next largest integer if P is not an integer,

    (c)   the total cost of the paint (C).

(3) The computer should print the following items *with appropriate labels:*

    (a)   the total area to be painted,

    (b)   the total number of gallons of paint to buy,

    (c)   the total cost of the paint.

Variation (*For Card Reader Use*)

Modify the program by using READ and DATA statements instead of INPUT statements.

Project 3 CLASSIFICATION OF TRIANGLES

The computer can be used to classify objects according to the data received. This problem calls for you to make a program based on the flow chart in Figure 6-9. The computer will request three positive numbers as input. It first checks to be sure that all of the numbers are positive. If so, it *interprets them as the lengths of the sides of a triangle.* If all three sides are equal, it classifies the triangle as "equilateral." If two sides are equal, the triangle is "isosceles." If no two sides are equal, the triangle is "scalene."

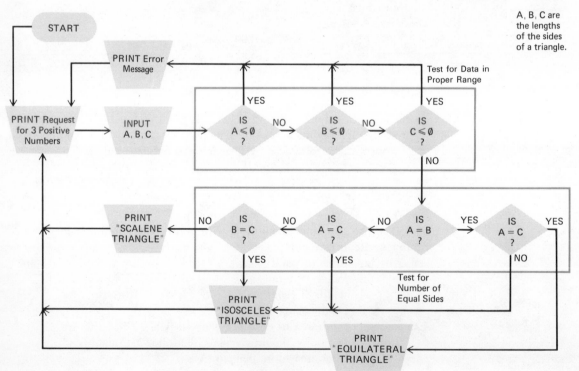

Figure 6-9 Project 3.

Project

Work through all of the steps of the flow chart. Make a program based on the flow chart. Run the program on the computer using your own data as input.

Variation (*For Facilities Using Card Readers*)

Use READ and DATA statements instead of INPUT statements. (Have at least three DATA statements each containing three numbers.) Set up a loop that will cause the computer to use all of the numbers in the DATA statements and to stop after all of the numbers have been processed.

Project 4 GRADE AVERAGING

Professor Mindboggle needs a program to average the test grades in his mathematics class. The grades will be given to the computer in DATA statements. As a signal that all grades have been read, the last "grade" is −1. The computer counts the grades (G) as they are read using N as the counter, calculates their sum (S), divides the final value of S by the final value of N, and prints the result. The flow chart of Figure 6-10 shows the steps of the program.

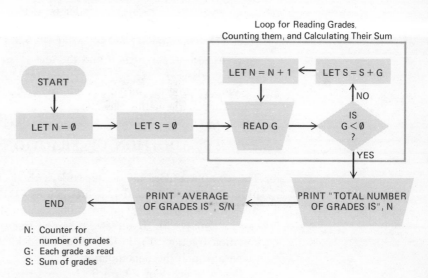

N: Counter for number of grades
G: Each grade as read
S: Sum of grades

Figure 6-10 Project 4.

Since Professor Mindboggle does not understand the computer, he has hired you to make the final program.

Project

Make a program based on Figure 6-10. Test the program with the following sample grades: 62, 93, 81, 40. Do not forget the phony "grade" of -1 after all of the grades have been entered.

Variation 1 (*Terminal Users Only*)

Modify the program so that the grades are entered as INPUT.

Variation 2

The professor also wants to know the range of grades. Modify your program so that it will print out the *highest* and *lowest* grades, with appropriate labels, as well as the *average* of the grades. (If you have trouble setting up this part of the program, study Figure 6-2.)

Variation 3

Professor Mindboggle follows a strict grading scale:

> 90–100 is A
> 80–89 is B
> 70–79 is C
> 60–69 is D
> Under 60 is F

Have the computer count and print the *total number of grades in each range* as well as the *highest* and the *lowest* grades and the *average* of all of the grades.

Run the program on your computer, using the last set of test grades from your mathematics class.

Project 5   ADDITION OF FRACTIONS

The computer does all of its work with decimal-type numbers. If we program it to add $\frac{2}{3}$ and $\frac{1}{7}$, it first changes these numbers from common fractions to a decimal-type notation before calculating the sum. In some cases we may wish to add two common fractions and print the sum as a common fraction. To do this we need to know the exact values of the numerator and the denominator.

The flow chart of Figure 6-11 shows the basic steps in a program to add common fractions. The numerators and denominators of the two fractions are read from DATA statements. (The computer recognizes them as N1, N2 and D1, D2.) The computer first checks to be sure that the

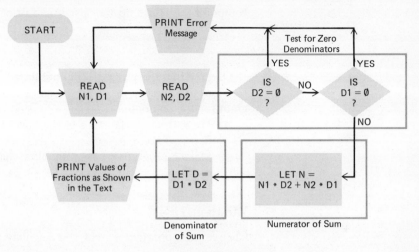

Figure 6-11 Addition of common fractions (Project 5).

denominators are not zero. Then it calculates the numerator (N) and the denominator (D) of the sum, using statements based on the rule

$$\frac{N1}{D1} + \frac{N2}{D2} = \frac{N1 * D2 + N2 * D1}{D1 * D2} = \frac{N}{D}$$

Finally, it prints out the result in the form

$$N1/D1 + N2/D2 = N/D$$

and reads new data. The program terminates when the computer runs out of data.

Note: If a teletype terminal is used, the READ statements can be changed to INPUT statements, and *semicolons* should be used in the PRINT statement instead of commas.

The PRINT statement should be similar to

PRINT N1, "/", D1, "+", N2, "/", D2, "=", N, "/", D

If the first few numbers on the data list are 3, 7, 4, 2, this statement will cause the computer to print

3	/	7	+	4
/	2	=	34	/
14				

If the semicolons are used instead of commas (by terminal users), the computer will close up the spaces and print

$$3 \quad / 7 \quad + 4 \quad / 2 \quad = 34 \quad / 14$$

Observe that the program does not instruct the computer to reduce the fractions to lowest terms. This operation is comparatively difficult to program for the computer.

(1) Work through the steps of the flow chart with the following data: 5, −1, 7, 2, 3, 0, 5, 9.

(2) Make a program based on the flow chart. Run your program on the computer.

Variation 1

Change your program so that you calculate and print the *product* of the fractions.

Variation 2

Change your program so that you calculate and print the *quotient* of the fractions.

Project 6   SIMULATION; COIN-TOSS GAME

Study the RND function before attempting this project.

This project is a gambling game simulation for *terminal users,* which pits the student (player) against the computer (banker).

At the beginning of the game, the banker welcomes the player and tells him that his original stake is $10. The player is instructed to *input* two numbers (D and B) each time the computer types a question mark. The first number, D, is the *decision:* Type 0 for a bet on HEADS, 1 for a bet on TAILS. Any other input results in an error message. The second number, B, is the *bet.* B can be any positive number less than or equal to the current value of the *player's stake* (S). A value of B that is less than or equal to zero is interpreted by the computer as a signal to terminate the game.

During play, the computer types a question mark as a signal that D and B should be typed as input. It first checks to see that B is in the proper range, then checks that D is either 0 or 1. At this point the program branches according to the value of D. In both branches the computer calculates a random number, RND(1). If RND(1) < 0.5, the toss is HEADS; if RND(1) ≥ 0.5, it is TAILS. The computer checks to see if the player won or lost, adjusts the stake accordingly, and prints the result and the new value of the stake. If the value of the stake is zero, it tells the player that he is broke and terminates the game. If the stake is not zero, the computer requests new input.

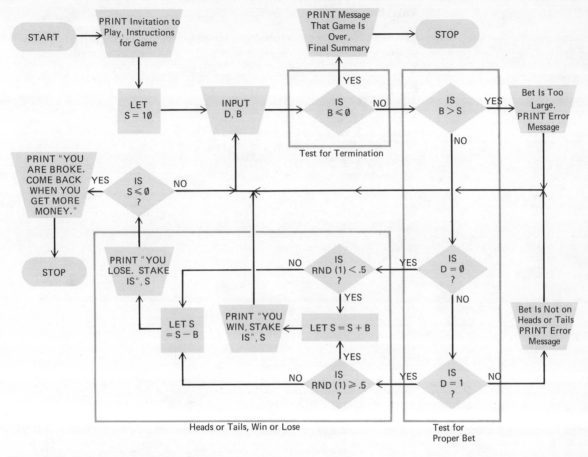

Figure 6-12 Coin-toss game (Project 6).

(1) Work through the steps of the flow chart in Figure 6-12 several times before you attempt to write the program. Make sure that you understand all of the steps.

(2) Write a program based on Figure 6-12. Add embellishments of your own to the PRINT statements that will improve the output. Have the computer print the *instructions to the player* (second box in the flow chart) as described in the second paragraph of this section. Run the program several times on the computer. Did you win or lose?

Variation 1

Set an upper limit of $500 that any player can win from the banker. Have the computer check each bet to be sure that the total winnings will not be more than $500 if the player wins. In the event that a player wins $500, have the computer print the message, "You have won the maximum amount; the game is over," and then terminate the game.

Variation 2

The flow chart (and the program based on it) can be improved by some simple changes. The fragment in Figure 6-13 begins with the box "IS B > S?" in the main flow chart. The checks to ensure that D is in the proper range are omitted—an illegal bet loses automatically. We calculate X = INT(2 * RND(1)). Then

$$X = 0 \quad \text{if} \quad \text{RND}(1) < 0.5$$
$$X = 1 \quad \text{if} \quad 0.5 \le \text{RND}(1) < 1$$
$$X = 2 \quad \text{if} \quad \text{RND}(1) = 1 \quad \text{(an event that almost never occurs)}$$

Since we can assume that X = 0 or X = 1 almost all of the time, we merely check to see if D = X. If so, the player wins. If not, he loses.

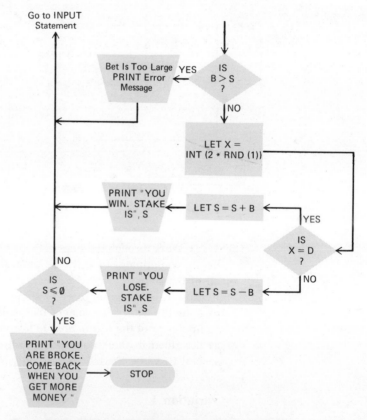

Figure 6-13 Portion of flow chart for Variation 2 (Project 6).

Make a new flow chart for the coin-toss game by incorporating the fragment in Figure 6-13 into the main flow chart. Work through the steps of the new flow chart two or three times to be sure that you understand it. Make a new program and run it several times on the computer.

Project 7 SIMULATION; DICE TOSSING

Study the INT and RND functions before attempting this section.

The computer can be used to simulate a number of different types of random events. In this project we simulate the tossing of a pair of dice.

A die is a cube with the numbers 1, 2, 3, 4, 5, 6 written in spots on the six sides. These sides are equally likely to turn up on a toss of an honest die. To simulate the tossing of a single die, we need to find a random way of selecting one of the numbers 1, 2, 3, 4, 5, 6 so that all of the numbers have an equally likely chance of being selected.

Recall that the random number function RND(1) causes numbers between 0 and 1 to be selected at random. *One-sixth* of these numbers are between 0 and $\frac{1}{6}$; *one-sixth* of them are between $\frac{1}{6}$ and $\frac{2}{6}$; *one-sixth* are between $\frac{2}{6}$ and $\frac{3}{6}$, and so on. To simulate the toss of a die, we shall let a random number between 0 and $\frac{1}{6}$ count as a toss of 1, a random number between $\frac{1}{6}$ and $\frac{2}{6}$ count as a 2, and so on. This can be accomplished directly by the expression

$$INT(6*RND(1))+1$$

As an example, suppose that RND(1) is in the range

$$2/6 \leq RND(1) < 3/6$$

Then

$$2 \leq 6*RND(1) < 3$$

so that

$$INT(6*RND(1))=2$$

and

$$INT(6*RND(1))+1=3$$

Thus, $INT(6*RND(1)) + 1 = 3$ whenever $2/6 \leq RND(1) < 3/6$. A similar argument shows that

$$INT(6*RND(1))+1=4$$

whenever $3/6 \leq RND(1) < 4/6$, and so on. The only problem that could occur is that perhaps once every billion or so times we might get RND(1) exactly equal to 1. In that case, INT $(6*RND(1)) + 1$ would equal 7. We can get around this problem by including a test statement in the program that causes the computer to recalculate INT $(6*RND(1)) + 1$ in the unlikely event it is equal to 7.

Thus, to simulate the tossing of a die, we need a pair of instructions similar to

$$
\begin{cases}
1\emptyset\emptyset & \text{LET } D1 = INT(6*RND(1)) + 1 \\
11\emptyset & \text{IF } D1 = 7 \text{ THEN } 1\emptyset\emptyset
\end{cases}
$$

These instructions cause the variable D1 to be set equal to one of the numbers 1 through 6, chosen at random.

To simulate the tossing of a pair of dice, we use two sets of such instructions, one for D1, the other for D2.

Project

Figure 6-14 outlines the steps in a program to record the outcomes of 100 tosses of the dice. Work through the flow chart two or three times, making up your own values for the random numbers. Then make a program and run it several times on the computer.

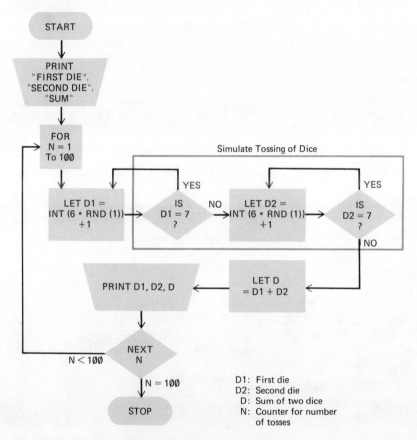

Figure 6-14 Dice toss simulation (Project 7).

Variation 1

During 100 tosses of a die, each of the numbers 1 through 6 should occur approximately 16 times. Modify the program so that the total number of times each of these numbers is tossed on the first die will be recorded and printed at the end of the run. For example, the final bit of output might be similar to the following:

```
SUMMARY OF TOSSES WITH FIRST DIE
1    TOSSED  2Ø   TIMES
2    TOSSED  13   TIMES
3    TOSSED  17   TIMES
4    TOSSED  22   TIMES
5    TOSSED  12   TIMES
6    TOSSED  16   TIMES
```

Variation 2

The total value of a toss of two dice is a number (denoted by D in the program) that is between 2 and 12. These numbers do not occur with the same frequency. For example, a *seven* occurs approximately $\frac{1}{6}$ of the time, but a *twelve* occurs only $\frac{1}{36}$ of the time. Modify the program so that the computer will print a final summary telling how many times each of these numbers has been tossed.

Project 8 RANDOM NUMBER LISTS

The Gasdrinker Engine Co. manufactures 2000 automobile engines each day. The company is required by contract to test at least 2 percent of these engines. Company officials interpret this contract to mean that Gasdrinker must test exactly 40 engines each day. The 40 test engines must be randomly selected from the 2000 engines in such a way that each engine has an equal chance of being selected. The company has hired Billy Wunderkind to make a computer program for the selection of the test engines.

Billy decides to make a daily list of integers randomly selected from the integers 1, 2, 3, . . . , 2000. His first problem is to change the random numbers produced by the computer into integers in the proper range. [Recall that RND(N) is always between 0 and 1.] An almost identical problem was solved in the dice-tossing project (Project 7). The pair of instructions

$$\begin{cases} 1\emptyset\emptyset & \text{LET } D = 1 + \text{INT}(6*\text{RND}(1)) \\ 11\emptyset & \text{IF } D = 7 \text{ THEN } 1\emptyset\emptyset \end{cases}$$

was used to produce integers randomly selected from the numbers 1, 2, 3, 4, 5, 6. He decides to use a similar pair of instructions to produce integers from the list 1, 2, 3, 4, . . . , 2000.

Billy decides that it will be easy to produce a list of 40 random integers each day, but he is concerned about the possible duplication of numbers. *How can duplicates on a list be avoided?* He thinks that duplicates will not be too frequent, but they will occur. To get around the problem, he decides to print 50 numbers for each day and use the first 40 distinct numbers. He knows that this is not a very good solution to the problem.

There is a second problem, not as important as the first. For convenience, the random numbers on each list should be in numerical order. The list order would then correspond to the assembly line order of the engines.

Project

Construct the "basic program" that will cause the computer to print five 50-number lists of random integers selected from the numbers 1, 2, 3, . . . , 2000. Include a title for the printout and individual titles for the five lists.

Variation (Difficult)

Read about "subscripted variables" in any standard textbook on BASIC. Use what you learn to modify your program so that

(1) exactly 40 numbers will be printed on each day's list, *with no duplicates;*

(2) the 40 numbers will be printed in *numerical order.*

Project 9 THE INSTALLMENT LOAN

Study the INT function before starting this project. Recall that for accuracy all monetary amounts should be expressed in *cents*. The expression

$$INT(X + 0.5)$$

can be used to round-off calculations to the nearest cent.

Joe Student, recently married, has purchased $1500 worth of furniture from the Friendly Furniture Co. Joe paid $500 down and promised to pay the remaining $1000 in equal monthly installments, with interest at the rate of 1.5 percent per month on the unpaid balance (18 percent true annual interest). Friendly Furniture has agreed to let Joe set the amount of the monthly payments. He can choose any monthly payment that he wishes to make provided that it is at least $20.

Joe is constructing a computer program that can be used to calculate the total interest he will pay and the total number of months needed to pay off the note at several different monthly payment amounts. He will enter the amount of the monthly payment as data.

The *amount* of the monthly payment will be entered in dollars and cents. An amount less than $20 is a signal to stop the program. An amount of $20 or more will be converted to cents. The computer will then set the *initial value of the principal* at 100,000 cents, the *initial value of the total interest* at 0, and the *initial value of the month counter* at 0. A loop will then be entered for the monthly calculations. For each month the computer will calculate the *interest*, add it to the *total interest* and to the *principal*, and will add 1 to the *month counter*. It then will check to see if the *principal* is less than or equal to the payment *amount*. If not, then it will subtract the *amount* from the *principal* and work through the loop again. If the *principal* is less than or equal to the *amount*, then the computer will recognize that this is the last payment and will print out (1) the *amount*, (2) *the number of months to pay off the loan*, and (3) the *total interest*, all with appropriate labels, the monetary amounts expressed in dollars and cents.

Joe has constructed a preliminary flow chart (Fig. 6-15) that outlines the basic steps in words.

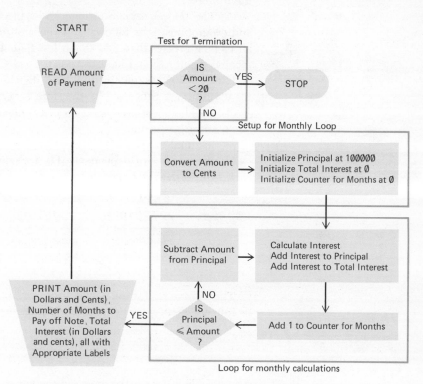

Figure 6-15   Basic steps for the installment loan problem (Project 9).

(1) Make a detailed flow chart from Figure 6-15 that breaks the steps down into those allowable in BASIC. Work through the steps of your flow chart several times.

(2) Make a program from your flow chart. Run the program on the computer with the following monthly payments:

$$\$20, \ \$30, \ \$40, \ \$50, \ \$100, \ \$150$$

Project 10 STATE INCOME TAX

Study the INT function before starting this project. Recall that for accuracy all monetary amounts should be expressed in cents. The expression

$$INT(X + 0.5)$$

can be used to round-off calculations to the nearest cent.

The State of Fetchumi has the following income tax regulations:

(1) All income, regardless of source, must be listed in the gross income.

(2) All medical expenses can be deducted; no other deductions are allowed.

(3) Each taxpayer has a $1000 exemption; each dependent gets a $700 exemption.

(4) The tax on *taxable income* (gross income *less* medical deductions and exemptions) is as follows: 1 percent of first $2000, plus 3 percent of all income greater than $2000 and less than $5000, plus 10 percent of all income greater than $5000.

Each taxpayer is identified by his 9-digit social security number. When his tax return is processed by the state, a DATA card is made up listing

(1) his identification (Social Security) number (I),

(2) his gross income (G),

(3) his total medical deductions (M),

(4) the number of dependents (D), including the taxpayer himself.

A typical data statement is

900 DATA   227361865,   20762.43,   572.65,   3
            └─────┬─────┘ └────┬────┘ └───┬──┘ └──┬──┘
             identification gross income medical number of
               number   deductions dependents

The income tax for the person listed on DATA card 900 can be calculated as follows. First, we calculate the *total exemptions:*

$$1000 + 2 * 700 = 1000 + 1400 = 2400$$

The *taxable income* is the *gross income* less *exemptions* less *medical deductions:*

gross income:		20,762.43
less exemptions:	−	2,400.00
		18,362.43
less medical deductions:	−	572.65
taxable income:		17,789.78

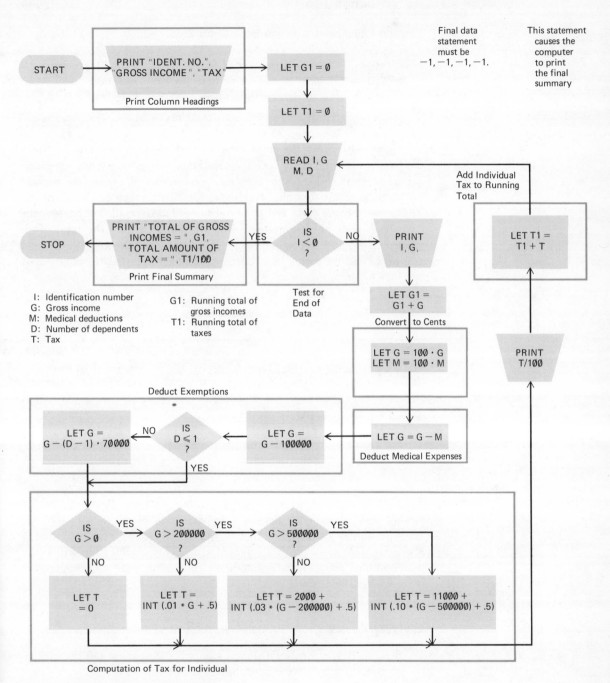

Figure 6-16 State income tax (Project 10).

The state income tax is calculated as follows:

$$
\begin{array}{rr}
1 \text{ percent of first } 2000 = & 20.00 \\
\textit{plus } 3 \text{ percent of next } 3000 = & 90.00 \\
\textit{plus } 10 \text{ percent of all over } 5000 = & 1278.98 \\
\hline
\text{total tax} & \$1388.98
\end{array}
$$

Figure 6-16 shows the steps used in calculating the income tax from the DATA statements. After each DATA statement is read, the tax is computed and a line is typed listing the identification number, the gross income, and the amount of tax. Running totals also are kept by the computer of the total gross income (G1) and the total tax (T1). After all data statements have been read, the computer prints a summary listing the total of all gross incomes and the total amount of income tax collected. As a signal that all of the tax forms have been processed, the *last* data statement is

$$9999 \quad \text{DATA} \; -1, \; -1, \; -1, \; -1$$

Project

Make a program based on Figure 6-16. Run the program with at least four DATA statements in addition to the "final" DATA statement that signals the computer that the data have ended. *Caution:* Be sure to use a dangling comma on the PRINT statement

$$\text{PRINT I, G,}$$

Variation

Modify the program so that the computer will check that I, $100 * G$, $100 * M$, and D are integers, that G and M are nonnegative, and that D is greater than or equal to 1 (unless $I < 0$). This modification will cause the computer to reject data statements containing many of the standard typographical errors.

SUGGESTIONS FOR FURTHER READING

1. Bell, E. T. *Men of Mathematics.* Chap. 5, "Greatness and the Misery of Man" (Pascal). New York: Simon & Schuster, 1937.
2. Eames, Charles, and Eames, Ray. *A Computer Perspective.* Cambridge, Mass.: Harvard University Press, 1973.
3. Hawkes, Nigel. *The Computer Revolution.* World of Science Library. New York: Dutton, 1972.

4. Kemeny, John. *Man and the Computer*. New York: Scribner, 1972.
5. Morrison, Philip, and Morrison, Emily. "The Strange Life of Charles Babbage," *Mathematics in the Modern World*, Chap. 8. Readings from *Scientific American*. San Francisco: Freeman, 1968.
6. Nolan, Richard L. *Introduction to Computing Through the BASIC Language*. New York: Holt, Rinehart & Winston, 1969.

7

NONLINEAR RELATIONSHIPS

7-1 EXAMPLES OF NONLINEAR RELATIONSHIPS

In our previous work with algebra (Chapter 5), we dealt mainly with *linear equations*—equations of form

$$y = mx + b$$

where m and b are constants. The graphs of these equations are lines in the xy-plane (Fig. 7-1). Most of the mathematical relationships that we encounter in practical problems, however, involve *nonlinear equations*—the graphs are not straight lines.

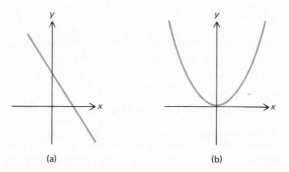

(a)   (b)

Figure 7-1 (a) Linear relationship. The graph is a line; (b) nonlinear relationship. The graph is a curve.

Example 1. Population Growth Figure 7-2(a) shows the population of the United States as tabulated in the census years since 1790. In Figure 7-2(b) the points are connected by line segments. In Figure 7-2(c) the points are approximated by a smooth curve. These graphs are not straight lines. Thus, the population growth relationship is nonlinear. The two graphs can be used to estimate the population at any time since 1790.

Example 2. Population Growth A population of bacteria will double in size over a fixed period of time, provided its environmental conditions do not change. A certain colony of a bacteria strain, which finds the human body a suitable habitat, is now residing in both lungs of George Octogenarian. At the present time, only 2 million bacteria are in his lungs. This number doubles every 6 hours. Assuming that George does not die or take antibiotics or change the environment for the bacteria in any other way, how many bacteria will live in his lungs after 24 hours?

Solution. The number of bacteria doubles every 6 hours:

Time	Number of Bacteria
Present	2 million
6 hours	4 million
12 hours	8 million
18 hours	16 million
24 hours	32 million

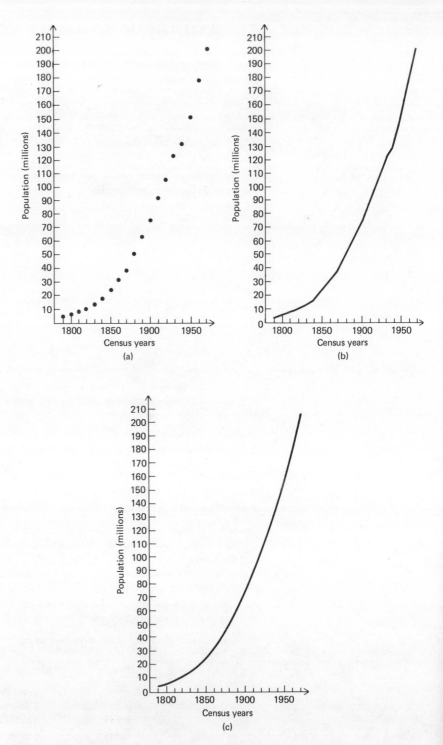

Figure 7-2 U.S. population in census years (Example 1). (a) Population in census years; (b) broken-line graph; (c) smooth curve that approximates the population.

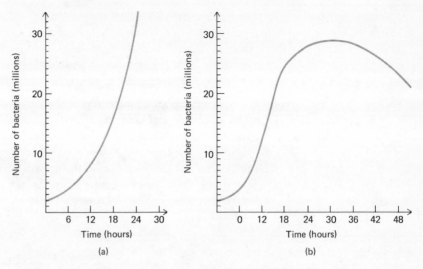

Figure 7-3 Population of bacteria (Example 2). (a) Typical growth curve during the early period of growth; (b) growth curve influenced by taking antibiotics.

Thus, at the end of 24 hours, approximately 32 million bacteria will live in poor old George.

Figure 7-3(a) pictures the situation graphically. We can use the graph to estimate the number of bacteria in George's lungs at any time. For example, at the end of 15 hours, there will be approximately 11.3 million bacteria.

Growth Rates

There are a number of different formulas for predicting future population sizes. All of these lead to different estimates. The simplest of the formulas is based on a constant growth rate. If the growth rate remains the same over a large number of years, then the size of the population can be predicted with great accuracy as in Example 2. The more complicated formulas attempt to predict the ways the growth rate will change and use the different growth rates to predict future population sizes.

Based on the high growth rates following the Civil War, it was predicted that the U.S. population would be 186 million by the year 1930 (a figure that was 50 percent too high).

Based on the low birthrates during the Depression, it was predicted that the U.S. population would be less than 160 million by the year 1970 (a figure that was 20 percent too low).

Any estimate of future population size should be treated with great suspicion. Unless we know which formula was used and which assumptions were made, we cannot tell if the estimate is a reasonable one or not. It may be as inaccurate as the ones made in 1870 and in 1930.

Remark: It is obvious that the situation discussed in Example 2 cannot continue indefinitely. In a very short time, the conditions for growth must change—George will either become ill and die or his body will produce antibodies that will attack the bacteria. Consequently, the doubling of the bacteria every 6 hours only holds for a brief time period—two or three days at the most. If George takes antibiotics at the beginning of the infection, then the bacterial growth follows a curve similar to the one in Figure 7-3(b).

A typical population curve is shown in Figure 7-4. The population increases more and more rapidly until it reaches one-half of its maximum possible value; then the growth rate begins to slow down and the population eventually stabilizes. This law appears to govern all populations—plant and animal. At the beginning of the growth period, there is abundant space and ample food for everyone. When the population nears its theoretical maximum, it is inhibited by scarcities of food, overcrowded conditions, and so forth.

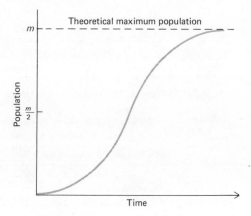

Figure 7-4 Population curve. Typical growth curve over a large period of time. The population increases rapidly until it reaches one-half of its maximum value. Then the rate of growth slows down and eventually becomes almost zero near the maximum population.

Example 3. Inflation The policies of the United States government encourage an inflation rate of approximately 2 percent per year. William Barrister, age 30, currently has a gross income of $20,000. Assume that the inflation rate will average 2 percent per year for the next 30 years. How much must William earn at age 60 in order to have an income equivalent to his current income?

Solution. Each year he must earn 2 percent more than the preceding year. To remain even in gross salary, he must earn the following amounts:

Example 3 is concerned only with gross salary, not net salary (after taxes). If we take the progressive nature of the federal income tax into account, then in 30 years of 2 percent yearly inflation (based on current tax rates) Barrister would have to earn about $45,000 per year in order to have an equivalent net income.

Year	*Salary*
Now	20,000
1	$20,000 + 0.02(20,000) = 20,400$
2	$20,400 + 0.02(20,400) = 20,808$
3	$20,808 + 0.02(20,808) = 21,224$

$$4 \qquad\qquad 21{,}224 + 0.02(21{,}224) = 21{,}648$$
$$\vdots \qquad\qquad\qquad\qquad \vdots$$
$$30 \qquad\qquad 35{,}517 + 0.02(35{,}517) = 36{,}227$$

Thus, William must earn more than \$36,000 per year when he is 60 years old in order to have an income equivalent to his current income of \$20,000 per year.

The graph in Figure 7-5 shows the salaries that William must earn each year. The points obtained in the above calculations were connected with a smooth curve.

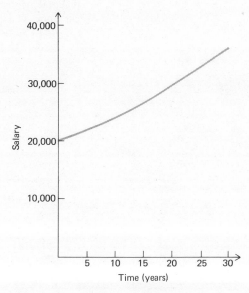

Figure 7-5 Salary curve (Example 3). This curve shows the cash income equivalent to a current income of \$20,000, assuming an inflation rate of 2 percent per year.

Example 4. Braking Distance The braking distance of a car is the shortest distance in which the car can stop after the brakes are applied. The braking distance depends on many factors, including the weight and speed of the car, the type and conditions of the brakes, and the amount of tread on the tires.

The actual braking distance of a car can only be determined experimentally. We brake the car at a convenient speed and measure the braking distance. We then can calculate the braking distances for various other speeds. The key factor is given in the following rule: *If the speed is doubled, the braking distance is multiplied by 4; if the speed is tripled, the braking distance is multiplied by 9, and so on.*

When the speed is *doubled,* the braking distance is *quadrupled.* When the speed is *tripled,* the braking distance is multiplied by *nine.*

A certain car has a braking distance of 20 feet at 20 miles per hour. Construct a graph showing the braking distance for all speeds up to 80 miles per hour.

Solution. We plot the braking distance *y* against the speed *x*. First, we calculate a few braking distances:

speed = 0:   If the car is at rest, the braking distance obviously is 0. Thus, the point $(0, 0)$ is on the graph.

speed = 20:   We are given that the braking distance is 20 feet at 20 miles per hour. The point (20, 20) is on the graph.

speed = 40:   If the speed is doubled, the braking distance is multiplied by 4. Thus, the braking distance at $x = 40$ miles per hour is $y = 4 \cdot 20 = 80$ feet. The point (40, 80) is on the graph.

speed = 60:   If the speed is tripled, the braking distance is multiplied by 9. Thus, the braking distance at $x = 60$ miles per hour is $y = 9 \cdot 20 = 180$ feet. The point (60, 180) is on the graph.

speed = 80:   The speed is double the speed at 40 miles per hour. Consequently, this braking distance is 4 times the braking distance at 40 miles per hour. Thus, the braking distance at $x = 80$ miles per hour is $y = 4 \cdot 80 = 320$ feet. The point (80, 320) is on the graph.

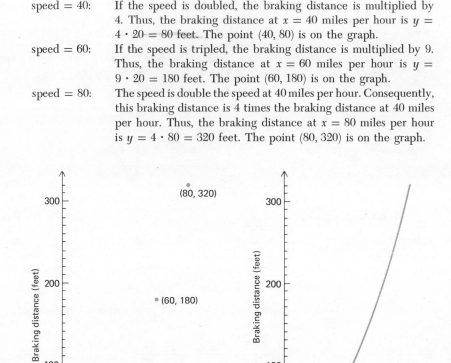

Figure 7-6   Braking distance (Example 4). (a) Points on the curve; (b) the braking distance curve. This curve is based on a braking distance of 20 feet at 20 miles per hour.

The points calculated above are shown in Figure 7-6(a). In Figure 7-6(b) the points are connected by a smooth curve, which can be used to estimate the braking distance at any speed up to 80 miles per hour. For example, at 70 miles per hour the braking distance is approximately 245 feet.

EXERCISES

1. Plot enough points to determine the shape of each of the following graphs. Then connect up the points with a smooth curve.

(a) $y = x^2$ (d) $y = 1/x$ ($x < 0$ only)

(b) $y = -\frac{1}{2}x^2$ (e) $y = 2^x$

(c) $y = 1/x$ ($x > 0$ only)   (f) $y = 3^x$

2. Make a copy of the population graph in Figure 7-2(c). Extend this graph to cover future years, *using the graph in Figure 7-2(c) as a pattern*. What is your estimate for the population of the United States in the year 2000? 2050? (See Example 1.)

3. Make a copy of the population graph in Figure 7-2(c). Extend this graph to cover future years, *using the graph in Figure 7-4 as a pattern*. What is your estimate for the population of the United States in the year 2000? 2050? (See Example 1.)

4. During recent years the United States government has not been successful in holding the inflation rate to 2 percent per year. In some years rates have been 10 percent or higher. Assume that the inflation rate will be 4 percent per year for the next 20 years. At the present time the average gross salary in the United States is approximately $300 per week. What must be the average salary in 20 years in order to have a gross salary equivalent to the present $300 per week? (See Example 3.)

5. Use the graph in Figure 7-6 to estimate the braking distance at the following speeds (Example 4):

(a) 10 miles per hour   (c) 50 miles per hour

(b) 30 miles per hour   (d) 90 miles per hour

6. Assume that your car has a braking distance of 4 feet at 10 miles per hour. Make a graph similar to Figure 7-4 showing the braking distance at all speeds up to 90 miles per hour (Example 4).

7-2 POLYNOMIALS; FACTORING

Algebraic expressions such as

$$3x^2 - 4x + 7$$
$$x + 2$$
$$5x^3 + x - 13$$

and

$$27x^{15} + 5x^{14} + 3x^9 - 7x^4 + 2x^2 - 5x + 7$$

are called *polynomials*. A polynomial is a sum of multiples of powers of x. In the first of the above examples, the polynomial $3x^2 - 4x + 7$ (which can be written as $3x^2 - 4x^1 + 7 \cdot x^0$) is a sum of multiples of x^2, x^1, and x^0. The second polynomial, $x + 2$ ($= x^1 + 2 \cdot x^0$), is a sum of multiples of x^1 and x^0. Each of the exponents of x is a nonnegative integer.

The highest power of x that occurs in a polynomial is called the *degree* of the polynomial. For example,

the polynomial $3x^2 - 4x + 7$ has degree 2;
the polynomial $x + 2$ has degree 1;
the polynomial $5x^3 + x - 13$ has degree 3;

and so on. First-degree polynomials are called *linear polynomials;* second-degree polynomials are called *quadratic polynomials;* third-degree polynomials are called *cubic polynomials;* fourth-degree polynomials are called *quartic polynomials,* and so on.

Example 1.
(a) The polynomial $3x - 4$ is linear. It has degree 1.
(b) The polynomial $x^2 + 5x + 3$ is quadratic. It has degree 2.
(c) The polynomial $17x^3 - 8x^2 + 0 \cdot x - 4$ is cubic. It has degree 3.
(d) The polynomial $x^4 + 15x^3 - 9x^2 + 7x - 2$ is quartic. It has degree 4.

On occasion we shall find it necessary to factor some simple polynomials. Given a quadratic polynomial such as

$$6x^2 + 31x + 35$$

we wish to find two linear polynomials which have the quadratic polynomial as their product. In this particular example, we might find by trial and error that

$$\begin{aligned}(3x + 5)(2x + 7) &= 3x(2x + 7) + 5(2x + 7) \\ &= 6x^2 + 21x + 10x + 35 \\ &= 6x^2 + 31x + 35\end{aligned}$$

so that

$$6x^2 + 31x + 35 = (3x + 5)(2x + 7)$$

The terms $3x + 5$ and $2x + 7$ are called the *factors* of $6x^2 + 31x + 35$.

The factoring that we shall do in this book will be minimal. We do not expect every reader to become proficient with the technique. We do, however, expect the reader to acquire enough of an understanding of the principles to be able to follow simple arguments in which factoring is used.

The basic techniques can best be illustrated by factoring a particular polynomial. For our example we factor the quadratic polynomial

$$x^2 - 5x + 6$$

If $x^2 - 5x + 6$ can be factored in a meaningful way, then each factor must be a linear polynomial. Let us suppose that the two factors are

$$ax + b \quad \text{and} \quad cx + d$$

where a, b, c, and d represent numbers that we must find. For example, it is conceivable that a might be 1 and b be 6. Then $ax + b$ would be $x + 6$.

If $ax + b$ and $cx + d$ are to be factors of $x^2 - 5x + 6$, then

$$(ax + b)(cx + d) = x^2 - 5x + 6$$

If we multiply the two expressions on the left and compare the coefficients with those of original polynomial, we find that

$$ac = 1 \quad \text{and} \quad bd = 6$$

This means that a must be a factor of 1, and b must be a factor of 6. If $a, b, c,$ and d are integers, then a must be either 1 or -1, and b must be one of the numbers $1, -1, 2, -2, 3, -3, 6,$ or -6.

Observe also that if we know a and b, then we can compute c and d. We have

$$ac = 1 \quad \text{and} \quad bd = 6$$

so that

$$c = \frac{1}{a} \quad \text{and} \quad d = \frac{6}{b}$$

Thus, for example, if $a = 1$ and $b = 6$, then $c = 1/1 = 1$ and $d = 6/6 = 1$. If that were the case, $ax + b$ would be $x + 6$, and $cx + d$ would be $x + 1$.

To actually find $ax + b$ and $cx + d$, we resort to trial-and-error methods. We select a pair of possible values of a and b, calculate the corresponding values of c and d, multiply $ax + b$ and $cx + d$ together and see if we get

$$x^2 - 5x + 6$$

for the product. If so, then we have found the factors; if not, then we try another pair of values for a and b.

Recall that a must be 1 or -1, that b must be $1, -1, 2, -2, 3, -3, 6,$ or -6, that

$$c = \frac{1}{a} \quad \text{and} \quad d = \frac{6}{b}$$

We now test several possible values of a and b.

(1) If $a = 1$ and $b = 1$, then $c = 1$ and $d = 6$. The possible factors are

$$ax + b = x + 1 \qquad \text{and} \qquad cx + d = x + 6$$

The product is

$$(x + 1)(x + 6) = x^2 + 7x + 6$$

Since the product is not equal to the original polynomial, then $x + 1$ and $x + 6$ are not factors of $x^2 - 5x + 6$.

(2) If $a = 1$ and $b = -1$, then $c = 1$ and $d = -6$. The possible factors are

$$ax + b = x - 1 \qquad \text{and} \qquad cx + d = x - 6$$

The product is

$$(x - 1)(x - 6) = x^2 - 7x + 6$$

Since the product is not equal to the original polynomial, then $x - 1$ and $x - 6$ are not factors of $x^2 - 5x - 6$.

(3) If $a = 1$ and $b = 2$, then $c = 1$ and $d = 3$. The possible factors are

$$ax + b = x + 2 \qquad \text{and} \qquad cx + d = x + 3$$

The product is

$$(x + 2)(x + 3) = x^2 + 5x + 6$$

Since the product is not equal to the original polynomial, then $x + 2$ and $x + 3$ are not factors of $x^2 - 5x + 6$.

(4) If $a = 1$ and $b = -2$, then $c = 1$ and $d = -3$. The possible factors are

$$ax + b = x - 2 \qquad \text{and} \qquad cx + d = x - 3$$

The product is

$$(x - 2)(x - 3) = x^2 - 5x + 6$$

Since the product is equal to the original polynomial, we have found the factors of $x^2 - 5x + 6$. These factors are $x - 2$ and $x - 3$.

As we can see from the above example, factoring involves a great deal of trial-and-error work. We list all of the possible algebraic expressions that could be factors, pair each of them with the other possible factors, then multiply the pairs of possible factors together. If one of the products is equal to the original expression, then we have a factorization.

Example 2. Factor $x^2 - 4x + 3$.

Solution. If $ax + b$ and $cx + d$ are the factors, then

$$(ax + b)(cx + d) = x^2 - 4x + 3$$
$$acx^2 + (ad + bc)x + bd = x^2 - 4x + 3$$

so that

$$ac = 1$$
$$ad + bc = -4$$
$$bd = 3$$

Then a must be a factor of 1 so that $a = 1$ or -1, and b must be a factor of 3 so that $b = 1, -1, 3,$ or -3. Furthermore,

$$c = \frac{1}{a}$$

and

$$d = \frac{3}{b}$$

If $a = 1$ and $b = 1$, then $c = 1$ and $d = 3$. The possible factors are $x + 1$ and $x + 3$. The product of these terms is

$$(x + 1)(x + 3) = x^2 + 4x + 3$$

which is not equal to the original expression.

If $a = 1$ and $b = -1$, then $c = 1$ and $d = -3$. The possible factors are $x - 1$ and $x - 3$. The product of these two terms is

$$(x - 1)(x - 3) = x^2 - 4x + 3$$

Since the product is equal to the original polynomial, then $x - 1$ and $x - 3$ are the factors.

Suppose we start with the polynomial $x^2 - 5$ and attempt to factor it as in the above examples. We shall find that the possible factors are $x - 1$, $x + 1, x + 5,$ and $x - 5$ and that none of these is a factor. Does this mean that the original polynomial cannot be factored? We shall see that it does not—there does exist a factorization. It is necessary, however, to have irrational numbers as some of the coefficients of the factors. In this particular problem, it can be shown that the factorization is

$$x^2 - 5 = (x - \sqrt{5})(x + \sqrt{5})$$

Recall that $\sqrt{5}$ is the positive square root of 5, the positive number which when squared is equal to 5. The two square roots of 5 are

$$\sqrt{5} \approx 2.236$$

and

$$-\sqrt{5} \approx -2.236$$

It is proved in more advanced courses that every polynomial is, in theory, a product of linear factors. The coefficients of these factors may not be real numbers, however. Even if the coefficients of the linear factors are real numbers, it may not be possible to calculate them exactly. We may be forced to use approximations to the actual coefficients.

This example is considered further in Example 4.

In general, the method that we used in Example 2 can be used to find all of the factors that have *rational coefficients*. If no such factors exist, it still may be possible to factor the expression into terms that have irrational coefficients. This is the situation with the polynomial $x^2 - 5$. The factors have irrational coefficients.

Differences of Two Squares

Polynomials of form $x^2 - a^2$, where a is a number, can be factored very easily. The factors are $x - a$ and $x + a$. This can be seen by multiplying the two factors together:

$$(x - a)(x + a) = x(x + a) - a(x + a)$$
$$= x^2 + ax - ax - a^2$$
$$= x^2 - a^2$$

Example 3. (a) $x^2 - 3^2 = (x - 3)(x + 3)$
(b) $x^2 - 16 = x^2 - 4^2 = (x - 4)(x + 4)$
(c) $x^2 - 25 = (x - 5)(x + 5)$

Example 4. Factor
(a) $x^2 - 5$
(b) $x^2 - 17$

Solution. (a) Write 5 as $5 = (\sqrt{5})^2$. Then

$$x^2 - 5 = x^2 - (\sqrt{5})^2 = (x - \sqrt{5})(x + \sqrt{5})$$

(b) Write 17 as $17 = (\sqrt{17})^2$. Then

$$x^2 - 17 = (x - \sqrt{17})(x + \sqrt{17})$$

EXERCISES

1. Identify the following polynomials as *linear, quadratic, cubic,* or *quartic.* What is the degree of each polynomial?
   (a) $x^2 - 7x + 2$
   (b) $3 - x + x^4 - 4x^3$
   (c) $\pi x^4 - 7x^3 + 3x + 2$
   (d) $x^4 + 2x^3 - x^4 + 5x^2 + 3x$
   (e) $ax + b$ (a, b are numbers)
   (f) $ax^3 + bx^2 + cx + d$ (a, b, c, d are numbers)

2. Factor the following polynomials.

(a) $x^2 - 5x$ (e) $x^2 - x - 12$
(b) $x^2 + 3x + 2$ (f) $15x^2 - 23x - 28$
(c) $x^2 + 3x - 10$ (g) $4x^2 - 15x - 4$
(d) $x^2 + 2x - 3$ (h) $x^2 + 2x - 63$

3. Factor the following differences of squares.

(a) $x^2 - 8^2$ (e) $y^2 - 7$
(b) $x^2 - 1$ (f) $y^2 - 3$
(c) $x^2 + (-81)$ (g) $t^2 - 8$
(d) $y^2 - 9$ (h) $t^2 - 13$

7-3 QUADRATIC EQUATIONS; SOLUTIONS BY FACTORING

An equation that can be written in the form

$$ax^2 + bx + c = 0$$

where a, b, and c are numbers, $a \neq 0$, is called a *quadratic equation*.
For example,

$$2x^2 - 3x + 7 = 0$$
$$x^2 + 5x - \pi = 0$$

and

$$3x^2 + 9x - 4 = 0$$

are quadratic equations. The equation

$$x^4 - 2x^2 + 4x - 7 = x^4 + 3x^2 - 8x + \pi$$

also is a quadratic equation because it can be rewritten as follows:

$x^4 - 2x^2 + 4x - 7 = x^4 + 3x^2 - 8x + \pi$ (original equation)
$x^4 - 2x^2 + 4x - 7 - x^4 - 3x^2 + 8x - \pi = 0$ (subtract the terms on the
$-5x^2 + 12x + (-7 - \pi) = 0$ right-hand side from both
sides of the equation)

Quadratic equations occur in a great many problems. They can be solved with ease, provided we can factor the left-hand side into two linear factors.

The key result needed to solve quadratic equations by factoring rests on the fact that *if a product of two numbers is zero, then at least one of the factors must be zero*. If $ab = 0$, where a and b are numbers, then a or b (or both) must be equal to zero. Stated another way, *if a and b are not zero, then their product ab cannot be zero*.

Example 1. Solve the equation

$$x^2 - 3x = 0$$

Solution. We can factor the left-hand side of the equation as

$$x(x - 3) = 0$$

The symbol x represents a number. Thus, we have a product of two numbers, x and $x - 3$, which is equal to zero. Since a product can be zero only in case one (or more) of the factors is zero, then we must have

$$x = 0 \quad \text{or} \quad x - 3 = 0$$
$$x = 0 \quad \text{or} \quad x = 3$$

POLYNOMIAL: X↑2 – 7*X + 5 = Ø
SOLUTIONS: X1 = .8Ø742,
X2 = 6.19258

There are several methods for approximating the solutions of a polynomial equation to any desired degree of accuracy. A number of these methods have been programmed for the computer. The solutions obtained by these programs are approximate.

The above argument shows that the only possible solutions of the equation are $x = 0$ and $x = 3$. In other words, we have limited the possible solutions to these numbers. No other numbers can be solutions. We now must substitute them into the left-hand side of the equation and see if they actually are solutions:

$$x = 0: \quad x^2 - 3x = 0^2 - 3 \cdot 0 = 0 - 0 = 0$$
$$x = 3: \quad x^2 - 3x = 3^2 - 3 \cdot 3 = 9 - 9 = 0$$

Thus, $x = 0$ and $x = 3$ are solutions of the equation.

Example 2. Solve the equation

$$x^2 - 2x = 2x - 3$$

Solution. We first rewrite the equation with all of the terms on the left-hand side, then factor it:

$$x^2 - 2x = 2x - 3$$
$$x^2 - 2x - 2x + 3 = 0 \quad \text{(subtract } 2x - 3 \text{ from both sides)}$$
$$x^2 - 4x + 3 = 0$$
$$(x - 3)(x - 1) = 0 \quad \text{(factor the quadratic)}$$

$x^2 - 7x + 5 = 0,$
$$x_1 \approx 0.80742$$
$$x_2 \approx 6.19258$$
$x^2 - 7x + 5$
$\approx (x - 0.80742)(x - 6.19258)$

Since a product can be zero only in case one of the factors is zero, we must have

$$x - 3 = 0 \quad \text{or} \quad x - 1 = 0$$
$$x = 3 \quad \text{or} \quad x = 1$$

There is a natural correspondence between the factors of a polynomial and the solutions of the corresponding polynomial equation. If we know the factors of the polynomial, we can solve the equation. Similarly, if we know the solutions of the equation, we can factor the polynomial.

The only possible numbers that can be solutions are $x = 3$ and $x = 1$. If we substitute these values into the left- and right-hand sides of the equation, we find that they actually are solutions. Therefore, the solutions are

$$x = 3 \quad \text{and} \quad x = 1$$

In order to apply factoring to the solution of equations, it is necessary to have an algebraic expression equal to zero. Thus, in Example 2 we had to rewrite the equation as

$$x^2 - 4x + 3 = 0$$

before we could solve it. We could not factor the two sides of the original equation and equate the factors, for the results would have been meaningless.

The procedure can be extended to algebraic expressions with any number of factors.

Example 3. Solve the equation

$$(x^2 - 7x)(x - 2)(x^2 + 5x + 6) = 0$$

Solution. We can factor the left-hand side further if we note that

$$x^2 - 7x = x(x - 7)$$

and

$$x^2 + 5x + 6 = (x + 2)(x + 3)$$

Thus,

$$(x^2 - 7x)(x - 2)(x^2 + 5x + 6) = 0$$
$$x(x - 7)(x - 2)(x + 2)(x + 3) = 0$$

Since at least one of these factors must be zero, we must have

$$x = 0 \quad \text{or} \quad x - 7 = 0 \quad \text{or} \quad x - 2 = 0 \quad \text{or} \quad x + 2 = 0 \quad \text{or} \quad x + 3 = 0$$
$$x = 0 \quad \text{or} \quad x = 7 \quad \text{or} \quad x = 2 \quad \text{or} \quad x = -2 \quad \text{or} \quad x = -3$$

It can be verified by substitution that all of these numbers are solutions of the original equation. Therefore, the solutions are

$$x = -3, -2, 0, 2, 7$$

Differences of Two Squares

As we saw in the preceding section, a difference of two squares can be factored:

$$x^2 - a^2 = (x - a)(x + a)$$

It follows that the equation

$$x^2 - a^2 = 0$$

can be solved by the procedures of this section.

Example 4. Solve the following equations

(a) $x^2 - 16 = 0$

(b) $x^2 - 5 = 0$

(c) $y^2 = 3$

(d) $(x + 1)^2 = 9$

Solution. (a)
$$x^2 - 16 = 0$$
$$(x - 4)(x + 4) = 0$$
$$x = 4 \quad \text{or} \quad x = -4$$

(b)
$$x^2 - 5 = 0$$
$$x^2 - (\sqrt{5})^2 = 0$$
$$(x - \sqrt{5})(x + \sqrt{5}) = 0$$
$$x = \sqrt{5} \quad \text{or} \quad x = -\sqrt{5}$$

(c)
$$y^2 = 3$$
$$y^2 - 3 = 0$$
$$y^2 - (\sqrt{3})^2 = 0$$
$$(y - \sqrt{3})(y + \sqrt{3}) = 0$$
$$y = \sqrt{3} \quad \text{or} \quad y = -\sqrt{3}$$

(d)
$$(x + 1)^2 = 9$$
$$(x + 1)^2 - 9 = 0$$
$$(x + 1)^2 - 3^2 = 0$$

Let $y = x + 1$. The equation becomes

$$y^2 - 3^2 = 0$$
$$(y - 3)(y + 3) = 0$$
$$y = 3 \quad \text{or} \quad y = -3$$

Therefore, since $y = x + 1$,

$$x + 1 = 3 \quad \text{or} \quad x + 1 = -3$$
$$x = 2 \quad \text{or} \quad x = -4$$

It can be verified by substitution that the numbers obtained actually are solutions of the equations.

The procedures used in this section will enable us to solve any equation that can be factored completely. We shall find, however, that some equations cannot be factored by the techniques that we developed in Section 7-2. For example, the left-hand sides of the equations

$$x^2 + 2x + 6 = 0$$

and

$$x^2 + 7x - 9 = 0$$

cannot be factored by these methods. We shall see later, however, that these equations do have solutions. (In fact, they can be factored by more advanced methods.) The basic conditions that separate equations into those that can be factored and those that cannot are concerned with the nature of the solutions. If the solutions of a quadratic equation are rational numbers, then it can be solved by a simple factorization of the type considered in Section 7-2. If the solutions are not rational, then a different type of factorization will be necessary. It may be necessary to rewrite the equation as a difference of two squares, or apply some other method.

EXERCISES

1. Solve the following equations.
   (a) $x(x - 4) = 0$ (e) $(x - 81)(2x + 5) = 0$
   (b) $(x + 3)x = 0$ (f) $(3x + 8)(5x + 7) = 0$
   (c) $(x + 1)(x - 5) = 0$ (g) $x(x - 1)(x - 3) = 0$
   (d) $(x + 7)(x + 9) = 0$ (h) $(x - 1)(x - 1)(2x + 3)(5x - 4) = 0$
2. Solve the following equations.
   (a) $x^2 - 3x + 2 = 0$ (e) $x^2 - 4x + 4 = 12 - 2x$
   (b) $x^2 + 8x + 7 = 0$ (f) $2x^2 - 2x = x^2 - 1$
   (c) $2x^2 - 9x - 5 = 0$ (g) $(x - 7)(x + 1) = -2(x + 5)$
   (d) $6x^2 + 7x - 3 = 0$ (h) $x^2 + 3x + 2 = -1 \cdot (x - 7)$
3. Use the factorization $x^2 - a^2 = (x - a)(x + a)$ to solve the following equations. [*Hint:* In (g) let $x = y + 1$.]
   (a) $x^2 - 4 = 0$ (f) $x^2 = 7$
   (b) $t^2 - 25 = 0$ (g) $(y + 1)^2 = 8$
   (c) $y^2 = 9$ (h) $(2z - 1)^2 = 49$
   (d) $u^2 = 16$ (i) $(x + 3)^2 = 8$
   (e) $v^2 = 18$ (j) $(2x + 5)^2 = 25$

7-4 FUNCTIONS; THE PARABOLA $y = ax^2$

There are many important nonlinear relationships in mathematics. The next few sections of this chapter will be devoted to a study of one of these relationships—that one defined by equations of form

$$y = x^2, \qquad y = 2x^2, \qquad y = 3x^2, \ldots$$

Observe that in each of the above equations, we can calculate the exact value y if we know x. Such a relationship is called a *function*. We say that *y is a function of x*. To be more precise, a mathematical function is a relationship involving two variables, x and y, which can be used to calculate y for a given value of x.

We have encountered several special functions in this book—the trigonometric functions sin, cos, tan, and so on, in Chapter 4, and the computer functions INT, RND, SQR, and ABS in Chapter 6.

Consider the relationship defined by

$$y = x^2$$

If we have any given value of x, we can calculate the corresponding value of y. For example,

$$\text{if } x = 2 \quad \text{then } y = 2^2 = 4$$
$$\text{if } x = -1 \quad \text{then } y = (-1)^2 = 1$$

Since we can calculate a value of y for any value of x, then the relationship

$$y = x^2$$

defines a mathematical function. The variable y is a function of the variable x.

THE PARABOLA

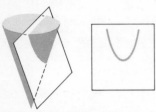

The ancient Greeks discovered that the parabola is the curve of intersection of a cone and a plane. If we slice through a cone with a plane parallel to one side of the cone, then the resulting curve is a parabola.

The Graph of $y = x^2$

We can graph the equation $y = x^2$ by plotting a large number of points and then connecting them with a smooth curve. For example,

$$\text{if } x = -3, y = x^2 = (-3)^2 = 9; \quad \text{point } (-3, 9)$$
$$\text{if } x = -2, y = x^2 = (-2)^2 = 4; \quad \text{point } (-2, 4)$$
$$\text{if } x = -1, y = x^2 = (-1)^2 = 1; \quad \text{point } (-1, 1)$$
$$\text{if } x = 0, y = x^2 = 0^2 = 0; \quad \text{point } (0, 0)$$
$$\text{if } x = 1, y = x^2 = 1^2 = 1; \quad \text{point } (1, 1)$$
$$\text{if } x = 2, y = x^2 = 2^2 = 4; \quad \text{point } (2, 4)$$
$$\text{if } x = 3, y = x^2 = 3^2 = 9; \quad \text{point } (3, 9)$$

The points are shown in Figure 7-7(a). The graph is in Figure 7-7(b).

The graph is a cup-shaped curve called a *parabola* that occurs naturally in many physical problems. This curve was first studied as a part of theoretical geometry by the ancient Greeks. It came as a surprise to the seventeenth-century physicists to find that the parabola is the graph of certain natural relationships. Before we consider any of these applications, we shall examine the graph in more detail.

The graph of the parabola $y = x^2$ has its greatest curvature near the origin. The farther we get away from the origin, the straighter the curve becomes. The point where the curvature is greatest (in this case, the origin) is called the *vertex* of the parabola (Fig. 7-8).

The y-axis is a *line of symmetry* for the parabola $y = x^2$. If we graph the portion of the curve on one side of the y-axis, we find that the part on the other side is a "mirror image" of the part that was graphed (Fig. 7-8). Observe that the vertex is on the line of symmetry.

If a is any nonzero constant, then the graph of the function

$$y = ax^2$$

SYMMETRY

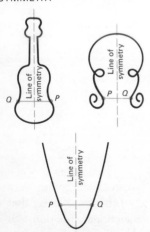

The curves are symmetric about the lines. For each point P on a curve there is a symmetric point Q. The line of symmetry is perpendicular to the line segment PQ and bisects that line segment.

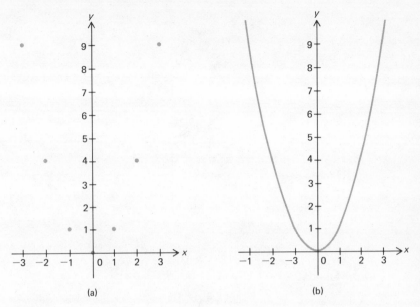

Figure 7-7 The parabola $y = x^2$. (a) Points on the graph; (b) the graph of $y = x^2$.

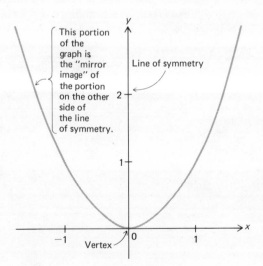

Figure 7-8 The parabola $y = x^2$.

is a parabola similar in shape to $y = x^2$. All of these parabolas have their vertices at the origin; all have the y-axis as a line of symmetry. If $a > 0$, the parabola "opens upward." If $a < 0$, it "opens downward." (See Fig. 7-9.)

Parabolas have a very simple shape. If we know the vertex and the line of symmetry, we can make a very accurate graph after plotting a relatively few points. It is best to plot the vertex, several points near the vertex (where

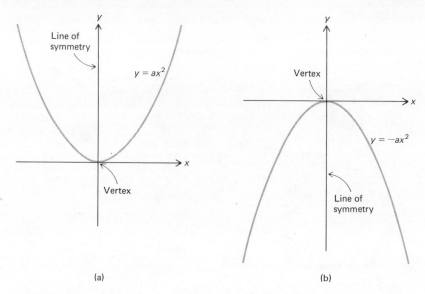

Figure 7-9   (a) The parabola $y = ax^2$ ($a > 0$); (b) the parabola $y = -ax^2$ ($a > 0$).

The trajectory of a cannonball is very nearly parabolic. It was discovered by Galileo in the seventeenth century that the path of a cannonball fired in a vacuum is a portion of a parabola. In actual practice, the parabolic path must be modified to account for wind speed and air resistance.

the curvature is greatest), and one or two points farther away from the vertex. The symmetry can be utilized in drawing the actual graph.

Example 1.   Sketch the graph of the parabola

$$y = 2x^2$$

Solution.   The function is of the form $y = ax^2$ with $a = 2$. Therefore, the vertex is at the origin; the y-axis is the line of symmetry. We first calculate the vertex and several points on the curve to the right of the y-axis:

$x = 0$:	$y = 2 \cdot 0^2 = 0$	vertex $(0, 0)$
$x = \frac{1}{4}$:	$y = 2 \cdot (\frac{1}{4})^2 = 2 \cdot \frac{1}{16} = \frac{1}{8}$	point: $(\frac{1}{4}, \frac{1}{8})$
$x = \frac{1}{2}$:	$y = 2 \cdot (\frac{1}{2})^2 = 2 \cdot \frac{1}{4} = \frac{1}{2}$	point: $(\frac{1}{2}, \frac{1}{2})$
$x = 1$:	$y = 2 \cdot 1^2 = 2$	point $(1, 2)$
$x = 2$:	$y = 2 \cdot 2^2 = 8$	point $(2, 8)$

Because of the symmetry, each of the points on one side of the y-axis has a corresponding symmetric point on the other side. For example, $(-\frac{1}{4}, \frac{1}{8})$ is the symmetric point corresponding to $(\frac{1}{4}, \frac{1}{8})$; similarly, $(-1, 2)$ is the symmetric point corresponding to $(1, 2)$. Figure 7-10(a) shows all of the points calculated above along with the corresponding points on the left side of the y-axis. Figure 7-10(b) shows the graph obtained by drawing a smooth curve through these points.

The graph of $y = 2x^2$ is "steeper" in shape than the graph of $y = x^2$. In general, the larger the value of a, the "steeper" the sides of the "cup." The closer the value of a to zero, the "shallower" the "cup." Several parabolas are shown on the same set of axes in Figure 7-11 for comparison.

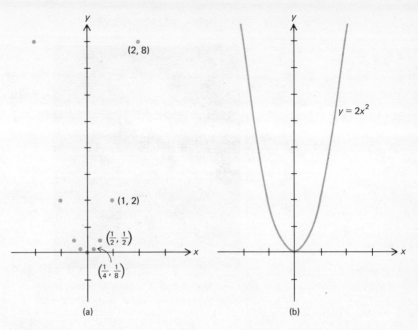

Figure 7-10 The parabola $y = 2x^2$ (Example 1). (a) Points on the graph; (b) the graph of $y = 2x^2$.

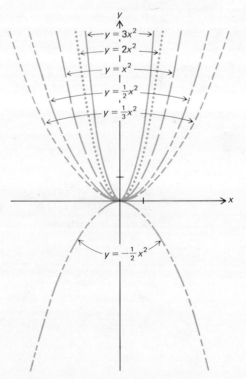

Figure 7-11 The parabola $y = ax^2$ ($a = -\frac{1}{2}$, $\frac{1}{3}$, $\frac{1}{2}$, 1, 2, 3). If $a > 0$, the parabola opens upward. If $a < 0$, the parabola opens downward. The larger the value of a, the "steeper" the sides of the parabola.

Photo from Boyer: *A History of Mathematics,*
1968, by permission of John Wiley & Sons, Inc.

In these formulas t is the number of seconds the object has been falling; v is the velocity in feet per second; d is the total distance it has fallen.

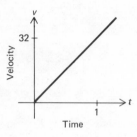

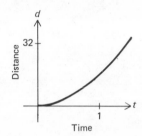

Galileo Galilei (1564–1642)

Galileo, the founder of the study of dynamics, was one of the outstanding physicists of the seventeenth century. During his life he studied the laws of motion that governed motion on the earth and in the heavens.

In order to study the motion of falling bodies, he rolled balls down inclined planes, thus slowing them sufficiently so that the short intervals of time could be accurately measured. He found that, contrary to popular opinion, the speed with which an object falls is independent of its weight—light objects fall at the same speed as heavy ones.

Neglecting air resistance, the speed of a falling body is determined by the length of time it has been falling. Galileo discovered that the laws of motion for an object falling in a vacuum (or one which has negligible resistance to air) are

$$v = 32t$$

and

$$d = 16t^2$$

Galileo extended his work to cover the trajectories of projectiles. He split the motion of a projectile into horizontal and vertical components and analyzed them separately. He was able to establish that the path of a projectile in a vacuum is a portion of a parabola. This result had a major significance for the martial arts in that it helped to establish artillery as a mathematical science.

Later in his life Galileo became interested in the laws of planetary motion. He built one of the world's first telescopes and used it to study the motion of the planets. Among other discoveries, he was the first person to observe the mountains of the moon, the rings of Saturn, and several of the moons of Jupiter. These discoveries led him to accept the Copernican theory of motion—that the earth and planets revolve about the sun.

In 1632 he published an astronomy book in which he attempted to prove the Copernican theory. Because this theory was in direct opposition to the views of the church, which held that the sun and planets revolve about the earth, he was arrested by the Inquisition. After several months of interrogation, he recanted his belief in the Copernican theory and was released.

Soon after his release from prison, he became blind and spent the remaining few years of his life in darkness, unable to pursue his beloved researches.

Square Roots

Every positive real number n has a square root. This is a number r such that

$$r^2 = n$$

In fact, every positive number n has *two* square roots—a positive one and its negative. We use the symbol $\sqrt{n}$ to indicate the positive square root of n.

For example,

$$\sqrt{4} = 2$$
$$\sqrt{5} \approx 2.236$$
$$\sqrt{6} \approx 2.449$$
$$\sqrt{7} \approx 2.646$$
$$\sqrt{8} \approx 2.828$$

and

$$\sqrt{9} = 3$$

Thus, the two square roots of 4 are 2 and -2; the two square roots of 9 are 3 and -3; the two square roots of 5 are $\sqrt{5} \approx 2.236$ and $-\sqrt{5} \approx -2.236$, and so on.

The graph of the parabola $y = x^2$ can be used to estimate square roots. It is necessary, of course, to have an accurate graph constructed on a good-quality graph paper.

The key to estimating square roots is the observation that if (a, b) is a point on the graph of $y = x^2$, then

$$a^2 = b$$

so that a is a square root of b.

To estimate the square root of 7, for example, we draw a horizontal line through the point $b = 7$ on the y-axis (Fig. 7-12). This line intersects the curve in two points, one on the left, one on the right of the y-axis. The point on the right has coordinates $(a, 7)$, where

$$a^2 = 7$$

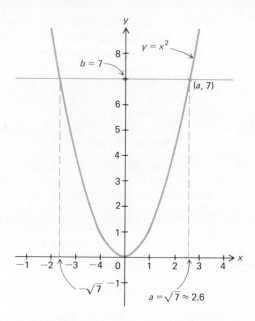

Figure 7-12 Estimation of square roots by graphs. The horizontal line through $b = 7$ intersects the parabola $y = x^2$ at the points (a, b), where $a^2 = 7$. If $a > 0$, then $a = \sqrt{7}$. If $a < 0$, then $a = -\sqrt{7}$.

Since the point is to the right of the y-axis, then $a > 0$ so that a is the positive square root of 7:

$$a = \sqrt{7} \approx 2.6$$

Similarly, the point on the left of the y-axis has x-coordinate equal to

$$-\sqrt{7} \approx -2.6$$

Reflection Property of Parabolas

Associated with every parabola is a point F, located on the axis of symmetry, called the *focus*. There is a remarkable reflection property of parabolas, involving the focus, that is responsible for many of the modern applications.

If we revolve a parabola about its axis of symmetry in three-dimensional space, it will sweep out a "cup-shaped" solid (Fig. 7-13). This solid is a bowl with a parabola for each of its cross sections. If we coat the inside of the bowl with silver, we obtain a *parabolic mirror*. When a light source is placed exactly at the focus, then each light ray is reflected out of the bowl parallel to the line of symmetry [Fig. 7-14(a)]. Thus, all of the light rays are concentrated in a beam the size of the original bowl. In other words, a true parabolic mirror makes a perfect spotlight. The beam of light (in theory, at least) never diffuses outward.

In actual practice it is impossible

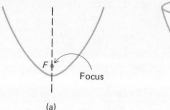

Figure 7-13 (a) The focus of a parabola; (b) if a parabola is revolved about its axis of symmetry, it generates a cup-shaped figure in three-dimensional space.

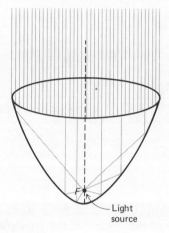

Figure 7-14 A parabolic reflector. Rays from a light source at the focus are reflected parallel to the line of symmetry.

to make a perfectly true parabolic mirror. The principle is utilized, however, in spotlights and automobile headlights in order to project beams great distances.

A similar principle is used in a parabolic dish antenna to collect radio signals from distant stars. The radio signals come toward the earth in parallel lines. They are collected on the surface of the antenna and "reflected" to the focus in order to get a stronger signal.

EXERCISES

1. Graph the following parabolas. In each case label the vertex and the line of symmetry.
   (a) $y = -x^2$ (d) $y = x^2/10$
   (b) $y = 3x^2$ (e) $y = -2x^2$
   (c) $y = x^2/3$ (f) $y = x^2/2$

2. (*Symmetry*) The following points are on a curve that is symmetric about the y-axis. To each point there corresponds another point directly across the y-axis. Plot and label the given point and the point that corresponds to it. Then sketch the graph.
   (a) $(1, 2)$ (d) $(3, 6)$
   (b) $(-2, 5)$ (e) $(-4, 7)$
   (c) $(0, 2)$ (f) $(-6, 8)$

3. Without looking at the book, carefully explain what is meant by a *square root*. Illustrate your explanation with two or three examples. Are we allowed to use the symbol $\sqrt{}$ for any square root?

4. Make an accurate graph of the parabola $y = x^2$ for values of x between 0 and 5. Use the graph to estimate the following numbers. Compare your answers with Table II at the end of the book.

(a) $\sqrt{16}$ (d) $\sqrt{3}$ (g) $\sqrt{8}$

(b) $\sqrt{144}$ (e) $\sqrt{0}$ (h) $\sqrt{17}$

(c) $\sqrt{1}$ (f) $\sqrt{4}$ (i) $\sqrt{24}$

5. Construct a computer program for calculating points on the parabola $y = ax^2$. Read the coefficient a from a DATA statement (or as INPUT). Have the computer print the coordinates of the points (with appropriate labels) for values of x between -3 and $+3$ in increments of 0.1. In other words, calculate y for $x = -3.0, -2.9, -2.8, \ldots, 2.9, 3.0$.

7-5   THE PARABOLA $y = ax^2 + bx + c$

The graph of the function

$$y = ax^2 + bx + c$$

is a parabola that is identical to the graph of

$$y = ax^2$$

in shape and orientation, but is shifted in location. If $a > 0$, the parabola opens upward; if $a < 0$, it opens downward. The larger the value of a, the "steeper" the "sides" of the parabola, and so on.

Figure 7-15 pictures the graphs of

$$y = x^2 \qquad \text{and} \qquad y = x^2 - 2x + 2$$

Observe that the two graphs are identical in shape and orientation.

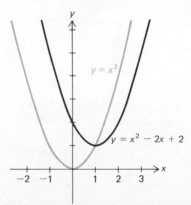

Figure 7-15 The graphs of $y = x^2$ and $y = x^2 - 2x + 2$ are identical in shape and orientation.

As we saw in Section 7-4, a parabola has a line of symmetry on which the vertex is located. We shall show how to find the line of symmetry and the vertex. We then will be able to make accurate drawings of parabolas by plotting a few points near the vertex and connecting them with a smooth curve.

The Parabola $y = ax^2 + bx$

As we shall see later in this section, the parabolas

$$y = ax^2 + bx \qquad \text{and} \qquad y = ax^2 + bx + c$$

have the same line of symmetry. Thus, we turn our attention to the first of these parabolas.

In order to find the line of symmetry of the parabola

$$y = ax^2 + bx$$

we observe that it is midway between the points where the parabola crosses the x-axis. [See Fig. 7-16(c).] Also observe that the parabola crosses the x-axis at the points on the graph where $y = 0$. Thus, we must find the values of x for which

$$y = ax^2 + bx = 0$$

These values are the solutions of the quadratic equation

$$ax^2 + bx = 0$$

Example 1.   (a) Find where the parabola

$$y = 3x^2 - 6x$$

crosses the x-axis.
   (b) Find the line of symmetry.
   (c) Find the vertex.
   (d) Graph the parabola.

Solution.   (a) The parabola crosses the x-axis at the points where

$$y = 3x^2 - 6x = 0$$

Thus, we must solve the quadratic equation

$$3x^2 - 6x = 0$$

We factor the equation

$$3x^2 - 6x = 0$$
$$x(3x - 6) = 0$$

$$x = 0 \quad \text{or} \quad 3x - 6 = 0$$
$$x = 0 \quad \text{or} \quad x = 2$$

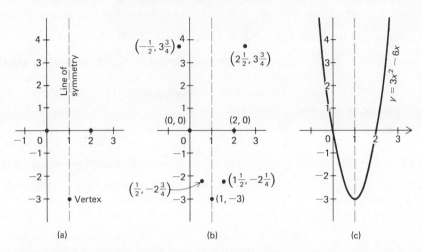

Figure 7-16 The parabola $y = 3x^2 - 6x$ (Example 1). (a) The line of symmetry is midway between the points where the parabola crosses the x-axis. The vertex is on the line of symmetry; (b) points on the graph; (c) the graph of $y = 3x^2 - 6x$.

The parabola crosses the x-axis at the points where

$$x = 0 \quad \text{and} \quad x = 2$$

(b) The line of symmetry is midway between the points where the parabola crosses the x-axis. Thus, the line of symmetry is the vertical line through the point where $x = 1$ [Fig. 7-16(a)].

(c) The vertex is on the line of symmetry. Thus, the x-coordinate of the vertex is $x = 1$. To find the y-coordinate, we substitute in the equation of the parabola:

$$x = 1: \quad y = 3x^2 - 6x = 3 \cdot 1^2 - 6 \cdot 1 = 3 - 6 = -3$$

The vertex is the point $(1, -3)$ [Fig. 7-16(a)].

(d) To sketch the graph we plot a few points near the vertex and connect them with a smooth curve:

$$x = -\tfrac{1}{2}: \quad y = 3(-\tfrac{1}{2})^2 - 6(-\tfrac{1}{2}) = \tfrac{3}{4} + 3 = 3\tfrac{3}{4} \quad \text{point } (-\tfrac{1}{2}, 3\tfrac{3}{4})$$
$$x = 0: \quad y = 3 \cdot 0^2 - 6 \cdot 0 = 0 \quad \text{point } (0, 0)$$
$$x = \tfrac{1}{2}: \quad y = 3 \cdot (\tfrac{1}{2})^2 - 6(\tfrac{1}{2}) = \tfrac{3}{4} - 3 = -2\tfrac{1}{4} \quad \text{point } (\tfrac{1}{2}, -2\tfrac{1}{4})$$
$$x = 1; \quad y = 3 \cdot 1^2 - 6 \cdot 1 = 3 - 6 = -3 \quad \text{vertex } (1, -3)$$
$$x = 1\tfrac{1}{2}: \quad y = 3(\tfrac{3}{2})^2 - 6(\tfrac{3}{2}) = 3 \cdot \tfrac{9}{4} - 9 = -2\tfrac{1}{4} \quad \text{point } (1\tfrac{1}{2}, -2\tfrac{1}{4})$$
$$x = 2: \quad y = 3 \cdot 2^2 - 6 \cdot 2 = 12 - 12 = 0 \quad \text{point } (2, 0)$$
$$x = 2\tfrac{1}{2}: \quad y = 3 \cdot (\tfrac{5}{2})^2 - 6 \cdot (\tfrac{5}{2}) = \tfrac{75}{4} - 15 = 3\tfrac{3}{4} \quad \text{point } (2\tfrac{1}{2}, 3\tfrac{3}{4})$$

The points are shown in Figure 7-16(b). The parabola is in Figure 7-16(c).

We can find the line of symmetry for the parabola

$$y = ax^2 + bx$$

by the argument used in Example 1. The parabola crosses the x-axis at the points where

$$ax^2 + bx = 0$$

We factor the equation as

$$x(ax + b) = 0$$

Then $x = 0$ or $x = -\dfrac{b}{a}$

(See Fig. 7-17.) Since the line of symmetry is midway between these two points, then it is the vertical line through the point

$$x = -\frac{b}{2a}$$

on the x-axis (Fig. 7-17). Since the vertex is on the line of symmetry, then its x-coordinate is

$$x = -\frac{b}{2a}$$

The y-coordinate of the vertex can be found by substituting the x-coordinate in the equation of the parabola.

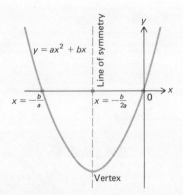

Figure 7-17 The parabola $y = ax^2 + bx$. The parabola intersects the x-axis at the origin and at the point $x = -b/a$. The vertex is located at the point where $x = -b/2a$.

Example 2. Find the vertex and the line of symmetry of the parabola

$$y = x^2 + 4x$$

Solution. If we compare the equation

$$y = x^2 + 4x$$

with the equation

$$y = ax^2 + bx$$

we see that

$$a = 1, \qquad b = 4$$

The line of symmetry is the vertical line through the point

$$x = -\frac{b}{2a} = -\frac{4}{2} = -2$$

The y-coordinate of the vertex is

$$y = (-2)^2 + 4(-2) = 4 - 8 = -4$$

The vertex is the point $(-2, -4)$. (See Fig. 7-18.)

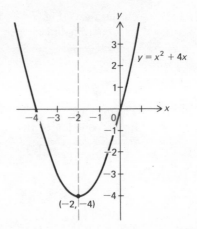

Figure 7-18 The graph of $y - x^2 + 4x$ (Example 2). Vertex: $x = -b/2a = -4/2 = -2$; $y = (-2)^2 + 4(-2) = 4 - 8 = -4$.

The Parabola $y = ax^2 + bx + c$

The graph of the equation

$$y = ax^2 + bx + c \qquad (a \neq 0)$$

is a parabola identical to the graph of $y = ax^2 + bx$ except that it is shifted c units in the vertical direction. If $c > 0$, it is shifted upward. If $c < 0$, it is shifted downward. The following example shows how this shift is accomplished.

Example 3. Sketch the graph of the parabola

$$3x^2 - 6x + 2$$

Solution. This graph can be obtained in two steps:
  (1) We sketch the graph of the related parabola

$$y = 3x^2 - 6x \qquad (\text{Example 1})$$

This parabola opens upward and has its vertex at the point $(1, -3)$ [Fig. 7-19(a)]. For any number x on the x-axis, the point $(x, 3x^2 - 6x)$ is on the graph of $y = 3x^2 - 6x$.

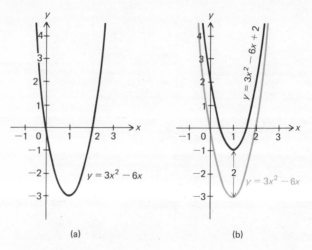

(a)   (b)

Figure 7-19 The graph of $y - 3x^2 - 6x + 2$ can be drawn by shifting the graph of $y = 3x^2 - 6x$ two units in the vertical direction (Example 3). (a) The graph of $y = 3x^2 - 6x$; (b) the graph of $y = 3x^2 - 6x + 2$.

  (2) We now sketch the graph of

$$y = 3x^2 - 6x + 2$$

For any number x on the number line, the point $(x, 3x^2 - 6x + 2)$ is on the graph. Observe that this point is exactly 2 units upward from the corresponding point on the parabola graphed in (1). Since we can do this for every point on the x-axis, this means that the graph of $y = 3x^2 - 6x + 2$ is an exact copy of the graph of $y = 3x^2 - 6x$, but is shifted 2 units in the vertical direction [Fig. 7-19(b)].

The procedure in Example 3 can be used to graph any equation of form

$$y = ax^2 + bx + c$$

where $a \neq 0$.

Construction of a Parabola by
Paper Folding

A parabola can be outlined on wax paper by the following process:

(1) Mark a line and a point on the wax paper.

(2) Fold the paper so that the line is directly over the point. Crease the wax paper. Refold and repeat the process several times.

(3) The shape of the parabola is outlined by the creases on the wax paper. Each of the crease lines is tangent to the curve.

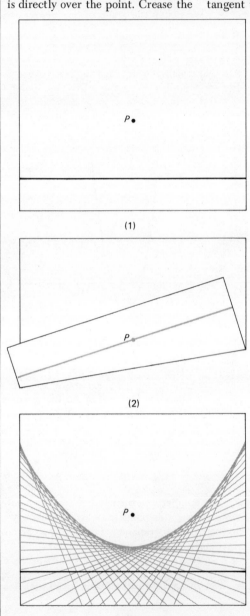

(1)

(2)

(3)

Figure 7-20 shows how the graphs of $y = ax^2$, $y = ax^2 + bx$, and $y = ax^2 + bx + c$ are related:

(1) The graph of $y = ax^2$ is a parabola with vertex at the origin. It opens upward if $a > 0$; opens downward if $a < 0$.

(2) The graph of $y = ax^2 + bx$ is identical to the graph of $y = ax^2$, but is shifted so that the vertex is at the point where $x = -b/2a$. This parabola passes through the origin.

(3) The graph of $y = ax^2 + bx + c$ is identical to the graph of $y = ax^2 + bx$, but is shifted upward c units. The vertex is at the point where $x = -b/2a$.

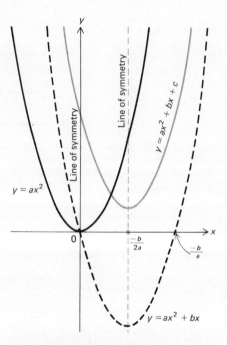

Figure 7-20 The graphs of $y = ax^2$, $y = ax^2 + bx$, and $y = ax^2 + bx + c$ are identical in size, shape, and orientation.

Quadratic Equations

Later in this chapter we shall develop a method for finding the exact solutions of all quadratic equations. This method is very specialized and cannot be used to solve other types of equations. It may be more practical to use graphs to solve equations. As an illustration we will use graphs to solve quadratic equations.

Observe that the solutions of

$$ax^2 + bx + c = 0$$

are those values of x where the parabola

$$y = ax^2 + bx + c$$

crosses the x-axis. (See Fig. 7-21.) These numbers are sometimes called the *roots* of the equation.

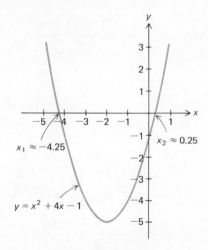

Figure 7-21   The solutions of the equation $x^2 + 4x - 1 = 0$ are the values of x where the parabola $y = x^2 + 4x - 1$ crosses the x-axis (Example 4).

Example 4.   (a) Sketch the graph of the parabola

$$y = x^2 + 4x - 1$$

(b) Find where the parabola crosses the x-axis.
(c) Find the approximate solutions of the equation

$$x^2 + 4x - 1 = 0$$

Solution.   (a) The graph is a parabola that opens upward. ($a = 1, b = 4, c = -1$.) Its vertex is at the point with

$$x = -\frac{b}{2a} = -\frac{4}{2} = -2$$

The y-coordinate of the vertex is

$$y = x^2 + 4x - 1 = (-2)^2 + 4(-2) - 1 = 4 - 8 - 1 = -5$$

The vertex is the point $(-2, -5)$. The graph is shown in Figure 7-21.

(b) We see from the graph that the parabola crosses the x-axis at the points

$$x_1 \approx -4.25 \qquad \text{and} \qquad x_2 \approx 0.25$$

(c) The solutions of the equation

$$x^2 + 4x - 1 = 0$$

are exactly those values of x that make

$$y = x^2 + 4x - 1$$

equal to zero. Thus, the solutions are the values of x where the graph of the parabola $y = x^2 + 4x - 1$ crosses the x-axis. It follows from (b) that the solutions of the equation $x^2 + 4x - 1 = 0$ are $x_1 \approx -4.25$ and $x_2 \approx 0.25$.

As a partial check on our work, we substitute these values into the equation. Since we have approximate values of the solutions, we shall not get the exact value of zero for the answer.

$$x \approx -4.25: \quad x^2 + 4x - 1 \approx (-4.25)^2 + 4(-4.25) - 1$$
$$\approx 18.06 - 17 - 1 \approx 0.06 \approx 0$$
$$x \approx 0.25: \quad x^2 + 4x - 1 \approx (0.25)^2 + 4(0.25) - 1$$
$$\approx 0.063 + 1 - 1 \approx 0.063 \approx 0$$

The procedure used in Example 4 can be used to approximate the real solutions of any quadratic equation

$$ax^2 + bx + c = 0$$

(1) We graph the parabola

$$y = ax^2 + bx + c$$

(2) The solutions of the equation $ax^2 + bx + c = 0$ are the values of x where the parabola crosses the x-axis. (Fig. 7-21).

There are three situations that can occur when we graph the parabola $y = ax^2 + bx + c$. Either the graph crosses the x-axis at two points *or* it touches the x-axis at one point *or* it lies completely on one side of the x-axis (Fig. 7-22). If the first case holds, then the equation $ax^2 + bx + c = 0$ has two real solutions x_1 and x_2 equally spaced on each side of the number $x = -b/2a$ [Fig. 7-22(a)]. If the second case holds, then the only real solution is $x_0 = -b/2a$, the x-coordinate of the vertex [Fig. 7-22(b)]. If the third case holds, then the equation has no real solution at all [Fig. 7-22(c)].

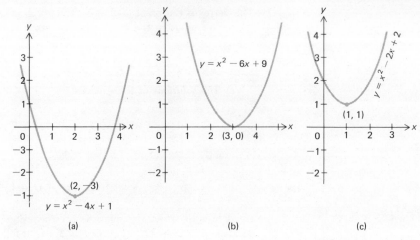

Figure 7-22 A quadratic equation $ax^2 + bx + c = 0$ has two real solutions, one real solution, or no real solution at all (Example 5). (a) The equation $x^2 - 4x + 1 = 0$ has two real solutions [Example 5(a)]; (b) the equation $x^2 - 6x + 9 = 0$ has one real solution [Example 5(b)]; (c) the equation $x^2 - 2x + 2 = 0$ has no real solution [Example 5(c)].

Example 5. Find the number of real roots of the following quadratic equations.
   (a) $x^2 - 4x + 1 = 0$
   (b) $x^2 - 6x + 9 = 0$
   (c) $x^2 - 2x + 2 = 0$

Solution. All of the related parabolas open upward.
   (a) $x^2 = 4x + 1 = 0$
The related parabola is $y = x^2 - 4x + 1$. The vertex is the point $(2, -3)$. Since the vertex is below the x-axis, the parabola must cross the x-axis at two points. The equation

$$x^2 - 4x + 1 = 0$$

has two real roots [Fig. 7-22(a)].
   (b) $x^2 - 6x + 9 = 0$
The related parabola is $y = x^2 - 6x + 9$. The vertex is the point $(3, 0)$. Since the vertex is on the x-axis, the parabola touches the x-axis only at this point. The equation

$$x^2 - 6x + 9 = 0$$

has only the solution $x = 3$ [Fig. 7-22(b)].
   (c) $x^2 - 2x + 2 = 0$
The related parabola is $y = x^2 - 2x + 2$. The vertex is the point $(1, 1)$. Since the vertex is above the x-axis and the parabola opens upward, then it lies completely on one side of the x-axis. The equation

$$x^2 - 2x + 2 = 0$$

has no real solution [Fig. 7-22(c)].

EXERCISES

1. Locate the line of symmetry and the vertex of each parabola.

   Work (a)–(d) by setting y equal to zero and factoring the resulting equation in order to find where the parabola crosses the x-axis.

   Work (e)–(h) by using the fact that the x-coordinate of the vertex is $x = -b/2a$.

   (a) $y = x^2 - 2x$ (e) $y = -x^2 + 4x$
   (b) $y = 2x^2 + 8x$ (f) $y = 3x^2 - 18x$
   (c) $y = x^2/4 + x$ (g) $y = x^2/2 - 3x$
   (d) $y = -x^2/6 - x/2$ (h) $y = -x^2/3 + 6x$

2. Plot a few points near the vertex of each parabola in Exercise 1, (a)–(h). Sketch the parabola. Label the vertex and the line of symmetry.

3. Graph the three parabolas on the same set of axes. Label the vertices and lines of symmetry.

   (a) $y = x^2$, $y = x^2 + 4x$, $y = x^2 + 4x + 2$
   (b) $y = -x^2$, $y = -x^2 + 6x$, $y = -x^2 + 6x + 1$
   (c) $y = 2x^2$, $y = 2x^2 - 8x$, $y = 2x^2 - 8x + 9$

4. Locate the vertex and the line of symmetry. Sketch the graph.

   (a) $y = x^2 + 2x - 1$ (d) $y = x^2 - 6x + 9$
   (b) $y = x^2 + 4x + 4$ (e) $y = x^2 + 2x + 3$
   (c) $y = -x^2 - 2x + 7$ (f) $y = -x^2 - 5x + 3$

5. Construct a flow chart for the steps involved in getting the following information about the parabola

$$y = ax^2 + bx + c$$

   (a) Decide if the parabola opens upward or downward.
   (b) Find the coordinates of the vertex.
   (c) Find the line of symmetry.

6. Make a computer program based on your flow chart in Exercise 5. Read a, b, c from a DATA statement. Print the coordinates of the vertex, the location of the line of symmetry, and several points on the parabola, all with appropriate labels. Decide if the parabola opens upward or downward and print that fact.

   (a) Run the program using your own data. Use at least three sets of data.
   (b) Make a rough sketch of the parabola based on the information from the computer.

7. Use graphs to determine the approximate solutions of the following equations.

   (a) $3x^2 + 12x - 5 = 0$ (d) $x^2 + 4x - 17 = 0$
   (b) $x^2 + 30x - 7 = 0$ (e) $x^2 - 15x + 7 = 0$
   (c) $2x^2 - 8x - 9 = 0$ (f) $x^2 - 5x + 8 = 0$

8. Each of the following equations can be obtained from a parabola $y = ax^2 + bx + c$ that opens upward by setting $y = 0$. Calculate the y-coordinate of the vertex of each parabola and use this number to decide

if the equation has one real solution, two real solutions, or no real solution at all. (*Hint:* If the y-coordinate of the vertex is positive, then the parabola lies completely above the x-axis.)

(a) $x^2 + 12x + 5 = 0$ (d) $x^2 - 8x + 16 = 0$
(b) $2x^2 - 7x + 7 = 0$ (e) $3x^2 - 5x + 3 = 0$
(c) $x^2 - 4x + 9 = 0$ (f) $x^2 + 3x + 2 = 0$

9. (a) Show that the vertex of the parabola $y = ax^2 + bx + c$, $a > 0$, is at the point

$$\left(\frac{-b}{2a}, \frac{-b^2 + 4ac}{4a} \right)$$

(*Hint:* Substitute the x-coordinate of the vertex into the equation $y = ax^2 + bx + c$.)

(b) Show that if $b^2 > 4ac$, then the equation $ax^2 + bx + c = 0$ has two real solutions; if $b^2 < 4ac$, the equation has no real solution; if $b^2 = 4ac$, the equation has exactly one real solution. (*Hint:* If $b^2 > 4ac$, show that the vertex of the parabola $ax^2 + bx + c$ is below the x-axis; so the parabola must cross the x-axis at two points.)

7-6 APPLICATIONS

Many practical problems can be solved by finding the vertices of certain parabolas. The clue to the solution lies in the following observation (Fig. 7-23):

(1) If a parabola opens downward, then the largest value of y at any point on the curve is at the vertex.

(2) If a parabola opens upward, then the least value of y at any point on the curve is at the vertex.

In this section we consider three typical applications.

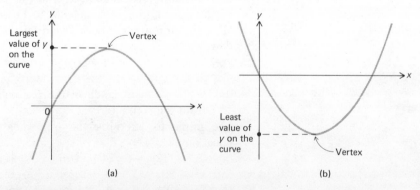

Figure 7-23 (a) If a parabola opens downward, the largest value of y on the curve is at the vertex; (b) if a parabola opens upward, the least value of y on the curve is at the vertex.

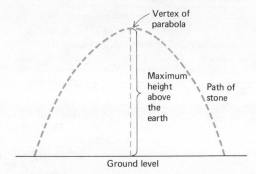

Figure 7-24 A stone thrown upward from the earth follows a parabolic path. The maximum height of the stone above the earth occurs at the vertex of the parabola.

For our first example, consider the discovery made by Galileo in the early seventeenth century: *A stone (or other object) thrown upward at an angle to the earth follows a parabolic path* (Fig. 7-24).

Example 1. A stone thrown upward at an angle to the ground follows the parabolic path

$$y = -\tfrac{1}{50}x^2 + 2x$$

(All distances are measured in feet.)
  (a) Find the maximum height of the stone above the ground.
  (b) Find where the stone will strike the ground.

Solution. (a) The equation is of form

$$y = ax^2 + bx \qquad \text{where} \quad a = -\tfrac{1}{50},\, b = 2$$

The graph is shown in Figure 7-25. The x-axis is drawn at ground level; the origin is the point at which the stone was thrown.

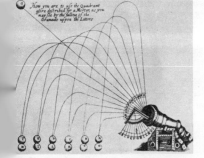

Diagram from The Complete Gunner (1672) by Thomas Venn. Photo courtesy of The British Library.

Venn quoted Galileo's work on ballistics, which established that the trajectories of cannonballs are parabolas. His drawings of the trajectories, however, incorrectly showed them to be different types of curves.

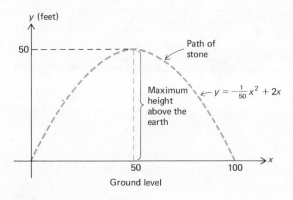

Figure 7-25 Example 1.

The maximum height above the ground occurs at the vertex of the parabola. As we saw in the last section, this is the point on the graph where

$$x = -\frac{b}{2a} = -\frac{2}{2(-\frac{1}{50})} = \frac{-2}{-\frac{1}{25}} = (-2) \cdot (-\frac{25}{1}) = 50$$

When $x = 50$, $y = -\frac{1}{50}x^2 + 2x = -\frac{1}{50}(50)^2 + 2(50) = -50 + 100 = 50$. The vertex is the point $(50, 50)$. Since the maximum height above the ground occurs at the vertex, the maximum height is 50 feet.

(b) The stone strikes the ground at the point where $y = 0$. If we set $y = 0$ in the equation of the parabola and solve for x, we obtain

$$-\tfrac{1}{50}x^2 + 2x = 0$$
$$(-\tfrac{1}{50}x + 2)x = 0$$
$$-\tfrac{1}{50}x + 2 = 0 \quad \text{or} \quad x = 0$$
$$x = 100 \quad \text{or} \quad x = 0$$

The solution $x = 0$ represents the point where the stone was thrown. Therefore, the stone strikes the ground at the point $x = 100$—one hundred feet from the point where it was thrown.

If an object is thrown straight up into the air, its height above the earth can be calculated by a simple formula. The formula expresses the height h as a function of time t. The graph of the function does not describe the actual path of the object; it shows the relationship between the height and the time.

Example 2.   A stone is thrown upward from the side of a bridge 84 feet above a river with an initial velocity of 100 feet per second. The height of the stone above the river after t seconds is given by the formula

$$h = -16t^2 + 100t + 84$$

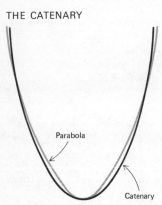

THE CATENARY

Contrary to popular opinion, a hanging cable does not assume the shape of a parabola. The actual shape is that of a curve called a *catenary*, which resembles a parabola.

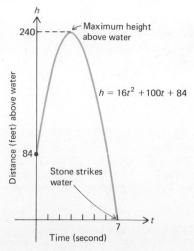

Figure 7-26   Height is plotted against time (Example 2). The graph does not show the actual path of the stone. (Different scales are used on the two axes.)

(a) What is the maximum height of the stone above the river?

(b) When does the stone strike the water?

Solution. (a) The graph of the parabola

$$h = -16t^2 + 100t + 84$$

is shown in Figure 7-26. The maximum value of h occurs at the vertex of the parabola. The vertex is at the point where

$$t = -\frac{b}{2a} = \frac{-100}{-32} = \frac{100}{32} = \frac{25}{8}$$

At this point we have

$$h = -16t^2 + 100t + 84$$

$$= -16 \cdot \left(\frac{25}{8}\right)^2 + 100\left(\frac{25}{8}\right) + 84$$

$$= \frac{-16 \cdot 625}{64} + \frac{2500}{8} + 84$$

$$= -\frac{625}{4} + \frac{1250}{4} + 84 = \frac{625}{4} + 84 = 156.25 + 84$$

$$\approx 240 \text{ feet}$$

If a hanging cable has equal weights distributed over its length, it assumes the shape of a parabola. Many suspension bridges have parabolic shapes.

Thus, the maximum height of the stone above the water is approximately 240 feet.

(b) We see from the graph that the stone strikes the water after 7 seconds. This is the time when h, the height above the water, is equal to zero.

The following problem shows how graphs can sometimes be used to find the largest possible values of quantities even when there is no physical law involved.

Example 3. The Thin Soup Co. finds that at a wholesale price of x cents per can, it can sell $(26 - x)$ million cans of Thin Tomato Soup each month. What selling price yields the largest revenue for the company?

Solution. Let y be the total revenue received by the company. At a price of x cents per can, it can sell $(26 - x)$ million cans. The total revenue is

$$y = \underbrace{x}_{\substack{\text{price} \\ \text{per can}}} \cdot \underbrace{(26 - x)}_{\substack{\text{number} \\ \text{of cans}}} = (26x - x^2) \text{ million cents}$$

The equation $y = 26x - x^2$ is of form $y = ax^2 + bx$, where $a = -1$, $b = 26$. The graph is shown in Figure 7-27.

The largest possible value of y occurs at the vertex. The x-coordinate of the vertex is

$$x = -\frac{b}{2a} = -\frac{26}{2(-1)} = \frac{26}{2} = 13$$

This selling price (13 cents per can) yields the maximum revenue for the company.

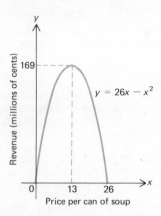

Figure 7-27 Example 3. (Different scales are used on the two axes.)

Parabolas can be used to solve some problems that have no apparent connection to an equation of form $y = ax^2 + bx$. This is the case with a large class of problems that require maximum values of quantities. The following example is typical of these problems. The main result that can be drawn from this example is that *a square encloses the largest possible area in a rectangle of fixed perimeter.*

Example 4.   Farmer Brown plans to enclose a rectangular pasture by fencing some unused land. He has bought 2400 feet of fencing. What is the largest area that he can enclose with the 2400 feet of fencing? What are the dimensions of this largest possible pasture?

Solution.   A diagram of the field is shown in Figure 7-28(a). If we let

$$x = \text{the base of the rectangle}$$
$$h = \text{the height of the rectangle}$$

and

$$y = \text{the area of the rectangle}$$

then $$y = xh$$

and $2x + 2h = 2400$ (the perimeter is equal to the length of fencing).

Our problem reduces to finding the largest possible value of $y = xh$ subject to the restriction

$$2x + 2h = 2400$$

We solve this last equation for h:

$$2x + 2h = 2400$$
$$2h = 2400 - 2x$$
$$h = 1200 - x$$

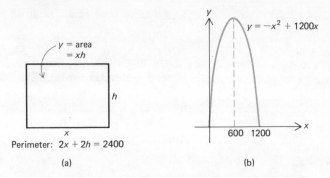

Figure 7-28 (a) x = length of field, h = width of field, area: $y = xh$, perimeter: $2x + 2h = 2400$; (b) the graph of $y = -x^2 + 1200x$. The maximum value of y occurs when $x = 600$ (Example 4). (Different scales are used on the two axes.)

If we substitute this expression for h into the equation $y = xh$, we obtain

$$y = xh$$
$$y = x(1200 - x)$$
$$y = -x^2 + 1200x$$

The equation $y = -x^2 + 1200x$ is of form $y = ax^2 + bx$ with $a = -1$, $b = 1200$. Its graph is the parabola in Figure 7-28(b). The largest possible value of y (the area) occurs at the vertex of the parabola.

The x-coordinate of the vertex is

$$x = -\frac{b}{2a} = -\frac{1200}{-2} = 600$$

Therefore, the largest possible area occurs when $x = 600$.

To find the value of h, we substitute $x = 600$ into the equation

$$h = 1200 - x$$
$$h = 1200 - 600$$
$$h = 600$$

The largest possible value of the area is

$$y = xh = 600 \cdot 600 = 360,000 \text{ square feet}$$

EXERCISES

1. Graph the following parabolas. Find the largest value of y on each of them.

(a) $y = -x^2$
(b) $y = -x^2/2 + 36x$
(c) $y = -x^2 - 14x$
(d) $y = -x^2/10 + 25x$

2. A projectile fired at an angle to the earth follows the parabolic path

$$y = -x^2 + 150x$$

where x and y are measured in meters.
   (a) Sketch the path of the projectile.
   (b) What is the maximum height above the earth?
   (c) Where does it strike the ground?
3. Work Exercise 2 on the assumption that the path of the projectile is

$$y = -\frac{x^2}{2} + 60x$$

4. A baseball struck from a height of 3 feet above the ground follows the parabolic path

$$y = -\tfrac{1}{100}x^2 + \tfrac{5}{4}x + 3$$

   (a) Determine the maximum height of the ball above the ground.
   (b) Find approximately where the ball will strike the ground.
5. A stone thrown straight up from ground level with an initial speed of 64 feet per second obeys the following law for height above the earth:

$$h = -16t^2 + 64t$$

where t is the elapsed time in seconds and h is the distance above the earth in feet.
   (a) Graph the equation $h = -16t^2 + 64t$.
   (b) Use the graph to determine where the stone strikes the ground.
   (c) Use the graph to find the maximum height above the ground.
6. A stone dropped from a bridge 144 feet above a river obeys the following formula for height above the river:

$$h = -16t^2 + 144$$

where t is the time in seconds and h is the distance above the river in feet. Construct the graph of the equation. Use the graph to find when the stone strikes the water.
7. Work Example 3 on the assumption that the Thin Soup Co. can sell

$$30 - \tfrac{3}{2}x$$

million cans of soup at a wholesale price of x cents per can. What value of x yields the largest possible total revenue? What is the maximum possible revenue?

8. The King Checker Co. can sell

$$15 - 2x$$

hundred boxes of its *Deluxe* checker sets each week at a selling price of $x per box. What value of x yields the largest possible gross revenue for the company?

7-7 ROOTS; FRACTIONAL EXPONENTS

Square roots have been an essential tool in mathematics for thousands of years. In most cases, rational approximations to these roots have been used.

We have considered square roots several times in this book. As we saw, every positive real number N has two square roots, $\sqrt{N}$ and $-\sqrt{N}$. There is an extensive theory of roots that we have only touched on thus far. Our results with square roots can be extended to cover roots of any order.

Every positive number N not only has a positive square root, but a positive cube root, a positive fourth root, a positive fifth root, and so on. These roots are defined as follows:

A is a *square root* of N provided $A^2 = N$
B is a *cube root* of N provided $B^3 = N$
C is a *fourth root* of N provided $C^4 = N$
D is a *fifth root* of N provided $D^5 = N$
$\ldots$

Example 1. Most sets of mathematical tables contain tables of roots. The following roots were calculated from values in the Rinehart Mathematical Tables:

(a) The positive cube root of 8 is 2. This holds because

$$2^3 = 8$$

(b) The positive fourth root of 81 is 3. This holds because

$$3^4 = 81$$

(c) The positive fourth root of 60 is approximately equal to 2.78316. This follows from the fact that

$$(2.78316)^4 \approx 60.0002 \approx 60$$

(d) The positive fifth root of 147 is approximately equal to 2.713086. This follows from the fact that

$$(2.713086)^5 \approx 147.0002 \approx 147$$

There are two standard symbols for the positive roots of a positive

NOTATIONS FOR ROOTS

A number of different notations have been used to indicate roots over the centuries. Most of them involved writing an "R" or similar symbol before the root. (The "R" stood for *radix*, the Latin word for root.) For example, Chuquet, the sixteenth-century French mathematician, wrote

R)² · 14 · m̄ · R)² 180

for the number we would write as

$$\sqrt{14 - \sqrt{180}}.$$

The modern symbol $\sqrt{\ }$, which is probably a variation of the letter *r*, was first used by the sixteenth-century German mathematicians.

number N. The symbols

$$\sqrt[k]{N} \quad \text{and} \quad N^{1/k}$$

both indicate the positive kth root of N.

Example 2.   (a) $\sqrt[2]{5} = \sqrt{5} = 5^{1/2} \approx 2.23607$

(b) $\sqrt[3]{8} = 8^{1/3} = 2$

(c) $\sqrt[4]{81} = 81^{1/4} = 3$

(d) $\sqrt[5]{147} = 147^{1/5} \approx 2.713086$

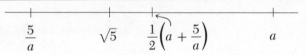

$$\frac{5}{a} \qquad \sqrt{5} \qquad \frac{1}{2}\left(a + \frac{5}{a}\right) \qquad a$$

The Square-Root Algorithm

The ancient Babylonians and Greeks had an interesting algorithm for calculating square roots. To calculate the square root of 5, for example, we choose a first approximation, call it a_1, and then define a_2, a_3, a_4, and so on, as follows:

$$a_2 = \frac{1}{2}\left(a_1 + \frac{5}{a_1}\right)$$

$$a_3 = \frac{1}{2}\left(a_2 + \frac{5}{a_2}\right)$$

$$a_4 = \frac{1}{2}\left(a_3 + \frac{5}{a_3}\right)$$

and so on.

It can be shown that the numbers a_1, a_2, a_3, a_4, and so on, form a sequence of better and better approximations to $\sqrt{5}$.

For example, if we choose $a_1 = 2$, then

$$a_2 = \frac{1}{2}\left(2 + \frac{5}{2}\right) = \frac{1}{2}\left(\frac{4+5}{2}\right) = \frac{1}{2}\left(\frac{9}{2}\right) = \frac{9}{4} = 2.25$$

$$a_3 = \frac{1}{2}\left(\frac{9}{4} + \frac{5}{9/4}\right) = \frac{1}{2}\left(\frac{9}{4} + \frac{20}{9}\right) = \frac{1}{2}\left(\frac{81+80}{36}\right) = \frac{161}{72} \approx 2.236$$

$$a_4 = \frac{1}{2}\left(\frac{161}{72} + \frac{5}{161/72}\right) = \frac{1}{2}\left(\frac{161}{72} + \frac{360}{161}\right)$$

$$= \frac{1}{2}\left(\frac{25{,}921 + 25{,}920}{11{,}592}\right) = \frac{51{,}841}{23{,}184}$$

$$\approx 2.236068$$

This last approximation is correct to five decimal places.

To understand the algorithm, observe that in every case the number $\sqrt{5}$ is located almost halfway between a and $5/a$. The number $\frac{1}{2}(a + 5/a)$, being the average of a and $5/a$, is exactly halfway between a and $5/a$. Thus, if a is an approximation to $\sqrt{5}$, then $\frac{1}{2}(a + 5/a)$ is a much better approximation.

Calculations on the Computer

Fractional exponents are almost essential when we calculate roots on the computer. There is an algorithm permanently stored in the computer that automatically computes the approximate value of $N^{1/k}$. We write the expression

$$N\!\uparrow\!(1/k)$$

to get the kth root of N.

Example 3. The following program causes the computer to print a brief table of square, cube, fourth, and fifth roots. The printout follows the program.

```
1Ø   REMARK * PROGRAM TO PRINT TABLE OF SQUARE, CUBE, FOURTH AND FIFTH
2Ø   REMARK * ROOTS OF NUMBERS 1,2,3, . . . ,1Ø
3Ø   PRINT "N","SQUARE ROOT","CUBE ROOT","FOURTH ROOT","FIFTH ROOT"
4Ø   PRINT
5Ø   FOR N=1 TO 1Ø
6Ø   PRINT N,N↑(1/2),N↑(1/3),N↑(1/4),N↑(1/5)
7Ø   NEXT N
8Ø   END
```

N	SQUARE ROOT	CUBE ROOT	FOURTH ROOT	FIFTH ROOT
1	1.	1.	1.	1.
2	1.41421	1.25992	1.18921	1.1487
3	1.732Ø5	1.44225	1.316Ø7	1.24573
4	2.	1.5874	1.41421	1.31951
5	2.236Ø7	1.7Ø998	1.49535	1.37973
6	2.44949	1.81712	1.565Ø8	1.43Ø97
7	2.64575	1.91293	1.62658	1.47577
8	2.82843	2	1.68179	1.51572
9	3.	2.Ø8ØØ8	1.732Ø5	1.55185
1Ø	3.16228	2.15443	1.77828	1.58489

Approximations from Graphs

We can use graphs to approximate higher-order roots much the same as we used the graph of the parabola $y = x^2$ to approximate square roots. We use the graph of $y = x^3$ to approximate cube roots, the graph of $y = x^4$ to approximate fourth roots, and so on.

The graphs of the functions $y = x^2$, $y = x^3$, $y = x^4$, $y = x^5$, and so on, have very simple shapes. There are, essentially, two basic shapes for these curves: The graphs of $y = x^2$, $y = x^4$, $y = x^6$, $y = x^8$, and so on, have one basic shape, similar to that of a parabola; the graphs of $y = x^3$, $y = x^5$, $y = x^7$, $y = x^9$, and so on, have the other shape (Fig. 7-29). These graphs

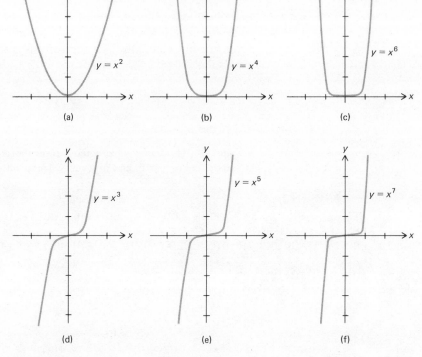

Figure 7-29 The graph of $y = x^n$ ($n = 2, 3, 4, 5, 6, 7$). (a) $y = x^2$; (b) $y = x^4$; (c) $y = x^6$; (d) $y = x^3$; (e) $y = x^5$; (f) $y = x^7$.

can be constructed by plotting a large number of points, then connecting the points with smooth curves.

To locate the kth roots of N, we draw a horizontal line through the point N on the y-axis. This line crosses the curve $y = x^k$ at the points where the x-coordinates are kth roots of N. The situation is pictured in Figure 7-30 for the cube and fourth roots. We see from the graphs that there is only one real cube root of 5, $\sqrt[3]{5} \approx 1.7$, but there are two real fourth roots of 5, approximately equal to 1.5 and -1.5.

Observe also in Figure 7-30 that negative numbers have real cube roots, but not fourth roots. For example, the horizontal line through $y = -2$ intersects the graph of $y = x^3$ at the point with $x \approx -1.3$ [Fig. 7-30(a)]. Thus,

$$\sqrt[3]{-2} \approx -1.3$$

There is no real fourth root of -2. The horizontal line through $y = -2$ does not intersect the graph of $y = x^4$ [Fig. 7-30(b)].

The general situation regarding roots of negative numbers, the number of roots, and so forth, is similar to the result for cube and fourth roots. The typical graph of $y = x^k$ (k odd) is shown in Figure 7-31(a), and the typical graph of $y = x^k$ (k even) is in Figure 7-31(b).

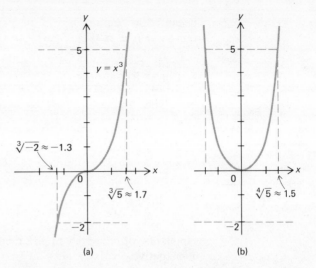

(a)   (b)

Figure 7-30 (a) There is one cube root of 5: $\sqrt[3]{5} \approx 1.7$. There is one cube root of -2: $\sqrt[3]{-2} \approx -1.3$; (b) there are two fourth roots of 5: $\sqrt[4]{5} \times 1.5$ and $-\sqrt[4]{5} \approx -1.5$. There is no real fourth root of -2. The horizontal line through $y = -2$ does not intersect the graph of $y = x^4$.

If k is *odd*, then every real number has exactly one real kth root. If N is *positive*, then $\sqrt[k]{N}$ also is positive. If M is *negative, then* $\sqrt[k]{M}$ also is negative [Fig. 7-31(a)].

If k is *even*, then every *positive* number N has two real kth roots, one positive and one negative. If M is negative, then M has no real kth root

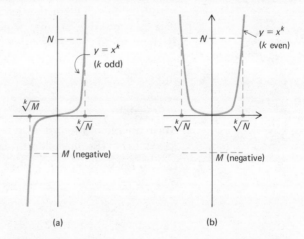

(a)   (b)

Figure 7-31 The graph of $y = x^k$ (k odd and k even). (a) The graph of $y = x^k$ (k odd). Every real number N has exactly one real kth root. If $N > 0$, then $\sqrt[k]{N} > 0$. If $M < 0$, then $\sqrt[k]{M} < 0$; (b) the graph of $y = x^k$ (k even). Every positive real number N has two real kth roots: $\sqrt[k]{N}$ and $-\sqrt[k]{N}$. Negative numbers do not have real kth roots.

because the graph of $y = x^k$ never intersects the horizontal line through the point $y = M$ [Fig. 7-31(b)].

Rational Exponents

FRACTIONAL EXPONENTS
Nicole Oresme, the fourteenth-century French priest-mathematician who became Bishop of Lisieux, was the first person to use fractional exponents to represent roots. Unfortunately, Oresme's notation was awkward, and his explanation was difficult to understand; so root signs were used for the next 200 years rather than fractional exponents.

Fractional exponents can be defined by the following rule

$$N^{a/b} = (\sqrt[b]{N})^a = (N^{1/b})^a$$

Example 4. (a) $3^{2/5} = (\sqrt[5]{3})^2 \approx (1.246)^2 \approx 1.553$
(b) $2^{3/2} = (\sqrt[2]{2})^3 \approx (1.414)^3 \approx 2.828$
(c) $16^{3/4} = (\sqrt[4]{16})^3 = 2^3 = 8$

The rules of exponents that we studied in Chapter 3 all hold with rational exponents, provided all of the numbers being raised to the powers are nonnegative.

LAWS OF EXPONENTS

$a^m \cdot a^n = a^{m+n}$

$(a^m)^n = a^{mn}$

$(ab)^m = a^m \cdot b^m$

$\left(\dfrac{a}{b}\right)^m = \dfrac{a^m}{b^m}$

Example 5. (a) $2^{1/2} \cdot 18^{1/2} = (2 \cdot 18)^{1/2} = (36)^{1/2} = 6$
(b) $(7^6)^{1/3} = 7^{6 \cdot 1/3} = 7^{6/3} = 7^2 = 49$
(c) $\dfrac{\sqrt{12}}{\sqrt{3}} = \dfrac{12^{1/2}}{3^{1/2}} = \left(\dfrac{12}{3}\right)^{1/2} = 4^{1/2} = 2$
(d) $(3 \cdot 2)^{-1/2} = \dfrac{1}{(3 \cdot 2)^{1/2}} = \dfrac{1}{\sqrt{6}}$

EXERCISES

1. Define in words what is meant by the following symbols. Do not actually try to calculate the numbers. (*Example:* $\sqrt[4]{2}$ is the fourth root of 2, the positive number which, when raised to the fourth power, is equal to 2.)

   (a) $\sqrt[2]{3}$ (e) $(-4)^{1/3}$
   (b) $7^{1/5}$ (f) $3^{1/9}$
   (c) $(-3)^{2/3}$ (g) $5^{1/73}$
   (d) $(-5)^{6/7}$ (h) $9^{1/80}$

2. Make a table of the first 10 powers of 2: $2^1 = 2$, $2^2 = 4$, $2^3 = 8$, . . . , $2^{10} = 1024$. Use the table to calculate the following roots.

   (a) $\sqrt[10]{1024}$ (d) $\sqrt[3]{512}$
   (b) $\sqrt[4]{256}$ (e) $\sqrt[7]{128}$
   (c) $\sqrt[3]{64}$ (f) $\sqrt[2]{256}$

3. Plot enough points to construct an accurate graph of $y = x^3$. [Use Fig. 7-30(a) as a guide.] Use your graph to approximate the following cube roots

   (a) $\sqrt[3]{3}$ (c) $\sqrt[3]{-7}$
   (b) $\sqrt[3]{4}$ (d) $\sqrt[3]{-5}$

4. Plot enough points to construct an accurate graph of $y = x^4$. [Use Fig. 7-30(b) as a guide.] Use your graph to approximate the following fourth roots, provided they exist as real numbers.

(a) $\sqrt[4]{0}$ (d) $\sqrt[4]{8}$

(b) $\sqrt[4]{12}$ (e) $\sqrt[4]{-8}$

(c) $\sqrt[4]{7}$ (f) $\sqrt[4]{-2}$

5. Use the rules of exponents to simplify the following expressions.

(a) $5^{3/2} \cdot 5^{1/2}$ (e) $2^{-1/2} \cdot 2^{3/2}$

(b) $28^{1/2} \cdot 7^{1/2}$ (f) $2^{1/2} \cdot 3^{3/2} \cdot 6^{-1/2}$

(c) $28^{1/2}/7^{1/2}$ (g) $4^{1/2} \cdot 9^{3/2}/(5^{1/2} \cdot 25^{1/4})$

(d) $3^{-0.5} \cdot 9^{0.25}$ (h) $7^{-0.3} \cdot 2^{-0.3}/14^{0.7}$

6. Modify the computer program in Example 3 so that the fifth, tenth, fifteenth, and twentieth roots of the numbers 1, 2, 3, . . . , 20 will be calculated and printed. Change the REMARK statements and Statements 30, 50, and 60 as appropriate.

Run the program on your computer.

7. There are two ways to calculate square roots on the computer—N↑(0.5) and SQR(N). In general, the SQR function gives a closer approximation to the square root than the expression N↑(0.5). The following program can be used to calculate both numbers, square them, and compare the closeness of the approximation.

Run the program on your computer using your own data. (The computer will abort the program when it runs out of data.]

```
10   REMARK * PROGRAM TO COMPARE SQR(N) AND N↑(.5)
20   PRINT "N","X=SQR(N)","X*X","Y=N↑(.5)","Y*Y"
30   READ N
40   LET X=SQR(N)
50   LET Y=N↑(.5)
60   PRINT N,X,X*X,Y,Y*Y
70   GO TO 30
80   DATA 5,.2,300,2000
90   END
```

7-8 COMPLEX NUMBERS

Several times in this section we have mentioned a special method for solving quadratic equations. There is a very old formula that can be used for this purpose. We shall develop the formula in Section 7-9. In this section we consider a few of the ideas behind it.

When the early mathematicians began to investigate square roots, they were aware that only nonnegative numbers have real square roots. There is no real number, for example, which can be a square root of -17. This follows from the fact that the square of any nonzero real number is positive. Thus, if x is a nonzero real number, then

$$x^2 > 0$$

so that x^2 cannot be equal to -17.

During the sixteenth century, the European mathematicians made a number of important discoveries in algebra. Certain of these discoveries caused them to think that it might make sense to consider symbols such as $\sqrt{-1}$, $\sqrt{-17}$, and so on. They were aware that these symbols could not represent "real" numbers, and they treated them only as abstract symbols. In some cases, however, they got intermediate results by using these symbols that could later be used to get final results involving "real" numbers.

Gradually these "imaginary" square roots came to be used more and more. During the eighteenth century, the symbol i became the standard abbreviation for $\sqrt{-1}$. Then

$$i = \sqrt{-1}$$

so that

$$i^2 = -1$$

Symbols such as $\sqrt{-9}$, $\sqrt{-17}$, and so on, were defined by using the symbol i:

$$\sqrt{-9} = i\sqrt{9} = 3i$$
$$\sqrt{-17} = i\sqrt{17}$$

and so on.

These numbers were called *imaginary* to distinguish them from the more familiar numbers that represented distances, which were then called *real* numbers.

Numbers such as $3 + 2i$, $5 - 7i$, $13i$, and so on, are called *complex* numbers. They consist of a real part added to an imaginary part. Observe that every real number also is a complex number (with a zero imaginary part) and that every imaginary number is a complex number (with a zero real part).

For example, $5 + 3i$ is a complex number. The real part is 5; the imaginary part is $3i$. The number

$$7 = 7 + 0 \cdot i$$

is a complex number. The real part is 7; the imaginary part is $0 \cdot i$. The

THE COMPLEX PLANE

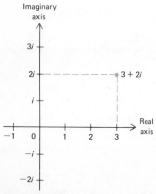

Complex numbers can be represented graphically as points in the plane. The x-axis is the "real" axis with the real numbers 1, 2, 3, and so on, on it. The y-axis is the "imaginary" axis with the numbers i, $2i$, $3i$, and so on. The complex number $3 + 2i$ is represented by the point (3, 2) in the plane.

number

$$13i = 0 + 13i$$

is a complex number. The real part is 0; the imaginary part is $13i$.

Addition and Subtraction

To add or subtract complex numbers, we regroup the numbers, combining the real parts and imaginary parts, then add or subtract within the groupings.

Example 1.
(a) $(3 + 2i) + (4 + 7i) = (3 + 4) + (2i + 7i)$
$$= 7 + 9i$$
(b) $(8i - 9) - (6i - 7) = 8i - 9 - 6i + 7$
$$= (8i - 6i) + (7 - 9)$$
$$= 2i - 2$$

Multiplication

For centuries imaginary numbers were considered to be mathematical oddities. Although they appeared to be necessary in mathematics, they seemed to lack physical significance. When electrical theory was developed, however, it was found that certain of the relationships that hold in electrical circuits can best be described by the use of complex numbers. The historical fact is that imaginary numbers, the tools needed to solve certain types of problems in physics, were invented many years before the problems were discovered.

To multiply complex numbers, we use the basic rules for multiplying algebraic expressions, replacing i^2 with -1 at the end of the calculations.

Example 2.
(a) $(3 - 7i)(4 + 2i) = 3(4 + 2i) - 7i(4 + 2i)$
$$= 12 + 6i - 28i - 14i^2$$
$$= 12 + 6i - 28i - 14(-1)$$
$$= (12 + 14) + (6i - 28i)$$
$$= 26 - 22i$$
(b) $(1 + 4i)(1 - 4i) = 1(1 - 4i) + 4i(1 - 4i)$
$$= 1 - 4i + 4i - 16i^2 = 1 - 16i^2$$
$$= 1 - 16(-1) = 1 + 16 = 17$$
(c) $(4i)^2 = 4i \cdot 4i = 16i^2 = 16(-1) = -16$

Conjugates

Observe in part (b) of Example 2 that the product

$$(1 + 4i)(1 - 4i) = 17$$

In this example the product of two nonreal complex numbers is a positive real number. A similar result can be established for any nonzero complex number. If we let $a + bi$ represent an arbitrary complex number where a and b are real numbers, then

$$(a + bi)(a - bi) = a^2 + abi - abi - b^2i^2$$
$$= a^2 - b^2(-1) = a^2 + b^2$$

which is a positive real number.

The complex number $a - bi$ is called the *conjugate* of $a + bi$. The product of a complex number and its conjugate is always a real number.

For example,

$$(2 + i)(2 - i) = 2^2 - i^2 = 4 - (-1) = 5$$
$$(3 + 5i)(3 - 5i) = 3^2 - 5^2 i^2 = 9 - 25(-1) = 34$$
$$(3 - 7i)(3 + 7i) = 3^2 - 7^2 i^2 = 9 - 49(-1) = 58$$

and so on.

Leonhard Euler
Library of Congress

Leonhard Euler (1707–1783)

$$e^{i\pi} = -1$$

Euler, a Swiss mathematician who lived much of his life in Russia, was the first person to use the symbol i for the square root of -1.

Euler made an extensive study of complex-valued functions, applying the principles of the calculus to them as well as to real-valued functions. The starting point was the identity

$$e^{ix} = \cos x + i \sin x$$

which established a relationship be-tween real-valued functions and complex-valued exponents. (The number e is an irrational number, approximately equal to 2.71828, that is important in higher mathematics. The number x is measured in *radians*, a special way of measuring angles.) Euler used his formula to establish the identity

$$e^{i\pi} = -1$$

which relates the four most important constants in mathematics: 1, i, e, and π.

Simplification of Fractions

The conjugate of a complex number can be useful in simplifying expressions involving fractions. We can change a fraction of two complex numbers into a single complex number by multiplying the numerator and the denominator by the conjugate of the denominator.

Example 3. Simplify

(a) $\dfrac{2 + 5i}{1 + i}$ (b) $\dfrac{6 - 7i}{3 - i}$

Solution. (a) We multiply both the numerator and the denominator by $1 - i$, the conjugate of the denominator:

$$\frac{2 + 5i}{1 + i} = \frac{(2 + 5i)(1 - i)}{(1 + i)(1 - i)} = \frac{2(1 - i) + 5i(1 - i)}{1(1 - i) + i(1 - i)}$$

$$= \frac{2 - 2i + 5i - 5i^2}{1 - i + i - i^2} = \frac{2 - 2i + 5i - 5(-1)}{1 - (-1)}$$

$$= \frac{2 - 2i + 5i + 5}{1 + 1} = \frac{7 + 3i}{2} = \frac{7}{2} + \frac{3}{2}i$$

(b) We multiply both the numerator and the denominator by $3 + i$, the conjugate of $3 - i$:

$$\frac{6 - 7i}{3 - i} = \frac{(6 - 7i)(3 + i)}{(3 - i)(3 + i)} = \frac{18 - 15i - 7i^2}{9 - i^2}$$

$$= \frac{18 - 15i + 7}{9 + 1} = \frac{25 - 15i}{10} = \frac{25}{10} - \frac{15}{10}i$$

$$= \frac{5}{2} - \frac{3}{2}i$$

Solution of Certain Equations

Complex numbers can be used to solve certain equations that cannot be solved with real numbers.

As an example, recall that the equation

$$x^2 = a^2$$

has the solution

$$x = a \qquad \text{and} \qquad x = -a$$

This result was obtained in Section 7-3 by rewriting the equation as $x^2 - a^2 = 0$ and factoring the left-hand side as $(x - a)(x + a)$. This result also holds if a is an imaginary number. For example, the equation

$$x^2 = -16$$

which we can write as

$$x^2 = (4i)^2$$

has the solutions

$$x = 4i \qquad \text{and} \qquad x = -4i$$

Example 4. Solve the equations
(a) $x^2 = -25$
(b) $x^2 + 17 = 0$
(c) $(x - 1)^2 = -9$

Solution. (a)
$$x^2 = -25$$
$$x^2 = 25i^2$$
$$x^2 = (5i)^2$$
$$x = -5i \quad \text{or} \quad x = 5i$$

(b)
$$x^2 + 17 = 0$$
$$x^2 = -17$$
$$x^2 = 17i^2$$
$$x^2 = (\sqrt{17}i)^2$$
$$x = -\sqrt{17}i \quad \text{or} \quad x = \sqrt{17}i$$

(c)
$$(x - 1)^2 = -9$$
$$(x - 1)^2 = (3i)^2$$

If we let $y = x - 1$, then the equation becomes

$$y^2 = (3i)^2$$

which has the solutions

$$y = -3i, \quad y = 3i$$

Since

$$y = x - 1$$

then

$$x - 1 = -3i \quad \text{or} \quad x - 1 = 3i$$
$$x = 1 - 3i \quad \text{or} \quad x = 1 + 3i$$

We frequently combine two solutions such as $x = 1 - 3i$, $x = 1 + 3i$ into the compact form

$$x = 1 \pm 3i$$

The symbol "$\pm$" is read *plus or minus*. The expression

$$x = 1 \pm 3i$$

then represents the two solutions $x = 1 + 3i$, $x = 1 - 3i$. Similarly, the symbol

$$\pm 4i$$

represents the two numbers $4i$, $-4i$.

EXERCISES

1. Calculate the following sums and differences.
   (a) $(5 + i) + (7 - 3i)$
   (b) $(2 + 7i) + (8 - 6i) + 4i$
   (c) $(3 - 11i) - (4i + 3)$
   (d) $(8 + 5i) - (17 + 16i)$

2. Calculate the following products.
   (a) $(5 + i)(7 - 3i)$
   (b) $(2 + 7i)(8 - 6i)$
   (c) $(3 - 11i)(4i + 3)$
   (d) $(8 + 5i)(17 + 16i)$
   (e) $(4i + 9)(3i - 6)$
   (f) $(2i - 1)(2i + 1)$

3. Write out the conjugate of each of the following complex numbers. Then multiply the number by its conjugate.
   (a) $2 + 7i$
   (b) $3i - 2$
   (c) $12 - 5i$
   (d) $-2i + 3$
   (e) 3
   (f) -12
   (g) $-6i$
   (h) $13i$

4. Simplify the following fractions. Write each fraction as a complex number.
   (a) $\dfrac{2 - i}{3 + i}$
   (b) $\dfrac{1 - i}{1 + i}$
   (c) $\dfrac{2i + 7}{3i + 5}$
   (d) $\dfrac{-i + 5}{i + 6}$
   (e) $\dfrac{12 + i}{12 - i}$
   (f) $\dfrac{6}{2 + i}$
   (g) $\dfrac{1}{i}$
   (h) $\dfrac{17 - 4i}{8i}$

5. Solve the following equations.
   (a) $x^2 = -36$
   (b) $x^2 + 1 = 0$
   (c) $4x^2 + 25 = 0$
   (d) $3x^2 + 17 = 0$
   (e) $(x + 2)^2 = -16$
   (f) $(2x + 1)^2 = -25$
   (g) $(3x - 2)^2 + 49 = 0$
   (h) $(4x + 7)^2 + 19 = 0$

7-9   "COMPLETING THE SQUARE"; THE QUADRATIC FORMULA

In the earlier parts of this chapter, we solved quadratic equations that could be factored. We now consider a general method that can be used to solve any quadratic equation.

The key to understanding the method is found in the solution of equations in which two squares are equal, such as

$$(x - 1)^2 = 3^2$$

If we let $y = x - 1$, the equation becomes

$$y^2 = 3^2$$

Special methods for solving quadratic equations were known to the ancient Babylonians 3000 years ago. All rules for the solution were remembered by "recipes," rather than by formulas. Mathematics students studied several examples worked by the same method. Supposedly this taught them the basic principles they needed to solve similar problems.

which has the solutions

$$y = \pm 3$$

Since $y = x - 1$, then

$$x - 1 = \pm 3$$
$$x = 1 \pm 3$$

The method (known as "completing the square") involves rewriting a quadratic equation in such a way that two squares are equal. We then solve the equation as in the above example.

Before we work through an example, let us recall that

$$(x + a)^2 = x^2 + 2ax + a^2$$

This result holds for any number a. For example,

$$(x + 1)^2 = x^2 + 2x + 1$$
$$(x + 2)^2 = x^2 + 4x + 4$$
$$(x - 7)^2 = x^2 - 14x + 49$$

and so on.

The Cubic and Quartic Equations

Not until the sixteenth century was a general method worked out for solving the cubic (third-degree) equation. (This was 3000 years after the quadratic equation had been solved by the Babylonians.)

In 1541 an impoverished Italian mathematician named *Nicolo Tartaglia* learned of a method for solving certain cubic equations. He began bragging of his abilities and was challenged to a contest by another mathematician. Tartaglia began to work in earnest on the problem and devised a general method to solve all cubic equations. When the contest was held, he was able to solve all of the problems posed by the other mathematician within two hours, thus completely vanquishing his opponent and greatly enhancing his own reputation.

Word of Tartaglia's success reached *Jerome Cardan* (*Geronimo Cardano*, 1501–1576), a prosperous physician-mathematician-scholar-astrologer-gambler. Cardan invited Tartaglia to his home, wined him, dined him, and flattered him. Finally, after a solemn oath by Cardan not to reveal his secret, Tartaglia told him his method.

Within a year Cardan published a book on algebra, in which he described Tartaglia's method in detail. Furthermore, the book contained more important results. Cardan's assistant, *Ferrari*, had extended Tartaglia's work to cover quartic (fourth-degree) equations as well. This last result completely eclipsed the work of Tartaglia.

Tartaglia devoted the rest of his life to an unsuccessful attempt to discredit Cardan and Ferrari, never finishing the major book on algebra that he had planned to write.

Suppose now that we have a quadratic equation and the left-hand side is

$$x^2 + 2x$$

What can we add to the left-hand side to make it a perfect square? If we look at the squares of $x + 1$, $x + 2$, and $x - 7$ that are listed above, we see that

$$(x + 1)^2 = x^2 + 2x + 1$$

Thus, if we add 1 to $x^2 + 2x$, we change it into $x^2 + 2x + 1$, which is $(x + 1)^2$.

Similarly, if we add 4 to $x^2 + 4x$, we change it into

$$x^2 + 4x + 4 = (x + 2)^2$$

If we add 49 to $x^2 - 14x$, we change it into

$$x^2 - 14x + 49 = (x - 7)^2$$

and so on.

We now consider two simple examples.

Example 1. Solve the quadratic equation

$$x^2 + 2x = 8$$

Solution. If we add 1 to both sides of the equation, the left-hand side is changed into a perfect square. Therefore,

$$x^2 + 2x = 8$$
$$x^2 + 2x + 1 = 8 + 1 = 9$$
$$(x + 1)^2 = 3^2$$
$$x + 1 = \pm 3$$
$$x = -1 \pm 3$$

The solutions are $x = -1 - 3 = -4$ and $x = -1 + 3 = 2$.

Example 2. Solve

$$x^2 + 4x - 5 = 0$$

Solution. We first rewrite the equation as

$$x^2 + 4x = 5$$

SOLUTIONS OF EQUATIONS

$$ax^2 + bx + c = 0$$

The *quadratic equation* was solved (positive solutions only) by the ancient Babylonians before 1500 B.C.

$$ax^3 + bx^2 + cx + d = 0$$

The *cubic equation* was solved by the Italian mathematician *Tartaglia* in 1541.

$$ax^4 + bx^3 + cx^2 + dx + e = 0$$

The *quartic equation* was solved by *Ferrari* (Cardan's assistant). The result was announced in Cardan's book *Ars magna* (1545).

$$ax^5 + bx^4 + cx^3 + dx^2 + ex + f = 0$$

It is not possible to solve the general equation of degree *five* or higher by a formula. This result was proved in 1824 by the Norwegian mathematician *Abel*. Abel was 19 years old at the time of his discovery.

Observe that if we add 4 to the left-hand side, it is a perfect square. Therefore, we add 4 to both sides of the equation:

$$x^2 + 4x = 5$$
$$x^2 + 4x + 4 = 5 + 4 = 9$$
$$(x + 2) = 3^2$$
$$x + 2 = \pm 3$$
$$x = -2 \pm 3$$

The solutions are

$$x = -2 + 3 = 1 \quad \text{and} \quad x = -2 - 3 = -5$$

The general problem that arises with this method is that of deciding what to add to both sides of the equation in order to make the left-hand side a perfect square. Observe that in both of the above examples, *we added the square of one-half the coefficient of x to both sides.*

If the left-hand side of the equation is of form

$$x^2 + Ax$$

where A is a number, we can always make the left-hand side into a perfect square by adding the square of one-half the coefficient of x. Thus, we must add

$$\left(\frac{A}{2}\right)^2 = \frac{A^2}{4}$$

to both sides of the equation to make the left-hand side a perfect square.

Example 3.
 (a) If the left-hand side of the equation is

$$x^2 + 6x$$

we must add $(\frac{6}{2})^2 = 3^2 = 9$ to both sides of the equation.
 (b) If the left-hand side of the equation is

$$x^2 + 20x$$

we must add $(\frac{20}{2})^2 = 10^2 = 100$ to both sides of the equation.
 (c) If the left-hand side of the equation is

$$x^2 + 5x$$

we must add $(\frac{5}{2})^2 = \frac{25}{4}$ to both sides of the equation.

Example 4. Solve

$$x^2 + 8x + 3 = 0$$

Solution.

$$x^2 + 8x = -3$$

We must add $(\frac{8}{2})^2 = 4^2 = 16$ to both sides of the equation:

$$x^2 + 8x = -3$$
$$x^2 + 8x + 16 = -3 + 16$$
$$(x + 4)^2 = 13$$
$$(x + 4)^2 = (\sqrt{13})^2$$
$$x + 4 = \pm\sqrt{13}$$
$$x = -4 \pm \sqrt{13}$$

The solutions are

$$x = -4 - \sqrt{13} \quad \text{and} \quad x = -4 + \sqrt{13}$$

If the coefficient of x^2 is different from 1, we must divide the equation by the coefficient of x^2 before proceeding further.

Example 5. Solve

$$2x^2 + 6x + 5 = 0$$

Solution. Since the coefficient of x^2 is different from 1, we divide the equation by that number:

$$2x^2 + 6x + 5 = 0$$
$$x^2 + 3x + \tfrac{5}{2} = 0$$
$$x^2 + 3x = -\tfrac{5}{2}$$

We now add $(\frac{3}{2})^2 = \frac{9}{4}$ to both sides of the equation.

$$x^2 + 3x = -\frac{5}{2}$$

$$x^2 + 3x + \frac{9}{4} = -\frac{5}{2} + \frac{9}{4} = -\frac{10}{4} + \frac{9}{4} = -\frac{1}{4}$$

$$\left(x + \frac{3}{2}\right)^2 = \left(\frac{1}{2}i\right)^2$$

$$x + \frac{3}{2} = \pm\frac{i}{2}$$

$$x = -\frac{3}{2} \pm \frac{i}{2}$$

The solutions are

$$x = -\frac{3}{2} + \frac{i}{2} = \frac{-3 + i}{2}$$

and

$$x = -\frac{3}{2} - \frac{i}{2} = \frac{-3 - i}{2}$$

The steps involved in completing the square are outlined in Figure 7-32.

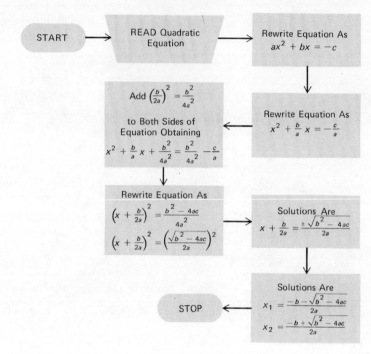

Figure 7-32 Flow chart for the solution of the quadratic equation $ax^2 + bx + c = 0$ by completing the square.

We start with the equation

$$ax^2 + bx + c = 0$$

where a, b, and c represent arbitrary numbers. The final solutions are

$$x = \frac{-b \pm \sqrt{b^2 - 4ac}}{2a}$$

Thus, there are two solutions

$$x = \frac{-b - \sqrt{b^2 - 4ac}}{2a} \quad \text{and} \quad x = \frac{-b + \sqrt{b^2 - 4ac}}{2a}$$

This result is known as the *quadratic formula*.

The Quadratic Formula. The solutions of

$$ax^2 + bx + c = 0$$

are

$$x = \frac{-b \pm \sqrt{b^2 - 4ac}}{2a}$$

We can use the method of completing the square or we can use the quadratic formula to solve any quadratic equation. If we use the formula, we must identify the numbers in the equation with the letters $a, b,$ and c in the formula. Recall that a is the coefficient of x^2, b is the coefficient of x, and c is the constant.

Example 6. Use the quadratic formula to solve
(a) $x^2 - 4x - 21 = 0$
(b) $2x^2 - x - 3 = 0$
(c) $2x^2 + 2x + 5 = 0$
(d) $3x^2 + 2x = 1 - 2x$

Solution. (a) $x^2 - 4x - 21 = 0$
In this equation $a = 1, b = -4, c = -21$. Then

$$x = \frac{-b \pm \sqrt{b^2 - 4ac}}{2a} = \frac{-(-4) \pm \sqrt{(-4)^2 - 4 \cdot 1 \cdot (-21)}}{2 \cdot 1}$$

$$= \frac{4 \pm \sqrt{16 + 84}}{2} = \frac{4 \pm \sqrt{100}}{2} = \frac{4 \pm 10}{2}$$

The solutions are

$$x = \frac{4 - 10}{2} = \frac{-6}{2} = -3$$

and

$$x = \frac{4 + 10}{2} = \frac{14}{2} = 7$$

(b) $2x^2 - x - 3 = 0$
In this equation $a = 2, b = -1, c = -3$. Then

$$x = \frac{-b \pm \sqrt{b^2 - 4ac}}{2a} = \frac{-(-1) \pm \sqrt{(-1)^2 - 4 \cdot 2 \cdot (-3)}}{2 \cdot 2}$$

$$= \frac{1 \pm \sqrt{1 + 24}}{4} = \frac{1 \pm \sqrt{25}}{4} = \frac{1 \pm 5}{4}$$

The solutions are

$$x = \frac{1 - 5}{4} = \frac{-4}{4} = -1$$

and

$$x = \frac{1 + 5}{4} = \frac{6}{4} = \frac{3}{2}$$

(c) $2x^2 + 2x + 5 = 0$

In this equation $a = 2$, $b = 2$, $c = 5$. Then

$$x = \frac{-b \pm \sqrt{b^2 - 4ac}}{2a} = \frac{-2 \pm \sqrt{2^2 - 4 \cdot 2 \cdot 5}}{2 \cdot 2} = \frac{-2 \pm \sqrt{4 - 40}}{4}$$

$$= \frac{-2 \pm \sqrt{-36}}{4} = \frac{-2 \pm 6i}{4}$$

The solutions are

$$x = \frac{-2 + 6i}{4} = \frac{2(-1 + 3i)}{2 \cdot 2} = \frac{-1 + 3i}{2}$$

and

$$x = \frac{-2 - 6i}{4} = \frac{2(-1 - 3i)}{2 \cdot 2} = \frac{-1 - 3i}{2}$$

(d) $3x^2 + 2x = 1 - 2x$

Rewrite the equation as

$$3x^2 + 4x - 1 = 0$$

Then $a = 3$, $b = 4$, $c = -1$. Therefore,

$$x = \frac{-b \pm \sqrt{b^2 - 4ac}}{2a} = \frac{-4 \pm \sqrt{4^2 - 4 \cdot 3 \cdot (-1)}}{2 \cdot 3}$$

$$= \frac{-4 \pm \sqrt{16 + 12}}{2 \cdot 3} = \frac{-4 \pm \sqrt{28}}{6}$$

The solutions are

$$x = \frac{-4 + \sqrt{28}}{6} \quad \text{and} \quad x = \frac{-4 - \sqrt{28}}{6}$$

These numbers can be simplified further if we observe that

$$\sqrt{28} = \sqrt{4 \cdot 7} = \sqrt{4}\sqrt{7} = 2\sqrt{7}$$

One solution is

$$x = \frac{-4 + \sqrt{28}}{6} = \frac{-4 + 2\sqrt{7}}{6} = \frac{\cancel{2}(-2 + \sqrt{7})}{\cancel{2} \cdot 3} = \frac{-2 + \sqrt{7}}{3}$$

The other solution is

$$x = \frac{-4 - \sqrt{28}}{6} = \frac{-4 - 2\sqrt{7}}{6} = \frac{\cancel{2}(-2 - \sqrt{7})}{\cancel{2} \cdot 3} = \frac{-2 - \sqrt{7}}{3}$$

EXERCISES

1. Solve the following quadratic equations by "completing the square."
   (a) $x^2 + 2x + 5 = 0$ (g) $5x^2 + 3x - 2 = 0$
   (b) $x^2 - 6x + 10 = 0$ (h) $x^2 - 4x - 8 = 0$
   (c) $x^2 - 3x + 2 = 0$ (i) $2x^2 + 6x + 5 = 0$
   (d) $2x^2 + 6x = -5 - x$ (j) $x^2 + 4x + 13 = 0$
   (e) $3x^2 + 9x + 6 = x + 1$ (k) $3x^2 + 5x + 2 = 0$
   (f) $14x^2 - 9x + 1 = 0$ (l) $8x^2 - 6x + 1 = 0$

2. Solve the equations in Exercise 1 by the use of the quadratic formula.

3. Suppose that a, b, and c are real numbers. Make a careful study of the terms in the quadratic formula. Show that
   (a) if $b^2 - 4ac > 0$, then the equation $ax^2 + bx + c = 0$ has two real solutions;
   (b) if $b^2 - 4ac = 0$, then the equation $ax^2 + bx + c = 0$ has just one real solution;
   (c) if $b^2 - 4ac < 0$, then the equation $ax^2 + bx + c = 0$ has two complex solutions which are conjugates of each other.

   Compare this result with Exercise 9 of Section 7-5.

7-10 VARIATION

In much of this chapter we have concentrated on topics related to quadratic equations. Such equations are only a small part of algebra. Almost every branch of mathematics has been developed as much as our work with equations that define parabolas and similar curves.

In this section we consider problems related to *variation*—the study of how one quantity changes with respect to other quantities. Many important problems can be expressed in terms of varying quantities.

Direct Variation

One type of variation was considered in Chapter 5—*direct variation*. We say that *y varies directly* with x if there is a constant $m \neq 0$ such that

$$y = mx \qquad \text{(direct variation)}$$

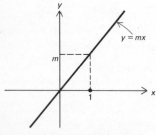

If y varies directly with x, then

$$y = mx$$

for a constant m. The graph of y as a function of x is a straight line through the origin with slope m.

For every change in x of 1 unit, there is a corresponding change in y of m units.

In other words, y varies directly with x provided y *is proportional to* x.

If y varies directly with x, then the quantities x and y have the number m as a constant ratio. If we know the value of y for one value of x, then we can calculate m. The value of m then can be used to calculate y for any value of x.

Example 1. It cost the Flatland Power Company $10,500 to construct 3 miles of high-tension power lines. Assume that the cost of construction is proportional to the distance. How much will it cost to construct 20 miles of power lines?

Solution. Let x be the number of miles and y be the total cost. Since y is directly proportional to x, then there is a constant m such that

$$y = mx$$

To find m we use the fact that $y = 10,500$ when $x = 3$:

$$y = mx$$
$$10,500 = m \cdot 3$$
$$3m = 10,500$$
$$m = \frac{10,500}{3} = 3,500$$

Thus, the equation relating x and y is

$$y = 3,500x$$

When

$$x = 20$$
$$y = 3,500 \cdot 20 = 70,000$$

In some problems y may vary directly with the *square* or the *cube* of x. In these cases the equation will be of the form

$$y = mx^2 \qquad \text{or} \qquad y = mx^3$$

where m is a constant.

Example 2. The *braking distance* of an automobile is the minimum distance required to stop after the brakes are applied. It can be shown that *the braking distance varies directly with the square of the speed.*

Joe Student measured that his car has a braking distance of 90 feet at 40 miles per hour. What is the braking distance at 80 miles per hour?

Solution. Let b represent the braking distance and s represent the speed. Since b is directly proportional to the *square* of s, there exists a constant m such that

$$b = ms^2$$

We know that $b = 90$ when $s = 40$. Therefore,

$$b = ms^2$$
$$90 = m \cdot 40^2$$
$$90 = 1600m$$
$$m = \frac{90}{1600} = \frac{9}{160}$$

Therefore, the actual equation relating s and b is (Fig. 7-33)

$$b = \frac{9}{160}s^2$$

When $s = 80$,

$$b = \frac{9}{160}s^2 = \frac{9}{160}(80)^2 = \frac{9}{160} \cdot 6400 = 360$$

The braking distance is 360 feet at 80 miles per hour.

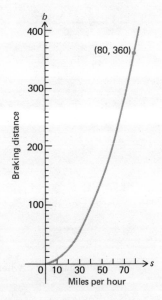

Figure 7-33 The braking distance varies with the square of the speed (Example 2). $b = \frac{9}{160} \cdot s^2$.

Joint Variation

If a quantity z varies directly with the *product* of x and y, we say that it varies *jointly* with respect to x and y.

In other words, z varies jointly with respect to x and y if there is a constant m such that

$$z = mxy \qquad \text{(joint variation)}$$

As with direct variation, if we know the value of z for particular values of x and y, we can calculate m, then use the value of m to calculate z for any values of x and y.

Example 3. The capacity of a tin can varies jointly with the height of the can and the square of the diameter of the can. A standard $\#303$ can is 3 inches in diameter, 4 inches high (neglecting the rim), and holds approximately 2 cups.

A standard $\#1\frac{1}{2}$ can is $3\frac{1}{4}$ inches in diameter and is $3\frac{1}{4}$ inches high. How much does it hold?

Solution. Let x be the height of a can (inches), y be the diameter (inches), and z be the capacity (cups). (See Fig. 7-34.) Since z varies jointly with x and y^2, then there is a constant m such that

$$z = mxy^2$$

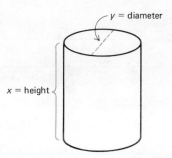

Figure 7-34 The capacity z varies jointly with the height x and the square of the diameter y (Example 3). $z = mxy^2$ $(m \approx \frac{1}{18})$.

We know that $z \approx 2$ when $x = 4$ and $y = 3$. Therefore,

$$z = mxy^2$$
$$2 \approx m \cdot 4 \cdot 3^2 = m \cdot 4 \cdot 9$$
$$2 \approx 36m$$
$$m \approx \tfrac{2}{36} = \tfrac{1}{18}$$

Therefore, the formula relating the capacity z to the height x and the diameter y is

$$z \approx \tfrac{1}{18}xy^2$$

When $x = 3.25$ and $y = 3.25$, we have

$$z \approx \tfrac{1}{18}xy^2$$

$$\approx \tfrac{1}{18}(3.25)(3.25)^2 \approx \frac{34.33}{18} \approx 2$$

Thus, a standard $\#1\frac{1}{2}$ can holds approximately 2 cups.

Inverse Variation

There is one other important type of variation that we consider. We say that y *varies inversely with* x if there is a constant m such that

$$y = \frac{m}{x} \qquad \text{(inverse variation)}$$

The graph of $y = m/x$ is worth examining. If we fix a value of $m > 0$, plot a large number of points for positive values of x, and connect them with a smooth curve, we get the graph in Figure 7-35(a). Observe that the larger the value of x, the closer y is to zero and, conversely, the closer x is to zero, the larger the value of y.

THE HYPERBOLA

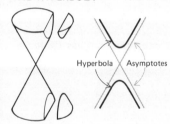

The graph of

$$y = \frac{m}{x}$$

is an example of a *hyperbola*. A typical hyperbola is shown in the figure. It is a two-branched curve obtained by slicing a double-napped cone with a plane parallel to the axis of the cone.

The original boundary lines of the cone are *asymptotes* to the hyperbola. As we trace along one of the branches of the curve, it gets closer and closer to one of the asymptotes, but never actually touches it.

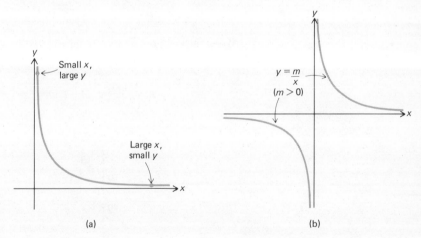

Figure 7-35 (a) The inverse variation graph $y = m/x$ (where m, x, and y are all positive); (b) the complete graph of $y = m/x$ ($m > 0$) consists of two distinct curves.

The complete graph of $y = m/x$ (for $m > 0$) is shown in Figure 7-35(b). The graph consists of two distinct branches. Most inverse variation problems involve only that branch of the graph with x and y both positive.

Many inverse variation problems occur in physics. The following problem is typical.

Example 4.   The pressure exerted by a gas on its container (at a fixed temperature) varies inversely with the volume of the container.

A certain quantity of compressed air, enclosed in an 8-cubic-foot container, exerts a pressure of 50 pounds per square inch on the container. This compressed air is pumped into a larger container with a capacity of 12 cubic feet. What pressure does it exert on the larger container?

Solution.   Let P represent the pressure and V the volume. Then

$$P = \frac{m}{V}$$

for some constant m. When $V = 8$, we know that $P = 50$. Thus,

$$P = \frac{m}{V}$$

$$50 = \frac{m}{8}$$

$$m = 8 \cdot 50 = 400$$

The actual equation relating the pressure and volume is

$$P = \frac{400}{V}$$

When the compressed air is pumped into the larger container, it has a volume of 12 cubic feet. The new pressure is (Fig. 7-36)

$$P = \frac{400}{12} = 33\frac{1}{3} \text{ pounds per square inch.}$$

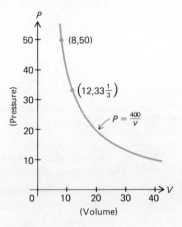

Figure 7-36   Example 4.

EXERCISES

1. Change each of the following equations to a sentence involving the word "varies." The symbols $x, y,$ and z represent varying quantities; a and b represent positive constants. (*Example:* The equation $y = 2x$ can be changed to "y varies directly with x.")
   (a) $y = 3xz$
   (d) $z = x/a$
   (b) $z = 2/x$
   (e) $y = bx^3$
   (c) $y = 5x$
   (f) $z = a/x^2$

2. It is known that $y = 5$ when $x = 2$. Express y as a function of x:
   (a) if y varies directly with x;
   (b) if y varies directly with x^2;
   (c) if y varies inversely with x;
   (d) if y varies inversely with x^2.

3. *The braking distance of a car varies directly with the square of the speed.* Determine the equation for the braking distance of each of the following cars. Then calculate the braking distance at 80 miles per hour. (See Example 2.)
   (a) Car A: braking distance of 80 feet at 30 miles per hour
   (b) Car B: braking distance of 70 feet at 40 miles per hour
   (c) Car C: braking distance of 25 feet at 20 miles per hour

4. *The value of a diamond of standard quality varies directly with the square of its weight.*
   A one-carat diamond sells for $1600. Calculate the values of the following diamonds of the same quality.
   (a) 2 carats
   (c) 8 carats
   (b) $\frac{1}{2}$ carat
   (d) $\frac{1}{4}$ carat

5. *The capacity of a tin can varies jointly with the height* (neglecting the rim) *and the square of the diameter.* In Example 3 we saw that a can 4 inches high and 3 inches in diameter holds approximately 2 cups.
   Calculate the approximate capacities (in *cups*) of the following size tin cans. In each case the height is measured by neglecting the rim.
   (a) *Soup can:* $3\frac{5}{8}$ inches high, $2\frac{1}{2}$ inches in diameter
   (b) *Seafood can:* $1\frac{13}{16}$ inches high, $3\frac{3}{8}$ inches in diameter

6. *The pressure exerted by a quantity of gas on its container varies inversely with the volume of its container* (Example 4).
   A certain quantity of gas, enclosed in a 4-cubic-foot container, exerts a pressure of 30 pounds per square inch.
   (a) What pressure would the gas exert if it were pumped into a 2-cubic-foot container?
   (b) What pressure would the gas exert if it were pumped into a 1-cubic-foot container?

7. One of the basic laws of astronomy was discovered by Kepler: *The time required for a planet to make one complete revolution about the sun varies directly with the square root of the cube of its average distance from the sun.*

(a) Express this law as an equation with x denoting the average distance from the sun and y denoting the length of time for one complete revolution.

(b) We know that the earth is approximately 93 million miles from the sun and makes one complete revolution in one year. Use this information to evaluate the constant in your equation.

(c) The planet Saturn is approximately 900 million miles from the sun. How long does it take Saturn to make one complete revolution of the sun?

7-11 EXPONENTIAL FUNCTIONS

The exponential function is one of the most important of the functions that occur in nature and society. This function occurs in problems concerning population growth, compound interest, investment growth, depreciation, inflation rates, and radioactive decay (the changing of radioactive substances such as uranium into stable substances such as lead).

To introduce this function, we consider a very simple idealized example.

A characteristic property of exponential functions is that the rates at which the functions change are proportional to the values of the functions. Thus, exponential functions occur in such diverse applications as *investment analysis* (rate of change of the value is proportional to the size of the investment), *population growth* (rate of change of the population is proportional to the size of the population), and *radioactive decay* (the rate at which a radioactive substance changes to lead is proportional to the amount of the substance).

Example 1. Bacteria of a certain strain are known to divide every hour, producing two bacteria for every one that previously existed.

A single one of these bacteria is breathed into George's lung (a favorable abode for such organisms).

(a) How many will live there at the end of 6 hours?

(b) How many at the end of t hours?

Solution. We make a table of values showing N, the number of bacteria at the end of t hours ($t = 1, 2, 3, 4, 5, 6$). We measure our time from the instant the single bacterium begins to live in the lung.

Time (hours)	Number of Bacteria in Lung
$t = 0$	$N = 1$
$t = 1$	$N = 2$
$t = 2$	$N = 4$
$t = 3$	$N = 8$
$t = 4$	$N = 16$
$t = 5$	$N = 32$
$t = 6$	$N = 64$

(a) At the end of 6 hours, 64 bacteria are in the lung.

(b) To find the number after t hours, observe that if

$$t = 1, \quad N = 2^1$$
$$t = 2, \quad N = 2^2$$
$$t = 3, \quad N = 2^3$$
$$t = 4, \quad N = 2^4$$

and so on. At the end of t hours, the number is

$$N = 2^t$$

HALF-LIFE

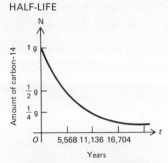

Years

The *half-life* of a radioactive substance is the length of time that it
takes one-half of the substance to
decompose into a different substance (usually lead). Radioactive
carbon (carbon-14), for example,
has a half-life of 5,568 years. After
5,568 years one-half of the original amount will remain; after an
additional 5,568 years one-fourth
will remain; after an additional
5,568 years one-eighth will remain, and so on.

Example 1 is a simplified version of what actually happens when bacteria
invade an organism. In a typical example, several million bacteria live in
the organism. Each of these will divide into two bacteria every hour (or
some other period of time). Furthermore, the times at which the bacteria
divide into two are scattered over the hour.

Example 2. Fifteen million bacteria live in an organism. Each of these divides
into two every hour.
  (a) How many will there be after 6 hours?
  (b) How many after t hours?
  (c) How many after $\frac{1}{2}$ hour?

Solution. We make a table of values:

Time (hours)	Number of Bacteria (millions)
$t = 0$	15
$t = 1$	$30 = 2 \cdot 15$
$t = 2$	$60 = 2^2 \cdot 15$
$t = 3$	$120 = 2^3 \cdot 15$
$t = 4$	$240 = 2^4 \cdot 15$
$t = 5$	$480 = 2^5 \cdot 15$
$t = 6$	$960 = 2^6 \cdot 15$

  (a) After 6 hours there are 960 million bacteria in the organism.
  (b) After t hours there are

$$N = 15 \cdot 2^t \text{ million}$$

bacteria in the organism. (See Fig. 7-37).

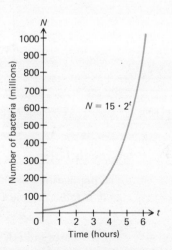

Figure 7-37 Number of bacteria after t hours (Example 2). Fifteen million bacteria at time $t = 0$. The
number doubles every hour. $N = 15 \cdot 2^t$.

(c) Because the bacteria divide continuously over each hour, we can use the formula

$$N = 15 \cdot 2^t$$

when $t = \frac{1}{2}$. The number of bacteria after $\frac{1}{2}$ hour is

$$15 \cdot 2^{1/2} = 15 \cdot \sqrt{2} \approx 15 \cdot (1.414) \approx 21.2 \text{ millions}$$

The function

$$N = 15 \cdot 2^t$$

is an example of an *exponential function*. If we use x and y for the variables, then the most general exponential function is

$$y = a \cdot b^x$$

where a and b are constants, $b > 0$. Examples of exponential functions are

$$y = 3^x$$
$$y = 5 \cdot 7^x$$
$$y = 8 \cdot (\tfrac{1}{2})^x$$

and so on.

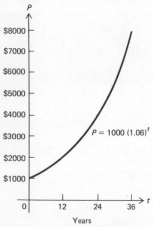

$P = 1000 \,(1.06)^t$

Years
Doubling time at 6-percent
compound interest is
12 years.

Compound Interest Rate (percent)	Approximate Doubling Time (years)
4	18
5	14
6	12
7	10
8	9

If an exponential function increases in size with time, then it doubles over a fixed period of time.

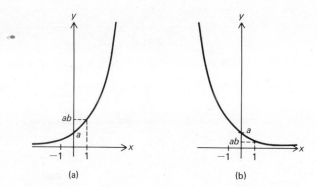

Figure 7-38 Graph of $y = a \cdot b^x$. (a) $b > 1$; (b) $b < 1$.

Growth Rates

The population growth of a city or country is usually expressed as a yearly *growth rate*. This rate is the ratio

$$\text{yearly growth rate} = \frac{\text{change in population over the year}}{\text{population at beginning of the year}}$$

and is frequently expressed as a percent.

For example, if the population of a city at the beginning of a year is 200,000 and at the end of the year is 204,000, then the increase in population is 4,000 and

$$\text{growth rate} = \frac{4,000}{200,000} = 0.02 = 2 \text{ percent}$$

In most cases a growth rate remains virtually constant over a period of several years. This usually holds even though the actual increase in population may vary from year to year. If we know the growth rate and the original population, we can calculate the increase and the size of the population after a year. Observe first that

increase in population = (growth rate) · (original size of population)

and that

population at end of year = (original population) + (increase)

For example, if the original population was 100,000 persons and the growth rate is 3 percent per year, then

increase of population = (growth rate) · (original size)
$$= (0.03)(100,000) = 3,000$$

and

population at end of year = (original population) + (increase)
$$= 100,000 + 3,000 = 103,000$$

We shall show that if the growth rate is constant for several years, then the population is an exponential function of time. To do this we let

P_t = the population at the end of t years

Then

P_0 = original population
P_1 = population at the end of 1 year
P_2 = population at the end of 2 years
P_3 = population at the end of 3 years

and so on. Suppose the growth rate is 0.03 per year. Then

$$P_1 = P_0 + \text{(increase in population)}$$
$$= P_0 + 0.03P_0 = (1.03)P_0$$
$$P_2 = P_1 + \text{(increase in population)}$$
$$= P_1 + 0.03P_1$$
$$= (1.03)P_1$$
$$= (1.03)(1.03)P_0 = (1.03)^2P_0$$
$$P_3 = P_2 + \text{(increase in population)}$$
$$= P_2 + 0.03P_2$$
$$= (1.03)P_2$$
$$= (1.03)(1.03)^2P_0 = (1.03)^3P_0$$

and so on. If we continue this argument, we find that the population after t years is

$$P_t = (1.03)^t P_0$$

A similar argument can be used for any constant growth rate.

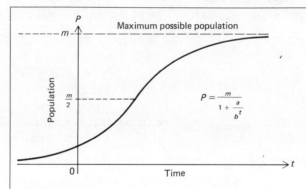

Population Growth

The assumption that the rate of change of a population is proportional to its size is reasonable over a short period of time. Over a long period, however, this condition cannot hold. No population can continue to increase at an ever-faster rate. Eventually inhibiting factors (lack of food or space, pollution, and so on) must curb the growth rate. Thus, for any given set of environmental conditions there is a maximum possible population that can be maintained.

If m is the maximum possible population, then both P (the size of the population) and $m - P$ (the potential for the population to increase) influence the rate of growth. It has been shown in experimental studies that in

many cases the rate of change is jointly proportional to P and $m - P$. Under those assumptions, the equation for the growth rate is

$$P = \frac{m}{1 + \dfrac{a}{b^t}}$$

where a and b are constants determined by the growth rate and the initial population.

Observe from the graph that the population increases more and more rapidly until $P = m/2$. At that time the growth rate slows down, and eventually the population levels off at a value slightly less than the value of m (the theoretical maximum).

Most of the exponential functions that must be calculated involve powers such as

$$(1.01)^3, \quad (1.02)^{15}, \quad (1.03)^7$$

and so on. Table IV at the end of the book contains the most common values of such exponential functions.

Example 3. The city of Fetchuni had a population of

$$P_0 = 200{,}000$$

persons five years ago and has had a growth rate of 2 percent per year since that time.
(a) What is the population after t years?
(b) What is the current population?

Solution. (a) Since the growth has been constant at 2 percent per year, then

$$P_1 = \text{population after 1 year} = (1.02)P_0$$
$$P_2 = \text{population after 2 years} = (1.02)^2 P_0$$
$$P_3 = \text{population after 3 years} = (1.02)^3 P_0$$

and so on. The general function for the population is

$$P_t = (1.02)^t P_0$$

(b) The current population is

$$P_5 = (1.02)^5 P_0$$
$$= (1.02)^5 (200{,}000)$$
$$\approx (1.104)(200{,}000) \qquad \text{(by Table IV)}$$
$$\approx 220{,}800$$

Exponential functions can be used to describe any type of function that has a constant rate of increase. For example, if money is invested at a fixed rate of interest per year, then the total value of the investment after t years is an exponential function of time.

Example 4. John Thrift invested \$1000 at 5 percent interest compounded yearly.
(a) What is the value of the investment after 4 years?
(b) What is the value of the investment after t years?

Solution. (a) Let P_t be the value of the investment after t years. Then

$$P_0 = 1000$$
$$P_1 = \text{value after 1 year} = P_0 + 0.05P_0 = (1.05)P_0$$
$$P_2 = \text{value after 2 years} = P_1 + 0.05P_1 = (1.05)P_1$$
$$= (1.05)(1.05)P_0 = (1.05)^2 P_0$$
$$P_3 = \text{value after 3 years} = P_2 + 0.05P_2 = (1.05)P_2$$
$$= (1.05)(1.05)^2 P_0 = (1.05)^3 P_0$$
$$P_4 = \text{value after 4 years} = P_3 + 0.05P_3 = (1.05)P_3$$
$$= (1.05)(1.05)^3 P_0 = (1.05)^4 P_0$$

Thus, the value after 4 years is

$$P_4 = (1.05)^4 P_0 = (1.05)^4 \cdot 1000 \approx (1.2155)(1000) \qquad \text{(by Table IV)}$$

$$\approx \$1215.50$$

(b) If we continue the argument in (a), we find that the value after t years is

$$P_t = (1.05)^t P_0$$

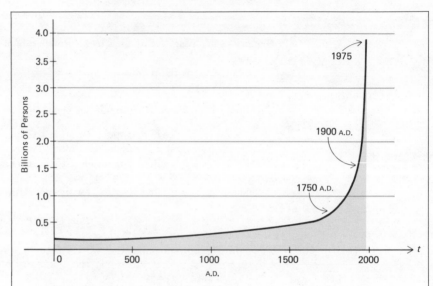

The World Population Explosion

The first person to fully grasp the mechanics of population growth was *Thomas Malthus,* an eighteenth-century Englishman. Malthus observed that it is natural for a population to increase exponentially, but that the production of food must increase at a slower rate. Thus, any period of explosive population growth must be followed by a period of famine unless food production is somehow increased to match the population increase. Since food supplies are limited by the amount of land available for farming, there must be eventual famine if the population continues to increase exponentially.

Malthus pointed out that wars, famine, and plagues are the natural companions of population growth and that these disasters help to correct the overcrowded conditions.

In our century it appears that Malthus' gloomy predictions may come true. Most economists predict large-scale famine in the underdeveloped countries during the next few years because of the rapid growth of population.

EXERCISES

1. Calculate the values of the exponential functions for the indicated values of x and t. Use Table IV for (i) and (j).

(a) $y = 3^x$; $x = 2, 3$

(b) $y = 5 \cdot 2^x$; $x = 0, 1, 5$

(c) $y = 4(\frac{1}{2})^x$; $x = 0, 2, 8$

(d) $y = 2 \cdot (\frac{1}{3})^x$; $x = -1, 0, 2$

(e) $y = 4 \cdot 3^x$; $x = -1, 1$

(f) $N = 2 \cdot 10^t$; $t = -2, 0, 2$

(g) $N = \frac{1}{2} \cdot (\frac{1}{3})^t$; $t = 1, 2$

(h) $N = (\frac{1}{10})^t$; $t = -2, 0, 1$

(i) $N = (1.02)^t$; $t = 3, 17$

(j) $N = (1.05)^t$; $t = 10, 20$

2. The population of a small city was 50,000 ten years ago. Since that time the population increased at the rate of 5 percent per year.

   (a) Express the population P as a function of time t (years).

   (b) What was the population after 3 years at this growth rate?

   (c) What is the current population (use Table IV)?

3. Mr. Moneybags has invested $100,000 in setting up small loan offices around the state. The average return on his investment is 1 percent per month, which is reinvested in other small loan offices.

   (a) What is the value of his investment after 1 year (use Table IV)?

   (b) What is the value of his investment after t years?

4. Houses in Bleaksburg have increased in value an average of 5 percent each year for the past ten years. Mr. Averageman bought a house for $20,000 ten years ago.

   (a) Express the value of the house t years after its purchase as a function of t.

   (b) Use Table IV to determine its current value.

5. Automobiles depreciate at the rate of 20 percent per year. (At the end of a year, the automobile is worth only 80 percent of its value at the beginning of the year.) Joe Schnook just bought a new Sleek-Zip sports car for $8000.

   (a) What will be its value in 3 years?

   (b) What will be its value in t years?

SUGGESTIONS FOR FURTHER READING

1. Cooley, Hollis R., and Wahlert, Howard E. *Introduction to Mathematics*, Chap. 6, 7. Boston: Houghton-Mifflin, 1968.

2. Eves, Howard. *An Introduction to the History of Mathematics*, 3rd ed., Chap. 7. New York: Holt, Rinehart & Winston, 1969.

3. Kline, Morris. *Mathematics for Liberal Arts*, Chap. 5, 13, 15. Reading, Mass.: Addison-Wesley, 1967.

4. Ore, Oystein. *Cardano, the Gambling Scholar*. New York: Dover, 1965.

8

TOPICS IN GEOMETRY

8-1 INTRODUCTION; ELEMENTARY CONSTRUCTIONS

Although the philosopher Plato made no important contributions to mathematics himself, his school at Athens was the world's center of mathematical studies for almost a century.

Plato firmly believed that all geometrical figures should be constructed by ruler-and-compass methods. It probably was his influence that caused the artificial emphasis on this technique among the ancient Greeks.

Plato was primarily concerned with the rigor of mathematical arguments—a rigor that he carried over to arguments in philosophy. He believed that all curves should be constructed in a rigorous way by means of lines and previously constructed curves, thereby ensuring that the theoretical curve had actually been constructed, not just an approximation to it.

Our mathematics has its roots in the work done by the ancient Greeks of the fifth to third centuries B.C. These men held deep beliefs in the perfection of art, philosophy, and nature and developed theories of nature and mathematics that incorporated those beliefs. Assuming that the point, the line, and the circle were perfect geometrical figures, they based much of their theory of nature on these figures—especially the circle.

The early Greek philosophers theorized that the earth is a sphere (a figure obtained by rotating a circle about its center), that the sun (a circular disk of light) the stars, and planets (points of light) move around the earth in circular orbits.

Because of their theories about perfection, which they believed were confirmed by observations of nature, the early Greeks used points, lines, and circles as the basic figures in their geometrical constructions. This is justified in an ancient story about the altar in the temple of Apollo at Delos. A delegation was sent to the temple to learn how to avoid the plague which was ravaging the countryside. They were informed by the oracle that Apollo wanted the altar, which was a perfect cube, to be doubled in size. Dutifully, the worshipers constructed a new altar twice as long, twice as wide, and twice as high, but to no avail—the plague struck anyway. They then were told by the oracle that they had not fulfilled Apollo's request. The new altar was eight times the volume of the original altar, whereas Apollo wanted one exactly double the volume. Furthermore, *the dimensions of the new cube must be worked out geometrically from the original cube, using only a straight edge and a compass. No other instruments could be used.* In other words, only points, lines, and circles could be used in the construction.

The Greeks actually attempted to construct all geometrical figures with a straight edge (for drawing lines between given points) and a compass (for drawing arcs and circles with one given point as center and another on the boundary). These restrictions were cultural, not mathematical. The Greeks limited themselves in this manner because they could not believe that a perfect subject, such as mathematics, was not based on perfect figures, such as lines and circles.

In the remainder of this section, we shall consider a few of the basic geometrical constructions that the Greeks performed. In our work we shall restrict ourselves to the ground rules that the Greeks established and shall use only a straight edge and a compass. Since the straight edge does not have markings and the compass can only be used to draw circles with one known point as center and another on the boundary, then lengths cannot be transferred directly. As we shall see, however, there is a rather complicated construction (*Construction 4*) that allows us to transfer lengths.

Construction 1. Given: points A and B on a line.

Construct: a line midway between A and B that is perpendicular to the given line.

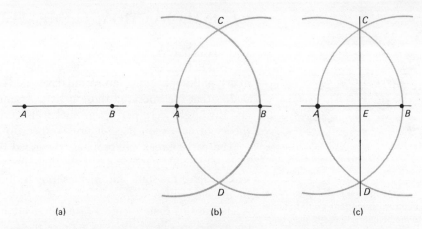

Figure 8-1 Construct the perpendicular bisector of a line segment (Construction 1). (a) Given: the line segment *AB;* (b) construct a circle with *A* as center and *B* on the boundary. Construct a second circle with center at *B* and *A* on the boundary; and (c) the line segment *CD* is perpendicular to *AB* and bisects *AB* at *E*.

Solution. (1) Draw two circles, one with *A* as center and *B* on the boundary, the other with *B* as center and *A* on the boundary. Let *C* and *D* be the points where the two circles intersect [Fig. 8-1(b)].

(2) Draw a line through points *C* and *D*. This line is perpendicular to *A* and *B*. It crosses line *AB* at point *E*, which is the midpoint between *A* and *B* [Fig. 8-1(c)].

Construction 2. *Given:* a line and a point *P* (which may or may not be on the line).

Construct: a line through *P* that is perpendicular to the given line.

Solution. (1) Choose a convenient point *A* on the line such that the line from *A* to *P* is not perpendicular to the given line. Draw a circle with *P* as center and *A* on the boundary. Let *B* be the other point where the circle intersects the given line [Fig. 8-2(b)].

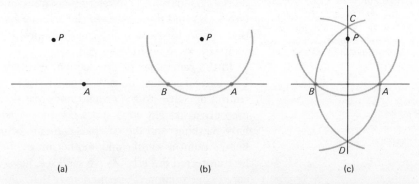

Figure 8-2 Construct a line segment through a given point perpendicular to a given line (Construction 2). (a) Given: a line and a point *P*. Choose point *A* on the line; (b) construct a circle with center at *P* and with *A* on the boundary. Let *B* be the other point where the circle intersects the line; and (c) use Construction 1 to construct the perpendicular bisector of the line segment *AB*.

(2) Complete the construction as in Construction 1. The line CD is perpendicular to the given line and passes through P [Fig. 8-2(c)].

Construction 3. *Given:* a line and a point P not on the line.
Construct: a line through P parallel to the given line.

Solution. (1) Use Construction 2 to construct a line PQ through P that is perpendicular to the given line [Fig. 8-3(b)].

(2) Use Construction 2 to construct a line through P that is perpendicular to the line PQ. This line is parallel to the given line [Fig. 8-3(c)].

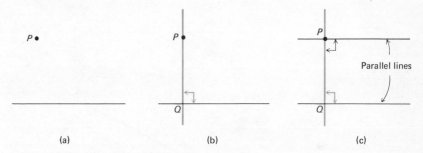

(a) (b) (c)

Figure 8-3 Construct a line through a given point parallel to a given line (Construction 3). (a) Given: a line and a point P not on the line; (b) construct a line through P perpendicular to the given line; and (c) construct a line through P perpendicular to the line constructed in (b).

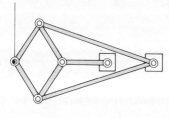
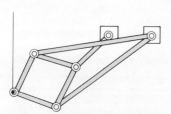
As we mentioned above, there is a construction that allows us to transfer lengths. (Recall that the straight edge and compass are not allowed to be used for this purpose.) The following construction, which uses only Constructions 1, 2, and 3, shows how lengths can be transferred. This construction is much more complicated than the first three constructions.

Construction 4. *Given:* line segment AB, a point P not on the line AB, and a line PQ.
Construct: a line segment on PQ, with one end at P, which has length equal to segment AB.

Solution. (1) Draw line PA.

(2) Draw line BR through B parallel to line PA (Construction 3). [See Fig. 8-4(b).]

(3) Draw PS through P parallel to line AB (Construction 3). Let T be the point where PS intersects BR [Fig. 8-4(c)]. Observe that the figure $PABT$ is a parallelogram. Thus, the opposite sides are equal. It follows that $|PT| = |AB|$.

(4) Draw a circle with P as the center and T on the boundary. Let V be the point where this circle intersects the given line PQ. Then

$$|PV| = |PT| \quad \text{and} \quad |PT| = |AB|$$

so that the distance $|PV|$ is equal to the distance $|AB|$ as desired [Fig. 8-4(d)].

Construction 4 shows that lengths of line segments can be transferred almost anywhere that we wish in the plane. The only defect in the con-

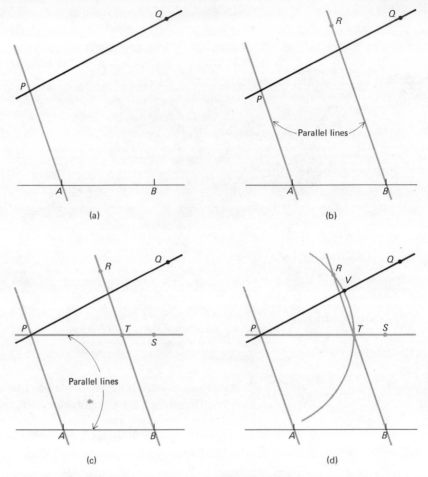

Figure 8-4 The transfer of lengths (Construction 4). (a) Given: line segments *AB* and *PQ*. Construct segment *AP;* (b) construct a line through *B* parallel to *AP;* (c) construct *PS* parallel to *AB*. Observe that |*PT*| = |*AB*|; and (d) construct a circle through *T* with center at *P*. Then |*PV*| = |*PT*| = |*AB*|.

struction is that the point *P* is not allowed to be on the line *AB*. If *P* is on the line *AB*, however, we can modify the construction in order to transfer the length |*AB*| to the point *P* (Exercise 8).

Once we know that lengths can be transferred, we can construct triangles with three given lengths as sides.

Construction 5. *Given:* a line *AB*, a point *P* on *AB*, and three line segments *CD, EF,* and *GH*.

Construct: a triangle with one vertex at *P*, one side on line *AB*, with the three sides equal in length to the three line segments *CD, EF,* and *GH*.

Solution. (1) Transfer the length |*CD*| to point *P*. Draw a line segment *PQ* with length equal to |*CD*| along line *AB*.

(2) Transfer the length |*EF*| to point *P*. Draw a circle of radius |*EF*| with *P* as center.

(3) Transfer the length $|GH|$ to point Q. Draw a circle of radius $|GH|$ with Q as center.

(4) Let R be one of the points where the two circles intersect. Then triangle PQR has sides of length (Fig. 8-5)

$$|PQ| = |CD|$$
$$|PR| = |EF|$$

and

$$|QR| = |GH|$$

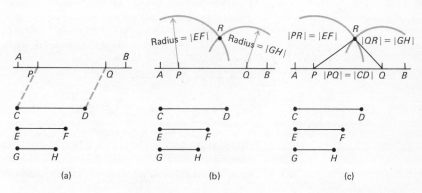

(a)   (b)   (c)

Figure 8-5 Construct a triangle with three given lengths for the sides (Construction 5). (a) Given: line segments *AB, CD, EF, GH*. Construct *PQ* equal in length to *CD*; (b) construct a circle with center at *P* and radius *EF*. Construct a circle with center at *Q* and radius *GH*; and (c) triangle *PQR* has sides equal in length to *CD, EF,* and *GH*.

Observe that Construction 5 cannot be carried out if length $|CD|$ is greater than the sum of the other two lengths because the two circles would not intersect. In general, a triangle can be constructed with three given lengths for the sides, provided each length is less than the sum of the other two lengths.

Three Famous Problems in Geometry
    The problem of doubling the volume of a cube is one of three famous problems tackled by the ancient Greeks. These three problems are (Fig. 8-6):
    (1) *Double the cube:* Construct a

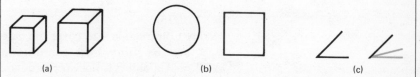

(a)   (b)   (c)

Figure 8-6 Three impossible ruler and compass constructions attempted by the ancient Greeks. (a) Double the cube. Construct a cube with double the volume of a given cube; (b) square the circle. Construct a square equal in area to a given circle; and (c) trisect the angle. Construct an angle one-third the size of a given angle.

cube that has volume double that of a given cube.

(2) *Square the circle:* Construct a square that has area equal to that of a given circle.

(3) *Trisect the angle:* Construct an angle that has measure exactly one-third that of a given angle.

All of the above constructions were to be carried out using only a straight edge and a compass.

The Greeks worked on the three problems for centuries without success. When the problems became known to the western world during the Renaissance, they were attacked with a new vigor—again without success. Finally, in the early nineteenth century, it was proved that the three constructions are impossible.

The proof of the impossibility of the constructions was based on a sophisticated use of higher algebra. It was shown that if the constructions can be carried out, then they cause logical contradictions—hence the impossibility. It is indicative of the spirit of modern mathematics that new results in algebra were developed in order to solve three ancient problems in geometry.

EXERCISES

Work through the steps of Constructions 1 through 5 in Exercises 1 to 5, using figures of your own that satisfy the given conditions. Be sure that your figures are not identical to the ones in the book.

1. Construction 1
2. Construction 2
3. Construction 3
4. Construction 4
5. Construction 5
6. Use Constructions 1, 2, and 3 as the basis for the following constructions.
    (a) *Given:* a line segment
        Construct: a square with the given line segment as one side
    (b) *Given:* a pair of line segments AB and AC that form an angle at A
        Construct: a parallelogram with AB and AC as two sides
    (c) *Given:* a parallelogram
        Construct: a rectangle with area equal to that of the parallelogram
7. Construct a triangle with the following lengths for the sides:
    (a) 2, 5, 6
    (b) 3, 4, 5
    (c) 9, 6, 13
8. Modify Construction 4 to cover the case where P is on the same line as A and B. [*Hint:* First construct points A' and B' such that $|A'B'| = |AB|$ and P is not on the same line as A' and B' (Fig. 8-7).]
9. Show that the problem of doubling the cube is equivalent to constructing the cube root of 2. (*Hint:* Let x be the length of one side of the original cube. Compute the length of one side of the new cube.)

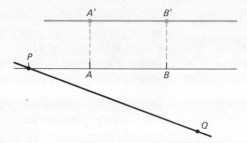

Figure 8-7 Exercise 8.

10. Show that the problem of squaring the circle is equivalent to constructing the square root of π. (*Hint:* Let x be the radius of the circle. Compute the length of one side of a square that has the same area as the circle.)

8-2 THE PYTHAGOREAN THEOREM I

One of the oldest geometrical problems is concerned with indirect measurement: *Given two sides of a right triangle, compute the length of the third side.*

The ancient Babylonians discovered that if a and b are the two sides of a right triangle and c is the hypotenuse, then

$$a^2 + b^2 = c^2 \qquad \text{(Pythagorean Theorem)}$$

(See Fig. 8-8.)

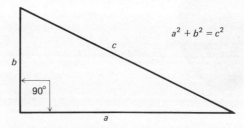

Figure 8-8 The theorem of Pythagoras. If a, b, and c are the sides of a right triangle, then $a^2 + b^2 = c^2$.

We believe that the Babylonians accepted the truth of the equation $a^2 + b^2 = c^2$ and did not prove it in a formal way. They did, however, study the consequences of the equation and used it to construct sophisticated trigonometric tables.

Apparently Pythagoras learned the fact that $a^2 + b^2 = c^2$ in his travels in the Persian world, puzzled over it, perhaps for years, and constructed a mathematical proof. His followers, the Pythagoreans, gave him full credit for the discovery of the result and its proof, hence the name that has become attached to it—the *Pythagorean Theorem*.

THE PYTHAGOREAN THEOREM
If a and b are the sides of a right triangle and c is the hypotenuse, then

$$a^2 + b^2 = c^2$$

Example 1. The best-known right triangle is the one with sides measuring 3, 4 and 5 units. Observe that

$$3^2 + 4^2 = 9 + 16 = 25 = 5^2$$

as predicted by the Pythagorean Theorem.

This triangle has been used by construction workers for centuries in order to build square corners. To make a large makeshift carpenter's square, simply nail together at the corners three pieces of wood measuring 3 feet, 4 feet, and 5 feet in length (Fig. 8-9).

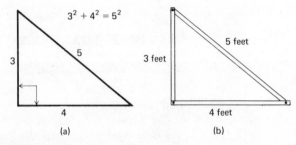

Figure 8-9 (a) The 3, 4, 5 triangle. $3^2 + 4^2 = 5^2$; (b) a makeshift carpenter's square based on the 3, 4, 5 triangle. (See Example 1.)

Example 2. A plot of land is shaped like a right triangle with one side equal to 80 feet and hypotenuse equal to 265 feet.
  (a) What is the length of the third side?
  (b) What is the area of the plot?

Solution. (a) Let a, b, and c represent the three sides of the triangle with $a = 80$ and $c = 265$ (Fig. 8-10). By the Pythagorean Theorem

$$a^2 + b^2 = c^2$$
$$80^2 + b^2 = 265^2$$
$$6{,}400 + b^2 = 70{,}225$$
$$b^2 = 70{,}225 - 6{,}400 = 63{,}825$$
$$b = \sqrt{63{,}825} \approx 252.64$$

The ancient Babylonians had full knowledge of the Pythagorean Theorem. This old tablet shows the value of $\sqrt{2}$ as 1.414222 (written in base-60 notation). The figure indicates that the square root of 2, which was calculated from an algebraic algorithm, is the length of the diagonal of the unit square.

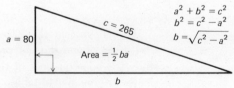

Figure 8-10 Example 2.

(b) The area is

$$\text{area} = \tfrac{1}{2}\,\text{base} \cdot \text{height}$$
$$= \tfrac{1}{2}ba$$
$$\approx \tfrac{1}{2}(252.64) \cdot 80 \approx 10{,}105.6 \text{ square feet}$$

Pythagorean Triangles

The ancient Pythagoreans were particularly interested in right triangles in which all three sides a, b, c are integers. Common examples of such "Pythagorean triangles" are the 3, 4, 5 triangle,

$$3^2 + 4^2 = 9 + 16 = 25 = 5^2$$

and the 5, 12, 13 triangle,

$$5^2 + 12^2 = 25 + 144 = 169 = 13^2$$

If we have one Pythagorean triangle, we can construct other triangles similar to it by multiplying each of the three sides by an arbitrary positive integer. For example, from the 3, 4, 5 triangle we can get the similar triangles

6, 8, 10	(multiply by 2)
9, 12, 15	(multiply by 3)
12, 16, 20	(multiply by 4)

and so on.

There is a method to generate a basic set of Pythagorean triangles. Apparently this method (or an equivalent one) was known to the ancient Babylonians. We choose two integers m and n, where $m > n$, and let

$$a = m^2 - n^2$$
$$b = 2mn$$
$$c = m^2 + n^2$$

Then

$$a^2 + b^2 = (m^2 - n^2)^2 + (2mn)^2$$
$$= (m^4 - 2m^2n^2 + n^4)$$
$$\qquad\qquad + 4m^2n^2$$
$$= m^4 + 2m^2n^2 + n^4$$
$$= (m^2 + n^2)^2$$
$$= c^2$$

Since

$$a^2 + b^2 = c^2$$

then a, b, c are the sides of a Pythagorean triangle.

After we generate the basic set of Pythagorean triangles by this method, we can multiply the triangles in the basic set by integers to obtain all possible Pythagorean triangles.

A few Pythagorean triangles in the basic set are listed in the following table:

m	n	a $(m^2 - n^2)$	b $(2mn)$	c $(m^2 + n^2)$
2	1	3	4	5
3	1	8	6	10
3	2	5	12	13
4	1	15	8	17
4	2	12	16	20
4	3	7	24	25
5	1	24	10	26
5	2	21	20	29
5	3	16	30	34
5	4	9	40	41

The Existence of Square Roots

The Pythagorean Theorem was indirectly responsible for a crisis in the mathematical theories of the ancient Greeks. In the early days of Greek mathematics, the concept of "number" apparently included only positive integers and fractions—no other types of numbers were known. It was only natural for distances to be considered as numbers somewhat as we identify the real numbers with distances and points on the number line.

When the Greeks applied the Pythagorean Theorem to the triangle with sides a, b, and c with $a = 1$ and $b = 1$, they computed that

$$c^2 = a^2 + b^2 = 1^2 + 1^2 = 2$$

so that c is a square root of 2. Thus, they were able to construct a length (represented by c) that was equal to a square root of 2 (Fig. 8-11). The results of this construction were accepted as part of the developing mathematical theory until, to their horror, one of the Pythagoreans found that there is no rational number whose square is equal to 2. In other words, *the square root of 2 is an irrational number*. This meant that lengths of line segments could not be identified with numbers, since numbers were restricted to being integers or fractions. The news was kept an official secret of the Pythagorean Society for a short period of time, but eventually was revealed to the other Greek mathematicians. This one discovery caused the Greeks to completely overhaul their concept of number. They reserved the word "number" for positive integers, dealt with distances geometrically, and developed an elaborate system of ratio and proportion to handle both rational numbers and irrational lengths of line segments.

Ancient Chinese proof of the Pythagorean Theorem. Photo courtesy of the British Library.

This figure from the *Chou Pei* manuscript (probably written in the Fourth Century B.C.) is the basis for an algebraic proof of the Pythagorean Theorem.

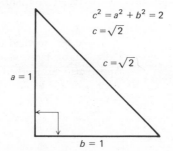

$$c^2 = a^2 + b^2 = 2$$
$$c = \sqrt{2}$$
$$c = \sqrt{2}$$
$$a = 1$$
$$b = 1$$

Figure 8-11 Construction of the square root of 2.
$c^2 = a^2 + b^2 = 1^2 + 1^2 = 2; c = \sqrt{2}$.

We take a more modern approach. We consider the real numbers to be represented by points on the number line. This line has no holes in it. Every real number can be represented by a point on the line, and every point on the line represents a real number. Since we can use the number line as a scale of measurement, then every distance can be represented by a positive real number. The existence of a triangle with one side equal to $\sqrt{2}$ causes us no difficulties. It merely establishes that the number 2 has a square root—a fact that might be difficult to prove otherwise.

There is a simple construction that can be used to show that *every* positive real number has a square root. The following example illustrates the construction for the square root of 7. We can replace the number 7 with any positive number different from 1, and the same construction will yield the square root.

Example 3. Show that there is a square root of 7. In other words, find a distance x (a point x on the number line) such that $x^2 = 7$.

Solution.   We perform the following construction (Fig. 8-12).

(1) Locate the point $(7 - 1)/2$ on the y-axis.

(2) Draw a circle with radius $(7 + 1)/2$, using the point $(7 - 1)/2$ on the y-axis as the center.

(3) Let x be the point where the circle intersects the positive x-axis.

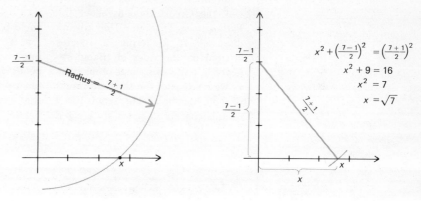

Figure 8-12   Construction of the square root of 7.

We shall show that $x^2 = 7$. Observe that the origin and the points x and $y = (7 - 1)/2$ form a right triangle with hypotenuse equal to $(7 + 1)/2$. By the Pythagorean Theorem

$$x^2 + \left(\frac{7 - 1}{2}\right)^2 = \left(\frac{7 + 1}{2}\right)^2$$
$$x^2 + 3^2 = 4^2$$
$$x^2 + 9 = 16$$
$$x^2 = 16 - 9$$
$$x^2 = 7$$

Thus, x is a square root of 7. As we mentioned above, the square root of any positive number can be constructed by this same procedure.

Tiling Problems

We have no way of knowing how the Pythagorean Theorem was first discovered. One theory is that the first special case of the theorem may have been seen in a pattern of floor tiles.

Figure 8-13 shows one of the popular patterns of tiles in the ancient world. In Figure 8-13(a) we have ac- cented the vertical and horizontal lines, in Figure 8-13(b) the diagonal lines. Figure 8-13(c) shows a triangle in the pattern with squares on its sides. *The total area of the two small squares is equal to the area of the square on the hypotenuse,* an example of the Pythagorean Theorem.

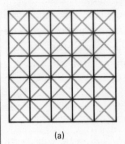
(a)

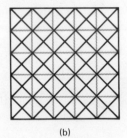
(b)

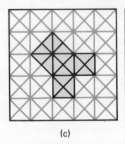
(c)

Figure 8-13 A special case of the Pythagorean Theorem revealed in a pattern of floor tiles. The square on the diagonal has area equal to the sum of the areas of the squares on the sides of the triangle.

Figure 8-13(c) is indicative of the way the Greeks thought about the Pythagorean Theorem. They always discussed it in terms of squares erected on the sides of triangles—never in terms of squares of numbers as we do. One of the standard proofs of the Pythagorean Theorem is based on this concept. Squares are erected on the three sides of a right triangle, then a line is drawn from the right angle perpendicular to the opposite side. This line splits the large square into two rectangles. The proof depends on showing that the two rectangles have areas exactly equal to the areas of the corresponding squares (Fig. 8-14).

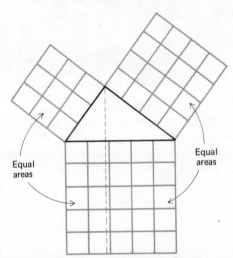

Equal areas

Equal areas

Figure 8-14 The Pythagorean Theorem. The classical proof of the Pythagorean Theorem is based on the fact that the square on the hypotenuse can be split into two rectangles equal in area to the two squares on the sides of the triangle.

EXERCISES

1. The following sets of numbers represent the sides of triangles. Which of the triangles are right triangles?

(a) 2, 3.5, 4.1   (d) 20, 23, 31
(b) 6, 8, 10   (e) 12, 13, 5
(c) 7, 12, 14   (f) 4, 7, 8

2. The following lengths are two of the sides of a right triangle a, b, and c, where c is the hypotenuse. Use the Pythagorean Theorem to calculate the third side.

(a) $a = 2, b = 5$ (d) $b = 7, c = 13$
(b) $a = 3, b = 4$ (e) $a = 2, b = 2$
(c) $a = 7, c = 9$ (f) $a = 5, c = 12$

3. Make a flow chart for the following operations.

   (a) Calculate the hypotenuse of a right triangle, given the lengths of the two sides.

   (b) Calculate the length of one side of a right triangle, given the hypotenuse and the length of the other side.

4. Make computer programs based on your flow charts in Exercise 3(a) and (b). Run the programs on the computer using your own data.

5. Work through a construction equivalent to the one in Example 3 to find the following square roots geometrically.

   (a) 5   (c) $\frac{1}{2}$
   (b) 13   (d) $\frac{2}{3}$

6. An ancient argument for the Pythagorean Theorem is based on the diagram in Figure 8-15.

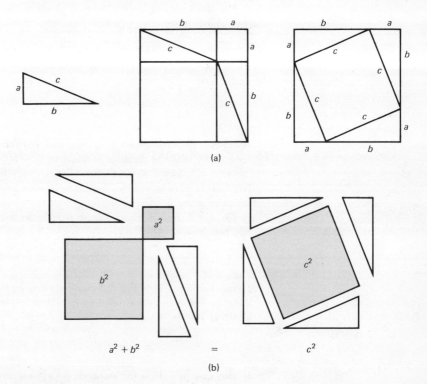

(a)

(b)

Figure 8-15 Exercise 6.

Start with a right triangle with sides a, b, and c. Cut two squares out of cardboard with sides of length $a + b$, and then cut the squares into squares and triangles as in Figure 8-15. Show that if the four triangles are removed from each figure, then the squares that remain in one figure have a total area of $a^2 + b^2$, while the square that remains in the other figure has an area of c^2. Thus, $a^2 + b^2 = c^2$.

8-3 THE PYTHAGOREAN THEOREM II; THE DISTANCE FORMULA

One of the most basic problems in coordinate geometry is that of determining the distance between two points in the xy-plane. This distance must be calculated rather than measured in order to get the exact value. In this section we use the Pythagorean Theorem to derive a formula for the distance between two points.

Before we attempt the distance formula, we must consider a special case:

If a and b are any two real numbers, then the distance between a and b on the number line is

$$|a - b|$$

We shall not attempt a formal proof of this fact, but we shall illustrate it in the following example.

Example 1. Show that the distance between a and b is $|a - b|$ in each of the following cases:
 (a) $a = 7, b = 3$
 (b) $a = 2, b = 5$
 (c) $a = 4, b = -3$
 (d) $a = -6, b = -2$

Solution. We shall measure the distance between a and b on the number line, then verify that the distance is equal to $|a - b|$.

(a) $a = 7, b = 3$. From Figure 8-16(a) we see that the distance between a and b is 4 units. Also, $|a - b| = |7 - 3| = |4| = 4$. In this case $|a - b|$ is the distance between a and b.

(b) $a = 2, b = 5$. From Figure 8-16(b) we see that the distance between a and b is 3 units. Also,

$$|a - b| = |2 - 5| = |-3| = 3$$

In this case $|a - b|$ is the distance between a and b.

(c) $a = 4, b = -3$. From Figure 8-16(c) we see that the distance between a and b is 7 units. Also,

$$|a - b| = |4 - (-3)| = |4 + 3| = |7| = 7$$

In this case $|a - b|$ is the distance between a and b.

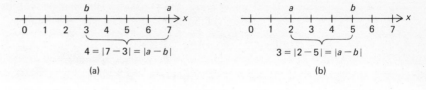

(a)

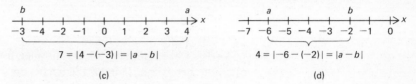

(c)   (d)

Figure 8-16 The distance between points a and b on the x-axis is $|a - b|$. (See Example 1.)

(d) $a = -6, b = -2$. From Figure 8-16(d) we see that the distance between a and b is 4 units. Also,

$$|a - b| = |(-6) - (-2)| = |-6 + 2| = |-4| = 4$$

In this final case, as well, $|a - b|$ is the distance between a and b.

A similar result holds for any two points on the y-axis. The distance between the points is the absolute value of the difference of their y-coordinates.

Before we derive the general distance formula, we shall work an example which illustrates the important steps.

Example 2.   Calculate the distance between the points $P_1(2, 1)$ and $P_2(4, 6)$.

Solution.   We plot the points P_1 and P_2 in the xy-plane. Then we draw lines parallel to the coordinate axes through P_1 and P_2 and form a right triangle. Let Q be the point where the two lines intersect (Fig. 8-17). Then Q is the point $Q(4, 1)$. The

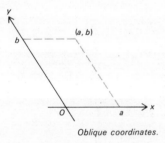

Oblique coordinates.

Points in the plane can be represented as well by oblique coordinates as by rectangular coordinates (perpendicular axes). If distances are to be calculated, however, it is almost essential that rectangular coordinates be used. Formulas can be derived for the distance between points represented by oblique coordinates, but these formulas are awkward and cumbersome to use.

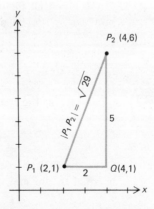

Figure 8-17 The distance between $P_1(2, 1)$ and $P_2(4, 6)$ is $|PQ| = \sqrt{29}$. (See Example 2.)

points P_1, P_2, and Q form a right triangle with two sides of length 2 and 5. The length of the hypotenuse is equal to the distance from P_1 to P_2. By the Pythagorean Theorem

$$|P_1P_2|^2 = 2^2 + 5^2 = 4 + 25 = 29$$

so

$$|P_1P_2| = \sqrt{29}$$

The general argument for the distance formula is similar to the above example. Let $P_1(x_1, y_1)$ and $P_2(x_2, y_2)$ be any two points in the plane. Draw lines through P_1 and P_2 parallel to the coordinate axes and let Q be the point where the lines cross (Fig. 8-18). Then P_1, P_2, and Q are the vertices of a right triangle.

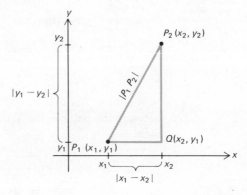

Figure 8-18 The distance formula. The distance between $P_1(x_1, y_1)$ and $P_2(x_2, y_2)$ is $|P_1P_2| = \sqrt{(x_1 - x_2)^2 + (y_1 - y_2)^2}$.

Observe that the distance between P_1 and Q is the same as the distance between x_1 and x_2 on the x-axis, which is equal to $|x_1 - x_2|$. Therefore,

$$|P_1Q| = |x_1 - x_2|$$

Similarly, the distance between Q and P_2 is the same as the distance between y_1 and y_2 on the y-axis, which is equal to $|y_1 - y_2|$. Therefore,

$$|QP_2| = |y_1 - y_2|$$

By the Pythagorean Theorem the square of the length of the hypotenuse is equal to the sum of the squares of the two sides:

$$|P_1P_2|^2 = |P_1Q|^2 + |QP_2|^2$$
$$= |x_1 - x_2|^2 + |y_1 - y_2|^2$$

THE DISTANCE FORMULA
The distance from $P_1(x_1, y_1)$ to $P_2(x_2, y_2)$ is

$$|P_1 P_2| = \sqrt{(x_1 - x_2)^2 + (y_1 - y_2)^2}$$

If we square the absolute value of a real number, we get the same result as if we square the number. Consequently, we can write the equation as

$$|P_1 P_2|^2 = (x_1 - x_2)^2 + (y_1 - y_2)^2$$

If we take the positive square root of both sides, we obtain

$$|P_1 P_2| = \sqrt{(x_1 - x_2)^2 + (y_1 - y_2)^2} \qquad \text{(The Distance Formula)}$$

Example 3. Use the distance formula to calculate the distances between the following pairs of points.
   (a) $P_1(2, 3)$, $P_2(5, -1)$
   (b) $P_1(1, -1)$, $P_2(-7, -6)$
   (c) $P(3, 7)$, $Q(2, -2)$

Solution. (a) $P_1(2, 3)$, $P_2(5, -1)$. [See Fig. 8-19(a).]

$$\begin{aligned}
|P_1 P_2| &= \sqrt{(x_1 - x_2)^2 + (y_1 - y_2)^2} \\
&= \sqrt{(2 - 5)^2 + [3 - (-1)]^2} \\
&= \sqrt{(-3)^2 + 4^2} = \sqrt{9 + 16} = \sqrt{25} \\
&= 5
\end{aligned}$$

The distance between P_1 and P_2 is 5 [Fig. 8-19(a)].

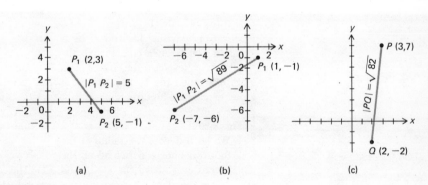

(a)   (b)   (c)

Figure 8-19 Example 3.

(b) $P_1(1, -1)$, $P_2(-7, -6)$. [See Fig. 8-19(b).]

$$\begin{aligned}
|P_1 P_2| &= \sqrt{[1 - (-7)]^2 + [(-1) - (-6)]^2} \\
&= \sqrt{(1 + 7)^2 + (-1 + 6)^2} \\
&= \sqrt{8^2 + 5^2} = \sqrt{64 + 25} = \sqrt{89}
\end{aligned}$$

The distance between P_1 and P_2 is $\sqrt{89}$.

(c) $P(3, 7)$, $Q(2, -2)$. [See Fig. 8-19(c).]

$$|PQ| = \sqrt{(3 - 2)^2 + [7 - (-2)]^2}$$
$$= \sqrt{1^2 + (7 + 2)^2}$$
$$= \sqrt{1^2 + 9^2} = \sqrt{1 + 81} = \sqrt{82}$$

The distance between P and Q is $\sqrt{82}$ [Fig. 8-19(c)].

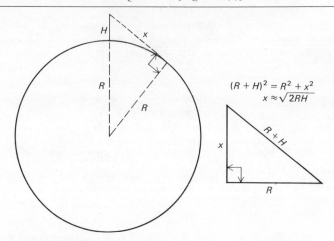

$$(R + H)^2 = R^2 + x^2$$
$$x \approx \sqrt{2RH}$$

Figure 8-20 The distance to the horizon. H = height of observer's eyes above the earth; R = radius of earth; x = distance to the visual horizon. $x \approx \sqrt{2RH}$.

The Distance to the Horizon

The Pythagorean Theorem can be used to calculate the distance from a person to his visual horizon. If he is standing on a flat surface, such as the salt flats of Nevada or the coastal plain of Texas, then the horizon occurs at the point where his direct line of sight passes over the curve of the earth (Fig. 8-20).

We let R denote the radius of the earth, H denote the height of the observer's eyes above the earth, and x denote the distance from the observer's eyes to the horizon. Observe in Figure 8-20 that a right triangle is formed with sides of length x and R and hypotenuse $R + H$. By the Pythagorean Theorem

$$x^2 + R^2 = (R + H)^2$$
$$x^2 + R^2 = R^2 + 2RH + H^2$$
$$x^2 = 2RH + H^2$$
$$x = \sqrt{2RH + H^2}$$

Thus, the exact distance to the horizon is

$$x = \sqrt{2RH + H^2}$$

Unless the observer is in a spaceship or high flying airplane, the value of H^2 is negligible when compared to $2RH$. If we omit it in the formula, we obtain the simpler formula

$$x \approx \sqrt{2RH}$$

The radius of the earth is approximately 4000 miles. For convenience, it is usually best to express the height H in miles as well.

Example. An observer is standing on a mountain 1000 feet above a flat plain. How far is the horizon?

Solution. $R = 4000$ miles, $H = 1000$ feet $\approx \frac{1}{5}$ mile. The distance from the observer to the horizon is

$$x \approx \sqrt{2RH} \approx \sqrt{2 \cdot 4000 \cdot \tfrac{1}{5}}$$
$$\approx \sqrt{1600} \approx 40 \text{ miles}$$

EXERCISES

1. Show that the distance from a to b on the x-axis is equal to $|a - b|$.
   (See Example 1.)
   (a) $a = 3, b = 8$ (c) $a = -5, b = -3$
   (b) $a = -1, b = 2$ (d) $a = 6, b = 1$

2. Use the distance formula to calculate the distances between the following pairs of points.
   (a) $(-2, -1), (1, 3)$ (d) $(4, 8), (6, 1)$
   (b) $(-1, 10), (4, -2)$ (e) $(3, -1), (-2, 6)$
   (c) $(-2, 0), (5, 1)$ (f) $(7, 2), (5, 5)$

3. Work through the steps in the derivation of the distance formula with the following points. In each case draw a diagram similar to Figure 8-17.
   (a) $P_1(1, 2), P_2(-3, -1)$
   (b) $P_1(-1, 1), P_2(6, 2)$
   (c) $P_1(1, 12), P_2(6, 0)$

4. Construct a computer program that can be used to calculate the distance between two points in the plane. The printout should list each pair of points and the distance between them. Test your program with the data in Exercise 2.

5. The following sets of points are the vertices of triangles. Use the Pythagorean Theorem to decide which of the triangles are right triangles.
   (a) $(1, 0), (2, 3), (-2, 1)$ (d) $(4, -2), (-4, 0), (-3, 4)$
   (b) $(-2, 1), (2, -1), (4, 2)$ (e) $(4, 2), (3, 4), (-4, -3)$
   (c) $(-1, -2), (-3, -1), (1, 2)$ (f) $(-3, -1), (2, 4), (14, -9)$

6. Calculate the approximate distance from an observer to the horizon, given H, the eye level above the earth.
   (a) $H = 5$ feet
   (b) $H = 2,000$ feet
   (c) $H = 10,000$ feet

7. Calculate the approximate distance from an observer on the moon to the horizon, given H, the eye level above the surface of the moon. $[R \approx 1,000$ miles.$]$
   (a) $H = 5$ feet   (b) $H = 2,000$ feet   (c) $H = 10,000$ feet

8-4 PERIMETERS AND AREAS OF PLANE FIGURES

The perimeter and area of a figure in the plane are two of its most important properties. Most readers will be surprised to learn, however, that these two concepts are quite difficult to define in an unambiguous manner. The correct definitions are usually not attempted until courses at the calculus level. For this reason we shall not attempt a formal definition of either term, but will consider the concepts intuitively.

Essentially, the perimeter of a figure is the total length of its boundary. If we could place a piece of string so that it exactly coincides with the

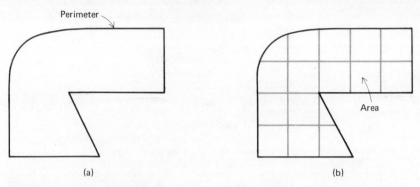

Figure 8-21 (a) The *perimeter* of a figure is the total length of the boundary; (b) the *area* of a figure is the total number of square units that could be cut from the figure.

boundary, then straighten out the string and measure it, we would find that the length of the string is equal to the perimeter [Fig. 8-21(a)].

We can think of the area of a figure as the number of square units we could cut up and reassemble in order to exactly cover the figure [Fig. 8-21(b)].

The areas and perimeters of certain figures can be obtained from simple formulas.

Rectangles

As we discussed in Chapter 4, a rectangle with base b and height h has *area* equal to the product of the base and height (Fig. 8-22).

$$\text{area} = \text{base} \cdot \text{height} = bh$$

The *perimeter* of the rectangle is equal to the total length of the four sides (Fig. 8-22).

$$\text{perimeter} = 2b + 2h$$

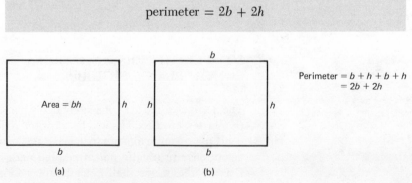

Figure 8-22 (a) The *area* of a rectangle is: area $= bh$; (b) the *perimeter* of a rectangle is: perimeter $= 2b + 2h$.

Example 1. (a) Express the area of one square foot in square inches.

(b) Express the area of one square yard in both square feet and square inches.

Solution. (a) A square foot is the area of a square that measures 12 inches on each side. The area [Fig. 8-23(a)] is

$$\text{one square foot} = 12^2 = 144 \text{ square inches}$$

(b) A square yard is the area of a square that measures 3 feet on each side. The area [Fig. 8-23(b)] is

$$\text{one square yard} = 3^2 = 9 \text{ square feet}$$

If we recall from (a) that each square foot measures 144 square inches, we see that

$$\text{one square yard} = 9 \text{ square feet} = 9 \cdot 144 = 1296 \text{ square inches}$$

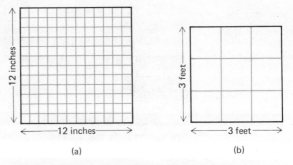

Figure 8-23 (a) 1 square foot = 144 square inches; (b) 1 square yard = 9 square feet. (See Example 1.)

Circles

A circle of radius R has a *perimeter* equal to

$$\text{perimeter} = 2\pi R$$

and an *area* equal to

$$\text{area} = \pi R^2$$

where $\pi \approx 3.14159$ (Fig. 8-24).

These results about circles are quite difficult to prove and usually are established in calculus courses. The basic idea behind the area formula is that of "filling up" the circle with rectangles.

In Figure 8-25(a) we have inscribed 48 rectangles in a circle of radius 1. The total area of the rectangles is 3.05, which is a little less than π,

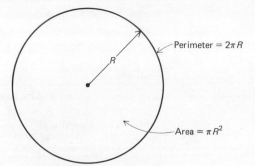

Figure 8-24   The circle. perimeter $= 2\pi R$;
area $= \pi R^2$.

the area of the circle. A similar idea is used to approximate the perimeter. We mark off a large number of points on the circumference and connect them with line segments [Fig. 8-25(b)]. The total length of the line segments is 6.25, which is a little less than 2π.

The constructions in Figure 8-25 only give approximations to the true area and perimeter. By choosing a large number of very small rectangles, we can make the total area of the rectangles as close to the area of the circle as we want, but we can never quite equal it. Similarly, we can make the total length of the line segments around the perimeter as close to the perimeter as we want, but be can never quite equal it. The exact determination of the area and the perimeter requires the use of the formulas that are derived in the calculus.

Example 2.   Calculate the area and the perimeter of each of the following circles correct to two decimal places.
  (a) radius $= 2$ inches
  (b) radius $= 3$ feet

Solution.   (a) $R = 2$ inches. The area and perimeter are calculated from the formulas:

$$\text{area} = \pi R^2 = \pi \cdot 2^2 = 4\pi \approx 4 \cdot (3.14159)$$
$$\approx 12.57 \text{ square inches}$$
$$\text{perimeter} = 2\pi R = 2 \cdot \pi \cdot 2 = 4\pi \approx 4(3.14159)$$
$$\approx 12.57 \text{ inches}$$

(b) $R = 3$ feet.

$$\text{area} = \pi R^2 = \pi \cdot 3^2 = 9\pi \approx 9 \cdot (3.14159)$$
$$\approx 28.27 \text{ square feet}$$
$$\text{perimeter} = 2\pi R = 2\pi \cdot 3 = 6\pi \approx 6 \cdot (3.14159)$$
$$\approx 18.85 \text{ feet}$$

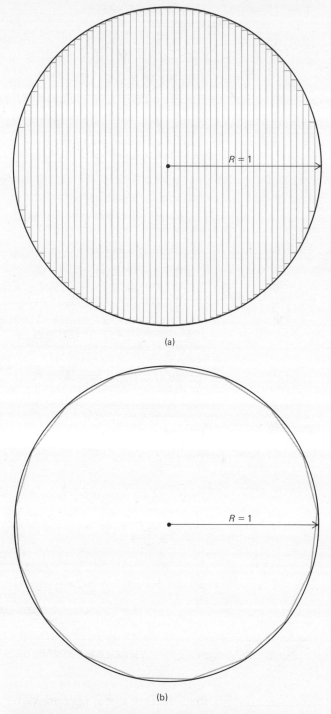

(a)

(b)

Figure 8-25 Approximations to the area and perimeter of a unit circle. (a) The area of the circle is approximately equal to the total area of the rectangles (3.05 square units); (b) the perimeter of the circle is approximately equal to the total length of the line segments ($\approx$ 6.25 units).

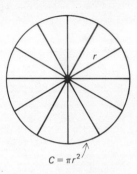

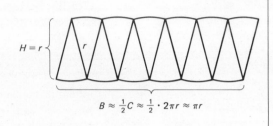

$C = \pi r^2$

$B \approx \frac{1}{2}C \approx \frac{1}{2} \cdot 2\pi r \approx \pi r$

The Area of the Circle

Some of the ideas behind the formula

$$A = \pi r^2$$

for the area of a circle can be understood by a careful study of the figure.

We cut the circle into a large number of equal pie-shaped wedges and rearrange them as shown in the diagram. The new figure is roughly shaped like a parallelogram with height $H = r$. The base of the figure is approximately equal to one-half of the circumference

$$B \approx \frac{1}{2} \cdot \text{circumference} \approx \frac{1}{2} \cdot 2\pi r \approx \pi r$$

Thus, the area of the circle is

area of circle = area of "parallelogramlike" figure
$$\approx BH \approx \pi r \cdot r \approx \pi r^2$$

Observe that the larger the number of wedges, the more the figure resembles a parallelogram, and the closer the product BH is to the actual value πr^2.

Triangles

The perimeter of a triangle with sides a, b, and c (Fig. 8-26) is simply

$$\text{perimeter} = \text{sum of the sides} = a + b + c$$

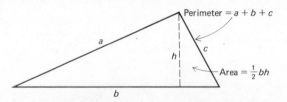

Perimeter $= a + b + c$

Area $= \frac{1}{2}bh$

Figure 8-26   The triangle. Perimeter $= a + b + c$, area $= \frac{1}{2} bk$.

TRIANGLE

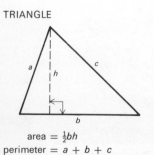

area $= \frac{1}{2}bh$
perimeter $= a + b + c$

If side b is designated the *base* and we let h be the *height* (the distance from the base to the opposite vertex), then the area is given by

$$\text{area} = \frac{1}{2} \text{base} \cdot \text{height} = \frac{1}{2}bh$$

The most natural dimensions to be given for a triangle are the lengths of the three sides. As we have seen, we can calculate the perimeter if these sides are given, but the area may be more difficult. If, in addition, we are given the measure of one of the angles, we can use the trigonometric functions to calculate the height, then use it in the calculation of the area.

Example 3. Triangle ABC has sides $b = 17$ feet, $c = 8$ feet, and angle $A = 47°$. Calculate the area [Fig. 8-27(a)].

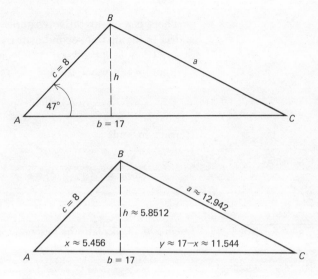

Figure 8-27 Example 3.

Solution. We first must calculate the height h. We see from Figure 8-27 that

$$\sin A = \frac{h}{8}$$

so that

$$h = 8 \sin A = 8 \sin 47° \approx 8 \cdot (0.7314) \approx 5.8512$$

Therefore,

$$\text{area} = \tfrac{1}{2}bh \approx \tfrac{1}{2} \cdot 17 \cdot (5.8512) \approx 49.74 \text{ square feet}$$

The perimeter of the triangle in Example 3 could be calculated, but the work is quite complicated. We would carry out the following steps [see Fig. 8-27(b)]:

(1) Calculate the base x of the small triangle on the left:

$$x = c \cos A \approx 8 \cdot (0.6820) \approx 5.456$$

(2) Use x to calculate y, the base of the triangle on the right:

$$y = b - x \approx 17 - 5.456 \approx 11.544$$

(3) Use y and h in the Pythagorean Theorem to calculate side a:

$$a^2 = h^2 + y^2 \approx (5.8512)^2 + (11.544)^2 \approx 167.5$$
$$a \approx \sqrt{167.5} \approx 12.943$$

The perimeter is

$$\text{perimeter} = a + b + c \approx 12.943 + 17 + 8 \approx 37.943$$

There is a comparatively simple formula discovered by Heron of Alexandria in the first or second century that can be used to calculate the area of any triangle if we know the lengths of its three sides. As above we let a, b, and c represent the lengths of the sides. We also let

$$s = \text{one-half the perimeter} = \tfrac{1}{2}(a + b + c)$$

Heron's formula is

HERON'S FORMULA

$$\text{area} = \sqrt{s(s - a)(s - b)(s - c)} \qquad \text{(Heron's Formula)}$$

$s = \frac{1}{2}(a + b + c)$

area = $\sqrt{s(s - a)(s - b)(s - c)}$

Example 4. Calculate the perimeter and the area of the triangle with sides

$$a = 13, \qquad b = 40, \qquad c = 45 \text{ inches}$$

Solution. The perimeter is

$$\text{perimeter} = a + b + c = 13 + 40 + 45 = 98 \text{ inches}$$

To use Heron's formula we let s equal one-half of the perimeter:

$$s = \tfrac{1}{2} \cdot (a + b + c) = \tfrac{1}{2} \cdot 98 = 49$$

The area is

$$\begin{aligned}
\text{area} &= \sqrt{s(s - a)(s - b)(s - c)} \\
&= \sqrt{49(49 - 13)(49 - 40)(49 - 45)} = \sqrt{49 \cdot 36 \cdot 9 \cdot 4} \\
&= \sqrt{7^2 \cdot 6^2 \cdot 3^2 \cdot 2^2} = 7 \cdot 6 \cdot 3 \cdot 2 = 252 \text{ square inches}
\end{aligned}$$

Quadrilaterals

A *quadrilateral* is a simple closed figure formed from four line segments. Squares, rectangles, trapezoids, and parallelograms are special types of quadrilaterals. Several quadrilaterals are shown in Figure 8-28.

It is a simple matter to measure the perimeter and to calculate the approximate area of any given quadrilateral. The perimeter is the sum of

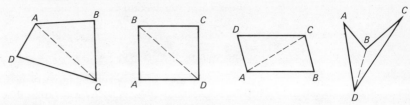

Figure 8-28 Quadrilaterals.

the lengths of the sides. The area can be found by decomposing the figure into two triangles (see Fig. 8-28) and calculating their areas separately.

There is no simple formula that can be used to calculate the area of every quadrilateral. In the special case where the figure is a trapezoid, a formula can be obtained quite easily by decomposing the figure into triangles. The reader will better appreciate the following argument for the area of a trapezoid if he actually constructs a figure out of cardboard and cuts it into the triangles.

The Area of a Trapezoid

The two parallel sides of a trapezoid are called its *bases*. The distance between the bases is called the *height*. If B_1 and B_2 are the lengths of the bases and H is the height, then the area is given by [Fig. 8-29(a)]

$$\text{area of trapezoid} = \frac{B_1 + B_2}{2} \cdot H$$

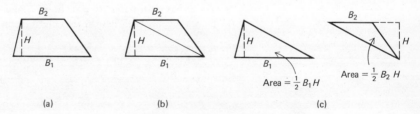

Figure 8-29 The area of a trapezoid is: area $= \frac{1}{2}(B_1 + B_2) \cdot H$.

TRAPEZOID

area $= \dfrac{b_1 + b_2}{2} \cdot h$

To establish the formula we decompose the trapezoid into two triangles as in Figure 8-29(b) and (c) and compute their areas separately. One triangle has base B_1 and height H; the other has base B_2 and height H. The area of the first triangle is

$$\tfrac{1}{2} B_1 H$$

the area of the second is

$$\tfrac{1}{2}B_2H$$

The sum of the areas is

$$\begin{aligned}\text{area of trapezoid} &= (\text{area of first triangle}) \\ &\quad + (\text{area of second triangle}) \\ &= \tfrac{1}{2}B_1H + \tfrac{1}{2}B_2H = \tfrac{1}{2}(B_1 + B_2)H\end{aligned}$$

This same formula was established by a different method in Exercises 5 and 6 of Section 4-2.

Area of a Parallelogram

PARALLELOGRAM

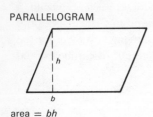

area = bh

In the special case of a parallelogram, the bases B_1 and B_2 are equal, say $B_1 = B_2 = B$. (See Fig. 8-30.) We obtain

$$\begin{aligned}\text{area of parallelogram} &= \tfrac{1}{2}(B_1 + B_2) \cdot H = \tfrac{1}{2}(B + B) \cdot H \\ &= \tfrac{1}{2}(2B) \cdot H = BH\end{aligned}$$

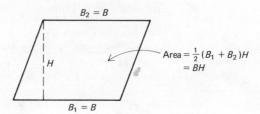

Figure 8-30   The area of a parallelogram is: area = BH.

Example 5.   Calculate the areas of the following figures.
   (a) Trapezoid with bases $B_1 = 3$, $B_2 = 7$, and height $H = 4$ [Fig. 8-31(a)].
   (b) Parallelogram with base $B = 3$ and height $H = 4$ [Fig. 8-31(b)].

Solution.   (a) area = $\tfrac{1}{2}(B_1 + B_2) \cdot H = \tfrac{1}{2}(3 + 7) \cdot 4$
          $= \tfrac{1}{2} \cdot 10 \cdot 4 = 5 \cdot 4 = 20$
   (b) area = $BH = 3 \cdot 4 = 12$

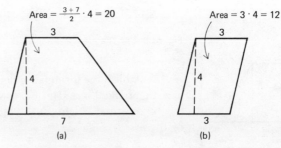

Figure 8-31 Example 5.

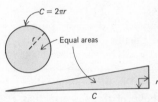

The area of a circle is equal to the area of a right triangle having the circumference of the circle for one side and the radius of the circle for the other side.

Archimedes (Third Century B.C.)

The Greeks stated mathematical relationships in terms of geometrical figures. If we let π denote the ratio of the circumference of a circle to diameter, then the circumference is

$$C = 2\pi r$$

where r is the radius. Archimedes established that

$$\begin{aligned}
\text{area of circle} &= \text{area of} \\
&\quad \text{right triangle} \\
&= \tfrac{1}{2} \cdot r \cdot C \\
&= \tfrac{1}{2} \cdot r \cdot 2\pi r \\
&= \pi r^2
\end{aligned}$$

As far as we know, Archimedes was the first person to prove this result, although its truth had been known to earlier mathematicians.

Archimedes

Archimedes of Syracuse (287–212 B.C.) was the greatest mathematician of the ancient world and one of the three or four greatest mathematicians who ever lived. According to the Roman historian Plutarch, he was a man possessed by mathematics, who was so engrossed in the subject that he forgot to eat or to bathe. Frequently his friends would forcibly take him to the public baths, where he would relax and study geometrical diagrams that he had drawn with his fingers in the ointment on his body.

Most of us have heard the story of how he ran naked through the streets shouting "Eureka! Eureka!" after discovering the basic principles of buoyancy while taking a bath. According to the story, he had been puzzling over a method to calculate the density of a metal. King Hiero suspected that a gold crown had been adulterated with silver and had asked Archi-medes to test it for purity. Archimedes realized that he could determine the volume of the crown by measuring the amount of water it displaced when immersed. By comparing the volume with the weight, he could determine the density. By comparing the density with the density of gold, he could determine if the crown were pure gold or not.

In addition to being an outstanding mathematician, he was an accomplished inventor. While in Egypt he invented the Archimedean screw, a hydraulic device used to raise water from wells and mines. He invented a system of levers and pulleys that enabled one man to launch a ship that the entire city of Syracuse had not been able to get to the water. As an old man in his seventies, he invented war machines to repulse the Romans during a seige of Syracuse. He designed large catapults that dropped rocks on ships far out in the sea, small catapults that hurled projectiles short distances, cranes that grasped ships and turned them over as they were landing, and a system of mirrors that focused the rays of the sun on ships and set them afire.

In spite of their success and the fame that they brought him, Archimedes looked upon his inventions with distain. According to Plutarch: ". . . He did not want to leave behind any writings on these subjects; he considered the construction of instruments, and, in general, every skill which is exercised for its practical use, as lowbrow and ignoble, and he only gave his efforts to matters which, in their beauty and their excellence, remain entirely outside the realm of necessity."

In his mathematics Archimedes was closer in spirit to modern workers than to the principles laid down by Plato. He worked on a large number of problems involving areas of plane

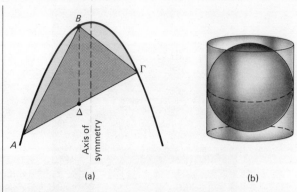

(a)   (b)

Typical problems solved by Archimedes. (a) The area of the parabolic segment is $\frac{4}{3}$ the area of the inscribed triangle. $\triangle$ is the midpoint of line segment $A\Gamma$. Line segment $\triangle B$ is parallel to the axis of symmetry; (b) the volume of a sphere is $\frac{2}{3}$ the volume of the circumscribed cylinder.

figures, surface areas, and volumes of solids. His mathematics was written in a rigorous, uncompromising style that made it difficult to understand and gave little hint of the ideas behind the proofs. For centuries it was thought that he had deliberately obscured his ideas in order to make them difficult for other mathematicians to comprehend.

In 1906 a lost manuscript written by Archimedes was discovered. This manuscript, "The Method," was devoted to an explanation of how Archimedes actually discovered the relationships that he later proved. In essence, his "method" involved the use of a number of the basic principles of the calculus—principles that were not rediscovered for almost 2000 years.

Archimedes' death is often taken to be symbolic of the new age that was dawning. After the long siege of Syracuse, the Romans captured the city. Marcellus, the Roman commander, had the greatest admiration for Archimedes and sent a soldier to bring the old man to him. When Archimedes delayed because of a theorem on which he was concentrating, the Roman soldier drew his sword and killed him.

EXERCISES

1. In each of the following cases, you are given sides and angles of a triangle with sides a, b, c and angles A, B, C labeled as in Figure 8-27. Compute the area of each triangle.
   (a) $a = 2, c = 5, B = 71°$
   (b) $a = 5, A = 30°, B = 50°, C = 100°$
   (c) $a = 20, b = 17, C = 10°$

2. Calculate the perimeter and the area of each of the following triangles. Use Heron's formula to calculate the area.
   (a) $a = 5, b = 5, c = 8$ (c) $a = 5, b = 3, c = 7$
   (b) $a = 65, b = 17, c = 80$ (d) $a = 8, b = 4, c = 6$

3. (1) Calculate the area and perimeter of each of the following circles correct to four decimal places.

   (2) Draw the circles to a convenient scale on good-quality graph paper. Estimate the area by counting the number of small squares and parts of squares enclosed in the circle. Compare this answer with the answer in part (1).

   (a) $R = 4$ inches   (c) $R = 11.12$ meters
   (b) $R = 7$ feet   (d) $R = 181.313$ centimeters

4. The three points are the vertices of a triangle in the xy-plane. Calculate the length of each side. Calculate the perimeter and (by Heron's formula) the area.

   (a) $A(0, 0), B(9, 12), C(16, -12)$ (c) $A(0, -6), B(12, 6), C(-5, 1)$
   (b) $A(-2, -3), B(3, -3), C(3, 9)$ (d) $A(1, 3), B(2, -4), C(-6, 4)$

5. The following points are the vertices A, B, C, D of a quadrilateral in the xy-plane. Draw each figure on good-quality graph paper. Be sure to use a large enough scale so that the lengths of the sides can be measured accurately.

   (1) Measure or compute the perimeter of each quadrilateral.

   (2) Decompose each quadrilateral into two triangles (as in Figure 8-28). Use Heron's formula on each of the two triangles and calculate the area of the quadrilateral.

   (a) $A(2, 2), B(4, 1), C(3, -1), D(-1, -2)$
   (b) $A(-1, 3), B(2, 2), C(5, 4), D(3, -1)$
   (c) $A(1, -2), B(1, 2), C(2, 3), D(4, 1)$
   (d) $A(1, 2), B(2, -1), C(3, 5), D(-4, 1)$

8-5 AREAS OF IRREGULAR FIGURES; AREAS UNDER CURVES

As we have seen, a polygon can be decomposed into triangles. The sides of these triangles can be measured and the areas computed by Heron's formula. Thus, in theory at least, we have a method which can be used to compute the areas of certain closed figures.

Unfortunately, the method described above is not a very practical one. A large number of calculations are required for a polygon with many sides, and the method does not work at all if parts of the boundary are curved. Thus, we need to consider alternate methods that can be used to compute the areas of plane figures, including those with many sides or with curved boundaries. A number of these alternate methods exist. All of them involve the use of accurate scale drawings or the exact coordinates of points on the boundaries. Many of the modern methods involve the use of the electronic computer.

One of the simplest methods to estimate the area of a figure is to superimpose a network of rectangles over it. This can be done quite easily if the figure is drawn to scale on graph paper. If the figure is irregular,

Greenwood Drive

0.66 acre

Surveyors have the practical problem of calculating areas of irregular figures. Traditionally, they superimposed rectangles and triangles with the same approximate area over a scale drawing and computed the areas of these figures.

Most surveys of large tracts of land have resulted in inaccurate estimates of area. It has been claimed that over half of the farms in this country have errors of 10 percent or more in their stated acreage.

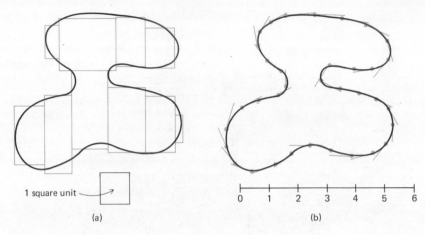

1 square unit

0	1	2	3	4	5	6

(a)   (b)

Figure 8-32 (a) Total area of figure $\approx$ total area of rectangles $\approx$ 16.6; (b) perimeter $\approx$ total length of line segments $\approx$ 26.8.

then the rectangles will not coincide with it exactly. Parts of the rectangles will be outside of the figure, and parts of the figure will not be covered by the rectangles. We try to construct the rectangles in such a way that the uncovered area of the figure is approximately equal to the total area of the portions of the rectangles that lie outside of the figure. If this is done, then the total area of the rectangles is a good approximation of the area of the figure.

Figure 8-32(a) illustrates this process with a particular example. Each rectangle is drawn so that its area outside the figure is approximately equal to the corresponding uncovered area in the figure. Since the combined area of the rectangles is approximately equal to 16.6 square units, then this is the approximate area of the figure.

There is a similar method that can be used to approximate the perimeter. First, we place a large number of points around the boundary. These points separate the boundary into short curves. From each of these points we draw a short line segment that is approximately equal to the length of the corresponding part of the boundary. These line segments do not have to connect or to terminate on the boundary of the figure. The total length of the line segments is approximately equal to the perimeter of the figure. [See Fig. 8-32(b).]

There are certain practical methods available for calculating areas of irregular figures. One of the most interesting involves the use of *uniform density paper*. This is a type of graph paper printed on a special cardboard, which is manufactured so that every square unit has the same weight as every other square unit. For purposes of our discussion, let us assume that each square centimeter weighs exactly 1 gram.

In use we make an accurate drawing of the figure on the uniform density paper, cut it out, and weigh it. If, for example, we find that it weighs 276.3 grams and we know that each square centimeter weighs 1 gram, then we know the area is 276.3 square centimeters (Fig. 8-33).

(a)   (b)   (c)

Figure 8-33 The weight of a figure is proportional to its area. (a) Draw the figure on uniform density paper; (b) cut it out; and (c) weigh it.

Uniform density paper is not very convenient to use, and in recent years its use has declined. A drafting machine called the *planimeter* has largely replaced it.

To use the planimeter we trace the arm around the boundary of the figure. At the end of one complete circuit of the boundary, the value of the area can be read from the machine. There is no mess, no trouble, and the answer is reasonably accurate.

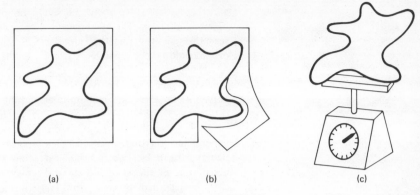

Photo from David Smith Collection.

Johann Kepler

Johann Kepler (1571–1630) made the first major advance in astronomy since the time of the ancient Greeks. Kepler, a religious mystic, firmly believed in the harmony of natural laws. It was inconceivable to him that God could have built the universe using haphazard methods. Just as musical harmony can be expressed by mathematical relationships, he believed that the natural harmony of celestial objects can be expressed by equations. Some of his arguments actually involved a mixture of music, theology, and mathematics.

During his early days as a mathematician, Kepler made a detailed study of mathematical curves. He was particularly interested in the ellipse, which he considered to be a distorted circle. By modifying the works of Archimedes, he was able to calculate the areas of sectors of ellipses.

As an astronomer who believed in the harmony of nature, Kepler was unable to accept the complicated "circle-within-circle" theory of planetary motion that was believed as an article of faith by all educated Europeans of the day. Searching for a simpler explanation, he studied the Copernican theory that the planets and the earth revolve about the sun in circular orbits. The recorded observations showed him that circular orbits could not be possible. A study of a hugh mass of data finally convinced Kepler that the planets revolve about the sun in ellipitical orbits.

Kepler was able to establish the general law relating the speed of a planet with its location in orbit at a given time of the year. He showed that the length of time required for the planet to move over an arc is proportional to the area of the sector of the arc of the ellipse with vertex at the sun.

It must be noted that Kepler's theory did not explain the movements of the planets any better than the old "circle-within-circle" theory. He believed in elliptical orbits solely because of the simplification that the new theory achieved. Consequently, his theory was not generally accepted during his lifetime.

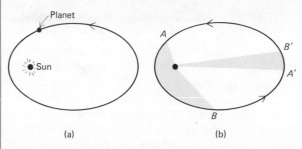

(a) A planet moves in an elliptical orbit around the sun. The sun is located at a point called the "focus" of the ellipse; (b) the shaded areas are equal. Kepler showed that it takes the planet the same length of time to travel from A to B as from A' to B'.

Perimeters and Areas of Similar Figures

At one point or another, almost all of us use scale drawings in our lives. Whenever we do this, we are working with a figure that is similar to the actual figure under consideration. (Recall from Chapter 4 that two figures are similar if they have identical shapes, but possibly differ in size and orientation.) We need to be able to interpret information about the scale drawing in terms of equivalent information about the similar figure.

If two figures are similar, there is a natural correspondence between their points and angles. In the discussion that follows, we let A, B, C, ... denote points on the smaller of the two figures and A', B', C', ... denote the corresponding points on the larger figure (Fig. 8-34).

There is a constant of proportionality that relates distances on similar figures. If the ratio of one pair of corresponding distances is k, then this same ratio holds for all corresponding distances. If $|A'B'| = k|AB|$, for example, then we must have

$$|A'E'| = k|AE|$$
$$|B'C'| = k|BC|$$

and so on. In Figure 8-34 every distance on the large figure is $k = \frac{3}{2}$ times the corresponding distance on the small figure.

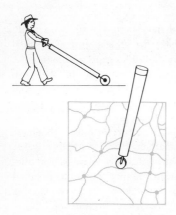

The most accurate measure of a perimeter is obtained by rolling a wheel around it and counting the number of revolutions of the wheel.

A penlike instrument with a small wheel on the end can be used to compute distances on a map or scale drawing. After the proper scale has been set to match the map scale, the wheel is rolled along the route marked on the map. The exact "map" distance can be read from the instrument.

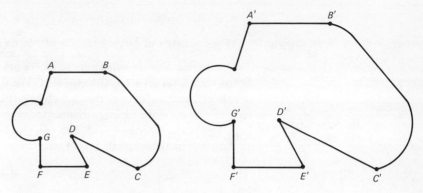

Figure 8-34 Corresponding points on similar figures.

If k is the constant of proportionality between two similar figures, then this same constant relates the perimeters of the two figures:

$$\text{perimeter of large figure} = k \cdot (\text{perimeter of small figure})$$

To show that this holds, we draw line segments around the boundary of the small figure with total length exactly equal to the perimeter. We next draw the corresponding line segments around the boundary of the large figure [Fig. 8-35(a) and (b)]. Then each line segment on the large figure is k times the length of the corresponding segment on the small figure.

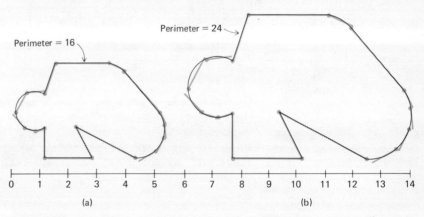

Figure 8-35 Perimeters of similar figures. Perimeter of large figure $= k \cdot$ (perimeter of small figure), where k is the constant of proportionality ($k = \frac{3}{2}$).

Thus,

$$
\begin{aligned}
\text{perimeter of large figure} &= \text{total length of line segments on large figure} \\
&= k \cdot (\text{total length of line segments on small figure}) \\
&= k \cdot (\text{perimeter of small figure})
\end{aligned}
$$

A similar argument can be used to show that

$$\text{area of large figure} = k^2 \cdot (\text{area of small figure})$$

To accomplish this we superimpose a network of rectangles over the small figure with total area equal to its area. We then draw the corresponding network of rectangles over the large figure [Fig. 8-36(a) and (b)]. The total area of the network of large rectangles is equal to the total area of the large figure. Furthermore, each large rectangle is equal to k^2 times the area of the corresponding small one. Thus,

$$\begin{aligned}\text{area of large figure} &= \text{total area of rectangles covering}\\ &\quad \text{large figure}\\ &= k^2 \cdot (\text{total area of rectangles}\\ &\quad \text{covering small figure})\\ &= k^2 \cdot (\text{area of small figure})\end{aligned}$$

1 square unit

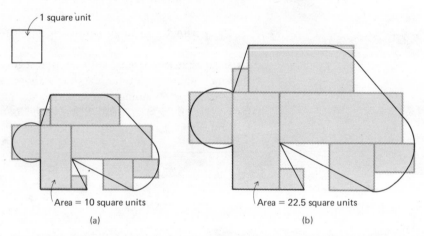

Area = 10 square units

(a)

Area = 22.5 square units

(b)

Figure 8-36   Areas of similar figures. Area of large figures $= k^2 \cdot (\text{area of small figure})$, where k is the constant of proportionality ($k = \frac{3}{2}$).

THE AREA UNDER A CURVE

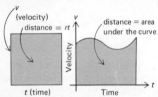

Any product of positive numbers can be thought of as an area. This seemingly trivial remark is the basis of many of the applications of higher mathematics which involve areas under curves. For example, if an automobile moves for a period of t seconds at a constant velocity of v feet per second, then the distance it travels is equal to the product vt—the area of the rectangle with base t and height v. If the velocity varies over the period of time, then the distance is equal to the area under the curve, which is the graph of v as a function of t.

Example 1.   Figure ABC in Figure 8-34 has perimeter 16 and area 10 square units. Figure $A'B'C'$ is similar to ABC with a constant of proportionality $k = \frac{3}{2}$. Calculate the perimeter and area of $A'B'C'$.

Solution.

$$\begin{aligned}\text{perimeter of } A'B'C' &= k \cdot (\text{perimeter of } ABC)\\ &= \tfrac{3}{2} \cdot 16 = 24 \text{ units (See Fig. 8-35)}\end{aligned}$$

$$\begin{aligned}\text{area of } A'B'C' &= k^2 \cdot (\text{area of } ABC)\\ &= (\tfrac{3}{2})^2 \cdot 10 = \tfrac{9}{4} \cdot 10 = \tfrac{90}{4} = 22.5 \text{ square units (Fig. 8-36)}\end{aligned}$$

The Area under a Curve

Many important problems in mathematics can be reduced to the single problem of determining the area under a curve. The curve is usually the graph of an equation involving x and y. By the "area under the curve" we mean the area of the region that is below the curve and above the x-axis.

In Chapter 9 we shall relate the area under a curve to problems in statistics. In this section we consider a different type of problem.

Figure 8-37 shows the velocity graph of an automobile over a short trip. It traveled for approximately 30 minutes at 20 to 30 miles per hour (in town), occasionally stopping for a traffic light, then speeded up to approximately 60 miles per hour until it reached its destination. The graph is drawn from one recorded by a special machine in the automobile.

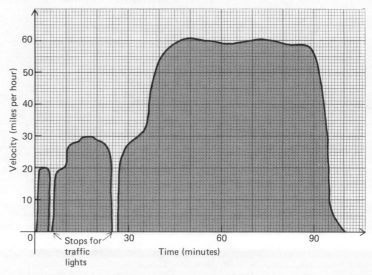

Figure 8-37 The area under the velocity graph is proportional to the distance traveled. Each small square represents $\frac{1}{60}$ mile. The total area is equal to 4295 small squares. The total distance is equal to $4295 \cdot \frac{1}{60} \approx 71.6$ miles.

A great deal of information can be obtained from this graph. We can, of course, read the velocity at any given instant directly from the graph. Just as important, the area under the graph is equal to the total distance the automobile traveled. The particular conversion for this graph (which depends on the scale) is

$$1 \text{ small square} = \tfrac{1}{60} \text{ mile} = 88 \text{ feet}$$

Since there are 4295 small squares under the curve, then the total distance is

$$\tfrac{1}{60} \cdot 4295 \approx 71.6 \text{ miles}$$

This same relationship between area and distance holds for any part of the trip. To see how far the automobile traveled during the first hour, for example, we count the small squares under the curve between $x = 0$ and $x = 60$ minutes. Since there are 2313 squares under the curve, then the automobile traveled

$$\tfrac{1}{60} \cdot 2313 \approx 38.6 \text{ miles}$$

during the first hour.

The relationship between area and distance pictured in Figure 8-37 is only one of many that involve areas under curves. A number of applications of mathematics to physical problems involve the determination of areas under curves at some point in the solution.

In recent years the electronic computer has been widely used to compute the approximate values of areas under curves. In the remainder of this section, we shall consider the ideas behind one of the simplest of the "area" programs for the computer.

In our discussion we shall restrict ourselves to the points on the curve in Figure 8-38. In that figure we have 13 points (x_1, y_1), (x_2, y_2), ..., (x_{12}, y_{12}), (x_{13}, y_{13}), where $x_1, x_2, \ldots, x_{13}$ are equally spaced along the x-axis. We wish to calculate the area under the curve between the vertical lines through x_1 and x_{13}.

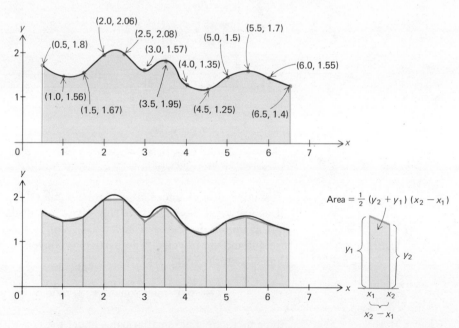

Figure 8-38 The total area under the curve $\approx$ the total area of the trapezoids ≈ 9.92. (a) Area under the curve; (b) area of the trapezoids.

Our first step is to connect up the points on the curve with line segments and to draw vertical lines through each of the 13 points. This gives us 12 trapezoids, which have a combined area approximately equal to the area under the curve [Fig. 8-38(b)].

Recall that the area of a trapezoid with bases B_1 and B_2 and height H is

$$\text{area of trapezoid} = \tfrac{1}{2}(B_1 + B_2)H$$

The first trapezoid has "bases" equal to y_1 and y_2 and "height" equal to $x_2 - x_1$. (Recall that the bases are the two parallel sides and the height is the distance between the bases.) Thus, the area of the first trapezoid is

$$\text{area of first trapezoid} = \tfrac{1}{2}(y_2 + y_1)(x_2 - x_1)$$

Similarly,

$$\text{area of second trapezoid} = \tfrac{1}{2}(y_3 + y_2)(x_3 - x_2)$$
$$\text{area of third trapezoid} = \tfrac{1}{2}(y_4 + y_3)(x_4 - x_3)$$
$$\cdots$$
$$\text{area of twelfth trapezoid} = \tfrac{1}{2}(y_{13} + y_{12})(x_{13} - x_{12})$$

The total area under the curve is given by

$$\text{area under curve} \approx \text{total area of trapezoids}$$
$$\approx \tfrac{1}{2}(y_2 + y_1)(x_2 - x_1) + \tfrac{1}{2}(y_3 + y_2)(x_3 - x_2)$$
$$+ \tfrac{1}{2}(y_4 + y_3)(x_4 - x_3) + \ldots + \tfrac{1}{2}(y_{13} + y_{12})(x_{13} - x_{12})$$

Although this expression is awkward to calculate, it is simple to program for the computer. Observe that the 12 terms have exactly the same form. Thus, we can set up a loop to evaluate each term and execute the loop 12 times. In order to set up the loop, we need a slightly different notation than in our example.

In each execution of the loop, we let X denote the x-term with largest subscript and, X1 denote the x-term with smallest subscript. Similarly, Y denotes the y-term with largest subscript and Y1 the y-term with smallest subscript. Thus, during the first execution of the loop, we have

$$X = x_2, \quad X1 = x_1, \quad Y = y_2, \quad Y1 = y_1$$

and we must evaluate

$$0.5 * (Y + Y1) * (X - X1) = 0.5(y_2 + y_1)(x_2 - x_1)$$

During the second execution of the loop, we have

$$X = x_3, \quad X1 = x_2, \quad Y = y_3, \quad Y1 = y_2$$

and we evaluate

$$0.5 * (Y + Y1) * (X - X1) = 0.5(y_3 + y_2)(x_3 - x_2)$$

The flow chart in Figure 8-39 shows the steps in the program. The program (including REMARK statements for clarity and an improved PRINT statement) follows with one run using the data from the points in Figure 8-38.

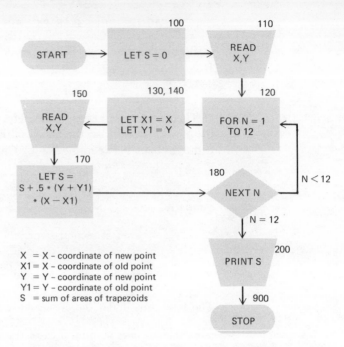

X = X – coordinate of new point
X1 = X – coordinate of old point
Y = Y – coordinate of new point
Y1 = Y – coordinate of old point
S = sum of areas of trapezoids

Figure 8-39 Flow chart for calculating the area under the curve in Figure 8-38 using trapezoids.

```
1Ø   REMARK * PROGRAM TO CALCULATE THE APPROXIMATE VALUE OF THE AREA
2Ø   REMARK * UNDER A CURVE GIVEN 13 POINTS (X,Y) ON THE CURVE. THE
3Ø   REMARK * PROGRAM IS BASED ON THE USE OF TRAPEZOIDS TO APPROXIMATE
4Ø   REMARK * THE AREA. FOR EACH TRAPEZOID X AND Y ARE THE COORDINATES
5Ø   REMARK * OF THE UPPER RIGHT HAND POINT, X1 AND Y1 ARE THE COORDINATES
6Ø   REMARK * OF THE UPPER LEFT HAND POINT. S IS THE RUNNING TOTAL OF
7Ø   REMARK * THE AREAS OF THE TRAPEZOIDS.
1ØØ  LET S=Ø
11Ø  READ X,Y
12Ø  FOR N=1 TO 12
13Ø  LET X1=X
14Ø  LET Y1=Y
15Ø  READ X,Y
16Ø  REMARK * ADD THE AREA OF THE TRAPEZOID
17Ø  LET S=S+.5*(Y+Y1)*(X−X1)
18Ø  NEXT N
19Ø  REMARK * PRINT VALUE OF S AFTER FINAL EXECUTION OF LOOP
2ØØ  PRINT "THE AREA UNDER THE CURVE IS APPROXIMATELY EQUAL TO",S
3ØØ  DATA .5,1.8,1,1.56,1.5,1.67,2,2.Ø6,2.5,2.Ø8,3,1.57,3.5,1.95,4
31Ø  DATA 1.35,4.5,1.25,5,1.5,5.5,1.7,6,1.55,6.5,1.4
9ØØ  END

THE AREA UNDER THE CURVE IS APPROXIMATELY EQUAL TO      9.92
```

EXERCISES

1. (1) Plot the following sets of points on graph paper. (Use one "large" square for each square unit.) Then draw a *smooth curve* through the points to form a simple closed figure.
   (2) Superimpose a network of rectangles over the figure with total area approximately equal to the figure. Use the rectangles to estimate the area. [See Fig. 8-32(a).]
   (3) Draw short line segments around the boundary with total length approximately equal to the perimeter. Use the line segments to estimate the perimeter. [See Fig. 8-32(b).]
   (a) $(0, 3)$, $(1, 4)$, $(2, 5)$, $(3, 6)$, $(4, 6)$, $(5, 5)$, $(4, 4)$, $(6, 3)$, $(6, 2)$, $(5, 0)$, $(4, -1)$, $(3, 0)$, $(2, 1)$, $(1, 2)$
   (b) $(5, 1)$, $(3, 2)$, $(2, 3)$, $(3, 5)$, $(2, 7)$, $(1, 8)$, $(2, 9)$, $(4, 10)$, $(5, 10)$, $(6, 9)$, $(7, 9)$, $(8, 8)$, $(8, 7)$, $(7, 6)$, $(6, 5)$, $(6, 4)$, $(6, 3)$

2. Draw a circle with radius $R = 3$ on graph paper.
   (a) Estimate the area by counting the squares on the graph paper. Compare your answer with the value obtained from the formula.
   (b) Estimate the perimeter by measuring short line segments marked on the boundary. Compare your answer with the value obtained from the formula.

3. If the planimeter is available, use it to measure the areas of the figures in Exercises 1 and 2.

4. Use uniform density paper to measure the areas of the figures in Exercises 1 and 2. (If uniform density paper is not available, you can make a good substitute by gluing graph paper to cardboard. Postal scales can be used to weigh the figures.)

5. You are given the perimeter and area of closed figure ABC. Figure $A'B'C'$ is similar to ABC with constant of proportionality k. Calculate the area and perimeter of figure $A'B'C'$.
   (a) perimeter of $ABC = 15.3$, area of $ABC = 12.1$; $k = 3$
   (b) perimeter of $ABC = 183$, area of $ABC = 276$; $k = \frac{5}{2}$
   (c) perimeter of $ABC = 21.32$, area of $ABC = 33.14$; $k = 80$

6. Make a velocity graph similar to Figure 8-37 for a short automobile trip. (Let one small unit on the x-axis represent 1 minute of time, one small unit on the y-axis represent 1 mile per hour of velocity. Then each small square unit represents $\frac{1}{60}$ of a mile of distance.) Measure the total area under the curve. Use the area to calculate the total distance traveled on the trip.

7. Modify the computer program at the end of the section so that N points (x, y) are used instead of 13 points, where N is the first number read from data. Use your program, with the appropriate value of N to calculate the areas under the following curves.
   (a) the parabola $y = x^2$ between $x = 0$ and $x = 2$ (N = 5)
   (b) the parabola $y = x^2 + 2$ between $x = 1$ and $x = 3$ (N = 3)

8-6 VOLUMES AND SURFACE AREAS OF SOLIDS

RECTANGULAR SOLID

volume = LWH
surface area = $2LH + 2HW + 2LW$

In this section we consider the volumes and surface areas of several solids in three-dimensional space. In contrast to our work in Section 8-4, we shall restrict ourselves to stating the appropriate formulas with a minimum of explanation. The reader should use this section primarily for reference.

Rectangular Solids

A *rectangular solid* has six sides, all of which are rectangles. If L is the length, W the width, and H the height (Fig. 8-40), then

$$\text{volume} = \text{length} \cdot \text{width} \cdot \text{height} = LWH$$

and

$$\text{surface area} = 2LH + 2HW + 2LW$$

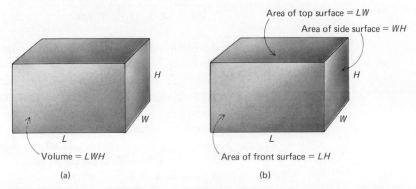

Figure 8-40   The rectangular solid. (a) volume = LWH; (b) surface = $2LH + 2LW + 2HW$.

Example 1. (a) Calculate the number of cubic inches in one cubic foot.
(b) Calculate the number of cubic centimeters in one cubic inch.

Solution. (a) One cubic foot is the volume of a cube that measures 12 inches in each dimension. [See Fig. 8-41(a).] Thus,

$$\text{one cubic foot} = \text{length} \cdot \text{height} \cdot \text{width} = 12 \cdot 12 \cdot 12$$
$$= 1728 \text{ cubic inches}$$

(b) One cubic inch is the volume of a cube that measures approximately 2.54 centimeters in each dimension. [See Fig. 8-41(b).] Thus,

$$\text{one cubic inch} = \text{length} \cdot \text{width} \cdot \text{height} \approx (2.54) \cdot (2.54) \cdot (2.54)$$
$$\approx 16.387 \text{ cubic centimeters}$$

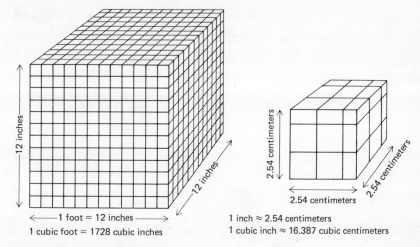

1 foot = 12 inches

1 cubic foot = 1728 cubic inches

1 inch ≈ 2.54 centimeters

1 cubic inch ≈ 16.387 cubic centimeters

Figure 8-41   Example 1.

SPHERE

volume = $\frac{4}{3}\pi R^3$

surface area = $4\pi R^2$

Spheres

A *sphere* can be formed by revolving a circle about its diameter. Every point on the surface of the sphere is R units from the center, where R is the radius of the original circle. (R is called the *radius* of the sphere.) The volume and surface area of a sphere are given by the formulas (Fig. 8-42):

$$\text{volume of sphere} = \frac{4}{3}\pi R^3$$

$$\text{surface area} = 4\pi R^2$$

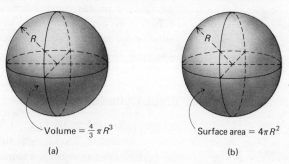

Volume = $\frac{4}{3}\pi R^3$ Surface area = $4\pi R^2$

(a)   (b)

Figure 8-42   The sphere. (a) volume = $\frac{4}{3}\pi R^3$; (b) surface area = $4\pi R^2$.

Example 2. Calculate the volume and surface area of a sphere of radius $R = 8$ inches.

Solution.

$$\text{volume} = \tfrac{4}{3}\pi R^3 = \tfrac{4}{3} \cdot \pi \cdot 8^3 \approx \tfrac{4}{3} \cdot (3.14159) \cdot 512$$
$$\approx 2144.66 \text{ cubic inches} \approx 1 \text{ cubic foot, } 416.66 \text{ cubic inches}$$
$$\text{surface area} = 4\pi R^2 = 4\pi \cdot 8^2 \approx 4 \cdot (3.14159) \cdot 64$$
$$\approx 804.25 \text{ square inch} \approx 5 \text{ square feet, } 84.25 \text{ square inches}$$

	Exact Value	Egyptian Value	Babylonian Value
Area of a rectangle	BH	BH	BH
Area of a triangle	$\tfrac{1}{2}BH$	$\tfrac{1}{2}BH$	$\tfrac{1}{2}BH$
Area of a circle	πR^2 ($\pi \approx 3.14159$)	$\tfrac{256}{81}R^2$	$3R^2$
Volume of a pyramid	$\tfrac{1}{3}B^2H$	$\tfrac{1}{3}B^2H$	$\tfrac{1}{2}B^2H$? (Surviving texts are not clear)

The Egyptians and Babylonians used approximation formulas for their calculations of area and volume. Some of these formulas were exact and others were not. As far as we know, no attempt was made to distinguish between exact and approximate results. The ancient Greeks were the first people to have accurate "formulas" for most of the common geometrical figures and solids. The Greeks based their work on theoretical arguments rather than on measurements.

RIGHT CYLINDER OR PRISM

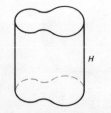

volume = (area of base) · height

lateral area = (perimeter of base) · height

Right Cylinders and Prisms

A *right cylinder* can be formed from a simple closed figure in a plane and a line segment of fixed length that is perpendicular to the plane. To form the cylinder, we construct a similar line segment at each point of the closed figure. This process causes a figure to be formed in three-dimensional space (Fig. 8-43).

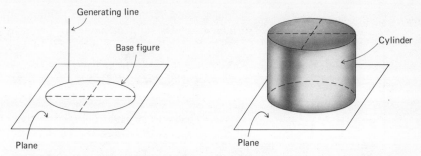

Figure 8-43 The cylinder. A cylinder can be formed by moving a line segment perpendicular to the plane of the base figure over every point of the base figure.

The original closed figure is called the *base* of the cylinder. (The opposite face also is called a base.) The original line segment is called the *generating line;* its length is called the *height* of the cylinder. The total surface area, not including the area of the two bases, is called the *lateral area* [Fig. 8-45].

Of the several cylinders shown in Figure 8-44, two are worthy of special attention. In Figure 8-44(a) the base figure is a circle. The corresponding cylinder is called a *right circular cylinder*. In (b) the base figure is a polygon. The corresponding cylinder is called a *prism*. Two unusual right cylinders are shown in Figure 8-44(c) and (d). The cylinder in (c) has a section of a parabola for its base; the one in (d) has an irregular base.

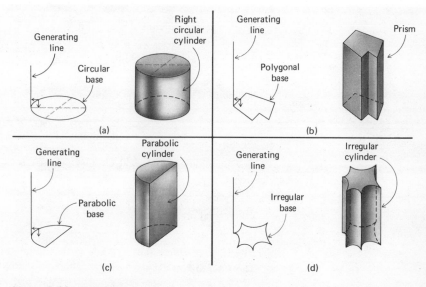

Figure 8-44 Cylinders.

RIGHT CIRCULAR CYLINDER

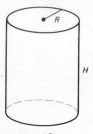

volume $= \pi R^2 H$
lateral area $= 2\pi RH$

The general formulas for the volume, the lateral area, and the total surface area of a right cylinder are given by

volume of right cylinder = (area of base) · (height)

lateral area of right cylinder = (perimeter of base) · (height)

total surface area of right cylinder = (lateral area) + 2 · (area of base)

Example 3. Use the general formulas to calculate the volume, surface area, and total area of a right circular cylinder of radius R and height H (Fig. 8-45).

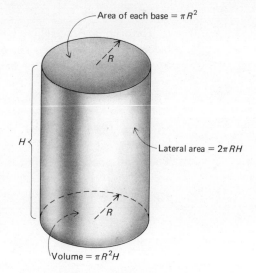

Figure 8-45 The right circular cylinder. Volume $= \pi R^2 H$; lateral area $= 2\pi RH$.

Solution. The base of the cylinder is a circle of radius R. Thus,

$$\text{area of base} = \pi R^2$$
$$\text{perimeter of base} = 2\pi R$$

The volume of the right circular cylinder is

$$\text{volume} = (\text{area of base}) \cdot \text{height} = \pi R^2 H$$

The lateral area is

$$\text{lateral area} = (\text{perimeter of base}) \cdot \text{height} = 2\pi RH$$

The total area is

$$\text{total area} = \text{lateral area} + 2 \cdot (\text{area of base})$$
$$= 2\pi RH + 2\pi R^2 = 2\pi R(H + R)$$

Example 4. Calculate the total surface area of a right circular cylinder with radius $R = 5$ centimeters and height $H = 3$ centimeters.

Solution. The area of the base is

$$\text{area of base} = \pi R^2 = \pi \cdot 5^2 = 25\pi \text{ square centimeters}$$

The lateral area is

$$\text{lateral area} = 2\pi RH = 2 \cdot \pi \cdot 5 \cdot 3 = 30\pi \text{ square centimeters}$$

The total surface area is equal to the lateral area *plus* the area of the two bases

$$\begin{aligned}
\text{total surface area} &= 30\pi + 2 \cdot 25\pi = 30\pi + 50\pi \\
&= 80\pi \approx 80(3.14159) \approx 251.33 \text{ square centimeters}
\end{aligned}$$

Example 5. Because of an accident at sea, 20,000 cubic feet of oil has been lost from a tanker. This oil has spread out into an oil slick that is approximately one-eighth of an inch thick. Estimate the total area covered by the oil slick.

Solution. The oil is approximately in the shape of an irregular right cylinder with height $H = \frac{1}{8}$ inch and volume $V = 20,000$ cubic feet. Recall that for any cylinder,

$$\text{volume} = (\text{area of base}) \cdot \text{height}$$

If we let x be the area of the base (the area covered by the oil slick), then

$$20,000 \text{ cubic feet} = x \cdot \tfrac{1}{8} \text{ inch}$$

In order to solve for x, the area of the base, we express the height in feet in order to have compatible units. Since

$$1 \text{ inch} = \tfrac{1}{12} \text{ foot}$$

then

$$\tfrac{1}{8} \text{ inch} = \tfrac{1}{8} \cdot \tfrac{1}{12} = \tfrac{1}{96} \text{ foot}$$

Thus, our equation becomes

$$\begin{aligned}
20,000 &= x \cdot \tfrac{1}{96} \\
x &= \frac{20,000}{\frac{1}{96}} = 20,000 \cdot \frac{96}{1} \\
&= 1,920,000 \text{ square feet}
\end{aligned}$$

To get a better idea of the area covered by the oil slick, observe that

$$1 \text{ square mile} = (5,280)^2 \text{ square feet} \approx 28,000,000 \text{ square feet}$$

Thus, the area is approximately

$$\frac{1,920,000}{28,000,000} \approx 0.07 \text{ square mile}$$

CONE OR PYRAMID

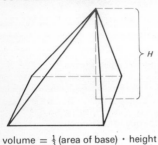

volume = ⅓ (area of base) · height

Cones and Pyramids

A cone can be formed by a process similar to the one used to generate a cylinder. We start with a closed figure in a plane (the *base*) and a point in space (the *vertex*) that is not in the plane of the base. The solid formed by constructing straight line segments from the vertex to each point of the base is called a *cone* (Fig. 8-46).

The perpendicular distance from the vertex to the plane of the base is called the *height* of the cone. The total surface area, not including the area of the base, is called the *lateral area* (Fig. 8-46).

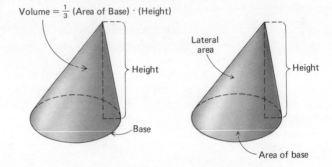

Figure 8-46 The cone. Volume = ⅓(area of base) · height.

Several cones are shown in Figure 8-47. If the base figure is a circle, the cone is called a *circular cone* [Fig. 8-47(a)]. If the vertex is directly above the center of the circle, the circular cone is called a *right circular cone* [Fig. 8-47(b)]. If the base is a polygon, the cone is called a *pyramid*.

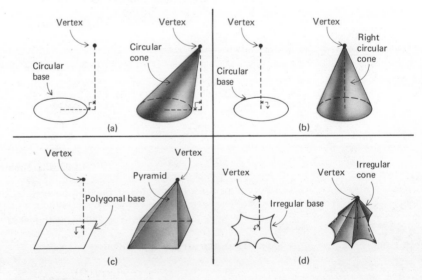

Figure 8-47 Cones.

The volume of a cone or pyramid is quite easy to calculate. It is equal to *one-third* of the volume of a cylinder with the same base and height. Thus,

$$\text{volume of cone} = \tfrac{1}{3} \cdot (\text{area of base}) \cdot \text{height}$$

Example 6. Calculate the volume of a right circular cone of radius $R = 3$ inches and height $H = 4$ inches.

Solution. The area of the base is

$$\text{area of base} = \pi R^2 = \pi \cdot 3^2 = 9\pi$$

The volume of the cone is

$$\begin{aligned}\text{volume} &= \tfrac{1}{3}(\text{area of base}) \cdot \text{height} \\ &= \tfrac{1}{3} \cdot 9\pi \cdot 4 = \tfrac{1}{3} \cdot 36\pi = 12\pi \\ &\approx 12 \cdot (3.14159) \approx 37.7 \text{ cubic inches}\end{aligned}$$

(See Fig. 8-48.)

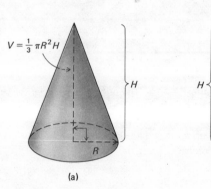

Figure 8-48 The right circular cone. (a) volume $= \tfrac{1}{3}\pi R^2 H$; (b) lateral area $= \pi R \sqrt{R^2 + H^2} = \pi R S$.

An argument similar to the one in Example 6 can be used to show that the volume of a right circular cone is

$$\text{volume of right circular cone} = \tfrac{1}{3}\pi R^2 H$$

The lateral area of a cone is much more difficult to calculate. We shall restrict ourselves to the special case of a right circular cone. In this special case, the formula is

$$\text{lateral area of right circular cone} = \pi R \sqrt{R^2 + H^2}$$

RIGHT CIRCULAR CONE

volume $= \tfrac{1}{3}\pi R^2 H$
lateral area $= \pi R S$
$= \pi R \sqrt{R^2 + H^2}$

If we let S denote the "slant height" of a right circular cone (S is the distance from a point on the circular base to the vertex), then by the Pythagorean Theorem

$$S = \sqrt{R^2 + H^2}$$

(See Fig. 8-48.) The formula for the lateral area can be expressed in terms of the radius of the base and the slant height as

$$\text{lateral area of right circular cone} = \pi RS$$

Example 7.   A right circular cone is constructed from a semicircle of radius 4 inches by bending the figure as in Figure 8-49.
   (a) Determine the slant height S and the lateral area of the cone.
   (b) Calculate the radius R and the height H of the cone.
   (c) Calculate the volume of the cone.

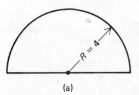
(a)

(b)

(c)

Figure 8-49   (a) The original semicircle; (b) bend the semicircle; and (c) form a cone. (See Example 7.)

Solution.   (a) The slant height S of the cone is equal to the radius of the original semicircle. Thus,

$$S = 4$$

The lateral area of the cone is equal to the area of the original semicircle. Thus,

$$\text{lateral area of cone} = \text{area of original semicircle}$$
$$= \tfrac{1}{2} \cdot \pi \cdot 4^2 = 8\pi \approx 25.13 \text{ square inches}$$

(b) To calculate the radius, recall that

$$\text{lateral area} = \pi RS \quad \text{(where } R \text{ is the radius of the cone)}$$
$$8\pi = \pi \cdot R \cdot 4 = 4\pi R$$
$$R = 2 \text{ inches}$$

To calculate the height, recall that

$$S = \sqrt{R^2 + H^2}$$

Thus,

$$S^2 = R^2 + H^2$$
$$H^2 = S^2 - R^2 = 4^2 - 2^2$$
$$= 16 - 4 = 12$$
$$H = \sqrt{12} \text{ inches}$$

(c) The volume is

$$\text{volume} = \tfrac{1}{3}\pi R^2 H = \tfrac{1}{3} \cdot \pi \cdot 2^2 \cdot \sqrt{12}$$
$$\approx \tfrac{1}{3}(3.14159) \cdot 4 \cdot (3.464) \approx 14.5 \text{ cubic inches}$$

The Volume of a Theoretical Wine Barrel

While making his studies of the heavens which led to the laws of planetary motion, Kepler became interested in the problem of calculating the volumes of irregularly shaped wine barrels. To approximate the shapes of the barrels, he rotated sections of ellipses, circles, and other curves about lines in their planes, thereby obtaining various solids of revolution. He extended the methods he had used to calculate areas to the calculation of the volumes of these solids.

Three of the 93 different solids studied by Kepler are shown in the figure. A *torus* (donut-shaped solid) is obtained by rotating an entire circle about a line that does not intersect it. Kepler obtained figures that he called "apples" and "lemons" by rotating arcs of a circle about their chords. If an arc is greater than half a circle, the solid is an "apple." If the arc is less than half a circle, it is a "lemon."

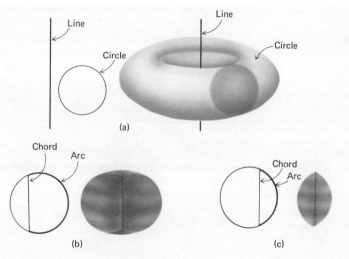

(a) Torus: Revolve a circle about a line that does not intersect it; (b) Kepler's "apple": revolve a major arc of a circle about its chord; (c) Kepler's "lemon": revolve a minor arc of a circle about its chord.

Summary of Formulas

$$L = \text{length}$$
$$H = \text{height}$$
$$W = \text{width}$$
$$R = \text{radius of base}$$
$$B = \text{area of base}$$
$$P = \text{perimeter of base}$$
$$S = \text{slant height of cone}$$

Figure	Area of Base	Volume	Lateral Area
Rectangular solid		LWH	
Sphere		$\frac{4}{3}\pi R^3$	$4\pi R^2$
Cylinder (or prism)		BH	PH
Right circular cylinder	πR^2	$\pi R^2 H$	$2\pi RH$
Cone (or pyramid)		$\frac{1}{3}BH$	
Right circular cone	πR^2	$\frac{1}{3}\pi R^2 H$	$\pi R\sqrt{R^2 + H^2} = \pi RS$

EXERCISES

1. Calculate the total surface areas and volumes of the following rectangular solids and boxes.
   (a) solid: length 5, width 3, height 2
   (b) solid: length 25 feet, width 17 feet, height 8 feet
   (c) box: length 2 feet, width 2 feet, height 3 feet
   (d) open-topped box: length 3 feet, width 2 feet, height 2 feet
   (e) open-topped box: length 2 feet, width 2 feet, height 2 feet
2. Calculate the surface areas and volumes of the following spheres.
   (a) $R = 7$ inches   (d) $R = 4,000$ miles (the earth)
   (b) $R = 5$ feet   (e) $R = 432,000$ miles (the sun)
   (c) $R = 8$ centimeters   (f) $R = 1,080$ miles (the moon)
3. Calculate the volumes, lateral areas, and total surface areas of the following figures.
   (a) right circular cylinder: $R = 3$, $H = 2$
   (b) right circular cylinder: $R = 7$ feet, $H = 20$ feet
   (c) right prism: base—a right triangle with sides 3, 4, 5; height—3.
   (d) right prism: base—a rectangle with sides 2 centimeters and 4 centimeters; height—14 centimeters
   (e) right circular cone: $R = 5$ inches, $H = 5$ inches
   (f) right circular cone: $R = 7$ inches, $H = 4$ feet
4. A right circular cone is formed by bending a quarter of a circle of radius 10 inches until the two straight sides coincide. Find the radius R, slant

height S, lateral area, and volume of the cone. (See Example 7 for a related problem.)

5. The Great Pyramid of Egypt originally had a 754.5-foot square base and a height of 481 feet.
   (a) Calculate the original volume of the Great Pyramid (in cubic yards).
   (b) The stones in the Great Pyramid have an average weight of approximately 125 pounds per cubic foot. Estimate the total weight of the structure.

6. The formula for the lateral area of a standard pyramid with square base of length x and height h is

$$\text{lateral area} = x\sqrt{x^2 + 4h^2}$$

How much paint would have been required to paint the Great Pyramid in its original form? (Assume that 1 gallon of paint covers approximately 500 square feet.)

7. A hot water heater contains an inner water tank, which is a right circular cylinder 20 inches in diameter and 3 feet high. The sides and top of the cylinder are covered by insulating material 1 inch thick.
   (a) Calculate the approximate volume of the inner tank in *gallons* (1 cubic foot ≈ 7.5 gallons).
   (b) Calculate the approximate volume of the insulating material.

8. An ancient stone monument was carved in the shape of a right circular cone with radius 10 feet and height 15 feet. An earthquake toppled the monument and caused the small end to break off 3 feet from the vertex. Calculate the approximate volume of each of the two remaining pieces.

8-7 MAXIMUM AREA AND VOLUME

Maximum-Area Problems

A farmer plans to fence in a rectangular pasture using 2000 feet of fencing. He needs to determine the dimensions of the pasture that will enclose the largest possible area. Having an experimental bent of mind, he calculates the area for several values of the length and width (Fig. 8-50).

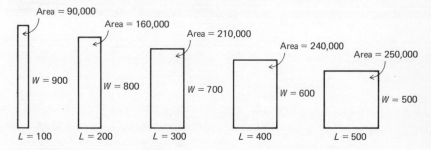

Figure 8-50 A rectangle of fixed perimeter has the maximum area if it is a square.

Length (L)	Width (W = 1000 − L)	Area (LW)
100 feet	900 feet	90,000 square feet
200 feet	800 feet	160,000 square feet
300 feet	700 feet	210,000 square feet
400 feet	600 feet	240,000 square feet
500 feet	500 feet	250,000 square feet
600 feet	400 feet	240,000 square feet

The farmer observes from the table of areas that the closer the length and width are to being equal, the greater is the area. He decides that having both dimensions equal to 500 feet yields the greatest possible area.

The conclusion made by the farmer is correct—*a square is the rectangle of maximum area that can be enclosed in a fixed perimeter.* As we saw in Section 7-4, this result can be established by an argument that involves a parabola of form $y = ax^2 - bx$. Similar results for more complicated figures can be obtained by the use of the calculus.

Mathematicians are not satisfied until they have general theories to explain their results. When they realized that a square has the greatest area of any rectangle with a fixed perimeter, they became interested in the same problem for other geometrical figures.

The Rectangle of Maximum Area

In the text it is claimed that the rectangle of perimeter 2000 feet which has the maximum possible area is a square.

To establish this result mathematically we let x denote the length and y the height of the rectangle. The perimeter is

$$2x + 2y = 2000$$

so that

$$2y = 2000 - 2x$$
$$y = 1000 - x$$

The area of the rectangle with length x and height y is

$$A = xy = x(1000 - x)$$
$$= -x^2 + 1000x$$

The graph of

$$A = -x^2 + 1000x$$

is a parabola with vertex at the point where $x = 500$. Since the largest possible value of A occurs at the vertex, then the area is a maximum when $x = 500$. At this value of x, we have

$$y = 1000 - x = 1000 - 500 = 500$$

Thus, the rectangle of maximum area is a square with each side equal to 500 feet.

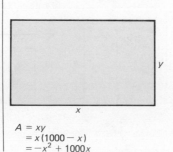

$A = xy$
$= x(1000 - x)$
$= -x^2 + 1000x$

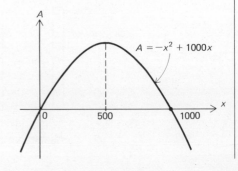

$A = -x^2 + 1000x$

The simplest polygon is the triangle. It was quickly established that *the equilateral triangle (three equal sides and three equal angles) has the greatest area of all triangles with the same perimeter.*

The next most complicated figure is the quadrilateral (four-sided closed figure). It was found that *the square (four equal sides and four equal angles) has the greatest area of all quadrilaterals with the same perimeter.*

Similar results were established for pentagons (five sides) and hexagons (six sides). In each case it was found that the *regular polygon* (the one with equal sides and equal angles) has the greatest area.

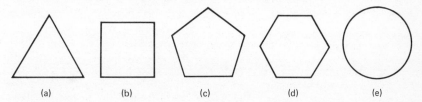

(a)   (b)   (c)   (d)   (e)

Figure 8-51 The greater the number of sides, the greater the area of a regular polygon of fixed perimeter. The circle has the greatest area of any figure with the same perimeter. (a) Equilateral triangle; (b) square; (c) regular pentagon; (d) regular hexagon; and (e) circle.

After studying a number of such cases, the obvious conjecture was made and then proved: *The regular polygon has the greatest area of all polygons with the same perimeter and same number of sides.*

The fact that the circle is the limiting case of regular polygons when the number of sides increases without bound was successfully exploited by Archimedes to estimate the value of the number π.

How do different regular polygons with the same perimeter compare in area? It is not difficult to show that the greater the number of sides, the greater the area. Thus, assuming the figures all have the same perimeter, the square has greater area than the equilateral triangle; the regular pentagon has greater area than the square; the regular hexagon has greater area than the regular pentagon, and so on.

As we see in Figure 8-51, the shape of a regular polygon approaches that of a circle as the number of sides increases. This limiting figure, the circle, has area greater than that of any other plane figure with the same perimeter. In other words, *the largest possible area that can be enclosed in a plane figure with a given perimeter is enclosed by a circle.* The closer a figure approximates the shape of a circle, the greater its area in relation to its perimeter.

An ancient story concerns Dido, the daughter of the king of the Phoenician city of Tyre, who ran away from home and landed on the North African shore of the Mediterranian Sea. Wanting to found a city, she paid a sum of money for "all the land that could be encompassed by a bull's hide." First, she cut the hide into very thin strips and tied them together to form a long leather string with which to "encompass the land." Because the shoreline would not have to be enclosed by the string, she chose a piece of land along the beach. She then had to determine the shape to lay out the leather string. After considerable thought, she decided that the most land would be enclosed by a semicircle. She laid out the string in that shape, enclosed the land, and built the city of Carthage there.

Dido's solution to the problem was the correct one—the maximum area that can be enclosed with a fixed perimeter along the seashore is bounded by a semicircle [Fig. 8-52(a)]. This can be established without too much difficulty if we first complicate the problem as in the following discussion.

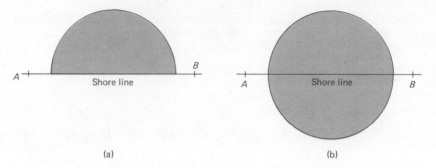

(a)   (b)

Figure 8-52 (a) Dido's problem: Enclose the maximum area along the shore line with a fixed perimeter; (b) the equivalent problem: Enclose the maximum area with twice the perimeter.

The problem faced by Dido is that of enclosing a maximum area on one side of shoreline AB with a string of length P [Fig. 8-52(a)]. This is equivalent to enclosing the maximum area on both sides of shoreline AB with a string of length $2P$ [Fig. 8-52(b)]. If we solve this last problem, we automatically have a solution of the original one. We know that the maximum area that can be enclosed with a perimeter of $2P$ is contained in a circle. Thus, the maximum area that can be enclosed with a perimeter of P on one side of the shoreline is contained in a semicircle.

Maximum Volume

Relationships similar to the ones discussed above hold between the *volume* and the *surface area* of a solid. We shall discuss the results for two special solids (Fig. 8-53).

The cube has the maximum volume of all rectangular solids with the same surface area. The closer a rectangular solid is to a cube, the greater its volume in relation to its surface area.

The sphere has the maximum volume of all solids with the same surface area. The closer the shape of a solid is to a sphere, the greater its volume in relation to its surface area.

The results about maximum area and maximum volume have been widely utilized in construction work. Many factories have been built with floor plans that are approximately square in order to enclose the largest possible area for the available money. A geodesic dome (a portion of a sphere) is used to enclose a large volume with a comparatively small surface area. It was claimed that the Pentagon building near Washington was built so as to enclose a larger area than would have been possible with a standard rectangular building of the same perimeter. (Critics have occasionally

The Pentagon building.

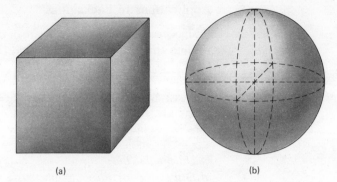

Figure 8-53 Solids of maximum volume. (a) The cube has the maximum volume of any rectangular solid with the same surface area; (b) the sphere has the maximum volume of any solid with the same surface area.

wondered if perhaps the five-pointed star on the generals' insignia might not have been more of an inspiration for the shape of the Pentagon. The actual increase in area caused by using a pentagon instead of a square is about 10 percent, whereas about 16 percent could have been added to the area by using a hexagon and 23 percent by using a circle.)

Minimum Perimeters and Surface Areas

The results concerning maximum area for a given perimeter and maximum volume for a given surface area can be considered in a different way. *Of all plane figures with a given area, the circle has the minimum perimeter. Of all solids with a given volume, the sphere has the minimum surface area.*

These geometrical properties can be seen in certain natural phenomena. To understand them, recall that every liquid has a *surface tension*. This is an elasticlike property of the surface of the liquid that causes it to tend toward a minimum surface area. Because the surface tensions of most liquids are very weak, the effects can be observed in the physical world only under special conditions.

If we carefully place a drop of oil in a bowl of very still water, the surface tension of the oil causes it to assume the shape in which it has the minimum perimeter—a circle. Ideally, the surface tension should cause the oil to assume a spherical shape, but the attraction of gravity is so much stronger than the surface tension that the oil spreads out in a circle. When other forces are present, they also influence the shape of the drop of oil. If we stir the water, for example, the forces acting on the oil cause it to break up into irregular shapes. When the water becomes still again, each droplet of oil assumes a circular shape. A similar result holds when a liquid is released in a gravity-free atmosphere such as the interior of a spacecraft. The surface tension is then the strongest force affecting the liquid; so it assumes a spherical shape.

Applications to Packaging

Businessmen have long been aware of the relationships involving maximum volume and minimum surface area. One of the main concerns has been to hold down costs, and this can be partially accomplished by using the minimum amount of material to enclose certain volumes. Thus, a cube uses the least material to enclose a given volume of any rectangular container. Obviously, the cube is not the best container for all products. Most tin cans, for example, are cylindrical. Even for this shape, the form that most closely resembles a cube is the most economical to use. *The cylinder with height equal to the diameter of the base encloses the greatest possible volume of any cylinder with that same surface area* [Fig. 8-54(b)].

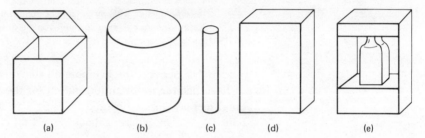

Figure 8-54 Packages designed for maximum capacity. (a) Cubical box; (b) cylindrical can with height equal to diameter. Some standard package designs are shown in (c), (d), and (e).

If we examine the canned goods in a grocery store, we shall find that approximately 90 percent of them have the diameter equal to the height—this being the shape that yields the minimum cost for the container.

When we examine the goods packaged in rectangular boxes, however, we find that a completely different situation exists. Almost none of the packages are cubical. Obviously these packages have been designed for some purpose other than minimizing costs.

There are several reasons for not using cubical containers for most products. One is convenience—cubes are hard to handle. (Most persons prefer a package that is not more than 3 or 4 inches deep.) A more important reason, however, is to fool the eye. When we see two rectangular objects, our minds automatically assume that the one that appears to be the largest actually has the largest volume. Thus, a box with a large frontal surface area appears larger than a box with a small frontal surface area, even if they hold the same amount. A standard cereal box, for example, appears to the eye to be much larger than a cube that encloses the same volume.

The design of packages to "fool the eye" is taken to extremes that almost qualify as deceptive merchandising. A cosmetics package is shown in Figure 8-54(e). A small rectangular bottle (enclosing a very small volume of perfume for the large frontal surface area) is contained in a much larger box, which has a false bottom and false top. The sides of the box are bent in at angles to give the illusion of depth. The picture presented to the eye is that of a much larger container of perfume than actually is found in the box.

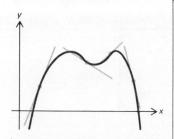

TANGENT LINES TO A CURVE
A tangent line to a curve is a line that touches the curve at a point and is the line that best approximates the curve in the vicinity of the point. In any very small neighborhood of the point, the tangent line is closer to the curve than is any other line.

Maximum—Minimum Problems and the Calculus

Problems involving maximum values of functions usually are solved by the use of the calculus. This branch of mathematics is concerned with two major problems, which can be interpreted geometrically as:

(1) Determine the equations of the tangent lines to a curve.

(2) Find a general method for calculating the area under a curve.

The part of the calculus that is concerned with tangent lines can be used to determine where a function has its maximum or its minimum value. The graph of the function

$$y = -x^2 + 2x + 3$$

for example, which is shown in the drawing, has its maximum value at the point where $x = 1$. Observe that the tangent line at this point is horizontal. In general, this same result holds for an arbitrary function: *If there is a tangent line to the graph of a function at the point where it has its maximum (or minimum) value, then that tangent line is parallel to the x-axis.*

To find where a function has its maximum (or minimum) value, we use the calculus to find the values of x for the points where its graph has horizontal tangent lines. We then examine these points to find where the maximum (or minimum) value of the function occurs.

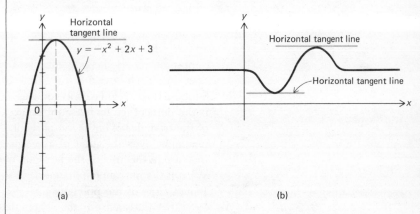

(a) The graph of $y = -x^2 + 2x + 3$ has a horizontal tangent line at the maximum value of y; (b) the tangent line to the graph of a function is horizontal at the point where the function has its maximum (or minimum) value.

EXERCISES

1. A farmer wishes to enclose a rectangular pasture with 700 feet of fencing. What dimensions give the maximum area of the field?
2. The farmer in Exercise 1 has decided to use an existing fence for one

boundary of the field. What dimensions now give the largest area of the field? [*Hint:* He only needs to enclose three sides of a rectangular field. Show that this problem is equivalent to enclosing a field twice the size with 1400 feet of fencing (Fig. 8-55).]

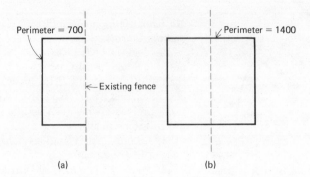

Perimeter = 700

Perimeter = 1400

←—Existing fence

(a)   (b)

Figure 8-55 (a) The farmer's problem. Enclose three sides of a rectangle so that the total area of the rectangle is a maximum; (b) the equivalent problem. Enclose the rectangle of maximum area with twice the fencing. (See Exercise 2.)

3. The following table lists the approximate areas of several regular polygons, each with sides of length *s*.

Figure	Approximate Area
Equilateral triangle	$0.433s^2$
Square	s^2
Regular pentagon	$1.72s^2$
Regular hexagon	$2.60s^2$

(a) Suppose each regular polygon has a perimeter of 60 inches. What area does each cover? (*Hint:* First find *s*.)

(b) Suppose each regular polygon has a perimeter of *P* units. What is the percentage increase in the area of the square over the area of the triangle? What is the percentage increase in the area of the pentagon over the area of the square? What is the percentage increase in the area of the hexagon over the area of the pentagon? What is the percentage increase in the area of a circle of perimeter *P* over each of the polygons?

4. We know that the square is the rectangle of minimum perimeter that encloses a given area. Show that there is no rectangle of maximum perimeter that encloses a given area. (*Hint:* Explain how a rectangle with perimeter greater than 1000 feet could be constructed with an area of 1 square inch. Then explain how rectangles of even larger perimeter could be constructed.)

5. *Puzzle problem.* A hiker plans to walk from point *A* to point *B* on the desert as shown in Figure 8-56. Because he is out of water, he must walk from *A* to *x* (at the river's edge), fill his canteen, and then walk to *B*. Where should point *x* be located in order to minimize the total

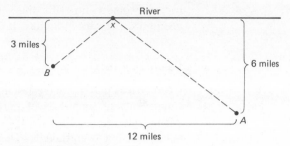

Figure 8-56 Exercise 5.

distance? (*Hint:* Draw a new figure locating point *B'* directly across the river from point *B*. Show that traveling from *A* to *B'* is mathematically equivalent to walking from *A* to *x* to *B*.)

8-8 GEOMETRY IN LIFE AND ART

Throughout history geometry has influenced society more than any other branch of mathematics. This has been accomplished through the practical

Figure 8-57 Ancient designs based on geometrical patterns. (a) American Indian pottery (1000–2000). Photo courtesy of School of American Research; (b) Peruvian textile pattern (1000–1450). Photo courtesy of the American Museum of Natural History.

(a)

(b)

use of geometry in daily life—through geometrical patterns in design, geometrical shapes in architecture, and geometrical principles in art.

Geometry in Design

Geometrical patterns have fascinated artisans from the earliest recorded times. Ancient fragments of pottery reveal repeated patterns based on geometrical figures (Fig. 8-57). The importance of geometry in designs has never lessened, even though styles have changed. In our society we do not use bold geometrical patterns on all of our pottery, but we do use such patterns (often disguised) in arrangements of floor and counter tiles, on wallpaper, and as the basis for advertising layouts (Fig. 8-58).

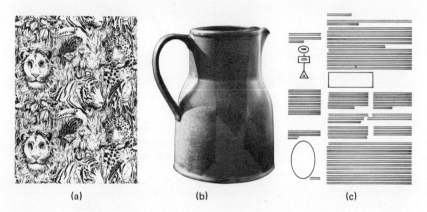

(a)   (b)   (c)

Figure 8-58   Modern designs based on geometrical patterns. (a) Wallpaper design based on intersecting sets of parallel lines. Photo courtesy of the Birge Company; (b) pottery design based on triangles. Photo courtesy of Iron Mountain Stoneware, by Nancy Patterson; and (c) design layout of book based on rectangles.

Geometry in Architecture

For both practical and aesthetic reasons, the most important applications of geometry to human life have been in the design of buildings and other structures. Every building, by conscious design or not, is a geometrical object. Ultimately, it must be planned from points, lines, planes, and curved surfaces. Furthermore, some of the simplest geometrical shapes are easy to construct, possess great strength, and can be assembled with similar shapes into larger structures. Thus, the most elementary geometrical figures have been utilized in the design of some of the world's most beautiful buildings.

The rectangular box, often topped with a pitched roof, has been the basic construction form throughout history. A majority of the world's buildings have been based on this simple plan.

In church architecture a circular design often has been incorporated into the basic rectangular form by means of arches and domes. This allows

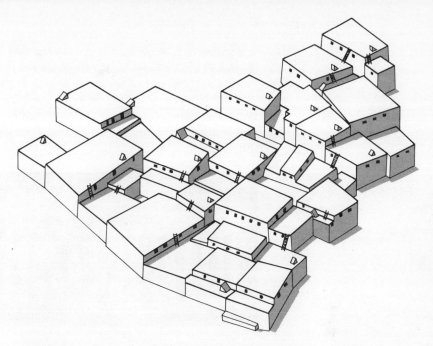

Figure 8-59 Reconstruction of the village of Çatal Hüyük (6000 B.C.). Çatal Hüyük (in modern Turkey) was one of the first villages established in the ancient world. The buildings were single rooms joined together at the walls. All entrances were at the roof level. The blank outer wall served as an effective barrier to invaders. (After J. Mellaart.)

(a) (b)

Figure 8-60 Model of the Hypostyle Hall, Egyptian temple at Karnak (about 1500 B.C.). The Egyptians used a post-and-lintel system of supports in their buildings. The posts or columns supported flat beams or stone slabs. This type of construction resulted in flat, rectangular buildings. No cement or mortar was used in the construction. The 66-foot-high columns in the Hypostyle Hall were held in place by the weight of the heavy stone ceiling slabs. The flat roof in the Hypostyle Hall required a large number of support columns, but not as many as were actually used. The dense groupings of the high columns had a religious significance for the worshipers, who were forced to follow narrow paths through the columns of the hall on their way toward the inner sanctuary. (a) Model of the Hypostyle Hall. Photo courtesy of The Metropolitan Museum of Art, Purchase, 1890, Levi Hale Willard Bequest; (b) post-and-lintel construction.

vast spaces to be enclosed without the necessity for close supporting pillars. [Compare the supports for the Egyptian temple at Karnak (Fig. 8-60) with the Hagia Sophia of Constantinople (Fig. 8-64).]

During the twentieth century, the curve has become popular in public buildings. This popularity is partly caused by the increased flexibility of design allowed by the use of reinforced concrete as a building material (Fig. 8-65).

Figure 8-61   Model of the Parthenon at Athens (447–432 B.C.). The Parthenon was built on a rectangular plan using a post-and-lintel construction. Photo courtesy of The Metropolitan Museum of Art, Purchase, 1890, Levi Hale Willard Bequest.

Figure 8-62   The Seagram Building, New York (1956–1958). The flat rectangular planes of the building reflect the rectangular rooms and corridors of the interior.

Figure 8-63 *Habitat,* an apartment complex at Expo 67, Montreal. The rooms are rectangular boxes, which are stacked together so as to give an irregular effect to the exterior of the buildings. The design reflects an attempt to orient each apartment to the optimum advantage of its occupants. Most buildings, by contrast, fit the individual rooms into modular frameworks dictated by the exteriors. Photo by George Carr.

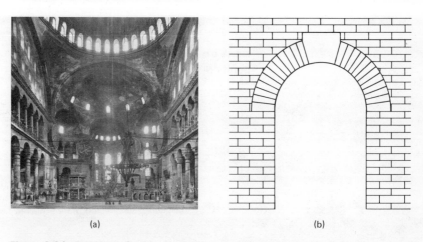

(a) (b)

Figure 8-64 The Hagia Sophia, Constantinople (532–537). The Hagia Sophia was one of the first Christian churches with circular vaults and arches dominating a rectangular base. The circular arch, a Babylonian invention, enables large openings to be enclosed with comparatively few supports. Arches were unknown to the Egyptians and were virtually ignored by the Greeks. The Romans, however, used them extensively. The dome and the vault are three-dimensional generalizations of the two-dimensional arch that allow vast open spaces to be enclosed. (a) Interior view of the Hagia Sophia. Photo courtesy of Marburg-Art Reference Bureau; (b) arch construction.

Figure 8-65 The Guggenheim Museum, New York. The modern use of reinforced concrete allows curved surfaces to be constructed almost as easily as flat planes.

Geometry in Painting

Many painters base their arrangements of elements on geometrical forms, but these forms are usually not noticed except by the experts. Some of these forms are based on the rectangular shape of the canvas. Every painting, for example, has several natural "centers of interest." The classical artist frequently arranged to have key elements of his paintings at these points (Fig. 8-66).

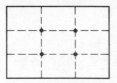 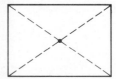

Figure 8-66 Centers of interest. Many of the classical painters either located important elements at these centers of interest or directed motion toward them.

Feelings of motion or of strength and serenity can be imparted by a conscious use of oblique lines. By directing apparent motion along a diagonal line, a feeling of true motion can be attained (Fig. 8-67). On the other hand, a triangular arrangement of major elements with the base of the triangle almost horizontal can cause a feeling of strength, serenity, or solidarity (Fig. 8-68).

Figure 8-67 Titian, *Christ Crowned with Thorns,* 1565. An illusion of motion is achieved by directing the action along the diagonal lines.

Figure 8-68 Cézanne, *Still Life.* A feeling of serenity is achieved by using a triangular (pyramidal) arrangement of elements. Photo courtesy of Musée du Louvre.

Perspective in Paintings

The problem of perspective has concerned artists for centuries. How can a three-dimensional scene be realistically represented on a two-dimensional canvas? The ultimate objective is an optical illusion where the picture appears to the viewer to be three-dimensional.

Figure 8-69   *Scribe Presenting Bible to St. Peter from Gero Codex,* 970. This tenth-century painting depicts the scene as if it were flat. The rules of perspective were unknown at the time, and distances were indicated by reducing figures in size. Photo courtesy of Marburg-Art Reference Bureau.

Figure 8-70. Jan van Eyck, *The Madonna with Chancellor Rolin,* 1435. The rules of perspective were developed during the Renaissance. All of the parallel lines that are directed away from the viewer cross at a point on the horizon. Photo courtesy of Alinari-Art Reference Bureau.

The basic rules of perspective were worked out in the fourteenth and fifteenth centuries. Figure 8-71 shows how the German artist Dürer achieved a three-dimensional effect. He viewed the scene from a fixed position (the eyepiece) through a sheet of glass. By marking and studying the apparent boundaries of objects on the glass, the artist was able to draw them as if they were seen from the original location of the eye. Eventually the rules of perspective were developed as the axioms and theorems of an abstract branch of mathematics called *projective geometry*.

One of the basic problems of perspective was that of finding where "parallel" lines intersect. In the real world, parallel lines that are not perpendicular to the viewer's line of vision appear to intersect on the horizon. It was necessary to draw them on canvas as if they would intersect, but the problem was complicated by the fact that several different sets

Figure 8-71 Albrecht Dürer, *Artist Drawing a Portrait of a Man*. Illustration from 1525 edition of textbook on *Perspective and Proportion*. This woodcut shows how Dürer made a practical study of perspective. The scene to be painted was viewed from a fixed position (the eyepiece) through a flat sheet of glass. The artist marked the outlines of objects on the glass, thereby recording their exact appearances from the position of the eyepiece. Photo courtesy of The Metropolitan Museum of Art, gift of Henry Walters, 1917.

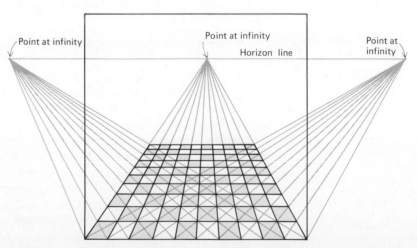

Figure 8-72 The points at infinity lie on the horizon line. This sketch, based on the pattern of floor tiles in van Eyck's painting *The Madonna with Chancellor Rolin* (Figure 8-70), shows how sets of ''parallel'' lines in a painting meet at points at infinity. The lines directed away from the viewer meet at the center point at infinity. The diagonals of the squares meet at points at infinity outside of the painting. The three points at infinity lie on the horizon line.

of parallel lines, running in different directions, had to be taken into account. One of the basic theorems in projective geometry establishes that each set of "parallel" lines in a painting has one common point of intersection and that these common points of intersection lie on a straight line. (See Fig. 8-72.)

The Pyramids

The pyramids of Egypt are some of the few major structures built with triangular sides. The largest of these pyramids is the size of a small mountain, containing 2,300,000 blocks of stone averaging 2.5 tons in weight. All of these blocks had to be dragged to the final site by slaves using only the most primitive machines.

The use of the pyramids for the burial of kings is well known. The cross section of the Great Pyramid shows the small amount of space allotted for this purpose in comparison to the great bulk of the stone blocks.

A recent (but not popular) theory is that the shape of the pyramids was determined by a secondary function. It is known that an early form of the pyramid was the *mastaba*, a large raised slab with sloping sides used as a burial chamber. It is thought that an obelisk was sometimes placed on a mastaba and that the entire construction was used as an observatory from the ground by sighting stars along the point of the obelisk. Periodically the mastaba was increased in height in order to obtain greater accuracy for the observations, resulting in the final form of the pyramids.

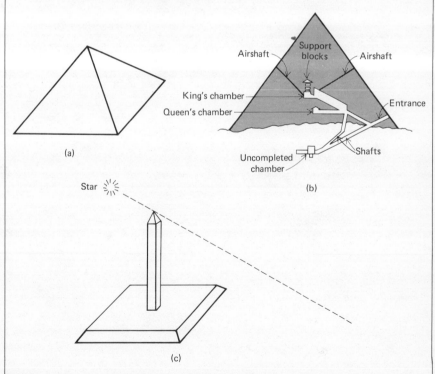

(a) The great pyramid; (b) cross-section of the great pyramid; and (c) possible use of a partially built pyramid as an observatory.

SUGGESTIONS FOR FURTHER READING

1. Bell, E. T. "Modern Minds in Ancient Bodies (Zeno, Eudoxus, Archimedes)," *Men of Mathematics*, Chap. 2. New York: Simon & Schuster, 1937.

2. Cooley, Hollis R., and Wahlert, Howard E. *Introduction to Mathematics*, Chaps. 8, 9. Boston: Houghton-Mifflin, 1968.

3. Kline, Morris. *Mathematics for Liberal Arts*, Chaps. 10, 12, 20. Reading, Mass.: Addison-Wesley, 1967.

4. Waerden, B. L. van der. *Science Awakening*. New York: Oxford University Press, 1961.

5. Weyl, Hermann. "Symmetry," in *World of Mathematics*, James R. Newman, ed., Vol. I, Part IV, Chap. 9. New York: Simon & Schuster, 1956.

9

INTRODUCTION TO STATISTICS

9-1 INTRODUCTION

The development of statistics began with a prosperous seventeenth-century London haberdasher named *John Graunt* (1620–1674). During that century the British government published periodic *Bills of Mortality* listing all of the deaths in each region and their causes. Graunt realized that London was changing in many ways, but he could not find an accurate way to determine how it was changing. Purely to satisfy his intellectual curiosity, he made a study of the death records contained in the *Bills of Mortality.* He was amazed to find that deaths from various diseases and accidents occurred with astonishing regularity in the populace as a whole, even though these events seemed to be completely random. For example, although there was no way to determine who would die of drowning in a given year, the total number of such deaths occurred with regularity and could be predicted with a high accuracy.

This regularity of apparently random occurrences spurred Graunt to make a more exhaustive study, including birth records as well as death records. In 1662 he published an influential book, *Natural and Political Observations on the Bills of Mortality,* in which he called for complete records of almost all phases of human endeavor. He called for vital statistics (birth, death, marriage, and divorce records), records of land transfers, land use, livestock, occupations, salaries, and ranks in society. In short, Graunt recognized the need for modern census. He went on to indicate how these records could be used to discover trends and to make predictions of the future.

Graunt's book was well received by the government and the scientific community, but not by the traditional philosophers and theologians who believed that births and deaths were determined by God and that mortals had no right to inquire into divine matters. In spite of these objections, a number of governments soon began to collect statistics and to require accurate record keeping for many phases of business and society.

Graunt was honored by the king of England by being named to the Royal Society. He was an active member for several years, but resigned in 1667.

We believe that one of his reasons for resigning from the Royal Society was a dispute over the authorship of the *Natural and Political Observations on the Bills of Mortality.* Many people of the day did not believe that Graunt, a shopkeeper, had written the book. They believed, instead, that Sir William Petty, an outstanding scientist and personal friend of Graunt, had written the book and published it under Graunt's name because of the controversy he knew it would cause.

In a sense Graunt's book was opposed to the spirit of reason that had influenced men since the Renaissance. The ancient Greek philosophers and their latter day European followers had believed that there were ultimate truths and that these could be discovered by reason. The theologians, on the other hand, believed that these truths could be learned through the church. Graunt suggested that both approaches were inadequate. He believed that it was necessary to investigate human conditions and that the collection of statistics was the proper way to begin the investigation.

After statistics are collected, they must be analyzed. The analysis involves the calculation of averages and measures of how the data are distributed about the average values. Trends are noted and projected into the future. Summaries are made up listing pertinent information in a form intelligible to the casual reader.

Modern society depends heavily on accurate and complete statistics of almost every kind. Government agencies regularly collect all of the statistics

The true foundation of theology is to ascertain the character of God. It is by the aid of Statistics that law in the social sphere can be ascertained and codified, and certain aspects of the character of God thereby revealed. The study of statistics is thus a religious service.

Florence Nightingale

Influential people of the seventeenth century believed that the collection of statistics was contrary to the wishes of God. By the nineteenth century workers in the field were claiming a religious significance for their results. The above quote of Florence Nightingale, the founder of nursing as a science, typifies this last attitude.

that are likely to be of interest to any major group. The raw data and the results derived from the data are made available to those who need them. The following example indicates how this information can be used.

The Montague County Planning Commission is preparing a proposal for a new hospital. As part of the report, the following information is needed:

(1) the total number of persons in each age group in the area served by the hospital (*state and national census reports*);

(2) population density figures—the hospital should be located so as to be as close as possible to the greatest number of persons (*census reports*);

(3) the total number of persons likely to need hospital care during each month (*this can be calculated from the population figures using statistics from similar hospitals*);

(4) the availability of trained medical personnel (*records of surveys conducted by regional medical associations*);

(5) highway use figures—the hospital should be located near a major highway that is not overcrowded with traffic; it would not do for ambulances to be regularly stalled in traffic (*state highway department traffic surveys*).

The statistics mentioned above are only a few of those that need to be considered in the preliminary planning stages of the hospital. As the plans become more advanced, many other statistics are needed. How much space should be devoted to a geriatrics ward, for example? How many heart patients are likely to need intensive care facilities? How many operations will be performed each week? These, and hundreds of other questions, can be answered by the proper use of statistics.

Our example with the hospital indicates only a few of the statistics that should be considered. Almost every major government or business decision is made after a careful study of all of the available statistics that seem pertinent. As we can see, the major uses of statistics are those that were first pointed out by John Graunt—to find out what the true situation is like and to project the findings into the future.

Topics for Discussion or Further Investigation

1. Discuss how statistics could be used in each of the following planning situations:
   (a) a new motel on an interstate highway;
   (b) the route of an expressway through a city;
   (c) a major new shopping center in the suburbs;
   (d) the number and types of automobiles to manufacture next year;
   (e) the location of a major-league baseball franchise;
   (f) the number and sizes of a new dress-style.
2. Suppose that you are an official with the U.S. Bureau of the Census. Explain how you would attempt to solve the following problems:
   (a) the counting of persons in very remote regions who have no official addresses (hermits in caves, for example);

(b) the counting of transients in hotels (these persons may be counted several times as they move around the country or not counted at all);

(c) the counting of migrant farm workers;

(d) persons who refuse to cooperate with the census enumerators because of religious convictions.

3. It has been estimated that each U.S. Census may be in error by as much as 5 percent. Use the latest census figures to estimate a range of values for the true size of the population of the United States. Can you give some reasons for this 5-percent possible error?

9-2 FREQUENCY DISTRIBUTIONS; GRAPHS

When we begin to collect statistics, we soon find that we must process the data so that they can be made intelligible. It does us no good to have thousands of numbers written on slips of paper (or stored in a computer's memory bank). The first step is usually a classification of data, followed by a frequency count.

The classification is a breakdown of classes or categories into which we decide (sometimes arbitrarily) to put the data. The frequency count is a count of the number of bits of data in each class. The completed list of frequency counts is known as the *frequency distribution*.

The data classes for a frequency chart can be set up in very arbitrary ways. In Example 1 we could have set up the classes as
375–399, 400–424, 425–449, . . ., as 300–399, 400–499, . . ., or in some other way.

Example 1. Forty-two college freshmen made the following scores on the Verbal S.A.T. (Scholastic Aptitude Test): 449, 490, 649, 610, 510, 640, 480, 624, 550, 580, 630, 590, 510, 591, 540, 390, 510, 617, 793, 419, 480, 450, 550, 540, 527, 589, 466, 466, 394, 590, 431, 706, 644, 438, 470, 500, 534, 676, 735, 660, 683, 466.

Make a frequency count based on a 50-point spread for the classes.

Solution. We set up our classes as follows: 351–400, 401–450, 451–500, 501–550, and so forth. The score 390 is listed in the 351–400 class; the score 419 is listed in the 401–450 class, and so on. We make a tally mark for each score, and at the end we count up the total number of scores in each class. The completed frequency distribution is shown in Table 9-1.

Table 9-1 **Frequency Distribution of 42 Verbal S.A.T. Scores**
(*Example 1*)

Class	Tally	Frequency
351–400	‖	2
401–450	卌	5
451–500	卌 ‖‖	8
501–550	卌 ‖‖‖	9
551–600	卌	5
601–650	卌 ‖	7
651–700	‖‖	3
701–750	‖	2
751–800	‖	1

It is obvious from Example 1 that the separation of data into classes was quite arbitrary. We could just as well have made the classes cover a 10-point range or even a 1-point range (each score being in its unique class) of S.A.T. scores. We chose a 50-point range for convenience. It allowed us to keep the table fairly short.

The frequency distribution chart gives us information in the form of numbers in a table. It usually is desirable to display the information on a graph. This is done by plotting the classes on the x-axis and the frequencies on the y-axis. We usually show only the portion of the x-axis that actually is needed. Because each class covers a wide range of values, we plot the frequencies above the *midpoint* of each class.

Figure 9-1 shows several graphs obtained from the data in Example 1. We base the graphs on the frequency distribution in Table 9-1. The first step is to plot the frequencies of the classes above the midpoints of the classes [Fig. 9-1(a)]. The class 551–600, for example, contains five items. Thus, we plot the point $(575, 5)$ above the point 575 on the x-axis. After we have plotted the points for the classes, we can proceed in two ways. We can connect up the points with *straight line segments, obtaining a broken-line graph,* as in Figure 9-1(b), or we can draw a *bar graph* as in Figure 9-1(c).

In some ways the bar graph is the most indicative of the true nature of the frequency distribution, because the area of each bar is proportional to the number of elements in that class. For example, the first bar in Figure 9-1(c) covers an area of two units corresponding to two items in the 351–400

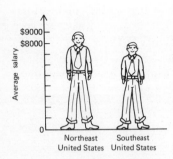

Beware of bar graphs with pictures of three-dimensional objects rather than bars. The graph in the figure compares the average salaries of persons in the northeastern and the southeastern United States. The actual percentage difference is 11 percent. The eye compares the apparent volumes of the two figures, however, and interprets the difference to be approximately 30 percent.

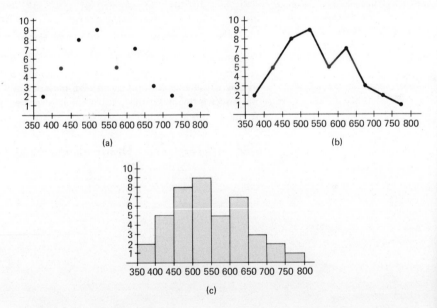

Figure 9-1 Distribution of verbal S.A.T. scores of 42 college freshmen (data from Example 1). (a) Points on the graph; (b) broken-line graph; and (c) bar graph.

class, while the second bar covers an area of five units corresponding to the five items in the 351–400 class.

There is a third type of graph that is widely used to represent certain types of data—the *pie graph*. To construct a pie graph, we split a circle into sectors that have areas proportional to the percentages of items in the various classes. The pie graph for the data in Example 1 is shown in Figure 9-2.

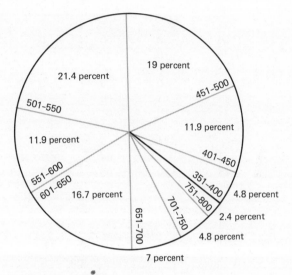

Figure 9-2 Pie graph (data from Example 1).

The pie graph is used mainly for final reports which involve relatively few classes. It is the best device for visualizing the proportion of rural-to-urban populations in our country or for explaining how the U.S. tax dollar is apportioned out to the major government agencies. The pie graph is not particularly helpful in conducting a statistical analysis. The broken-line graph and bar graph are used in the actual analysis, and the pie graph is used to show the final results. Because of its usefulness in analysis, we shall concentrate on the bar graph in this chapter.

Example 2. The following list shows the grades on a mathematics test. Make two frequency counts, one using one grade to each class (that is, show the number of persons who earned each grade), the other using a 10-point range of grades (61–70, 71–80, and so on). Which frequency distribution gives the best information about the distribution of grades?

Make a bar graph and a broken-line graph based on the distribution into the 10-point classes.

Grades: 75, 71, 51, 90, 84, 61, 65, 73, 37, 55, 87, 82, 91, 58, 67, 57, 78, 61, 100, 72, 32, 77, 64, 88, 50, 81, 43, 57, 54, 93, 62, 74, 73, 79, 63.

Solution. The grades range from 32 to 100. The first frequency chart must contain each of these numbers. It is shown in Table 9-2. The second frequency chart (Table

Table 9-2   **Frequency Distribution of 35 Grades** (*Example 2*)

Grade	Frequency	Grade	Frequency	Grade	Frequency	Grade	Frequency
32	1	49	0	66	0	83	0
33	0	50	1	67	1	84	1
34	0	51	1	68	0	85	0
35	0	52	0	69	0	86	0
36	0	53	0	70	0	87	1
37	1	54	1	71	1	88	1
38	0	55	1	72	1	89	0
39	0	56	0	73	2	90	1
40	0	57	2	74	1	91	1
41	0	58	1	75	1	92	0
42	0	59	0	76	0	93	1
43	1	60	0	77	1	94	0
44	0	61	2	78	1	95	0
45	0	62	1	79	1	96	0
46	0	63	1	80	0	97	0
47	0	64	1	81	1	98	0
48	0	65	1	82	1	99	0
						100	1

9-3) contains the seven classes 31–40, 41–50, 51–60, and so on. The bar graph and broken-line graph for the second chart are in Figure 9-3.

Table 9-3 (and the associated graphs) gives the more useful information, even though it is not as exact as that in Table 9-2. We can use Table 9-3 to determine how the grades are actually distributed. We see that the 71–80 class is the largest class of grades and that the sizes of the classes decrease as we move away from the largest class in either direction. It would seem natural to consider the 71–80 class to be the "average" class and to assign an "average" grade, such as a "C," to the grades in that class.

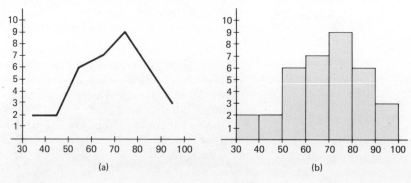

Figure 9-3 Distribution of 35 test grades (Example 2). (a) Broken-line graph; (b) bar graph.

Table 9-3   **Frequency Distribution
of 35 Grades in
10-point Classes**
(*Example 2*)

Class	Frequency
31–40	2
41–50	2
51–60	6
61–70	7
71–80	9
81–90	6
91–100	3

The Dishonest Bar Graph •

Although the standard bar graph gives the most accurate visual display for certain types of data, it also can be modified to give highly misleading impressions.

The employees of the Honestman Dice Works are considering their third strike of the year. To counter the union propaganda, the company newsletter published the bar graph in Figure 9-4(a).

This graph shows the average weekly salary of workers at Honestman ($160) and the average weekly salary of workers in the United States ($158). These two figures are almost identical. By cutting off almost all of the lower portions of the two bars, however, the company graph gives the misleading visual impression that the Honestman salary is half again as large as the average U.S. salary. At least this graph represents an improvement over the graphs Honestman used in the past—some of these graphs did not even show the scale on the vertical axis.

The union newsletter immediately published an analysis of Honestman's graph showing how it was intended to deceive. To counter the impression left by the "bad" graph, it also published a graph showing a comparison of the number of workers in the nation with average or above-average salaries (60 percent) and the number of workers at Honestman with average or above-average salaries (40 percent). To prove that the graph is honest, the vertical scale (0 to 70 percent) is prominently displayed.

Unfortunately, the union graph is just as misleading as the company graph. By changing the *width* of the two bars, the union manages to give the impression that proportionately only half as many Honestman employees have average salaries as in the nation as a whole.

As a matter of interest, both of the graphs in Figure 9-4 are virtually worthless. In each case, unrelated statistics, which seem to be related, are compared. To be meaningful, both comparisons should be with workers in related industries, not with workers in the nation as a whole. The union graph is particularly bad since it gives no information about how far the company wages are below "average."

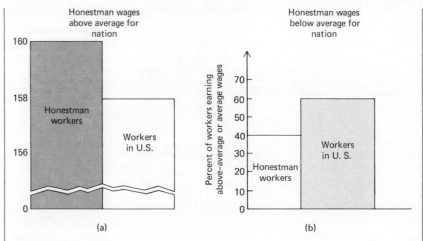

Figure 9-4   Misleading bar graphs used by company and union in Honestman wage dispute. (a) Company newsletter; (b) union newsletter.

EXERCISES

1–3. (a)   Make frequency distributions for the sets of data. Separate the data into classes as indicated.

   (b)   Make broken-line graphs based on the frequency distributions.

   (c)   Make bar graphs based on the frequency distributions.

1. The 33 freshmen enrolled in a mathematics class had the following scores on their Mathematics S.A.T. test:

   440, 589, 450, 700, 440, 600, 430, 690, 660, 589, 650, 702, 390, 449, 649, 510, 480, 550, 630, 510, 540, 510, 793, 480, 450, 527, 540, 466, 394, 431, 438, 500, 534. Separate the data into the classes 301–350, 351–400, 401–450, and so on.

2. The students in Exercise 1 made the following grades on their first test:

   45, 51, 84, 74, 70, 87, 93, 50, 32, 56, 87, 25, 78, 48, 41, 56, 78, 88, 56, 63, 87, 75, 60, 71, 63, 100, 35, 17, 58, 76, 82, 61, 57.

   Separate the data into the classes 21–30, 31–40, 41–50, and so on.

3. The final grades for the students in Exercise 1 are as follows:

   D, D, B, C, C, B, A, C, D, D, D, B, F, A, C, D,
   C, B, C, D, B, C, D, D, B, B, F, D, A, F, C, C.

   Separate the grades into the classes A, B, C, D, F.

4. Construct a flow chart for a computer program that will cause the computer to print a frequency distribution of test scores with appropriate labels. Assume that the scores all are between 1 and 100. Let the classes be 1–10, 11–20, 21–30, and so on.

5. Make a computer program based on the flow chart in Exercise 4. Run the program on the computer using the following data:
   (1) the data in Example 2;
   (2) the data in Exercise 2;
   (3) the class grades on your last mathematics test.
6. Find examples of bar graphs and broken-line statistical graphs in current news magazines. Are the graphs honest or do they tend to mislead the casual observer?
7. Prepare two accurate (but possibly misleading) bar graphs based on your current financial situation. One graph is to be used to convince the local credit union that you are a good risk for a loan. The other is to be used to convince your parents that you need additional money. The same basic data must be used for both graphs.

9-3 AVERAGES

The most important single bit of information we can get from a set of data is the average. We speak of the *average* of a set of grades, the *average* salary in the United States, the *average* dress size worn by women in New York. As we shall see, there are several different types of average that can be computed from a given set of numbers. The averages mentioned above are examples of the three most useful averages—the *mean*, the *median*, and the *mode*.

The Mean

HAVARIA
One of the earliest uses of the average was for a simple form of insurance for goods shipped by sea. Frequently, part of a load would be damaged or lost at sea. When the ship arrived at its destination, a pro rata share of the cargo that arrived safely was assigned to the owner of the lost or damaged cargo. The Latin word *havaria*, which meant the damage to cargo at sea, is the root word for our word *average*.

The common average that most of us have used since childhood is the *mean*. It is calculated by adding up the numbers in a set and dividing the sum by the number of items.

Example 1. A student's grades in five tests are 82, 76, 91, 88, and 94. His mean average is

$$\text{mean} = \frac{\text{sum of grades}}{\text{number of grades}} = \frac{82 + 76 + 91 + 88 + 94}{5} = \frac{431}{5} = 86.2$$

(See Fig. 9-5.)

The mean also can be calculated from frequency tables, but the final answer is not 100 percent accurate. The inaccuracy is caused by the fact

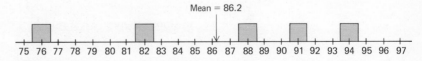

Figure 9-5   Example 1.

that the frequency table lists the number of items in each class rather than the individual items so that we no longer know the exact value of each item. The usual procedure is to assume that each element in a class has the average value for that class. In the following example, we assume that each item in the 31–40 class has the value 35, that each item in the 41–50 class has the value 45, and so on.

Example 2. The test grades of a mathematics class are shown in Table 9-4. Calculate the mean.

Table 9-4   Example 2

Class	Frequency	Class Sum
31–40	2	70
41–50	2	90
51–60	6	330
61–70	7	455
71–80	9	675
81–90	6	510
91–100	3	285
Sums:	35	2415

$$\text{mean} = \frac{\text{sum of grades}}{\text{number}} = \frac{2415}{35} = 69$$

Most users of statistics do not have access to the original data. These users must use average values of data classes (as in Example 2) when calculating the mean. Raw data collected by the U.S. Census Bureau, for example, must be kept confidential and not released to the public. Such a huge amount of data is collected in the census that even if it could be released, it would not be practical for anyone to use it in its original form.

Solution. We count each grade in the 31–40 class as if it were 35, each grade in the 41–50 class as if it were 45, and so on. We must calculate the sum of the grades and divide it by the number of grades.

There is a simple shortcut that can help us to avoid errors. Rather than add up the grades one at a time, we calculate the sum for each class and then add the class sums. The first class contains two grades of 35 for a class sum of 70. The second class contains two grades of 45 for a class sum of 90. This computation is carried out in the column headed "Class Sum."

The total of the grades is 2415—the total of the class sums. The number of grades is 35—the total of the frequencies. Thus,

$$\text{mean} = \frac{\text{sum of grades}}{\text{sum of frequencies}} = \frac{2415}{35} = 69$$

The mean average is 69 (Fig. 9-6).

As we mentioned above, this is not the exact value that we would have computed as the mean from the raw data. The original list of grades from which the frequency chart was prepared is in Example 2 of Section 9-2. If we add up those grades, we would find that

$$\text{mean (exact value)} = \frac{\text{sum of grades}}{\text{number of grades}} = \frac{2405}{35} \approx 68.7$$

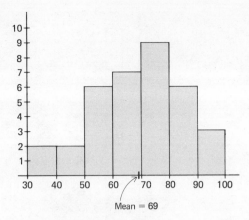

Figure 9-6 Example 2.

which is very close to the value 69 computed from the frequency table. In most cases the value that we compute from the frequency table is so close to the exact value of the mean that no significant error results.

The Median

When we speak of an "average," we usually have the mean in mind. There are cases, however, when the mean is completely misleading. The following example shows how the mean may give a false impression when a few items of data are completely out of line with the remaining data.

Example 3. Workers at the Honestman Dice Works are on strike, charging that the company pays substandard wages. The president, John N. Honestman, has just published an auditor's report showing that the average weekly salary is $227.76, which is considerably above the average for workers in related industries.

Honestman's report is accurate. The total weekly payroll of $11,160 is paid to 49 persons, resulting in a mean salary of

$$\text{mean salary} = \frac{11,160}{49} = \$227.76$$

The manner in which the payroll is distributed, however, is much more informative than is the mean. The 46 employees each earn $160; the plant manager (Bill Honestman) earns $800; the vice-president (John N. Honestman, Jr.) earns $1,000; the president (John N. Honestman) earns $2,000. Thus, we have the following payroll distribution (see Fig. 9-7):

46 workers at $160	=	$7,360
1 manager at $800	=	800
1 vice-president at $1,000	=	1,000
1 president at $2,000	=	2,000
Total: 49 employees		$11,160

It is easy to see that the average salary of the *workers*, who are the ones on strike, is $160 per week, not $227.76.

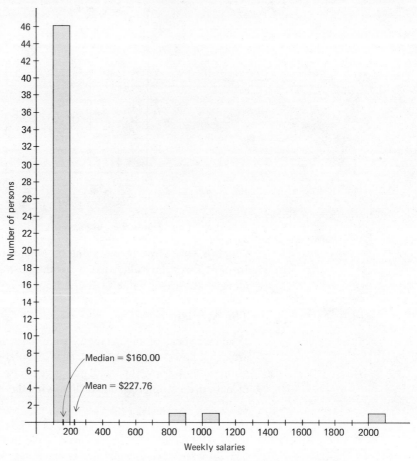

Figure 9-7 Average salaries at the Honestman Dice Works. In this example the mean gives a misleading impression because a few items of data are completely out of line with the other data (Example 3).

The *median* is a different type of average which is used in situations like the one described in Example 3. The median of a set of data is the midpoint of the set. In Example 3 the median is the salary of the 25th man from the bottom of the list (who also is the 25th man from the top). Thus, the median salary paid by the company is $160.

The median is one of the easiest averages to calculate. We list the items in order and count to the middle item on the list. This item is the median. If there are two middle items (as will be the case with an even number of data items), then the median is the average (mean) of the two middle items.

Example 4. (a) Nine test grades are

$$37, 61, 62, 62, 78, 79, 81, 100, 100$$

The median is 78, the middle grade [Fig. 9-8(a)].

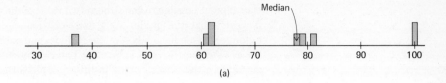

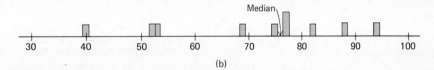

Figure 9-8 (a) The median of 37, 61, 62, 62, 78, 79, 81, 100, 100 is 78; (b) the median of 40, 52, 53, 69, 75, 77, 77, 82, 88, 94 is (75 + 77)/2 = 76. (See Example 4.)

(b) Ten test grades are

$$40, 52, 53, 69, 75, 77, 77, 82, 88, 94$$

The median is the average of the two middle grades [Fig. 9-8(b)]:

$$\text{median} = \frac{75 + 77}{2} = 76$$

When the data are separated into classes, we can find the approximate value of the median by counting the elements in each class. This gives us the class in which the median is located. If we need to be more exact, we can follow the process described in the following example.

Example 5.   The test grades from Example 2 are listed in Table 9-5.
(a) Locate the class that contains the median grade.
(b) Estimate the median grade.

Table 9-5   Example 5

Class	Frequency
31–40	2
41–50	2
51–60	6
61–70	7
71–80	9
81–90	6
91–100	3
Sum:	35

Solution.   (a) The median is the 18th grade from the lowest grade. This grade is the lowest grade in the 71–80 class. If we only need a range of values for the median, we could list the entire class. Thus, the 71–80 class is the *median class*.
(b) Knowing the median class only gives us a possible range of values for the median grade. Without knowing how the grades are distributed in the median

class it is not possible to know the actual value of the median. It is usually *assumed*, however, that the grades are evenly distributed in the median class. In this example, the median class contains 9 grades and covers a range of 10 points. Thus, we assume that there is a $\frac{10}{9} \approx 1.1$ point difference between the grades in the class. Since the median grade in this example is the lowest grade in the median class, we give it an approximate value of 72.

The Mode

In certain settings neither the median nor the mean gives information that is very useful. It may be more natural to consider a third type of average called the *mode*. If, when we construct a frequency chart, one class contains many more elements than any other class, then the large class is called the *modal class*. The average (mean) value of the class is called the *mode*. The mode is typically used for large quantities of data which are naturally grouped into a few classes, such as data on clothing sizes.

Example 6.   Men's sport shirts are made in four standard neck sizes: *Small* (14–$14\frac{1}{2}$), *Medium* (15–$15\frac{1}{2}$), *Large* (16–$16\frac{1}{2}$) and *Ex-large* (17–$17\frac{1}{2}$). A recent inventory at the Bleaksburg Men's Emporium shows that during the month of May, the following numbers of sport shirts were sold:

Shirt Size	Number Sold
Small	263
Medium	541
Large	180
Ex-large	53

The mode for the shirt customers of the store is *Medium*. Far more shirts were sold in that size than any other size (Fig. 9-9).

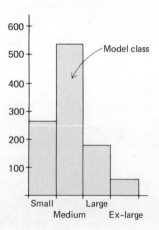

Figure 9-9 The modal class. The mode is Medium (Example 6).

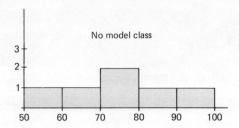

Figure 9-10 No modal class. One class is larger than the others, but not large enough to establish it as the mode.

The mode is not established unless an unusually large number of items are in one class. As an example, suppose that the grades of six students on a test are

$$53, 61, 71, 78, 85, 96$$

If we make a frequency chart based on a 10-point spread of grades, we get the following distribution (Fig. 9-10):

Class	Frequency
51–60	1
61–70	1
71–80	2
81–90	1
91–100	1

The 71–80 class does contain more elements than any other class, but not enough to establish it as the modal class.

In many cases it may not be obvious which average to use. This is especially true when the three averages give answers that vary widely. We may have to rule out one or more of the averages because they do not seem to fit our common sense notions about what the "average" value should be. Also, we should recall that the mode is used when a large number of items are naturally grouped into a few classes and that the median is used in preference to the mean if a few exceptional cases could change the mean drastically.

Example 7. A neighborhood consists of 25 houses. Three were built about the time of the Revolutionary War and have been restored—these are approximately 200 years old. One house is 30 years old; five are 20 years old; six were built 8 years ago by a developer; two are 5 years old; one is 2 years old; seven are 1 year old.

(a) What is the mean age? The median age? The modal age?

(b) Which of the averages should we use?

Solution. We make a frequency chart listing only the ages that occur in the neighborhood.

Age	Frequency	Class Sum
200	3	600
30	1	30
20	5	100
8	6	48
5	2	10
2	1	2
1	7	7
Sums:	25	797

(a) The mean is

$$\text{mean} = \frac{\text{sum of the ages}}{\text{number of houses}} = \frac{797}{25} \approx 31.9$$

The median is the 13th house built. Thus, the median age is 8.

$$\text{median} = 8$$

Three of the classes contain approximately the same number of items, all three being much larger than the other classes. Thus, there is no well-defined mode. If, however, we were forced to choose a modal age, we would have to choose 1 year because more of the houses are 1 year old than any other age.

Thus, we have three completely different averages:

$$\text{mean} = 31.9$$
$$\text{median} = 8$$
$$\text{mode} = 1 \quad \text{(if the mode is defined at all)}$$

(b) *Which average should we use?* As we pointed out above, the mode does not seem to be too promising in this example. Thus, we can rule out the mode as the best average to use. Observe that the main reason for the wide discrepancy between the median and the mean is the existence of the three very old houses. Since these exceptional cases can influence the mean so much, then the median is the most meaningful average.

EXERCISES

1. (a)   Think of one example each in which the three averages should be used.
   (b)   Think of one example each in which the three averages should not be used.
2. Which average would you use in the following examples? Why?
   (a)   the average number of children in families in the United States;
   (b)   the average number of beds in hospitals in the United States;
   (c)   the average hat size of men;
   (d)   the average weight of women aged 20–30;

    (e)   the average daily temperature;

    (f)   the average percent of sky covered by clouds;

    (g)   the average salary of industrial workers in the United States;

    (h)   the average price of a fifth of rum in Miami.

3. The Gasdrinker Automobile Co. advertises that the average gas mileage of 100 of its test automobiles is 18 miles to the gallon. Would you be more impressed if this were the mean, the median, or the mode? Explain.

4. Calculate the mean and the median of each of the following sets of quiz grades. Is there a mode?

    (a)   6, 10, 10, 4, 8, 6, 8, 8, 0, 6

    (b)   2, 4, 10, 9, 5, 10, 8, 9, 6, 9, 7, 1, 0, 7, 8

Use the frequency distributions to find the mean and the median for the data sets in Exercises 5 and 6. Is there a modal class?

5. The data in Exercise 1, Section 9-2.

6. The data in Exercise 2, Section 9-2.

7. Modify your computer program in Exercise 5 of Section 9-2 so that the computer will print the mean and median class. If you can figure how to do it, have it decide when there is a modal class and print the modal value as well.

9-4 DISPERSION ABOUT THE MEAN

The mean is the most common average that is used. Just knowing the mean, however, does not give us much information. We also need to have some idea of how the data are distributed about the mean. For example, each of the data sets

$$0, 1, 9, 10 \quad \text{and} \quad 4, 5, 5, 6$$

has a mean of 5, but the data are distributed about the mean in completely different ways. In the first set, the data are all far from the mean. In the second set, they are clustered closely about it.

    It is useful to know the range of values of the data. In the first of the sets, the range of values is 0 to 10; in the second set, it is 4 to 6.

    In many situations we can consider the mean to be an "ideal" value and the actual data to be imperfect attempts to reach the mean. The situation is analogous to a rifleman shooting at a target. The bull's eye represents the ideal shot. Most of the shots miss the bull's eye, but not by too much. If we "average" the locations of all of the shots, we should obtain a location close to the bull's eye. If the shots are tightly clustered about the bull's eye, we say the man is a marksman. If they are widely dispersed, we know that either he is a poor shot or that the rifle is no good. Similarly, if we know how data are dispersed about the mean, we can make judgments about the validity of using the mean as an average or about the way the data were collected or analyzed. If the data are tightly grouped about the

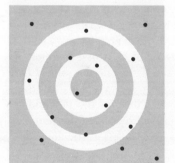

Good dispersion and poor dispersion of shots about bull's eye.

The bull's-eye target furnishes one of the best examples of dispersion. If all of the shots are clustered about the center, then the dispersion is small and the rifleman is a marksman.

mean, then the mean is a good average. If they are widely dispersed, then it may be worthless. We look for a single number that will measure how the data in a set are dispersed.

The Standard Deviation

We shall soon discuss the most common measure of dispersion about the mean—the *standard deviation*. Before we do, however, we introduce some terminology. If x is a typical item of data, then it is customary to let $\bar{x}$ denote the mean. Observe that the difference $x - \bar{x}$ is a measure of how far the data item x is from the mean $\bar{x}$. This number $x - \bar{x}$ is called the *deviation* of x from the mean.

For example, if the data consist of the six quiz grades 2, 3, 6, 7, 8, 10, then

$$\bar{x} = \text{the mean} = \frac{2 + 3 + 6 + 7 + 8 + 10}{6} = \frac{36}{6} = 6$$

The deviations are

$$2 - \bar{x} = 2 - 6 = -4 = \text{deviation of 2 from the mean}$$
$$3 - \bar{x} = 3 - 6 = -3 = \text{deviation of 3 from the mean}$$
$$6 - \bar{x} = 6 - 6 = 0 = \text{deviation of 6 from the mean}$$
$$7 - \bar{x} = 7 - 6 = 1 = \text{deviation of 7 from the mean}$$
$$8 - \bar{x} = 8 - 6 = 2 = \text{deviation of 8 from the mean}$$
$$10 - \bar{x} = 10 - 6 = 4 = \text{deviation of 10 from the mean}$$

Observe that some of the deviations are positive and some are negative. The deviation $x - \bar{x}$ is positive if x is greater than the mean $\bar{x}$. It is negative if x is less than $\bar{x}$. Observe also that the sum of the deviations is zero.

We are interested in obtaining a single number that describes all of the deviations. At first glance, it might appear that the average (mean) of the individual deviations would be a good measure of the total amount of deviation. If we think about it, however, we see that since the sum of the deviations is zero, then we always get zero for the average. Another approach is to use the average of the absolute values of the deviations. This would be a good measure of the total deviation except that this number is difficult to work with. To get around these problems, statisticians work with the squares of the deviations.

The actual measure of deviation that is used is a number called the *standard deviation*. It is *the square root of the average (mean) of the squares of the deviations*. The standard deviation is usually denoted by the symbol σ (Greek letter sigma).

As we see from the definition, the computation of the standard deviation is quite involved. We must perform the following operations in order as shown in the flow chart of Figure 9-11:

(1) Calculate the *mean*.
(2) Calculate the *deviations* of the data from the mean.

We often deal with dispersion in a purely practical way by means of graphs or maps. The map shows the locations of robberies that police think were attempted by the same person. The high density of robberies in this one part of the city makes it a likely place for future robberies.

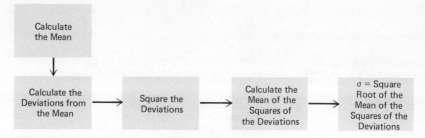

Figure 9-11 Flow chart for the standard deviation.

(3) *Square* the deviations.
(4) Calculate the *mean* of the squares of the deviations.
(5) Calculate the *square root* of the mean of the squares of the deviations.

Example 1. Calculate the standard deviation of the following set of quiz grades: 2, 3, 6, 7, 8, 10.

Solution. We make the following chart:

Data (x)	Deviations $(x - \bar{x})$	Squares $(x - \bar{x})^2$
2	-4	16
3	-3	9
6	0	0
7	1	1
8	2	4
10	4	16
Totals: 36		46

Mean: $\qquad \bar{x} = \dfrac{36}{6} = 6$

Standard deviation: $\sigma = \sqrt{\dfrac{46}{6}} \approx \sqrt{7.67} \approx 2.77$

The first step is the calculation of the mean. Since the six data items sum to 36, then the mean is

$$\bar{x} = \frac{36}{6} = 6$$

The second step is the calculation of the individual deviations (second column):

$$x_1 - \bar{x} = 2 - 6 = -4$$
$$x_2 - \bar{x} = 3 - 6 = -3$$

and so on (Fig. 9-12).

The third step is the squaring of the deviations (third column). The fourth step is the calculation of the mean of the squares:

$$\text{mean of squares} = \frac{16 + 9 + 0 + 1 + 4 + 16}{6} = \frac{46}{6} \approx 7.67$$

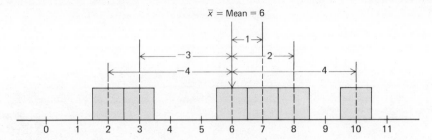

Figure 9-12 Deviations from the mean (Example 1). These deviations are used to calculate the standard deviation.

The final step is the calculation of the standard deviation:

$$\sigma \approx \sqrt{7.67} \approx 2.77$$

The mean and the standard deviation furnish a quick summary of the data. Most users of statistics do not have access to the original data. They must use information about the data that was obtained by others. A trained analyzer of statistics can obtain almost as much information from knowing the *range of data values,* the *size of the data set,* the *mean* and the *standard deviation* as by examining the original data.

Now that we can calculate the standard deviation, what do we do with it? It is proved in advanced courses in statistics that the standard deviation measures how closely the data are clustered about the mean. If σ is small (when compared with the range of values), then most of the data must be grouped within a narrow range about the mean. If σ is large, then a sizable number of the data must be far from the mean. To be more exact, in every case at least 75 percent of the data must be located within two standard deviations of the mean, and at least 88 percent of the data must be within three standard deviations of the mean.

As an example, recall the two data sets 0, 1, 9, 10 and 4, 5, 5, 6 that we mentioned at the beginning of this section. Each of these has a mean of 5. The standard deviations are approximately equal to 4.5 and 0.7, respectively. The large standard deviation of the first set indicates that the data are distributed so far from the mean that the mean may be worthless as an indication of the values of the data. The small standard deviation of the second set indicates that the data are tightly grouped about the mean.

Example 2. In Example 1 we calculated the mean and the standard deviation of the six items of data

$$2, 3, 6, 7, 8, 10$$

We found that $\bar{x} = 6$ and $\sigma \approx 2.77$. Then

$$\bar{x} - \sigma \approx 3.23 \qquad \text{and} \qquad \bar{x} + \sigma \approx 8.77$$

In this case, there are three items of data between 3.23 and 8.77. Thus, 50 percent of the data are within one standard deviation of the mean.

Similarly,

$$\bar{x} - 2\sigma \approx 6 - 5.54 = 0.46 \qquad \text{and} \qquad \bar{x} + 2\sigma \approx 6 + 5.54 = 11.54$$

so that 100 percent of the data are within two standard deviations of the mean.

Recall that our earlier claim was that at least 75 percent of the data must be within two standard deviations of the mean. In this case, we have far more of the data in that range than was predicted.

Figure 9-13 illustrates the situation that occurs with large and small standard deviations. In each case, the area under any part of the curve is proportional to the number of data in that particular range. In Figure 9-13(a) the standard deviation is large. Corresponding to this large standard deviation, the data are widely dispersed over a large range of values. In Figure 9-13(b) the standard deviation is small, and the data are grouped within a narrow range of values about the mean.

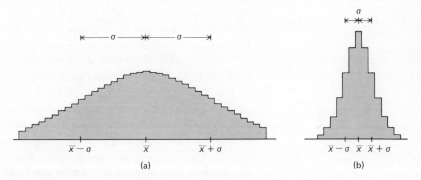

Figure 9-13 (a) Large standard deviation. The data are widely dispersed about the mean; (b) small standard deviation. The data are tightly grouped about the mean.

Example 3. A statistical study shows that the mean weight of 1000 college women is $\bar{x} = 127$ pounds with a standard deviation of 17 pounds. Make an estimate of the number of women who weigh between 93 and 161 pounds.

Solution. At least 75 percent of the women must have weights between $\bar{x} - 2\sigma$ and $\bar{x} + 2\sigma$. We calculate

$$\bar{x} - 2\sigma = 127 - 2 \cdot 17 = 127 - 34 = 93$$
$$\bar{x} + 2\sigma = 127 + 2 \cdot 17 = 127 + 34 = 161$$

Thus, at least 750 of the 1000 women must weigh between 93 and 161 pounds.

The estimate obtained in Example 3 is a very crude one. Actually, about 950 of the women are in that weight range. The weight distribution happens to be an example of one of the most studied distributions in statistics—the normal distribution. As we shall see in Section 9-6, we can obtain a much better estimate of the data distribution if we know that the data are distributed normally. Regardless of the type of distribution, however, at least 75 percent of the data items in any data set must be between $\bar{x} - 2\sigma$ and $\bar{x} + 2\sigma$.

As we can see, the minimum estimates for the amounts of data within two or three standard deviations of the mean that we mentioned above are the worst possible good estimates that can be obtained since they must hold for every data set. As we shall see in Section 9-6, many data distributions that occur in nature are distributed normally. Those have approximately 67 percent of the data items within one standard deviation of the mean and approximately 95 percent within two standard deviations.

The calculation of a standard deviation usually leads to horrendous computational problems. Our examples have been chosen for ease of computation, but imagine the work that would be involved in calculating the standard deviation of a data set containing thousands (or even millions) of items.

Statistics is one of the areas where the electronic computer has had its greatest impact. It is a simple matter to program the computer to calculate the standard deviation of a large data set. This means that the staggering work involved in many statistical problems can be left to the computer, freeing the statisticians for the more creative aspects of the discipline.

The computer is able to calculate many other statistical quantities in addition to standard deviations. We get a practical demonstration of its abilities in this field every presidential election. The election returns are analyzed statistically as they are received by the networks. The computer performs millions of calculations as it analyzes the returns and, when the trends are clear, makes predictions almost instantly. The theory for the analysis has been developed for many years. The electronic computer is the modern tool that allows the theory to be put to practical use. In the following section, we shall use the computer to calculate the standard deviation of a large data set.

EXERCISES

Calculate the mean $\bar{x}$ and the standard deviation σ of the data set in Exercises 1–6. (Use Table II at the end of the book for the square roots.) Calculate $\bar{x} - \sigma$, $\bar{x} + \sigma$, $\bar{x} - 2\sigma$, and $\bar{x} + 2\sigma$. What percent of the data are within one standard deviation of the mean? Verify in each case that at least 75 percent of the data are within two standard deviations of the mean.

1. Data: 1, 4, 5, 5, 7, 9.
2. Data: 10, 5, 6, 3, 5, 4.
3. Data: 6, 5, 4, 3, 5, 4, 0, 5, 6, 3.
4. Data: 2, 2, 5, 5, 5, 4, 6, 7, 5, 6.
5. Data: 2, 3, 0, 0, 1, 9, 10, 10, 10, 9.
6. If you have access to a desk calculator, use the data in
   (a) Exercise 1, Section 9-2;
   (b) Exercise 2, Section 9-2;
   (c) Example 1, Section 9-2

For Exercises 7 and 8 use the fact that in any set of data at least 75 percent of the data items are within two standard deviations of the mean and at least 88 percent are within three standard deviations.

7. The test grades of 120 students have a mean of 72 and a standard

deviation of 12. Estimate the number of students with test grades between 48 and 96.

8. The mean height of college men is 70 inches with a standard deviation of 2.5 inches. Estimate how many out of a sample of 1000 college men are between 62.5 and 77.5 inches in height.

9-5* A COMPUTER PROGRAM FOR THE STANDARD DEVIATION

In this section we develop a computer program for the calculation of the standard deviation. This program enables us to use the computer for all of the hard work. We simply give it the program and the data.

There are two problems that arise in the program. Because we need to use the same program with data lists of varying length, we must let the computer know when all of the data have been read. We can accomplish this by telling the computer in advance how many items to read *or* by having a "cue" number at the end of the data to signal the computer that all of the data have been read. We choose the last method. We put the "cue" statement

$$999 \quad \text{DATA} \quad 999999$$

just before the END statement. When the computer reads this number from the data list, it will be instructed to stop reading data and to compute the mean or the standard deviation.

The other problem is more basic to the program. Recall the steps in computing the standard deviation:

(1) *Add* and *count* the data.
(2) Calculate the *mean*.
(3) Calculate the *deviations* from the mean.
(4) Calculate the *mean* of the squares of the deviations.
(5) Calculate the *square root* of the mean of the squares of the deviations.

Our problem is that we must read completely through the data list in order to calculate the mean [Step (2)]; then we must read through it a second time in order to calculate the deviations [Step (3)]. This means that we either store the data as we read it the first time and then reread it from storage or we must find a way to actually reread the DATA list.

In certain of the computer languages we have no choice—we must store the data and then reread it later. In most dialects of BASIC, however, a simple statement is available that allows us to reread the original data list. The statement RESTORE causes the computer to start over and reread the data list from the first item of data. The following example illustrates how this statement is used.

Example 1. What is the output of the following program?

```
50   REM-SAMPLE PROGRAM USING THE RESTORE STATEMENT
100  READ A,B
110  LET S=A+B
120  PRINT A,B,"SUM  =",S
130  RESTORE
140  READ X,Y
150  LET P=X*Y
160  PRINT X,Y,"PRODUCT  =",P
170  DATA 5,13
180  END
```

Solution. When the computer executes Statement 120, it prints

| 5 | 13 | SUM = | 18 |

When it executes Statements 130 and 140, it goes back to the beginning of the data list and reads X = 5, Y = 13. When it executes Statement 160, it prints

| 5 | 13 | PRODUCT = | 65 |

Thus, the total output consists of the two lines

| 5 | 13 | SUM= | 18 |
| 5 | 13 | PRODUCT= | 65 |

The Flow Chart

Figure 9-14 is a flow chart showing the steps involved in computing the standard deviation.

It is desirable to have a list of the data, the range of values, and the value of the mean, as well as the value of the standard deviation. The variable N is used to count the data items; S is used to store the sum of the data; T is used to store the sum of the squares of the deviations. These are the variables that we need to calculate the mean and the standard deviation and to print these numbers along with the data. The listing of the range of values is left for the reader (Exercise 4).

At the beginning of the program, the computer sets the variables N (counter for data), S (sum of data), and T (total of squares of deviations) equal to zero. Next, it prints the word "DATA:" (followed by a semicolon or comma to conserve space on the printout). It then reads a number (X)

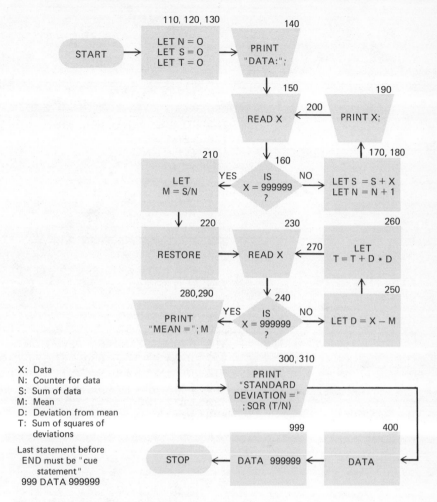

Figure 9-14 Detailed flow chart for the standard deviation.

from the data list, adds it to the sum (S), adds 1 to the counter (N), and prints the number on the permanent list.

As soon as the "cue" number 999999 is read, the computer stops reading, computes the mean, resets the data list to the beginning, and then works through a loop to compute the sum of the squares of the deviations. When it reads each number (X) from the data list, it computes the deviation (D) and then adds D^2 to T (the running total of the squares of the deviations).

When the computer encounters the "cue" number 999999 the second time, it prints the mean, calculates and prints the standard deviation, and stops work.

The following program is based on the flow chart. One run (using the data given in the program) is shown with the printout.

```
100   REM-MEAN AND STANDARD DEVIATION
110   LET N=Ø
120   LET S=Ø
130   LET T=Ø
140   PRINT "DATA:";
150   READ X ←
160   IF X=999999 THEN 21Ø
170   LET S=S+X
180   LET N=N+1
190   PRINT X;
200   GOTO 15Ø
21Ø   LET M=S/N
220   RESTORE
230   READ X ←
24Ø   IF X=999999 THEN 28Ø
25Ø   LET D=X−M
26Ø   LET T=T+D*D
27Ø   GOTO 23Ø
28Ø   PRINT
290   PRINT "MEAN =";M
3ØØ   PRINT
31Ø   PRINT "STANDARD DEVIATION =";SQR(T/N)
400   DATA  41,37,29,83,87,51,57,62,1Ø,93,1ØØ,1ØØ,98,41
41Ø   DATA  77,62,85,84,73,11,81,94,85,91,63,71,73
42Ø   DATA  65,73,78,68,5Ø,61,77,89,94,2Ø,53,67,72
43Ø   DATA  1ØØ,7Ø,67,63,72,88,71,62,57,65,64,57,83
999   DATA  999999
1ØØØ  END
```

```
DATA:  41   37   29   83   87    51   57   62   1Ø   93   1ØØ
      1ØØ   98   41   77   62   85    84   73   11   81   94   85
       91   63   71   73   65   73    78   68   5Ø   61   77   89
       94   2Ø   53   67   72   1ØØ   7Ø   67   63   72   88   71
       62   57   65   64   57   83
MEAN = 68.3962

STANDARD DEVIATION = 21.1984
```

EXERCISES

1. Work through each step of the program in Example 1 until you understand the effect of the RESTORE statement.
2. Work through each step of the main program, using the data statements

$$400 \quad \text{DATA} \quad 0, 5, 10$$
$$999 \quad \text{DATA} \quad 999999$$

Can you predict what the printout will be?

3. Run the main program of this section on your computer, using the data provided with the program. Compare your printout with the printout following the program.
4. Modify the main program of this section so that the range of values (highest and lowest data numbers) will be printed with an appropriate label. Run the program with the data provided in the program in the text. (Example 1 of Section 6-6 may be helpful.)
5. Run the main program using the class grades from your last test as data. Does the standard deviation show that the grades are tightly grouped around the mean?

9-6 NORMAL DISTRIBUTIONS

During the seventeenth century, the "error" curve received a good deal of attention from statisticians. This is the graph of the frequency distribution that results whenever a large number of measurements are made of an object using very precise instruments. If the measuring instruments are very delicate, then two measurements of the object will probably result in different measures of its size. If we make a very large number of measurements, then the bar graph of the frequency distribution of the measurements is similar to the one in Figure 9-15(a).

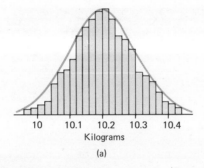
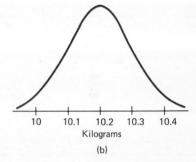

Figure 9-15 Measurements of the breaking strength of wire. (a) Bar graph; (b) the error curve.

It is customary to "round-off" the bar graph and replace it with a smooth curve through the points. This results in the typical error curve shown in Figure 9-15(b).

It was found in later years that the bell-shaped error curve is the typical curve when a manufacturing process is tested for uniformity. The following discussion indicates how the curve is used.

The Strong Wire Co. manufactures a special very thin wire used in space exploration. This wire is supposed to hold a weight of 10 kilograms without breaking. In theory it should hold more than the 10 kilograms, but in actual practice, due to imperfections in the metal alloy and minor variations in the manufacturing process, some samples are weaker than the standard and some are stronger.

The company tests a small sample when each spool is wound. The sample is subjected to a strain test, and the actual weight required to break it (measured to the closest tenth of a kilogram) is recorded.

At the end of the day, the data are tabulated; a frequency distribution is made; and the mean $\bar{x}$ and the standard deviation σ are calculated. The output is acceptable, provided 95 percent of the samples could hold at least 10 kilograms without breaking. From a practical standpoint, this means that $\bar{x} - 2\sigma$ must be greater than or equal to 10. If $\bar{x} - 2\sigma < 10$, the entire output is rejected.

The bell-shaped frequency curves (Figs. 9-15 and 9-16) are examined carefully. Each curve should have a sharp peak at the mean (indicating a narrow range of breaking values) as in Figure 9-16(a). If a curve is flattened out as in Figure 9-16(b), it is a sign that the wire lacks uniformity. Usually this means that the machines are out of adjustment.

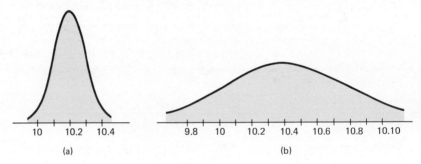

Figure 9-16   Error curves for the breaking strength of wire. (a) Good error curve. Most samples break at a strain of slightly more than 10 pounds; (b) poor error curve. The samples break over a wide range of strains. The machines are out of adjustment.

The area under any part of a frequency curve is proportional to the number of data items in that range. This is pictured in Figure 9-17 where approximately 20 percent of the area under the curve is in the range $10.1 \leq x \leq 10.2$. Thus, approximately 20 percent of the wire samples break in the range 10.1 kilograms $\leq x \leq$ 10.2 kilograms.

In 1833 the pioneer Belgian statistician *L. A. J. Quetelet* (1796–1874) made a study of the distributions of measurable human characteristics—height, weight, arm length, and so on. He found that the distribution of each characteristic fitted the error curve almost perfectly. This result has since been verified for hundreds of other measurements on all types of plants and animals. Within a given species, the distribution of almost any measurable characteristic fits the standard error curve almost exactly (Fig. 9-18).

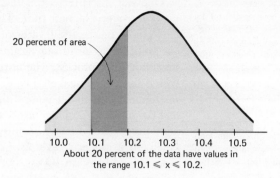

20 percent of area

10.0 10.1 10.2 10.3 10.4 10.5
About 20 percent of the data have values in
the range 10.1 ≤ x ≤ 10.2.

Figure 9-17 Area under the error curve. The total area is proportional to the total number of data items. The area between 10.1 and 10.2 is proportional to the number of data items with values between 10.1 kilograms and 10.2 kilograms.

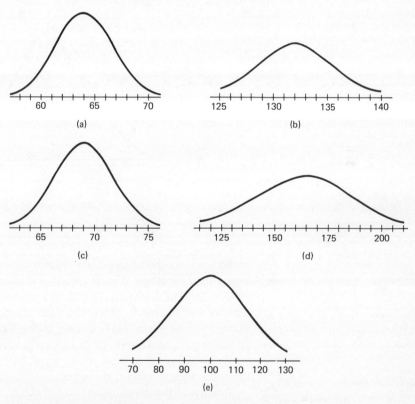

Figure 9-18 Normal distribution curves. (a) Heights of women; (b) weights of women; (c) heights of men; (d) weights of men; and (e) intelligence (as measured by IQ tests).

The basic assumptions made by Quetelet have been criticized by many statisticians. Quetelet assumed that if a number can be found that describes a natural phenomenom, then that number has a significance in nature. For example, if the mean height of college men is found to be 70 inches, then Quetelet would have interpreted this to mean that all college men should be 70 inches tall and that heights different from 70 inches are caused by errors.

Quetelet interpreted this discovery to mean that there is an ideal or *norm* for every human trait. The variability of individuals was seen by Quetelet to be the result of errors in the reproduction process—just as the variability of wire in our earlier example was the result of errors in the manufacturing process. The norm for each trait was the central figure at which the curve peaked. [Thus, from Figure 9-18 we see that the norm for height (men) is 69 inches, the norm for weight (men) is 165 pounds, and the norm for intelligence is an IQ of 100.] Quetelet defined the "average" person to be one who fitted all of the norms. Thus, in the sense of Quetelet, a modern "average" man would be 69 inches tall, weigh 165 pounds, and have an IQ of 100.

Because of the discovery of the norms for measurable characteristics, the bell-shaped curves also are known as *normal distribution curves*. When we deal with measurement, they still are called *error* curves.

The normal distribution curves all have the characteristic bell shape. They peak at the mean value and are symmetric about it. Thus, the mean, the median, and the mode all coincide. Consequently, when we speak of the "average" of data that are normally distributed, there is no ambiguity. The three common averages are equal to one another (Fig. 9-19).

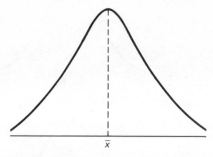

$\bar{x}$

Figure 9-19 If data are normally distributed, then the *mean, median,* and *mode* all are equal to $\bar{x}$.

"Coxcomb" figure used by Florence Nightingale to represent statistics.

Statistics were used by social reformers in nineteenth-century England to break down established prejudices. Foremost among these users was Florence Nightingale, who used statistics to such an extent in her arguments for medical reform that she was known as "the passionate statistician."

Some of the bell-shaped normal distribution curves peak in a sharper curve than do others (Fig. 9-20). This is a direct consequence of the size of the standard deviation. The smaller the standard deviation, the sharper the peak at the mean; the greater the standard deviation, the flatter the curve near the mean. Many human characteristics have a fairly flat curve, indicating that a wide range of variations is common. Others have a sharp peak, indicating that most of the variations are minor and the data are clustered tightly around the norm.

Because of the great importance of the normal-distribution curve, it has been studied extensively and in great depth. By calculating the area under the curve, the proportions of the data within certain ranges about the mean

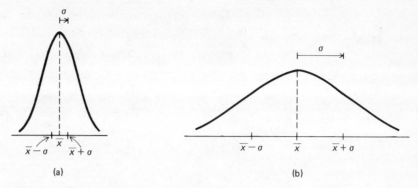

Figure 9-20 (a) Small standard deviation. The data are tightly clustered about the mean; (b) large standard deviation. The data are widely dispersed about the mean.

have been determined. In particular, it is known that

(1) a total of 68.3 percent of the data are within one standard deviation of the mean (between $\bar{x} - \sigma$ and $\bar{x} + \sigma$);

(2) a total of 95.4 percent of the data are within two standard deviations of the mean (between $\bar{x} - 2\sigma$ and $\bar{x} + 2\sigma$);

(3) a total of 99.7 percent of the data are within three standard deviations of the mean (between $\bar{x} - 3\sigma$ and $\bar{x} + 3\sigma$). (See Fig. 9-21.)

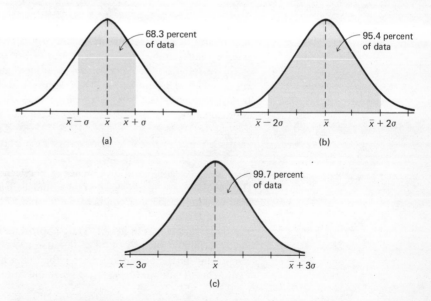

Figure 9-21 Areas under the normal distribution curve. (a) 68.3 percent of the data are within one standard deviation of the mean; (b) 95.4 percent of the data are within two standard deviations of the mean; and (c) 99.7 percent of the data are within three standard deviations of the mean.

The following example shows how the standard deviation can be used to determine the distribution of data within certain range.

Example 1.   Measurements of college men reveal a mean height of 70 inches with a standard deviation of 2.5 inches. A certain university has 10,000 men enrolled. Assume that they are representative of the entire population of college men.
   (a) How many of the men are between 67.5 and 72.5 inches in height?
   (b) How many of the men are between 65 and 67.5 inches in height?
   (c) How many are over $77\frac{1}{2}$ inches in height?

Solution.   In this example $\bar{x} = 70$ and $\sigma = 2.5$. (See Fig. 9-22.)

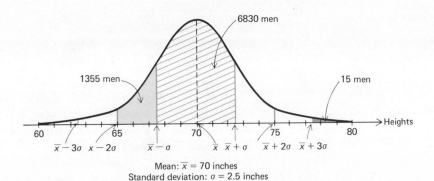

Mean: $\bar{x} = 70$ inches
Standard deviation: $\sigma = 2.5$ inches

Figure 9-22   Normal distribution curve showing the heights of 10,000 college men (Example 1).

   (a) A total of 68.3 percent of the men are between $\bar{x} - \sigma = 67.5$ and $\bar{x} + \sigma = 72.5$ inches in height. Thus, approximately 6830 of the men are between 67.5 and 72.5 inches tall.
   (b) A total of 95.4 percent of the men are between $\bar{x} - 2\sigma = 65$ and $\bar{x} + 2\sigma = 75$ inches in height. Thus, approximately 9540 men are in this range. In (a) we found that 6830 men are between 67.5 and 72.5 inches tall. Thus, approximately

$$9540 - 6830 = 2710$$

men are between 65 and 67.5 inches or between 72.5 and 75 inches tall.
Half of these are in the lower range, and half are in the upper range. Thus, approximately 1355 men are between 65 and 67.5 inches in height.
   (c) A total of 99.7 percent of the men are between $\bar{x} - 3\sigma = 62.5$ and $\bar{x} + 3\sigma = 77.5$ inches in height. Thus, approximately 9970 of the men are in this range. It follows that about 30 of the men are outside of the 62.5- to 77.5-inch range. Half of these men are shorter than 62.5 inches, and half are taller than 77.5 inches. Thus, approximately 15 men out of the thousand are taller than 77.5 inches.

   An analysis similar to the one in Example 1 can be performed on any data that are normally distributed. For simplicity, we considered only the ranges of values bounded by $\bar{x} - 3\sigma$, $\bar{x} - 2\sigma$, $\bar{x} - \sigma$, $\bar{x}$, $\bar{x} + \sigma$, $\bar{x} + 2\sigma$, and

$\bar{x} + 3\sigma$. Statisticians use elaborate tables that give the percentage of data over any range. For example, the tables can be used to show that 1.2 percent of the data are between $\bar{x} + 1.5\sigma$ and $\bar{x} + 1.6\sigma$, in case anyone is interested in that range of values.

The use of the normal distribution curve is so deeply ingrained in our society that we often are suspicious of any data that are not normally distributed. We should realize, however, that many statistics are not distributed normally.

The following statistics are not distributed normally. (This does not mean that they are abnormal.)

(1) *Income.* Far more persons are in the lower income brackets than are in the upper income brackets.

(2) *Death Rates.* The death rate is much higher for older persons than for younger persons.

(3) *Literacy Rate* (plotted against age). After a certain age, the percentage of persons who can read and write does not change appreciably.

(4) *Cloud Cover.* If we measure the percent of sky covered by clouds at 12:00 noon, we shall find that most of the days either have a very small percentage or a very high percentage of cloud cover. The mean may be 50 percent, but very few days actually have 50 percent cloud cover.

The word "normal" has a technical meaning in statistics that is different from its usual meaning. It refers to *norms* (averages), not to desirable traits. The fact that a person is far from the norm with respect to a trait does not mean that he has a defect that should be corrected or that he is abnormal.

Sir Francis Galton (1822–1911)

As a young man Galton studied medicine and mathematics but left his studies at the age of 22 when he received an inheritance, and set off to travel the world. On the basis of his explorations in South Africa, he was made a Fellow of the Royal Society at the age of 34.

Galton was a cousin of Charles Darwin, and he became interested in the applications of genetics to natural selection. This interest grew into a study of the problems inherent in the modern policy of allowing the unfit to survive and reproduce. Pursuing this study to its extreme, Galton became one of the founders and the chief developer of the science of eugenics—the study of the improvement of mankind through selective breeding.

In order to improve the human race, Galton found that it was first necessary to describe it accurately. This called for the large-scale collection of statistics, which he achieved in his Anthropometric Laboratory, a part of the South Kensington Science Museum. For a small fee anyone could register and have accurate measurements taken of his height, weight, span of arms, keenness of sight, breathing power, and so on. These statistics were kept on file for the subject to use for later comparison and for use by the laboratory.

Once he had collected the statistics, Galton was faced with the problem of interpreting them. This led him to develop the idea of *correlation*—the study of how two statistical quantities vary with respect to each other. His work with this and other statistical concepts became the basis for many of the applications of statistics to biology.

During much of his life Galton was obsessed with the collection of statistics. He counted the number of fidgets and yawns at public meetings in an attempt to establish an "index of boredom," observed the changes in color

as faces flushed with excitement during horse races, studied the average
number of children of heiresses, the
common properties of identical twins,
and the general inheritance of physical and mental traits. In addition, he
introduced graphical and statistical
methods into meteorology and made
a "beauty map" of England. (When
he traveled through the various cities
and towns, he classified each girl that
he passed as "attractive, indifferent,
or repellent," then used the average
numbers to rank the localities.)

Galton was primarily a social re-
former. Although his collections of
statistics were based on his own interests, he slanted all of his work toward the improvement of the human
race. As a consequence, he has been
criticized for lacking the dispassion-
ate love of knowledge for its own sake
that often is considered to be the
hallmark of a great scholar or scientist.

EXERCISES

1. In your opinion, which of the following statistics should be distributed
   normally?
   (a)   size of ears;
   (b)   musical ability;
   (c)   the death rate at each age due to cancer;
   (d)   the number of families with x number of children;
   (e)   the number of families who move x times during the year;
   (f)   the scores on the Scholastic Aptitude Test;
   (g)   the number of books sold each year in the United States.
2. The following measurements were obtained when a sixth grade class
   measured the length of a piece of wood. Draw the error curve for the
   data and use it to estimate the actual length.

26.5 centimeters	1 person
26.7 centimeters	1 person
26.9 centimeters	2 persons
27.0 centimeters	3 persons
27.1 centimeters	5 persons
27.2 centimeters	5 persons
27.3 centimeters	4 persons
27.4 centimeters	3 persons
27.5 centimeters	3 persons
27.8 centimeters	1 person

3. A statistical study shows that the mean weight of college women is 127
   pounds with a standard deviation of 17 pounds. A certain college has
   1000 women students.
   (a)   How many weigh between 110 pounds and 144 pounds?
   (b)   How many weigh less than 110 pounds?
   (c)   How many weigh less than 93 pounds?
   Compare your answers with Example 3 of Section 9-4.
4. The mean IQ of individuals in the United States is 100 with a standard
   deviation of 15.

(a) What percentage of the population has an IQ over 115?

(b) What percentage has an IQ over 130?

(c) What percentage has an IQ over 145?

5. The state legislature has passed a law requiring the 4 million residents of the state to take IQ examinations. All persons with IQ under 70 are to be confined to mental institutions. At present, the state has facilities for 2000 of these persons.

(a) How many additional spaces must be provided for the new inmates?

(b) Is this a good law? Were the statistical data properly considered by the legislature before passing the law?

6. A characteristic that is normally distributed in a large population may not be distributed normally in a small sample. Often the sample contains two or more distinct groups in which the characteristic is distributed differently. Figure 9-23 shows the final distribution of grade averages in a mathematics class. What tentative conclusions would you make about the composition of the class?

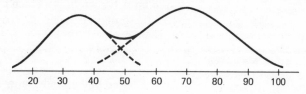

Figure 9-23 Exercise 6.

9-7 SOME USES (AND MISUSES) OF STATISTICS

Sampling

The normal distribution curve is widely used in sampling procedures. Suppose, for example, that we wish to know the average weight of U.S. women in the age group 20–29. It would not be feasible to attempt to weigh all of the women in that age group in the United States. Thus, we choose a sample of the women, weigh them, and draw conclusions about the average weight in the entire United States.

The sampling process is the greatest possible source of error in making the study. It is necessary to have a sample that is both *large* and *representative* of the entire population under study. Within these limits the sample should be chosen in a completely random way. If the sample is too small (say, it only includes a dozen women), then the final results may not be valid. In the same way, if the sample is not representative or is not chosen randomly, then the final results may not mean anything except for the women in the sample population. (Think of the results if all the women are members of a national weight-reducing club.)

After the sample is chosen and the statistics are collected, then the mean and the standard deviation are calculated. There is still a chance of error, because the total sample may not have been truly representative in spite of the attempts to make it so. Thus, the sampling procedure should be repeated several times. If all of the means and the standard deviations are in basic agreement, then the mean and the standard deviation are calculated from the entire data set. The final mean and the standard deviation are assumed to hold for the entire population.

Consumer Testing

One of the major uses of sampling is in the testing of consumer products. The most typical example is the testing of light bulbs. Every box of light bulbs has a label stating the number of hours that the bulbs should burn (typically 750 hours). It is not possible to test the individual bulbs that will be sold. If they are burned for 750 hours before being sold, then they will almost certainly burn out within a few hours of use in the home.

A sample of each day's output of light bulbs is selected and tested. This sample must be carefully chosen so that the outputs of each machine are tested, these outputs being chosen in a random way. (It would not do to test exactly every 100th bulb from a machine, for example, because a malfunction might cause every 5th bulb to be defective. If every 100th bulb were tested, then either the defective ones would never be found or all of the tested bulbs would show up as defective.) The results of the tests are tabulated; the mean and the standard deviation are calculated. If the mean is greater than or equal to 750 hours and the standard deviation is small, then the bulbs can be sold. If the mean is less than 750 hours or the standard deviation is very large, then the entire output of bulbs should be rejected.

Fraudulent Testing

Consumer testing is one area in which statistical fraud is common. The fraud usually is perpetrated by deliberately ignoring unfavorable statistics or by choosing samples that are not representative or are too small to give meaningful results. The following example illustrates one of the methods.

The Brite-Glo Toothpaste Company claims that 80 percent of its customers had fewer cavities after using Brite-Glo than with their former toothpaste. This figure was obtained by the following process.

Two hundred persons were selected at random to test Brite-Glo. Their dental records were examined, and the average number of cavities over a six-month period was computed for each person. The subjects were then split into 20 teams of 10 persons each. Each team used Brite-Glo for six months; then the number of cavities was recorded for each person and compared with that person's average number.

With any small group of persons, one of three things may happen. Either the members of the group will have fewer cavities or more cavities or about the same number. In a few test groups, a majority had fewer cavities; in a few groups, the majority had more cavities; in most groups, there was no substantial change in the number of cavities.

The company found that in *one* 10-person group, 8 persons had fewer cavities; 1 had the same number; and 1 had more cavities. They kept the report from this team and "filed" the other reports where they could never be found. Thus, the "80 percent fewer cavities" figure was obtained by ignoring the actual results and keeping only one small, nonrepresentative sample out of the larger sample.

In recent years the Federal Trade Commission has required that complete statistics be turned over to their statisticians for analysis before claims such as those made by Brite-Glo can be used in advertising. Since we have used toothpaste for our example, we should mention that toothpastes endorsed by the American Dental Association (ADA) have been tested extensively. Exhaustive statistical studies must be conducted under ADA supervision before the society gives its endorsement.

Correlations

A major use of statistics is to establish correlations between quantities that may, or may not, seem to be related. A significant correlation exists if one quantity changes in a consistent way when the other quantity changes. Some standard correlations that have been established are between IQ scores and scholastic achievement, between the Dow-Jones average and the "health" of the U.S. economy, between the price of grain and the price of eggs.

Correlations are of special value when changes in one quantity can be used to predict changes in the other. Thus, S.A.T. scores can be very useful in determining college admissions. The scores are fairly good predictors of future academic progress.

One of the major errors made with correlations is to assume that because two factors have a significant correlation one of the factors has an influence on the other one. This may well be the case, but the correlation does not establish it. For example, one study found a positive correlation between the average salary paid to Presbyterian ministers in New England and the price of rum in Jamaica. No one would ever claim, however, that the salaries paid to Presbyterian ministers have any effect on the price of rum or, conversely, that the price of rum has any effect on ministers' salaries. The truth is that the world economic situation influenced both the salaries of the ministers and the price of rum.

A number of important correlations have been made—some of which are quite controversial. One of the most famous is the correlation between cigarette smoking and lung cancer. It appears that a definite correlation exists. Opponents of smoking claim that this correlation means that smoking cigarettes causes cancer. The tobacco industry, on the other hand, claims that the correlation has never been established and, even if it does exist, it does not *prove* that smoking causes lung cancer—some other factor may actually cause it.

TESTING OF DRUGS
Each new drug must be tested for effectiveness and safety. In general, both the reactions to an infection and the responses to a drug vary widely with individuals. Thus, the medical statistician must design careful experiments to test the drug that will take these varying factors into account.

Scientific theories are usually validated by statistical methods. Until recently there were two competing theories for the creation of the universe. The "continuous creation" theory held that matter was being continually created in interstellar space. The "big bang" theory held that all matter was once contained in a huge dense ball that exploded about 15 million years ago. By comparing the average number of quasars near the edge of the universe with the average number near the Milky Way galaxy, the "continuous creation" theory was ruled out as a possibility.

Scatter Diagrams and Correlations

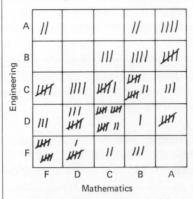

The *scatter diagram* is a device for comparing two varying quantities. The above scatter diagram compares the mathematics and engineering grades of 104 freshman students. (If a student received an A in mathematics and a B in engineering, his grade is recorded as a tally in the appropriate box.) Observe that there is an apparent relationship between the grades. In general, a student who did well in one subject also did well in the other, and a student who did poorly in one did poorly in the other.

Francis Galton developed his concept of *correlation* from elementary ideas similar to the ones illustrated in the above box. In the modern theory an *index of correlation* is established between two varying quantities. The index is a number between -1 and $+1$. An index near $+1$ indicates that one quantity generally increases when the other increases, and vice versa. An index near -1 indicates that one quantity generally decreases when the other increases, and vice versa. An index near 0 indicates a randomness which shows no actual relationship between the quantities.

Grading "on a Curve"

We have discussed the bell-shaped normal distribution curve several times in this chapter. Most human characteristics seem to be normally distributed in the population. It is a mistake to assume, however, that all types of human activity can be fitted to the normal distribution curve. Furthermore, small samples usually are not normally distributed even when the characteristic is normally distributed in the entire population. The obvious reason for this is that the samples are usually not representative of the total population.

Academic tests furnish a good illustration of the above remarks. It appears that the grades on most tests fit a normal distribution curve, provided the tests are given to a very large group of students. Thus, if we know the mean and the standard deviation, we can make a reasonably "good" distribution of grades by fitting them to the curve. The usual procedure is to use one standard deviation for each letter grade with the mean midway in the "C" range (Fig. 9-22). This gives us the following grade scale:

F range: below $\bar{x} - 3\sigma/2$ (7 percent)
D range: $\bar{x} - 3\sigma/2$ to $\bar{x} - \sigma/2$ (24 percent)
C range: $\bar{x} - \sigma/2$ to $\bar{x} + \sigma/2$ (38 percent)
B range: $\bar{x} + \sigma/2$ to $\bar{x} + 3\sigma/2$ (24 percent)
A range: above $\bar{x} + 3\sigma/2$ (7 percent)

In a class of 100 students with grades that are normally distributed, there are 7 F's, 24 D's, 38 C's, 24 B's, and 7 A's.

Several problems arise with fitting grades to the normal distribution curve. First, no one has ever established that grades either are, or should be, normally distributed, although these two assumptions are held to be matters of faith by some educators. Furthermore, even if grades are normally distributed for a very large class, they almost surely will not be for a small class. In addition, some teachers do not use the normal distribution curve at all, but claim that they are "grading on a curve" when they preassign the distribution of grades as 7 percent A, 24 percent B, 38 percent C, 24 percent D, and 7 percent F. This last practice is at best a very dubious method for deciding the grades of individual students.

The author heard of one professor who followed a strict "curve" for the grades in all of his classes. This worked out quite well for his 200-student freshman lecture sections, but caused a great deal of anguish for him (and for his students) in a small senior seminar. The seminar was taken by only three students, all very good in the subject, who had nearly perfect results on their seminar projects. The professor puzzled over the grades for two weeks before he finally assigned one D, one C, and one B, the closest he could fit the grades to the normal distribution curve.

EXERCISES

1. Find an additional example of how statistics can be used to mislead rather than inform. (The book *How to Lie with Statistics* has a wealth of such information.)

2. Explain why we need random samples of populations if surveys are to be meaningful.

3. Which of the following would come close to being random samples of the population?
   (a) every 10th person who passed a street corner;
   (b) every 50th name in the telephone directory;
   (c) every 10th person in line at an airline's ticket counter;
   (d) every 10th person encountered on a college campus.

4. At least three methods can be used to assign grades in a mathematics class: (1) Assign grades on the basis of the student proficiency in the subject, using some type of absolute standard. (2) Assign grades by using the mean and the standard deviation as explained in the text. (3) Preassign the distribution of grades so that a certain number of students receive F, a certain number receive D, and so on.
   (a) Discuss the merits and the failings of each system.
   (b) Obtain a copy of the scores of your last mathematics test. Compare the distributions of letter grades on the test for the three different systems.

SUGGESTIONS FOR FURTHER READING

1. David, F. N. *Games, Gods and Gambling*. New York: Hafner, 1962.
2. Huff, Darrell. *How to Lie with Statistics*. New York: Norton, 1954.
3. *Statistical Abstract of the United States*. Bureau of the Census, U.S. Department of Commerce.
4. Weaver, Warren. *Lady Luck*. Anchor Books. New York: Doubleday, 1963.

The following can be found in *World of Mathematics*, James R. Newman, ed. New York: Simon & Schuster, 1956.

5. Galton, Francis. "Classification of Men According to Their Natural Gifts," in *World of Mathematics*, Vol. 2, Part VI, Chap. 2.
6. Graunt, John. "Foundations of Vital Statistics," in *World of Mathematics*, Vol. 3, Part VIII, Chap. 1.
7. Malthus, Thomas Robert. "Mathematics of Population and Food," in *World of Mathematics*, Vol. 2, Part VI, Chap. 3.

10

PROBABILITY

10-1 EMPIRICAL PROBABILITY

Probability theory is concerned with expectations and likelihoods. When we say that the probability of rain is 40 percent, we mean that in the past rain has occurred on 40 percent of the days with similar climatic conditions. When we say that the probability of a 20-year-old person living to age 70 is 0.48, we mean that in the past 48 percent of the persons who reach the age of 20 have lived to age 70.

These two examples illustrate the concept of *empirical probability*—a branch of statistics concerned with making predictions from past performances. In order to make these predictions, large quantities of statistical data must be collected and analyzed. Another branch of probability theory is known as *a priori probability*. In this branch, which we shall consider in Section 10-2, underlying assumptions about events are used to calculate the probabilities that the events might occur.

Empirical probability began to be developed shortly after John Graunt's book in which he pointed out the regularity of deaths from certain causes. Eighteenth-century businessmen quickly realized that they could make money by predicting how long the average person would live. They studied mortality tables and formed companies to sell life insurance. These men realized that they could not predict when any one person would die, but they also knew that this information was not really necessary. They only needed to know the average length of time that persons lived. Knowing that fact they could calculate how much to charge for the premiums. The following simplified example shows how the process worked.

Empirical probability theory was originally developed to solve problems in insurance and pensions. It is now an indispensable part of all large-scale planning for the future, both in business and government.

Example 1. Suppose the average person who reaches age 20 will live to age 60. How much should be charged in monthly premiums on a $1000 life insurance policy sold to a 20-year-old person if the company is to neither make money nor lose money on the policy. (Neglect the effects of investments, interest earned for the company, and inflation.)

Solution. The average person will live an additional 40 years. Thus, for the company to break even, it must charge

$$\frac{1000}{40} = \$25 \text{ per year}$$

This reduces to a little over $2.08 per month.

At this rate the company will neither lose money nor make money. (Actually, it will lose a small amount because there is no payment for the work involved in writing up the policy, filing it, collecting premiums, and so on.) Once the company knows the "break-even" rate, however, it can modify the premium rate so as to cover expenses and make a profit.

The early insurance companies made extensive statistical studies of birth and death rates. The information was compiled in *mortality tables*, which told how many persons out of a standard population (say 1000 persons) would die each year. These tables also listed the probability of dying at

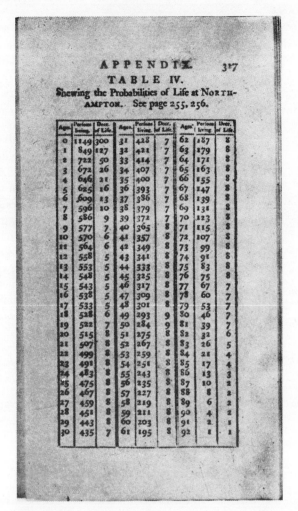

Figure 10-1 Eighteenth-century mortality table used by the Equitable Life Insurance Company. R. Price, *Observations on Reversionary Payments,* 1771.

a given age and the life expectancy at each age. Figure 10-1 shows a page from an early mortality table. Out of 515 persons alive at age 20, only 284 were alive at age 50.

In spite of their careful records and extensive calculations, all of the early insurance companies went bankrupt. The reason was that the life and death statistics kept by the governments were quite inaccurate. No one at the time (except the insurance companies) realized that accurate record keeping of vital statistics was important.

One of the first successful insurance companies was the Equitable Life Insurance Company of England. Its mortality tables also were inaccurate, but in the company's favor. The average policyholder lived much longer than predicted, so that more was paid in premiums than paid out by the company. In fairness to the company, we must report that it constantly revised its mortality tables so as to reflect the true death rate. Within a few years, the error in the tables was completely eliminated.

An interesting postscript shows how incorrect statistics can work to either advantage or disadvantage. Knowing that Equitable was successful, the British government borrowed the company's mortality tables and used them to establish its pension plan for civil service workers. Now the statistics worked to the disadvantage of the government. Because the tables predicted earlier deaths than actually occurred, much more was paid out in pension benefits than had been predicted, causing a severe strain on the British Treasury.

In the remainder of this section, we shall consider some examples of empirical probability based on mortality tables. In the rest of the chapter, we shall develop some of the elementary results concerned with a priori probability.

The *probability* that an event will have a successful outcome is the ratio of the total number of successful outcomes to the total number of events:

$$p = \frac{\text{total number of successful outcomes}}{\text{total number of events}}$$

Approve	27 percent
Disapprove	49 percent
Undecided	24 percent

How many people, representative of the nation as a whole, must be questioned in a poll before the results can be applied to the entire population with a 90-percent probability of being accurate to within 3 percentage points?

There is a close connection between probability and statistics. Whenever we apply statistics to a sample different from the original one, we must ask the probability that the results will be valid.

In probability theory the word "event" has a much broader meaning than in ordinary conversation. It can refer to a person in the population, to a day, or to an occurance. The word "successful" also has a broader meaning. Any outcome in which we are interested may be called "successful," even an outcome such as death, which is not ordinarily considered to be a success.

As an example, suppose that, on the average, 284 persons live to age 50 out of every 515 who reach age 20. (See Fig. 10-1.) In this case, the "events" are the persons alive at age 20, and the "successful outcomes" are the persons still alive at age 50. Thus, the probability that a person aged 20 would live to age 50 is

$$p = \frac{\text{number of persons alive at age 50}}{\text{number of persons alive at age 20}} = \frac{284}{515} \approx 0.55$$

We would expect approximately 55 percent of the persons alive at age 20 to live to age 50.

The above figure does not mean that exactly 55 percent of every group of people aged 20 will live to age 50. There is no way to predict how long any one person will live or how long the members of any small group will live. If we start with a group of 10 persons, for example, it is possible that all may die at age 21 in an accident or at age 22 in an epidemic.

On the other hand, all 10 may live to be more than 100 years old. The probability figures are only valid with large populations, and they must constantly be revised.

The mortality rate has changed drastically over the past hundred years. In 1850 the average newborn baby boy was expected to live only to the age of 38. By 1956 such a baby could expect to live to the age of 67.

To a large extent, empirical probabilities depend on local conditions. The figures 284 and 515 that we used above were obtained from an eighteenth-century English mortality table. We would not expect those figures to hold for twentieth-century America. Advances in medicine and higher standards of living have constantly increased the probability that the average person will reach an advanced age. (See Table 10-1.)

Table 10-1 **Life Expectancy at Birth and at Age 40 at Various Times in United States History.**

| Year | Life Expectancy | | | |
| | At Birth | | At Age 40 | |
	Male	Female	Male	Female
1850°	38.3	40.5	27.9	29.8
1880°	41.74	43.50	28.86	30.29
1900	48.23	51.08	27.74	29.17
1920	56.34	58.53	29.86	30.94
1940	62.81	67.29	30.03	33.25
1956	67.3	73.7	31.6	36.7

° Massachusetts only.

Source: Adapted from *Historical Statistics of the United States, Colonial Times to 1957,* U.S. Bureau of the Census.

Table 10-2 is a modern United States mortality table. It tells the number of persons alive each year out of an initial sample of 1 million persons alive at age 1, the number of persons who die each year, the probability of living through the year, and the life expectancy (the number of additional years a person of a given age can expect to live).

The increased life span of the average person is creating new problems for society. A few individuals always have lived to advanced ages, but not enough to create distinct social pressures. As the length of life continues to increase and the birthrate to decrease, the United States will increasingly become a nation of middle-aged persons. It is estimated that all of our established patterns of work, retirement, pensions, and the like, will have to be revised. Probability theory, based on accurate statistics, will be a major tool for planning the changes.

Example 2. Use Table 10-2 to calculate the probability that a person of age 20 will live to age 50 and to age 70.

Solution. Out of the original sample, 951,483 were alive at age 20, and 810,900 of these persons remained alive at age 50. Thus, the probability of a person aged 20 living to age 50 is

$$p = \frac{810,900}{951,483} \approx 0.85$$

About 85 percent of the persons alive at age 20 live to age 50.

Similarly, the probability that a person alive at 20 will live to age 70 is

$$p = \frac{454,548}{951,483} \approx 0.48$$

Only about 48 percent of the persons alive at age 20 live to age 70.

Table 10-2 **Modern Mortality Table**

x	l_x	d_x	p_x	$\overset{\circ}{e}_x$	x	l_x	d_x	p_x	$\overset{\circ}{e}_x$
0	1 023 102	23 102	.977 42	62.33	**50**	810 900	9 990	.987 68	21.37
1	1 000 000	5 770	.994 23	62.76	51	800 910	10 628	.986 73	20.64
2	994 230	4 116	.995 86	62.12	52	790 282	11 301	.985 70	19.91
3	990 114	3 347	.996 62	61.37	53	778 981	12 020	.984 57	19.19
4	986 767	2 950	.997 01	60.58	54	766 961	12 770	.983 35	18.48
5	983 817	2 715	.997 24	59.76	**55**	754 191	13 560	.982 02	17.78
6	981 102	2 561	.997 39	58.92	56	740 631	14 390	.980 57	17.10
7	978 541	2 417	.997 53	58.08	57	726 241	15 251	.979 00	16.43
8	976 124	2 255	.997 69	57.22	58	710 990	16 147	.977 29	15.77
9	973 869	2 065	.997 88	56.35	59	694 843	17 072	.975 43	15.13
10	971 804	1 914	.998 03	55.47	**60**	677 771	18 022	.973 41	14.50
11	969 890	1 852	.998 09	54.58	61	659 749	18 988	.971 22	13.88
12	968 038	1 859	.998 08	53.68	62	640 761	19 979	.968 82	13.27
13	966 179	1 913	.998 02	52.78	63	620 782	20 958	.966 24	12.69
14	964 266	1 996	.997 93	51.89	64	599 824	21 942	.963 42	12.11
15	962 270	2 069	.997 85	50.99	**65**	577 882	22 907	.960 36	11.55
16	960 201	2 103	.997 81	50.10	66	554 975	23 842	.957 04	11.01
17	958 098	2 156	.997 75	49.21	67	531 133	24 730	.953 44	10.48
18	955 942	2 199	.997 70	48.32	68	506 403	25 553	.949 54	9.97
19	953 743	2 260	.997 63	47.43	69	480 850	26 302	.945 30	9.47
20	951 483	2 312	.997 57	46.54	**70**	454 548	26 955	.940 70	8.99
21	949 171	2 382	.997 49	45.66	71	427 593	27 481	.935 73	8.52
22	946 789	2 452	.997 41	44.77	72	400 112	27 872	.930 34	8.08
23	944 337	2 531	.997 32	43.88	73	372 240	28 104	.924 50	7.64
24	941 806	2 609	.997 23	43.00	74	344 136	28 154	.918 19	7.23
25	939 197	2 705	.997 12	42.12	**75**	315 982	28 009	.911 36	6.82
26	936 492	2 800	.997 01	41.24	76	287 973	27 651	.903 98	6.44
27	933 692	2 904	.996 89	40.36	77	260 322	27 071	.896 01	6.07
28	930 788	3 025	.996 75	39.49	78	233 251	26 262	.887 41	5.72
29	927 763	3 154	.996 60	38.61	79	206 989	25 224	.878 14	5.38
30	924 609	3 292	.996 44	37.74	**80**	181 765	23 966	.868 15	5.06
31	921 317	3 437	.996 27	36.88	81	157 799	22 502	.857 40	4.75
32	917 880	3 598	.996 08	36.01	82	135 297	20 857	.845 84	4.46
33	914 282	3 767	.995 88	35.15	83	114 440	19 062	.833 43	4.18
34	910 515	3 961	.995 65	34.29	84	95 378	17 157	.820 12	3.91
35	906 554	4 161	.995 41	33.44	**85**	78 221	15 185	.805 87	3.66
36	902 393	4 386	.995 14	32.59	86	63 036	13 198	.790 63	3.42
37	898 007	4 625	.994 85	31.75	87	49 838	11 245	.774 37	3.19
38	893 382	4 878	.994 54	30.91	88	38 593	9 378	.757 00	2.98
39	888 504	5 162	.994 19	30.08	89	29 215	7 638	.738 56	2.77
40	883 342	5 459	.993 82	29.25	**90**	21 577	6 063	.719 01	2.58
41	877 883	5 785	.993 41	28.43	91	15 514	4 681	.698 27	2.39
42	872 098	6 131	.992 97	27.62	92	10 833	3 506	.676 36	2.21
43	865 967	6 503	.992 49	26.81	93	7 327	2 540	.653 34	2.03
44	859 464	6 910	.991 96	26.01	94	4 787	1 776	.629 00	1.84
45	852 554	7 340	.991 39	25.21	**95**	3 011	1 193	.603 79	1.63
46	845 214	7 801	.990 77	24.43	96	1 818	813	.552 81	1.37
47	837 413	8 299	.990 09	23.65	97	1 005	551	.451 74	1.08
48	829 114	8 822	.989 36	22.88	98	454	329	.275 33	.78
49	820 292	9 392	.988 55	22.12	99	125	125	.000 00	.50
x	l_x	d_x	p_x	$\overset{\circ}{e}_x$	x	l_x	d_x	p_x	$\overset{\circ}{e}_x$

x = year

l_x = number of persons alive at the beginning of the year

d_x = number of persons who die during the year

p_x = probability of living through the year

$\overset{\circ}{e}_x$ = life expectancy for persons alive at the beginning of the year

Source: From *Rinehart Mathematical Tables*, Holt, Rinehart and Winston.

EXERCISES

1. Use Table 10-2 to calculate the probability that a person of the given age will live an additional 10 years:
   (a)   20 years old
   (b)   30 years old
   (c)   60 years old
   (d)   70 years old

2. Use Table 10-2 to calculate the premiums that should be charged on a $1000 life insurance policy taken out at each of the following ages if the insurance company is to break even on the policy. (Neglect the cost of writing and maintaining the policy, the effect of inflation, and the money that the company can make from the policy.)
   (a)   20 years of age
   (b)   30 years of age
   (c)   60 years of age

3. In Example 2 and Exercise 1 we neglected the effects of inflation and investments that the life insurance company can make. Explain the joint effect of these factors.

4. Life insurance companies have been a major force in developing better public health laws. In addition, they have disseminated free health information and have provided services for public health projects. Explain why this is good business for the companies. Give statistical as well as general reasons for your answers. Assume that the average life span has been increased five years because of the efforts of the insurance companies.

5. Social security benefits are paid to most people at age 65. Use Table 10-2 to estimate the percentage of persons who will receive these benefits. What is the average length of time that they will be paid?

6. Most people think that their life expectancy remains the same over their lives. This is not the case. Statistically, persons who have proved their hardiness by staying alive tend to live proportionately longer than persons in the general population. Show that this is the case by comparing the life expectancies at age 10, age 40, and age 70 in Table 10-2.

7. Make two graphs from Table 10-1, one by plotting the life expectancy at birth against the year, the second by plotting the life expectancy at age 40 against the year. How do you account for the difference in the two graphs?

10-2   THE MATHEMATICS OF GAMBLING

Probability theory has always been closely associated with games of chance. The first book that touched on the subject was written by *Jerome Cardan* (1501–1576), the Italian physician-mathematician-gambler-astrologer. Cardan, one of the most colorful scholars ever to work in mathematics, supported himself as a professional gambler for several years while he and his family lived in the Milan poorhouse. Later he made an intensive study

PIERRE FERMAT
(1601–1665)
Fermat, a lawyer by profession, devoted most of his leisure time to mathematical research. In addition to his work on probability theory, he was a co-inventor of analytic geometry and solved several problems related to the invention of the calculus. His main work was in establishing the modern theory of the branch of mathematics known as the theory of numbers.

Fermat published almost nothing during his lifetime. He did correspond regularly with the friar Martin Mersenne, however, who notified the other European mathematicians of Fermat's discoveries.

of gambling games, which he published under the title *Liber de ludo aleae* (*Book on Games of Chance*). Cardan was primarily interested in strategies and in techniques to expose cheating, but he did include methods for "figuring the odds."

The modern development of probability theory began in 1654 with a correspondence between *Pascal* (the philosopher-theologian who invented one of the first adding machines) and *Pierre Fermat* (one of the great creative mathematicians of all times). The correspondence concerned the Chevalier de Méré, a French nobleman and gambler. The Chevalier had made a modest fortune betting that he could toss at least one 6 out of 4 tosses of a die. Eventually no one would bet against him; so he changed the game and bet that he would throw at least one double 6 out of 24 tosses of two dice. He soon lost his fortune and appealed to Pascal to find out what went wrong. Pascal and Fermat solved the problem and went on to establish probability theory as a branch of mathematics.

Photo from G. Cardano, The Great Art, translated by T. Richard Witmer, by permission of the M.I.T. Press, Cambridge, Mass.

Jerome Cardan (1501–1576)

Geronimo Cardano (Jerome Cardan), the illegitimate son of an Italian nobleman, was denied admission to the medical school in Milan for several years because of his birth. When finally admitted he achieved a brilliant record only to be denied permission to practice medicine, again because of his birth. Several years of bad fortune forced him to take refuge in the local poorhouse with his family, where he supported himself as a gambler. He continued his medical and mathematical studies during this period and gained an excellent reputation as a scholar. When the local medical society finally recognized his abilities and admitted him to practice, he became one of the most noted and wealthy physicians in Europe.

Unfortunately, Cardan's public recognition came hand in hand with personal tragedy. All of his sons were wastrals and gamblers, in constant trouble with the law. When one of his sons murdered his wife, Cardan cut off the son's ears as punishment. This act enraged the local populace, and Cardan had to flee and ask protection from the Pope. Because of his mathematical ability, he soon became the Pope's personal astrologer. The son was later executed, bringing further sadness to Cardan.

Cardan was embroiled in one controversy after another. As we related in Chapter 7, when Nicolo Tartaglia discovered the general method for solving cubic equations (those of form $ax^3 + bx^2 + cx + d = 0$), Cardan invited him into his home, flattered him,

and treated him like an honored guest until Tartaglia, under a promise of strict secrecy, revealed the method of solution. Within a year Cardan had published the method in his book on algebra, ruining the hopes of Tartaglia of publishing his own book. Tartaglia devoted the rest of his life to proving that Cardan had stolen the secret from him, but was never able to muster popular support to his cause.

Cardan, who prided himself on being the best astrologer in Italy, finally got into trouble with the church authorities when he published a horoscope of Jesus Christ. He was arrested by the Inquisition on a heresy charge and spent several months in prison. After his release, he never regained his former prominence.

A number of stories have been told about this man. Even the ones that are untrue can be used to show facets of his character. One such story concerns his death. Late in his life Cardan cast his own horoscope and found that he was scheduled to die on a certain day. He got his affairs in order, but when the day came, it was obvious to him that he would live. According to the story, he committed suicide to protect his reputation as the country's best astrologer.

Games of chance have been a favorite pastime for thousands of years. Ancient Egyptian and Babylonian games had moves determined by throws of the knucklebones of an animal. Over the centuries, the bones were squared off to crude rectangular shapes on which numbers were carved to make dice.

The basic concepts established by Pascal and Fermat involved *equally likely events*. [Examples of equally likely events are tossing a coin (*heads* or *tails*) and tossing an honest die (1, 2, 3, 4, 5, or 6).] They decided that if all possible outcomes are equally likely, then the probability that any one of them will occur is

$$p = \frac{1}{\text{number of possible outcomes}}$$

For example, if there are n possible outcomes of an event and these are equally likely, then the probability that any particular one will occur is

$$p = \frac{1}{n}$$

Example 1. (a) There are two equally likely outcomes when we toss a coin. Thus, the probability of tossing *heads* is

$$p = \tfrac{1}{2}$$

Similarly, the probability of tossing *tails* is

$$q = \tfrac{1}{2}$$

(b) The probability of tossing 4 with an unloaded die is $\tfrac{1}{6}$. There are six equally likely outcomes. The probability of any one of them is $\tfrac{1}{6}$.

The principle that equally likely events have equal probabilities is the most basic of five principles formulated by Pascal and Fermat. In most cases we are interested in calculating a probability when the outcomes are not equally likely. The other four principles involve techniques to use equally likely outcomes to calculate the probabilities of events that are not equally likely.

> *Probability Principle 1:* Equally likely events have equal probabilities.

In most of our formulations, we speak of "favorable" or "successful" outcomes. These are merely the outcomes that yield the result in which we are interested. These outcomes may be most unfavorable by any other standard. If, for example, we wish to know the probability that a man will die of cancer or that the United States will engage in a major nuclear war, then favorable events are "dying of cancer" or "engaging in nuclear war."

Tree Diagrams

A figure known as a tree diagram can help us list all of the possible outcomes of an event. The tree diagram in Figure 10-2(a) pictures the two possible outcomes of tossing a coin. The diagram in Figure 10-2(b) pictures the six possible outcomes of tossing a die.

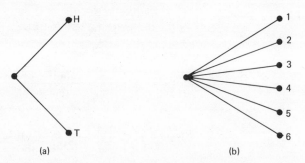

(a)   (b)

Figure 10-2 Tree diagrams. (a) The possible outcomes of a toss of a coin; (b) the possible outcomes of a toss of a die.

probability $= \frac{3}{7}$ odds are 4 to 3.

The odds in a fair game are determined by the ratio of the chances to win to the chances to lose.

If the probability of winning a game is $p = \frac{3}{7}$, then, on the average, there are three chances to win out of seven possible outcomes. If we subtract, we find that there are four chances to lose out of the seven possible outcomes. Thus, the player should be given four-to-three odds in order to make the game fair.

Example 2. A game is played by tossing a single die. You win if the toss is a 1 or 2.

(a) What is the probability of winning? Of losing?

(b) What odds should you be given to make the game fair?

Solution. (a) We see from the tree diagram in Figure 10-3 that there are six equally likely possible outcomes, that two of them result in a win for you, and four result in a loss. The probability of winning is

$$p = \frac{\text{number of favorable outcomes}}{\text{number of possible outcomes}} = \frac{2}{6} = \frac{1}{3}$$

The probability of losing is

$$q = \frac{\text{number of unfavorable outcomes}}{\text{number of possible outcomes}} = \frac{4}{6} = \frac{2}{3}$$

(b) Since it is twice as likely that you will lose as win (4 chances to lose compared to 2 chances to win), you should be given 2 to 1 odds in order to make the game fair.

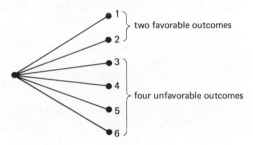

Figure 10-3 The probability of a favorable outcome is $p = \frac{2}{6} = \frac{1}{3}$. The probability of an unfavorable outcome is $q = \frac{4}{6} = \frac{2}{3}$.

Several basic principles can be established by arguments similar to the one in Example 2. The first is a direct extension of our previous work.

> *Probability Principle 2:* If an event has n equally likely outcomes and m of these are favorable, then the probability of a favorable outcome is
>
> $$p = \frac{\text{number of favorable outcomes}}{\text{number of possible outcomes}} = \frac{m}{n}$$

Example 3. A single card is drawn from a standard 52-card deck.

(a) What is the probability the card will be a Heart?

(b) What is the probability the card will be black Jack?

Solution. (a) There are 52 equally likely possible outcomes. Since there are 13 hearts in the deck, then 13 of these outcomes are favorable for drawing a Heart.

The probability of drawing a Heart is

$$p = \frac{\text{number of favorable outcomes}}{\text{number of possible outcomes}} = \frac{13}{52} = \frac{1}{4} = 0.25$$

(b) Since there are two black Jacks in the deck, then there are two ways of drawing one of them. The probability of drawing a black Jack is

$$p = \frac{\text{number of favorable outcomes}}{\text{number of possible outcomes}} = \frac{2}{52} = \frac{1}{26} \approx 0.038$$

When we compute a probability, we must calculate the ratio

$$p = \frac{\text{number of favorable outcomes}}{\text{number of possible outcomes}}$$

Because every favorable outcome is a possible outcome, then the numerator can be no larger than the denominator. Also observe that the numerator must be a number greater than or equal to zero. Thus, the fraction

$$p = \frac{\text{number of favorable outcomes}}{\text{number of possible outcomes}}$$

must be a number in the range

$$0 \leq p \leq 1$$

Furthermore, the value of p can only be 0 if the number of favorable outcomes is zero (that is, the event is impossible) and can only be 1 if the number of favorable outcomes is equal to the number of possible outcomes (the event is certain). Observe that the closer p is to 1, the greater the proportion of successful outcomes and the more likely the event. The closer p is to 0, the more unlikely the event.

Suppose we know that the probability that a successful event will occur is p. It can be shown that the probability the event will not be successful is

$$q = 1 - p$$

Consider the tree diagram in Figure 10-4. This lists n possible outcomes for an event, m of these are successful, the other $n - m$ are not successful. The probability of a successful event is

$$p = \frac{\text{number of successful outcomes}}{\text{number of possible outcomes}} = \frac{m}{n}$$

and the probability of an unsuccessful event is

$$q = \frac{\text{number of unsuccessful outcomes}}{\text{number of possible outcomes}} = \frac{n - m}{n}$$

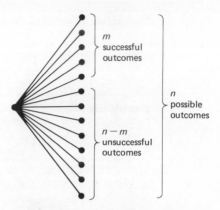

Figure 10-4 If m out of n possible outcomes are successful, then the probability of a successful outcome is $p = m/n$ and the probability of an unsuccessful outcome is $q = ((n - m)/n$. It follows that

$$p + q = \frac{m}{n} + \frac{n - m}{n} = \frac{n}{n} = 1$$

If we add p and q, we get

$$p + q = \frac{m}{n} + \frac{n - m}{n} = \frac{m + (n - m)}{n} = \frac{\cancel{m} + n - \cancel{m}}{n} = \frac{n}{n} = 1$$

Thus,

$$p + q = 1$$

and so

$$q = 1 - p$$

We summarize the above results in the following principle.

Probability Principle 3: If p is the probability that an event will occur and q is the probability that it will not occur, then

$$0 \leq p \leq 1$$

and

$$q = 1 - p$$

If $p = 1$, the event is a certainty. If $p = 0$, it is impossible. The closer p is to 1, the more likely the event, and the closer p is to 0, the less likely.

Example 4. (a) If the probability of rain is $p = 0.40$ (40 percent), then the probability of no rain is

$$q = 1 - p = 1 - 0.40 = 0.60 \text{ (60 percent)}$$

(b) The probability that a 50-year-old man will live during the next year is $p = 0.98768$. The probability that he will die during the year is

$$q = 1 - p = 0.01232$$

All of our work in this section has been with mutually exclusive outcomes. Two outcomes are *mutually exclusive* provided either could occur, but both could not occur simultaneously. If an event could occur by either of two mutually exclusive outcomes, then we can add their probabilities to obtain the probability of the event.

For example, suppose we toss a die. If it turns up either 2 or 5, we win. Since the two events are mutually exclusive (the die cannot turn up 2 and 5 simultaneously), then we can add their probabilities to obtain the probability of winning:

probability of winning = probability of throwing 2 or 5
= (probability of throwing 2) + (probability of throwing 5)
$= \frac{1}{6} + \frac{1}{6} = \frac{2}{6} = \frac{1}{3}$

This result can be stated as the following principle.

Probability Principle 4: Suppose an event can occur in two mutually exclusive ways which have probabilities p_1 and p_2, respectively. The probability that the event will occur is equal to the sum of the two probabilities:

$$p = p_1 + p_2$$

This result can be extended to any number of mutually exclusive outcomes.

When we apply Probability Principle 4, we must be certain that the two events are mutually exclusive. It is a common mistake to assume that unrelated events are mutually exclusive. The following example shows that unrelated events may not be mutually exclusive because they both could occur simultaneously.

Example 5. Henry Henpeckked plans to play golf Saturday unless the weather is bad or his wife wants him to help clean the house. He reads from the weather forecast that the probability of rain on Saturday is 30 percent. He knows from past experience that there is a 40-percent probability that his wife will want him to work.

He estimates that the probability that he will not be able to play golf is

$$p = \text{(probability of rain)} + \text{(probability wife wants him home)}$$
$$= 0.30 + 0.40 = 0.70$$

Thus, he calculates that the probability of playing golf is

$$q = 1 - p = 1 - 0.70 = 0.30$$

Is this a correct estimate?

Solution. Henry's estimate is not correct. According to Probability Principle 4, we only add probabilities when the events are mutually exclusive. Since it is possible that Henry's wife will request his help and it also will rain, the two events are independent, but are not mutually exclusive. We shall consider this example further in Section 10-3.

Divinations

In the ancient world, a multiplicity of conflicting gods was believed to control major events. Actions taken to appease one god would offend others. We believe that chance was used to determine a course of action that would avoid the responsibility of offending the gods who were displeased. Lots were drawn to determine who would go into battle; the guilty were detected by throwing knucklebones, and so on. The oracles of ancient Greece made predictions by tossing dice, by reading the arrangements of entrails, and by other methods involving pure chance.

When Christianity replaced the pagan religions in the Western world, a serious attempt was made to stop the reliance on chance in daily life. Unfortunately, the divinations which were banned by the church reappeared in the guise of devil worship. Even within the church a few methods of divination have persisted. Today some people make hard decisions by opening the Bible randomly, reading the first verse that strikes their eyes, and interpreting it in light of the problem that they need to solve.

EXERCISES

1. (a) State the four probability principles of this section.
   (b) Define the terms "equally likely outcomes" and "mutually exclusive outcomes."
2. Draw tree diagrams to illustrate all of the possible outcomes of the following situations:
   (a) Guessing the sex of an unborn baby;
   (b) Drawing a light bulb from a bag containing four bulbs, one of which is defective (*Hint:* Identify the bulbs with the numbers 1, 2, 3, 4, where 1, 2, 3 are good and 4 is defective);
   (c) Choosing a street at random at an intersection of three streets.
3. Use the tree diagrams in Exercise 2 to find the probabilities:
   (a) The probability that a soon-to-be-born baby will be a girl.
   (b) A light bulb is drawn from a bag containing four bulbs, one of which is defective. Find the probability that the bulb is good.
   (c) Three streets meet at an intersection. Two of the streets are under construction. A street is chosen at random. What is the probability that the street is under construction?
4. An army officer will be either assigned to a station in Germany or to a station in the Philippines or he will remain in the United States. He has been told that the probability of being assigned to the Philippines is $\frac{1}{3}$, and the probability of being assigned to Germany is $\frac{1}{4}$.
   (a) What is the probability of being assigned overseas?

(b) What is the probability of being assigned to a station in the United States?

(c) The officer wants to be assigned to Germany or to the United States. What is the probability that he will be assigned to one of these two countries?

5. A single card is drawn from a standard 52-card deck. What is the probability of each of the following events?

(a) The card is a spade.

(b) The card is not a spade.

(c) The card is either a heart or a diamond.

(d) The card is a king.

(e) The card is either the jack of hearts or the king of diamonds.

6. A student calculated from the mortality tables (Table 10-2) that the probability that a person aged 20 would live to age 30 is 0.97176. He then calculated that the probability that a person aged 30 would live to age 40 is 0.95537. He concluded that the probability that a person aged 20 would live to age 40 is $0.97176 + 0.95537 = 1.92713$. What, if anything, is wrong with his argument?

7. A student, attempting to find the probability of throwing a 7 with a pair of dice, made the following argument: There are 11 possible outcomes of a throw of 2 dice. We can throw 2, 3, 4, 5, 6, 7, 8, 9, 10, 11, or 12. Since only one of these outcomes is 7, then the probability of throwing 7 is

$$p = \frac{\text{number of favorable outcomes}}{\text{number of possible outcomes}} = \frac{1}{11}$$

What, if anything, is wrong with his argument?

10-3 PROBABILITIES OF INDEPENDENT EVENTS

Probability theory deals with general events, never with specific ones. Thus, there is no way to determine if *heads* will actually occur on the next toss of a coin, or to determine which automobile drivers will be killed in accidents tomorrow. We can, however, use probability theory to predict the approximate number of *heads* that will occur in 1000 tosses of a coin, or the approximate number of drivers who will be killed on the highways tomorrow.

The laws of probability have been called a kind of faith. They are unprovable on the one hand, but immutable on the other.

As the reader has noticed, we dealt with only the simplest probability problems in Section 10-2. In order to solve more complicated problems, we need to discuss the fifth of the probability principles formulated by Pascal and Fermat. This last principle is the most powerful of the five that we consider. With its use we can solve a large number of relatively complicated problems. It concerns the probability that two independent events will both occur, two events being independent provided neither has any influence on the other.

Before we make a formal statement of the principle, we consider a simple example.

Example 1. A game consists of flipping a coin and rolling a single die. You win if the coin is *heads* and the die is 6. What is the probability of winning?

Solution. The tree diagram in Figure 10-5 pictures the situation. This diagram is more complicated than the ones in the preceding section, because it involves

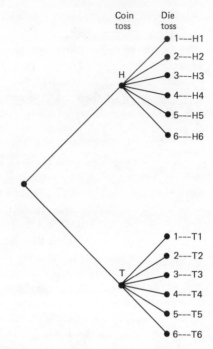

Figure 10-5 The possible outcomes of a toss of
a coin followed by a toss of a die (Example 1).

When we say that the probability of a coin toss resulting in *heads* is $\frac{1}{2}$, we do not mean that *heads* will occur exactly one-half the time. We mean, instead, that over a large number of tosses we expect to obtain *heads* about one-half of the time.

The results of previous experiments do not influence the outcomes of succeeding ones. Even if *heads* have occurred 50 times in a row, the probability of *heads* on the next toss is $\frac{1}{2}$. A number of fortunes have been lost by not understanding this principle.

two events. We assume that the coin is flipped first, then the die is rolled. By tracing from the starting point along the various branches of the "tree," we can list all possible outcomes of the game. For example, if we trace along the top line of the diagram, we have the outcome in which the coin toss is *heads*, and the die toss is 1. (This outcome is abbreviated as H1.) If we trace along the bottom lines, we have the outcome *tails* and 6 (abbreviated as T6).

The tree diagram in Figure 10-5 lists the 12 equally likely possible outcomes. Only one of these is H6. Thus, the probability of winning the game is

$$p = \frac{\text{number of favorable outcomes}}{\text{number of possible outcomes}} = \frac{1}{12}$$

Observe that this probability is the product of the individual probabilities of the two events. The probability of obtaining *heads* on the coin toss is $\frac{1}{2}$; the probability of a 6 on the die toss is $\frac{1}{6}$. The probability of both events happening is

$$p = \tfrac{1}{2} \cdot \tfrac{1}{6} = \tfrac{1}{12}$$

If we study the tree diagram in Figure 10-5, we can see why the product of the individual probabilities is involved. The probability of $\frac{1}{2}$ for *heads* is related to *two* main branches corresponding to the two outcomes of the coin toss. The probability of $\frac{1}{6}$ for the die toss is related to *six* small branches on each main branch. Thus, the probabilities of $\frac{1}{2}$ and $\frac{1}{6}$ lead to 12 small branches on the "tree," which, in turn, causes a probability of $\frac{1}{12}$ for each outcome.

We now are ready to state the last of the probability principles that we shall consider.

> *Probability Principle 5:* Suppose that two events are completely independent of each other. Let p_1 be the probability that the first event will occur and p_2 the probability that the second will occur. The probability that both events will occur is the product
>
> $$p = p_1 p_2$$
>
> This result can be extended to any number of independent events.

Example 2. A man and his wife plan to have three children.
  (a) What is the probability that all will be girls?
  (b) What is the probability that two will be boys and one a girl?
  (c) What is the probability that two will be of the same sex and one of the opposite sex?
(Assume that it is equally likely that a child will be a boy or a girl.)

Solution. (a) The probability that the first child will be a girl is $p_1 = \frac{1}{2}$. After the first has been born, the probability that the second will be a girl is $p_2 = \frac{1}{2}$. After the second child has been born, the probability that the third will be a girl is $p_3 = \frac{1}{2}$. Since these three events are independent, the probability that all three will be girls is the product of the probabilities:

$$p = p_1 p_2 p_3 = \tfrac{1}{2} \cdot \tfrac{1}{2} \cdot \tfrac{1}{2} = \tfrac{1}{8}$$

Thus, in about one-eighth of all such families, we find that all three children are girls.

The tree diagram in Figure 10-6 pictures the situation. Observe that there are eight equally likely outcomes and that only one yields a family of three girls. Thus, again we see that the probability of three girls is

$$p = \frac{\text{number of favorable outcomes}}{\text{number of possible outcomes}} = \frac{1}{8}$$

(b) The probability that two of the children will be boys and one will be a girl can be found from the tree diagram in Figure 10-6. Observe that three of the possible outcomes yield families of two boys and one girl (GBB, BGB, BBG). Thus, the probability of two boys and one girl is

$$p = \frac{\text{number of favorable outcomes}}{\text{number of possible outcomes}} = \frac{3}{8}$$

We would expect a three-child family to consist of two boys and one girl in three-eighths of all such families.

(c) The probability that two of the children will be of one sex and the other

FAMILY PLANNING

The probability that a three-child family will consist of three girls is $\frac{1}{8}$. Many a couple, having two girls and wanting a boy, will decide to have a third child, reasoning that the probability of the third child being a boy must be $\frac{7}{8}$.

The parents' reasoning is completely wrong. Once it is known that two of the children are girls, then most of the possible outcomes used in establishing the $\frac{1}{8}$ figure can be ruled out. The third child is an event that is completely independent of the other two events. Thus the probability that it will be a boy is $\frac{1}{2}$.

First child ↓ Second child ↓ Third child ↓

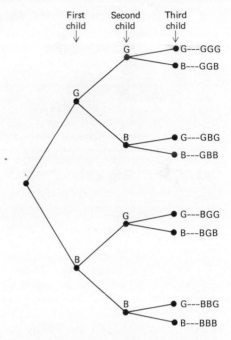

Figure 10-6 The possible distributions of children in a three-child family (Example 2).

of the opposite sex can be found in several different ways. If we use the tree diagram in Figure 10-6, we see that there are six ways we can have such a distribution: GGB, GBG, GBB, BGG, BGB, BBG. Thus, the probability of such a family is

$$p = \frac{\text{number of favorable outcomes}}{\text{number of possible outcomes}} = \frac{6}{8} = \frac{3}{4}$$

A second method allows us to use the result of part (b) to solve this problem. We found that the probability of two boys and a girl is $\frac{3}{8}$. Similarly, the probability of two girls and a boy is $\frac{3}{8}$. Since these events are mutually exclusive, then the probability of two children of one sex and one of the opposite sex

= (probability of two boys and one girl) + (probability of two girls and one boy)
$= \frac{3}{8} + \frac{3}{8} = \frac{6}{8} = \frac{3}{4}$

We would expect 75 percent of all three-child families to have two children of one sex and one of the opposite sex.

We now return to Example 5 of Section 10-2.

Example 3. Henry Henpeckked plans to play golf on Saturday unless the weather is bad or his wife wants him to clean the house. He knows that the probability of bad weather is 0.30 and the probability that his wife will want him to clean the house is 0.40. What is the probability that he will be able to play golf?

Solution. We first find the probability that both the weather will be good and Henry's wife will want him to clean the house. This probability, when added to the probability of rain, will yield the probability that he will not be able to play golf.

Assume that his wife's request will be independent of the weather. The probability that the weather will be good is $1 - 0.30 = 0.70$. The probability that his wife will ask him to clean the house is 0.40. The probability that both events will occur is the product

$$p = (0.70)(0.40) = 0.28$$

The probability that Henry will be forced to remain at home is

probability of staying home
  = (probability of rain) + (probability of good weather, but wife wants him home)
  = $0.30 + 0.28 = 0.58$

The probability that Henry will be able to play golf is

probability of playing golf = $1 - $ (probability of staying home)
                            = $1 - 0.58 = 0.42$

Thus, Henry has a 42-percent chance of playing golf.

If a very large number of possible outcomes exist for a combination of operations, then a tree diagram may not be feasible. In that case, we often represent the outcome as the various parts of a rectangular box as in Figure 10-7.

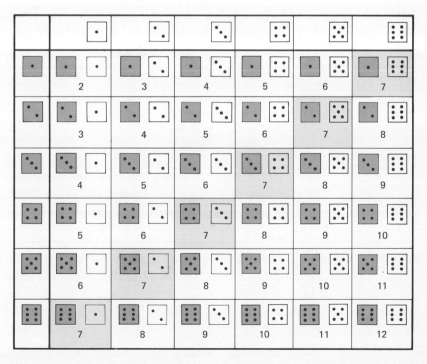

Figure 10-7 The 36 equally likely outcomes for a toss of two dice (Example 6).

Figure 10-7 pictures the possible outcomes of a toss of two dice. In order to distinguish between them, one of the dice is black the other is white. The outcomes of the black die are listed vertically, the outcomes of the white die horizontally. Since most problems with dice involve the sum of the numbers on the two dice, then this sum is listed in each compartment of the box. For example, a black 5–white 2 combination has a total value of 7.

Example 4.   What is the probability of throwing a 7 with a pair of dice?

Solution.   Observe that Figure 10-7 shows 36 possible outcomes of a throw of two dice. These 36 outcomes are equally likely. There is no reason to favor a black 4–white 3 combination over a white 4–black 3 combination, for example. Thus, the probability of any given combination is $\frac{1}{36}$.

Six of the entries in the table have a total value of 7. These are B1–W6, B2–W5, B3–W4, B4–W3, B5–W2, and B6–W1. Thus, the probability of a 7 is

$$p = \frac{\text{number of favorable outcomes}}{\text{number of possible outcomes}} = \frac{6}{36} = \frac{1}{6}$$

We can expect to throw a 7 one-sixth of the time.

EXERCISES

1. Draw tree diagrams to illustrate all possible outcomes of the following events:
   (a)   the tossing of four coins (assume the coins are tossed one at a time);
   (b)   the tossing of a die followed by the tossing of a coin;
   (c)   having two children;
   (d)   having four children, the first being a boy.
2. A coin is tossed twice. What is the probability of the following events? (Illustrate with a tree diagram.)
   (a)   both *heads*
   (b)   both *tails*
   (c)   one *heads*, one *tails*
3. A two-person game is played by tossing two coins. If they match (both *heads* or both *tails*), then one player wins. If they do not match, the other one wins. What is the probability that the coins will match?
4. Birthdays seem to be distributed evenly in the population. What is the probability of each of the following events? (Ignore the effect of leap years. Assume that each year has 365 days.)
   (a)   A man, chosen at random, has his birthday on December 25.
   (b)   A man and his wife, chosen at random, both have their birthdays on December 25.
   (c)   A man and his wife, chosen at random, have their birthdays on the same day.
5. A die is tossed three times.
   (a)   What is the probability that all three numbers are less than or equal to 3?

(b)   What is the probability that at least one of the three numbers is greater than 3. [*Hint:* First work part (a).]

6. Use Figure 10-7 to calculate the probabilities of throwing the following numbers with two dice.

(a)   11   (d)   5
(b)   12   (e)   10
(c)   6   (f)   any double (1–1, 2–2, 3–3, and so on)

7. A game consists of rolling two dice. If the sum is 2, 3, 4, 5, 10, 11, or 12, player A wins. If it is 6, 7, 8, or 9, player B wins. Would you rather be player A or B? Explain.

8. Approximately 1 percent of the people in the United States are named Smith. Two persons are chosen at random. Calculate the following probabilities:

(a)   Both are named Smith.
(b)   Neither is named Smith.
(c)   At least one is named Smith.

Explain why the probabilities in (a), (b), and (c) sum to more than 1.

9. The probability that a person chosen at random will die in a certain type of accident is 1/100,000. What is the probability that a man and his brother will both die in accidents of this type? Does the low probability mean that this could never happen?

10-4   SOME PROBABILITY PROBLEMS

Two events may be independent even though the probability of the second event depends on knowing the outcome of the first. To explain this apparent contradiction, recall that events are independent provided neither has an influence on the other. If the first event has occurred before the second and we know its outcome, then we can calculate the probability of the second event using the outcome of the first as one of the conditions for the probability of the second event.

Some gambling games are primarily games of skill rather than luck. Poker, for example, has been described as "a gambling game played with money in which cards are used to help determine the outcome." Professional gamblers who play these games use extensive tables of probabilities to help them decide when to bet and when to drop out.

Other games depend almost entirely on luck. These range from games such as coin tossing and craps, in which the player has almost even chances of winning, to "games" such as slot machines and lotteries, in which the chances are very low of winning as much as bet.

Example 1.   We draw a card from a standard 52-card deck, keep it, and draw again. What is the probability that both cards will be spades?

Solution.   Since there are 13 spades in the deck, the probability that the first will be a spade is

$$p_1 = \tfrac{13}{52} = \tfrac{1}{4}$$

If the first card is a spade, there are 12 spades left in the deck. Thus, the probability that the second card is a spade is $P_2 = \tfrac{12}{51}$. The probability that both cards are spades is

$$p = p_1 p_2 = \tfrac{1}{4} \cdot \tfrac{12}{51} = \tfrac{3}{51} = \tfrac{1}{17}$$

Example 2.   We draw a card from a standard deck, keep it, and draw again. What is the probability that both cards are of the same suit?

Solution. The first probability is that of drawing a card from the deck: $p_1 = 1$. After the first card is drawn, there are 51 cards left, 12 in the same suit as the first card. The probability of drawing one of these cards is

$$p_2 = \tfrac{12}{51} = \tfrac{4}{17}$$

The probability that both cards are in the same suit is

$$p = p_1 p_2 = 1 \cdot \tfrac{4}{17} = \tfrac{4}{17}$$

Probability principle 5 (independent events) is frequently used to calculate probabilities indirectly. It may be very easy to calculate the probability that an event will not occur, then subtract this number from 1 in order to calculate the probability that the event will occur. After the following example, we consider the original problem worked out for the Chevalier de Méré by Pascal and Fermat.

Example 3. Two cards are drawn from a standard deck. What is the probability that at least one card is a spade?

Solution. We first calculate the probability that neither card is a spade. Observe that the problem is essentially the same if the two cards are drawn at the same time or if one is drawn and then the other. In order to keep the discussion simple, we shall discuss the problem as if one card were drawn and then the other.

Let q_1 be the probability that the first card is not a spade. Then

$$q_1 = \tfrac{39}{52} = \tfrac{3}{4}$$

If the first card is not a spade, then the remaining part of the deck contains 13 spades and 38 nonspades. Thus, the probability that the second card is not a spade after the first card is found to be a nonspade is

$$q_2 = \tfrac{38}{51}$$

The probability that both cards are nonspades is

$$q = q_1 q_2 = \tfrac{3}{4} \cdot \tfrac{38}{51} = \tfrac{19}{34}$$

If p is the probability that at least one card is a spade, then

$$p = 1 - q = 1 - \tfrac{19}{34} = \tfrac{34}{34} - \tfrac{19}{34} = \tfrac{15}{34} \approx 0.44$$

Example 4. (a) What is the probability of rolling at least one 6 in four tosses of a die?

(b) What is the probability of rolling at least one double 6 in 24 tosses of a pair of dice?

Solution. (a) It is much easier to compute the probability that we shall not obtain a 6 on four rolls of a die. The probability that we shall not obtain a 6 on the first roll is $\tfrac{5}{6}$. After the first roll, the probability that we shall not obtain a 6 on the second roll is $\tfrac{5}{6}$, and so on, for the other two rolls. The probability that none

EXPECTATION

The mathematical expectation of a game furnishes us with the best way of evaluating its fairness. To calculate a player's expectation, we multiply the probability of each possible outcome by the financial effect of that outcome on the player. The sum of these products is the expectation.

If a player bets $1.00 that *heads* will be thrown on a toss of a coin, then he wins $1.00 if *heads* is thrown (probability $\tfrac{1}{2}$) and loses $1.00 if *tails* is thrown (probability $\tfrac{1}{2}$). His expectation is

$$E = \underbrace{\tfrac{1}{2}}_{\substack{\text{Probability} \\ \text{of } heads}} \cdot \underbrace{(1.00)}_{\substack{\text{Effect of} \\ heads}} +$$

$$+ \underbrace{\tfrac{1}{2}}_{\substack{\text{Probability} \\ \text{of } tails}} \cdot \underbrace{(-1.00)}_{\substack{\text{Effect of} \\ tails}} = 0$$

The value of the expectation is the average amount the player should expect to win or lose each time he plays. The expectation of zero in the coin-toss game shows that it is completely fair.

of the rolls is a 6 is

$$q = \frac{5}{6} \cdot \frac{5}{6} \cdot \frac{5}{6} \cdot \frac{5}{6} = \frac{5^4}{6^4} = \frac{625}{1296} \approx 0.4823$$

The probability that at least one 6 is rolled is

$$p = 1 - q \approx 1 - 0.4823 = 0.5177$$

Observe that this probability slightly favors the throwing of at least one 6.

(b) If we refer to Figure 10-7, we see that the probability of throwing a double 6 on one throw of a pair of dice is $\frac{1}{36}$. Thus, the probability of not throwing a double 6 on any given throw is $\frac{35}{36}$. By an argument similar to the one in (a), the probability of not throwing a double 6 in 24 tosses of the dice is

$$q = (\tfrac{35}{36})^{24}$$

This number can be computed (using the electronic computer if necessary) as

$$q = (\tfrac{35}{36})^{24} \approx 0.5086$$

The probability of throwing at least one double 6 is

$$p = 1 - q \approx 1 - 0.5086 \approx 0.4914$$

This probability slightly favors not throwing a double 6.

Thus far we have concentrated on probabilities involving cards and dice because these probabilities illustrate the basic principles and are relatively easy to compute. Many meaningful applications, however, involve problems in genetics, biology, business and government planning.

Example 5. A State Department spokesman once claimed that the probability of the United States being drawn into a major war in any given year is 4 percent. He then drew the conclusion that it was certain that the United States would be involved in at least one major war every 25 years. Is this a valid conclusion from his assumption?

Solution. We first calculate the probability q that the United States will not be involved in a major war over a 25-year period. Then $p = 1 - q$ is the probability of at least one war.

If the probability of war in any one year is 0.04, then the probability of no war is 0.96. Thus, the probability of no war the first year is 0.96.
The probability of no war for two successive years is

$$(0.96)(0.96) = (0.96)^2$$

The probability of no war for three years in succession is

$$(0.96)(0.96)(0.96) = (0.96)^3$$

If we continue this line of argument, we find that the probability of no war for 25 successive years is

$$q = (0.96)^{25}$$

This number can be calculated on the computer as

$$q \approx 0.3604$$

Thus, the probability of at least one war in the 25-year period is

$$p = 1 - q \approx 1 - 0.3604 \approx 0.6396$$

While this probability is high, it is not even close to the figure $p = 1$, which is needed for certainty.

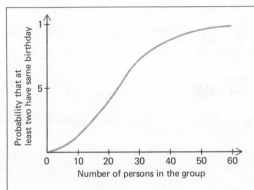

Chart showing the probability that at least two persons in a group will have the same birthday.

The Birthday Problem

A group of students, chosen at random, are seated in a room. *What is the probability that at least two of them have the same birthday?* Obviously, the answer depends on the number of students. If only two or three are in the group, then it is highly unlikely that any two have the same birthday, while if more than 365 are in the group, it is certain that at least two of them have the same birthday. We would expect that if a large number of students are present, then the probability is high that at least two will have the same birthday.

A related, and more interesting, problem is the following: *How many students must we have in the room before it is likely that at least two have*

the same birthday? The surprising answer to this problem is that only a relatively few persons need to be present. If 25 persons are present, then the probability is about 0.5. If 30 persons are present, then the probability is almost 0.7. If 50 persons are present, the probability is almost 1.

To see why this result holds, we consider the case where only two or three persons are in the room. For simplicity, we ignore leap years and assume that every year has 365 days.

If only two persons are in the room, the probability that their birthdays are *different* is

$$q = 1 \cdot \frac{364}{365}$$

We obtain this result by first lining the persons up in order and noting that after the first person is chosen, there are 364 possible days in which the second person could have a birthday different from the birthday of the first person.

If three persons are in the room, a similar argument shows that the probability that their birthdays are different is

$$q = 1 \cdot \frac{364}{365} \cdot \frac{363}{365}$$

If 10 persons are in the room, then the probability that all of the birthdays are different is

$$q = 1 \cdot \frac{364}{365} \cdot \frac{363}{365} \cdot \frac{362}{365} \cdot \frac{361}{365} \cdot$$

$$\cdot \frac{360}{365} \cdot \frac{359}{365} \cdot \frac{358}{365} \cdot \frac{357}{365} \cdot \frac{356}{365}$$

$$\approx 0.88$$

Thus, in this case the probability that at least two birthdays are the same is

$$p = 1 - q \approx 1 - 0.88 \approx 0.12$$

A similar argument shows that if 30 persons are in the room, then the probability that all of the birthdays are different is

$$q = 1 \cdot \underbrace{\frac{364}{365} \cdot \frac{363}{365} \cdot \ldots \cdot \frac{336}{365}}_{(30 \text{ terms})}$$

$$\approx 0.29$$

and the probability that at least two birthdays are the same is

$$p = 1 - q \approx 1 - 0.29 \approx 0.71$$

Other values are given below:

Number of Persons	Probability That at Least Two Have the Same Birthday
10	0.12
25	0.57
30	0.71
50	0.97
60	0.99

EXERCISES

1. A card is drawn from a standard 52-card deck, replaced, and a new card is drawn. What is the probability of the following events?
   (a) Both cards are black (clubs or spades).
   (b) The first card is black; the second is red.
   (c) One card is black; the other is red.
   (d) The two cards are the ace and king of spades.
   (e) The ace of spades is drawn twice.
   (f) The same card is drawn twice.
2. Work Exercise 1 on the assumption that the first card is not replaced.
3. The number 5 plays a special role in the game of *parchesi*. Two dice are thrown and the number on either die or their sum can be used. Thus, on a throw of 5–2, the 5 can be used. What is the probability of obtaining a 5 in parchesi? (Use Fig. 10-7.)
4. Two dice are rolled in the game of *craps*. If the roll is a 7 or 11, the player wins. If it is 2, 3, or 12, he loses. If any other number is rolled, the player continues to roll until he rolls a 7 (loses) or the same number as on the first roll (wins).
   (a) What is the probability of winning on the first roll?
   (b) What is the probability of losing on the first roll?

    (c)   What is the probability of losing on the second roll?

    (d)   What is the probability of losing on one of the first two rolls? (Use Fig. 10-7.)

5. A computerized war game is based on two assumptions:
    (1) Only one-half of the missiles penetrate enemy defenses;
    (2) Missiles are so inaccurate that only one-third of those that get through enemy defenses actually reach the proper targets.

    One player has just fired 300 missiles at enemy targets. How many of these are expected to hit the proper targets?

6. Most wild animals that mate in captivity have a very low chance of reproducing. A type of wild fox reproduces only 1 litter for every 20 matings in captivity. Zoologists have arranged for five pairs of foxes to mate.
    (a)   What is the probability that no litters will be produced?
    (b)   What is the probability that at least one litter will be produced?

10-5* APPLICATIONS TO GENETICS

A major application of probability theory is in the field of genetics—the study of inherited traits. All such characteristics are determined by pairs of *genes* received from the parents. Most of these traits are determined by several genes; a few, however, are determined by a single pair of genes, one from each parent. In this section we shall consider only those characteristics that are caused by a single pair of genes.

The majority of genes have two basic forms of intensity. Either they are *dominant* or *recessive*. If a person has a dominant gene for a trait, then that trait will appear regardless of the gene that it is paired with. Recessive genes must be paired with identical genes before they have any effect on the individual.

Eye color in humans is usually caused by a single pair of genes. Brown eyes are dominant, and blue eyes are recessive. Let B denote a gene for brown eyes and b a gene for blue eyes. If an individual has the BB combination or the Bb combination of genes, his eyes will be brown. If he has the bb combination, they will be blue.

When a child is conceived, it receives one gene from each parent for each inherited trait. If the father has the Bb combination for brown eyes, then the child could receive either the B gene or the b gene. These two genes occur in the sperm cells with equal frequency. Thus, the probability of receiving a B gene from the father is $\frac{1}{2}$, and the probability of receiving a b gene is $\frac{1}{2}$. The probabilities of receiving particular genes from the mother are likewise equal to $\frac{1}{2}$.

Example 1. A man and his wife both have the Bb genetic endowment for brown eyes. What is the probability that a child will have brown eyes? Blue eyes?

Solution.   The four equally likely genetic combinations are pictured in Figure 10-8. There are four possible combinations of genes (in each combination the gene from the father is shown first, the one from the mother is shown second): BB, Bb, bB, and bb. Three of these four combinations result in brown eyes; one results in blue eyes. Thus, the probability of a child having brown eyes is $\frac{3}{4}$. The probability of blue eyes is $\frac{1}{4}$.

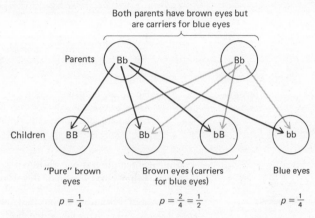

Figure 10-8 If both parents are brown-eyed but are carriers for blue eyes, then, on the average, one-fourth of their children will have "pure" brown eyes, one-half will be brown-eyed carriers for blue eyes, and one-fourth will have blue eyes (Example 1).

If an individual has one dominant and one recessive gene, he possesses the dominant characteristic and is a *carrier* for the recessive characteristic. In Example 1 both of the brown-eyed parents are carriers for blue eye color. On the average one-fourth of their children will have "pure" brown eyes (BB combination); one-half will have brown eyes but will be carriers for blue eyes (Bb combination); and one-fourth will have blue eyes (bb combination).

Example 2.   If one parent has the BB combination for brown eyes and the other has blue eyes, then all children will be brown-eyed carriers of the blue-eyed trait (Fig. 10-9).

Example 3.   If one parent has the Bb combination for brown eyes and the other has the bb combination for blue eyes, then on the average one-half of the children will have blue eyes and one-half will have brown eyes. The brown-eyed children all are carriers for blue eyes (Fig. 10-10).

We have used eye color—a harmless trait—to illustrate the basic principles of genetics. Other genes, which obey the same laws of genetics, can cause serious abnormalities. Fortunately, most of these "bad" genes are recessive.

Example 4.   Hereditary muscular atrophy is a condition, caused by a single pair of recessive genes, in which the muscles gradually waste away until the victim is completely helpless. Death usually results in a few years.

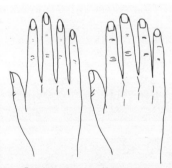

Female hand   Male hand
Long index finger   Short index finger

Many inherited traits are sex limited. The genes causing these traits are dominant in one sex and recessive in the other. Two harmless genes of this type in humans are the one that causes pattern baldness and the one that determines the length of the index finger.

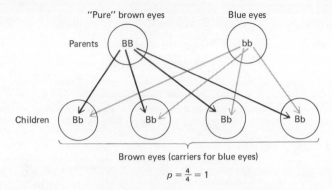

Figure 10-9 If one parent has "pure" brown eyes and the other has blue eyes, then all of their children will have brown eyes and be carriers for blue eyes (Example 2).

A normal couple has one child with muscular atrophy. What is the probability that their other child also will develop the condition?

Solution. Let N denote a "normal" dominant gene and m denote a recessive gene for muscular atrophy. Since both parents are normal and a child has the condition, then both parents must be carriers.

Once we know that the parents have the Nm combination, we can compute that the probability of any child having muscular atrophy is $\frac{1}{4}$. The probability that he is a carrier is $\frac{1}{2}$. The probability that he is completely free of the gene is $\frac{1}{4}$ (Fig. 10-11).

Example 5. The couple in Example 4 plans to have three additional children.

(a) What is the probability that all three will be completely normal (NN combination)?

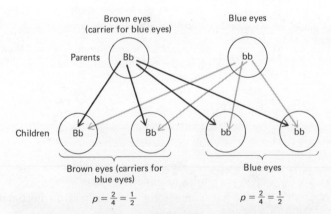

Figure 10-10 If one parent has brown eyes but is a carrier for blue eyes and the other has blue eyes, then, on the average, one-half of their children will have blue eyes and one-half will have brown eyes and be carriers for blue eyes (Example 3).

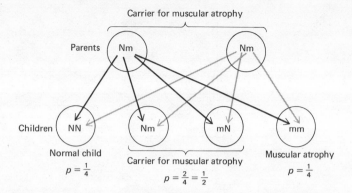

Figure 10-11 If both parents are carriers for hereditary muscular atrophy, then the probability that a child will develop the condition is $\frac{1}{4}$, the probability that it will be completely normal is $\frac{1}{4}$, and the probability that it will be a carrier for the trait is $\frac{1}{2}$ (Example 4).

(b) What is the probability that all three will appear normal, but that at least one will be a carrier for muscular atrophy (Nm combination)?

(c) What is the probability that at least one of the three children will have muscular atrophy (mm combination)?

Solution. (a) The probability that any given child will be completely normal is $\frac{1}{4}$. Since this event must occur independently three times in succession in order to have three completely normal children, then the probability of that event is

$$\tfrac{1}{4} \cdot \tfrac{1}{4} \cdot \tfrac{1}{4} = (\tfrac{1}{4})^3 = \tfrac{1}{64} \approx 0.016$$

Thus, in only 1 out of every 64 such families will all three of the children be completely normal.

(b) The probability that a child will *appear* normal is $\frac{3}{4}$. The probability that all three children will appear normal is

$$(\tfrac{3}{4})^3 = \tfrac{27}{64} \approx 0.42$$

Many of the very serious congenital defects, such as mongolism, are chromosome abnormalities not genetic defects. Genes are strung together like beads on long threads called chromosomes. At conception these chromosomes pair up, one member of each pair from each parent, thereby matching the genes. Occasionally an extra chromosome will be acquired. This usually results in a physical abnormality, coupled with severe mental retardation, and sterility.

We found in (a) that in only 1 out of 64 families will all three children be normal. Also, 27 out of 64 families have three children that appear normal. Thus, in the other 26 families at least one child is a carrier for the condition. It follows that the probability that all three children will appear normal, but at least one will be a carrier, is

$$\tfrac{26}{64} \approx 0.41$$

(c) Since the probability is only $\frac{27}{64}$ that none of the three children will develop muscular atrophy, then the probability is

$$1 - \tfrac{27}{64} = \tfrac{64}{64} - \tfrac{27}{64} = \tfrac{37}{64} \approx 0.58$$

that at least one of the three children will develop the condition.

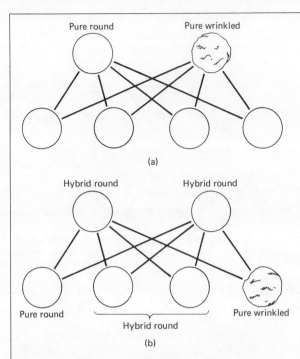

(a) First generation. All offsprings are hybrid round; (b) Second generation. Three-quarters of the offsprings are round; one-quarter are wrinkled.

GENES WITH VARIABLE EFFECTS

Many genes are neither pure dominant nor pure recessive. Their effects are variable. For example, if N and M denote two variable genes that pair together, then an NN combination has one trait; an MM combination has another trait; while an MN combination combines the two traits.

Mendel was fortunate in choosing traits to study that were caused by single pairs of genes, each gene being either dominant or recessive. If he had chosen other traits, the analysis would have been so complicated that he probably would have failed.

Gregor Mendel

Gregor Mendel (1822–1884), an obscure Bavarian monk, was the first person to understand the genetic basis for inheritance. Mendel was aware of the difficulties that had hampered biologists in their attempts to solve this difficult problem; so he made a preliminary study to isolate problems in technique that had contributed to the previous failures. He decided that earlier researchers had made a mistake in concentrating on the whole organism—it was much too complicated; so he concentrated on one trait at a time. He also decided that a simple organism should be studied—one that had a number of pure strains, that could be fertilized artificially, and that could be reproduced through several generations in a relatively short period of time. His choice was the common garden pea.

Mendel selected peas from pure breeding strains which exhibited seven different traits, including

wrinkled or smooth seeds,
brown or white seeds,
green or yellow seedpods.

Plants from the various strains were cross-bred, and their seeds were studied and counted. When plants from wrinkled seeds were cross-bred with plants from smooth seeds, all of the hybrid plants had smooth seeds. When the hybrid seeds were planted and cross-bred with one another, the seeds from the resulting plants were one-fourth wrinkled and three-fourths smooth. When these seeds were planted and cross-bred, the plants that came from parent plants with smooth seeds had three-fourths smooth seeds and one-fourth wrinkled seeds. The plants that came from parent plants with wrinkled seeds had only wrinkled seeds.

After a program of extensive cross-breeding in which all seven of the traits were studied for several generations, Mendel formulated the basic laws of inheritance. Each trait is determined by a single pair of *genes*, one received from each parent. Certain of the genes, such as the one for smooth seeds, are *dominant*. If such a dominant gene is present, it determines the trait regardless of the gene it is paired with. Other genes, such as the one for wrinkled seeds, are *recessive*. Both genes must be of the same type before recessive genes can determine the trait. All of this analysis was based on the mathematical structure of the results—the microscopes of the day were not capable of distinguishing objects as small as genes.

Mendel presented his conclusions to a small local German scientific society, where they were completely ignored. Not one person who heard the report or who saw the resulting paper realized that Mendel had discovered the single most important law of nature. Thirty-four years later (16 years after Mendel's death), his work was rediscovered and received the attention that it was due.

Eugenics

Eugenics is the science devoted to the improvement of the human race by selective breeding. One of the aims of eugenics is the elimination of hereditary defects by not allowing persons with the defects or known carriers of the defects to mate. Dominant defects can be eradicated from the population in only a few generations by this restriction, but recessive defects are much more difficult to eliminate. It has been estimated that thousands of generations of selective breeding would be necessary before most recessive defects could be eradicated. The major problem is that carriers for a trait cannot usually be spotted until they have already had a defective child. The following example indicates why the process would be so slow.

Example 6. One out of every thousand persons is a carrier for a certain genetic defect.

(a) What is the probability that a known carrier of the defect will marry a carrier?

(b) In what proportion of families in the population are both parents carriers?

Solution. (a) If we assume that mates are selected in a random manner (with respect to the trait), then the probability that a man who is a known carrier will marry a woman who is a carrier is

$$1 \cdot \frac{1}{1,000} = \frac{1}{1,000}$$

Thus, one out of every thousand carriers will marry a carrier.

(b) The probability that a man and wife selected at random are both carriers is

$$\frac{1}{1,000} \cdot \frac{1}{1,000} = \frac{1}{1,000,000}$$

Thus, only one out of every million families is likely to produce offspring with the trait.

Consider the problem of ridding the population of the hereditary defect described in Example 6. Out of 20 million couples of the childbearing age, only about 20 will produce defective offspring. Even if these couples could be induced to stop having children, there are approximately 20,000 other families in which a normal person is married to a carrier. (Recall that one out of every thousand persons is a carrier.) Since one-half of the children produced in these 20,000 families will be carriers, then the total proportion of carriers in the population will not change appreciably.

We are fortunate that most hereditary defects are recessive, because there are very many such defects in the general population. Some of these, such as muscular atrophy, cause severe damage to the victims and major heartache to their families. Others, such as nearsightedness, are comparatively mild and can be corrected by medical technology. It has been estimated that the average person is the carrier for at least eight recessive genetic defects.

The prevalence of "bad" genes in the population is responsible for the prohibitions against incest. Almost every society forbids marriage between close relatives. It is easy to see why such marriages frequently produce defective children. If a man has a particular "bad" gene, then the probability that a first cousin also has it is $\frac{1}{8}$. If he and the cousin both have the gene, then the probability of their having a defective child is $\frac{1}{4}$. Thus, if first cousins mate, the probability that a child will have any given defect for which the father is a carrier is

$$\frac{1}{8} \cdot \frac{1}{4} = \frac{1}{32}$$

If we assume that the average person has eight "bad" genes, then it can be computed that the probability of a defective child is

$$1 - (\tfrac{31}{32})^8 \approx 1 - 0.78 \approx 0.22$$

when first cousins marry.

Natural Selection and Eugenics

The theory of natural selection, developed by *Charles Darwin* (1809–1882), holds that the unfit in nature are constantly being eliminated by environmental factors. In most animal populations, this occurs among the very young so that individuals with genetic defects tend to be removed before they can breed and pass on the defects to their offsprings.

Man is the one animal to reverse the effects of natural selection. Every effort is made to keep a child alive even at the risk of allowing him to pass serious genetic defects on to his offsprings. This is good for the immediate generation, but causes monu-mental problems for future ones.

The science of eugenics, which is based on the use of genetics to elimi-nate defects from human beings, was founded by *Sir Francis Galton*, a cou-sin of Darwin. Proponents of eugenics claim that man has a much higher percentage of deleterious genes than other animals. It is estimated that the average person has four lethal reces-sive genes (causing death if paired with similar genes) and eight reces-sive genes that cause defects. Eugeni-cists claim that unless we stop in-creasing the average number of genes that cause defects, the day will come when a constant high level of medical technology will be needed to keep everyone alive.

EXERCISES

1. Albinism, the absence of color in the skin, hair, and eyes, is a recessive trait. George, who is normal and was born to normal parents, has an albino brother.
   (a)   What is the probability that George's parents are both carriers for albinism?
   (b)   What is the probability that George is a carrier for albinism?
2. George, who has an albino brother (see Exercise 1), plans to marry Mary, who also is normal but has an albino brother. What is the probability that they would have an albino child? (*Hint:* First find the probability that both George and Mary are carriers for albinism.)
3. Mendel's original experiments were with varieties of peas. One experi-ment involved the crossing of round peas with deeply wrinkled peas. The first-generation peas were all round. When these hybrid peas were planted and the second-generation peas were harvested, he obtained 5474 round peas and 1850 wrinkled peas.
   (a)   Which gene is dominant and which is recessive?
   (b)   Suppose the second-generation wrinkled peas were planted. What would you predict for the harvest?
4. The Rh factor in human blood is caused by a dominant gene R. If the gene is present, then the blood is Rh+; if the gene is absent, the blood is Rh−. A man and his wife are both Rh+ but are carriers for Rh−. What is the probability that their child will be Rh+? Rh−?
5. Blood factors are often used in helping to decide questions of disputed parentage. A college woman has brought suit against a college man charging that he is the father of her child. The man and woman both have Rh− blood. The child has Rh+. Is the man the father of the child? (See Exercise 4.)

10-6   ELEMENTARY COUNTING PRINCIPLES; PERMUTATIONS

At one point or another every probability problem requires us to count the number of possible outcomes or the number of favorable outcomes of an event. In our previous examples, these numbers were small and could actually be counted. Some of the examples that follow have very large numbers of possible outcomes—so large that weeks or months might be required to count them. In this section we consider some methods that allow us to calculate the number of possible outcomes rather than count them one by one.

The Fundamental Counting Principle

If a coin is tossed three times, there are eight possible outcomes. This is demonstrated in the tree diagram in Figure 10-12. In theory, a similar tree diagram can always be constructed when two or more operations are carried out in sequence. The number of possible outcomes of the first operation determines the number of main branches. The number of possible outcomes of the second operation determines the number of small branches on each of the main branches. Thus, the total number of small branches, which corresponds to the total number of ways the two operations can be carried out, is equal to the *product* of the number of possible outcomes for the first operation and the number of possible outcomes of the second operation.

The Fundamental Counting Principle is closely related to the probability principle that we multiply the probabilities of two independent events to obtain the probability that both will occur. Most textbooks, for example, use the Fundamental Counting Principle to prove the probability principle.

This line of argument can be developed into the following principle.

> *Fundamental Counting Principle:* Suppose an event can occur in m distinct ways and a second event, after the first has been completed, can occur in n distinct ways. Then the total number of ways in which the two events can occur is the product mn.

The Fundamental Counting Principle can be extended to cover any number of events. In each case, we form the product of the number of ways the events can be performed.

Example 1. A building has three entrances and two stairways. In how many possible ways (entrance–stairway) can a person go from the outside of the building up to the second floor?
(a) Use a tree diagram.
(b) Use the Fundamental Counting Principle.

Solution. (a) The three entrances into the building give us three main branches of the tree. The two stairways give two small branches on each of the main branches for a total of six small branches (Fig. 10-13). Thus, there are six ways to go into the building and up to the second floor.

(b) Since there are *three* possible outcomes for the first operation (getting into the building) and *two* possible outcomes for the second operation (getting to the second floor), then there are

$$3 \cdot 2 = 6$$

possible ways to perform the two operations in sequence.

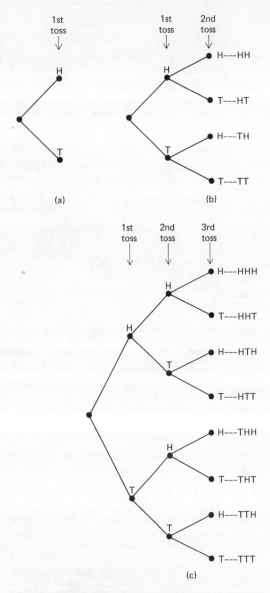

Figure 10-12 Tree diagram illustrating the eight possible outcomes for three tosses of a coin. (a) One toss; (b) two tosses; and (c) three tosses.

The Fundamental Counting Principle can be used when the total number of operations is so large that it would not be possible to actually make a tree diagram.

Example 2. The cryptographic department of the CIA is making a master list of five-letter code words to be used in secret communications. In order to make the code words easily recognizable (and thereby avoid certain common errors by

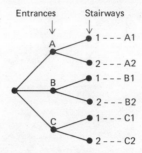

Figure 10-13 Tree diagram showing the six possible routes from the street to the second floor of the building (Example 2).

the operators who transmit the messages), the first letter must be a consonant, the second a vowel (A, E, I, O, U), the third a consonant, the fourth a vowel, and the fifth a consonant. Thus, for example,

$$\text{QAROR, BABAB, and ZUTIN}$$

are proper code words, but AUBOK is not.

How long will the completed list of code words be?

Solution. The number of code words in the completed list can be determined by counting the number of possible ways a code word can be formed. Recall that there are 21 consonants and 5 vowels. A code word can be formed in five stages:

(1) Choose a first letter (consonant): 21 ways
(2) Choose a second letter (vowel): 5 ways
(3) Choose a third letter (consonant): 21 ways
(4) Choose a fourth letter (vowel): 5 ways
(5) Choose a fifth letter (consonant): 21 ways

It follows from the Fundamental Counting Principle that there are

$$21 \cdot 5 \cdot 21 \cdot 5 \cdot 21 = 5^2 \cdot 21^3 = 231{,}525$$

possible ways that a code word can be formed. Thus, the completed list will contain 231,525 code words.

Example 3. The CIA has decided that such an extensive list of code words is not needed. (See Example 2.) They will keep the five-letter code words that do not have duplicated letters, but will delete all others. For example, code words such as QAROR and BABAB will be deleted; code words such as ZUTIN will be retained.

How many code words will be in the new list?

Solution. The problem is easier to solve if we consider it in a slightly different light. Rather than deleting the code words with duplicated letters, we make a new list in which the first, third, and fifth letters are consonants, the second and fourth are vowels, and no letter is repeated. Again there are five stages in the construction of a code word:

(1) Choose a first letter (consonant): 21 ways
(2) Choose a second letter (vowel): 5 ways
(3) Choose a third letter (consonant): 20 ways
(4) Choose a fourth letter (vowel): 4 ways
(5) Choose a fifth letter (consonant): 19 ways

(The third letter must be a consonant different from the first letter. Thus, there are only 20 choices. Similarly, there are only 19 choices for the fifth letter because it must be a consonant different from the ones used for the first and third letters.)

It follows from the Fundamental Counting Principle that there are

$$21 \cdot 5 \cdot 20 \cdot 4 \cdot 19 = 159{,}600$$

code words in the revised list.

Permutations

How many different arrangements can we make with a set of objects? For example, we might have the 26 letters of the alphabet or the four symbols ♠, ♡, ◇, ♣. For simplicity, we first consider the case where we have four symbols. (See Fig. 10-14.) When we form an arrangement of the symbols, we have four basic steps:

(1) Choose the first symbol: 4 ways
(2) Choose the second symbol: 3 ways
(3) Choose the third symbol: 2 ways
(4) Choose the fourth symbol: 1 way

Figure 10-14 The 24 permutations of the four symbols ◇ ♡ ♣ ♠. There are four possible ways the first symbol can be chosen. After the first symbol is selected, there are three possible ways the second symbol can be chosen. After the first two symbols are selected, there are two possible ways the third symbol can be chosen. After the first three symbols are selected, there is only one way the final one can be chosen.

The total number of ways we can arrange the four symbols is

$$4 \cdot 3 \cdot 2 \cdot 1 = 24 \text{ ways}$$

If we consider the problem of arranging the letters of the alphabet, we find that there are 26 basic steps:

(1) Choose the first letter: 26 ways
(2) Choose the second letter: 25 ways
(3) Choose the third letter: 24 ways

. . . .

(24) Choose the twenty-fourth letter: 3 ways
(25) Choose the twenty-fifth letter: 2 ways
(26) Choose the twenty-sixth letter: 1 way

Thus, there are

$$26 \cdot 25 \cdot 24 \cdots 3 \cdot 2 \cdot 1$$

possible arrangements of the 26 letters of the alphabet. The three dots in the expression represent the integers 23, 22, 20, and so on, that are not actually listed.

An arrangement of objects is called a *permutation* of the objects. Thus, we see that there are $4 \cdot 3 \cdot 2 \cdot 1 = 24$ permutations of four objects and $26 \cdot 25 \cdot 24 \cdots 3 \cdot 2 \cdot 1$ permutations of 26 objects.

Factorials

Expressions such as $5 \cdot 4 \cdot 3 \cdot 2 \cdot 1$ and $26 \cdot 25 \cdot 24 \cdots 3 \cdot 2 \cdot 1$ occur with such frequency when we count arrangements of objects that we have invented a special symbol to represent them. The use of the symbol enables us to avoid a great amount of writing.

The symbol $n!$ (read "n-factorial") represents the product of all of the positive integers from 1 to n. Thus,

The number of permutations of a large set of objects is so huge that it is awkward to write and to work with. For this reason we use symbols such as $n!$, P_n, and so on, to represent these numbers until we actually have to work out a numerical answer.

$$1! = 1$$
$$2! = 1 \cdot 2 = 2$$
$$3! = 1 \cdot 2 \cdot 3 = 6$$
$$4! = 1 \cdot 2 \cdot 3 \cdot 4 = 24$$
$$5! = 1 \cdot 2 \cdot 3 \cdot 4 \cdot 5 = 120$$
$$6! = 1 \cdot 2 \cdot 3 \cdot 4 \cdot 5 \cdot 6 = 720$$

and so on.

As we saw above, there are 4! permutations of the set ♠, ♡, ◇, ♣, and 26! permutations of the letters of the alphabet. In general, there are exactly $n!$ permutations of a set of n objects. This can be established by an argument similar to the one we used for the letters of the alphabet. There are n steps involved in finding the number of permutations of n objects.

    (1) Choose the first object: n ways
    (2) Choose the second object: $n - 1$ ways
    (3) Choose the third object: $n - 2$ ways

. . . .

    $(n - 2)$ Choose the $n - $2nd object: 3 ways
    $(n - 1)$ Choose the $n - $1st object: 2 ways
    (n) Choose the nth object: 1 way

Thus, there are

$$n \cdot (n - 1) \cdot (n - 2) \cdots 3 \cdot 2 \cdot 1 = n!$$

permutations of n objects.

Although the first few factorials are relatively small, they get large very rapidly. For example, 26!, the number of permutations of the letters of the alphabet, is a number greater than

$$4 \cdot 10^{26} = 400{,}000{,}000{,}000{,}000{,}000{,}000{,}000{,}000$$

There is one further convention with factorials. It is standard to define 0! to be the number 1. The reason for this definition, which does not seem to be consistent with our other definition and which appears strange when we first encounter it, is that it simplifies a large number of formulas that we shall encounter later.

Thus,

$$0! = 1$$

and, if n is greater than 0,

$$n! = 1 \cdot 2 \cdot 3 \cdots (n - 1) \cdot n,$$

the product of all the positive integers that are greater than or equal to 1 and less than or equal to n.

EXERCISES

1. Calculate the following factorials.
    (a) 3!   (c) 10!   (e) 1!
    (b) 8!   (d) 0!   (f) 9!
2. Calculate the following ratios. (Make the work as easy as possible. Rewrite the factorials as products and cancel before multiplying.)
    (a) 7!/6!   (d) 9!/7!
    (b) 12!/11!   (e) 5!/3!2!
    (c) $(n + 1)!/n!$ (f) 8!/5!

3. Make a flow chart for the operation of calculating factorials.

4. Make a computer program based on your flow chart in Exercise 3. Have the computer ask for N (a nonnegative integer) as INPUT. Print out the value of N! with an appropriate label. Print an error message if N is negative or is not an integer.

5. (a) Make a tree diagram to illustrate the possible outcomes of the tossing of two dice. (*Hint:* Assume the dice are tossed one at a time.)

   (b) Use the Fundamental Counting Principle to count the number of possible outcomes of the tossing of two dice. (*Hint:* Throws such as 5-2, 3-4, and 2-5 should be counted separately, even though they sum to the same number.)

   (c) Use the tree diagram to find the probability that the sum of the numbers on the two dice will be 5.

6. A maze used to test learning abilities of rats in a psychology class has three points where the rats must make a choice of directions. At the beginning a rat has a choice of three directions. After he has chosen one of these and has progressed two feet into the maze, he has a choice of two directions. After he has progressed two feet farther, he has a choice of two other directions.

   (a) Make a tree diagram listing the possible routes of the rat through the maze.

   (b) Use the Fundamental Counting Principle to count the number of paths through the maze.

   (c) Assume that the rat makes a "blind" choice each time. What is the probability that it will choose a specified·path through the maze?

7. A combination lock has 12 numbers on the dial. How many different three-number combinations exist if no number can be repeated in a combination?

8. State automobile license plates have three letters followed by a three-digit number. (The number must be a proper three-digit number. Numbers such as 007 are not allowed.) How many different license plates can be manufactured?

9. How many different five-letter code words can be formed if the first, third, and fifth letters are vowels, the second and fourth are consonants, and no letter can be repeated?

10. A Little League baseball team has 18 players, all of whom can play all positions. In how many ways can nine players be selected for a starting team? (*Hint:* First select a pitcher, then a catcher, and so on.)

10-7 COMBINATIONS AND PERMUTATIONS

In some problems we need permutations made from a few elements of a larger set. For example, we might be interested in the total number of five-letter "words" that could be made from the letters of the alphabet if no letter is repeated. In this example, there would be five basic steps:

(1) Choose the first letter: 26 ways
(2) Choose the second letter: 25 ways
(3) Choose the third letter: 24 ways
(4) Choose the fourth letter: 23 ways
(5) Choose the fifth letter: 22 ways

It follows from the Fundamental Counting Principle that the total number of five-letter permutations that can be formed from the letters of the alphabet is

$$26 \cdot 25 \cdot 24 \cdot 23 \cdot 22 = 7{,}893{,}600$$

This number is denoted by the symbol $P_5{}^{26}$. We can write this product in a more compact form by first multiplying it by $21!/21!$, obtaining

$$P_5{}^{26} = 26 \cdot 25 \cdot 24 \cdot 23 \cdot 22 \cdot \frac{21!}{21!} = \frac{26!}{21!}$$

In general,

$$P_m{}^n = \frac{n!}{(n-m)!}$$

is the number of permutations of m objects that can be formed from a set of n objects.

For example,

$$P_3{}^5 = \text{the number of permutations containing three}$$
$$\text{objects that can be formed from a set of five objects}$$
$$= \frac{5!}{2!} = \frac{5 \cdot 4 \cdot 3 \cdot 2 \cdot 1}{2 \cdot 1} = 5 \cdot 4 \cdot 3 = 60$$

and

$$P_3{}^7 = \text{the number of permutations containing three objects that can be}$$
$$\text{formed from seven objects}$$
$$= \frac{7!}{(7-3)!} = \frac{7!}{4!} = \frac{7 \cdot 6 \cdot 5 \cdot 4 \cdot 3 \cdot 2 \cdot 1}{4 \cdot 3 \cdot 2 \cdot 1} = 7 \cdot 6 \cdot 5 = 210$$

Table III at the end of this book is a table of factorials. This table can be used to compute the value of $P_m{}^n$ when n and m are large. For example,

$$P_5{}^{26} = \frac{26!}{21!}$$

Figure 10-15 The number of permutations of four objects taken two at a time is

$$P_2^4 = 4 \cdot 3 = \frac{4 \cdot 3 \cdot 2 \cdot 1}{2 \cdot 1} = \frac{4!}{2!}$$

From Table III we find that

$$26! \approx (4.03) \cdot 10^{26} \quad \text{and} \quad 21! \approx (5.11) \cdot 10^{19}$$

Thus,

$$P_5^{26} = \frac{26!}{21!} \approx \frac{(4.03) \cdot 10^{26}}{(5.11) \cdot 10^{19}} \approx \frac{(4.03)}{(5.11)} \cdot \frac{10^{26}}{10^{19}} \approx (0.789) \cdot 10^7$$

$$\approx (7.89) \cdot 10^6 \approx 7,890,000$$

Example 1. A seventh-grade home room teacher, finding that none of her students want to be class officers, decides to select the officers by lottery. The names of the 30 students are to be put in a box and 3 names drawn. The first person will be president, the next vice-president, the third secretary-treasurer.

(a) In how many ways can the class offices be assigned?

(b) Billy secretly wants to be an officer, but is embarrassed to say so. What is the probability that he will be president?

Solution. (a) The essential problem is to determine the number of permutations of 30 objects taken 3 at a time. This number is

$$P_3^{30} = \frac{30!}{(30 - 3)!} = \frac{30!}{27!} = \frac{30 \cdot 29 \cdot 28 \cdot 27!}{27!}$$

$$= 30 \cdot 29 \cdot 28 = 24,360$$

Thus, there are 24,360 ways the offices can be assigned.

(b) Since there are 30 names in the box at the first drawing, then the probability that Billy's name will be drawn for president is $p = \frac{1}{30}$.

There is an alternate way of solving Billy's problem that can also be used in more complicated problems. Suppose that Billy's name is drawn first. Then the other two offices can be filled in exactly P_2^{29} ways. (There are 2 remaining names to be drawn out of 29 names.) We found in (a) that there are P_3^{30} ways that the names can be drawn. Since P_2^{29} of these are favorable for Billy, then the probability that his name will be drawn first is

$$p = \frac{\text{number of favorable outcomes}}{\text{number of possible outcomes}} = \frac{P_2^{29}}{P_3^{30}}$$

$$= \frac{29 \cdot 28}{30 \cdot 29 \cdot 28} = \frac{1}{30}$$

Combinations

In a very large number of probability problems, we must count the number of possible sets of certain size that can be formed from a large set of objects (the number of five-card hands that can be dealt from a 52-card deck, for example). In a problem of this type, we are concerned only with the sets of objects and not with the order. If two sets contain the same objects, they are the same regardless of whether the order is the same. The following example illustrates the problem with a small set of objects.

Example 2. How many three-object sets can be formed from the five letters a, e, i, o, u?

Solution. We list all of the possible three-object sets:

$$\{a, e, i\} \quad \{a, i, o\} \quad \{a, o, u\} \quad \{e, i, o\} \quad \{i, o, u\}$$
$$\{a, e, o\} \quad \{a, i, u\} \qquad\qquad \{e, i, u\}$$
$$\{a, e, u\} \qquad\qquad\qquad\qquad \{e, o, u\}$$

There are 10 different sets that can be formed.

Observe that we do not list such sets as $\{e, i, a\}$ and $\{e, a, i\}$. These two sets are identical to the set $\{a, e, i\}$ which was listed.

A set of three objects formed from a set of five objects is called a *combination of the objects taken three at a time*. We use the symbol

$$\binom{5}{3} = 10$$

to indicate the *number* of combinations of five objects taken three at a time.

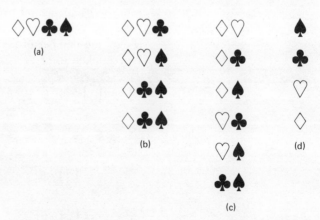

Figure 10-16 The combinations of four objects taken four at a time, three at a time, two at a time, and one at a time.

(a) $\binom{4}{4} = 1$; (b) $\binom{4}{3} = 4$; (c) $\binom{4}{2} = 6$; and (d) $\binom{4}{1} = 4$.

In general, the symbol

$$\binom{n}{m}$$

where $1 \leq m \leq n$ is used for the *number of combinations of n objects taken m at a time*. This number counts the total number of m-object subsets that can be formed from a large set containing n objects.

There is an alternate way that we could have solved Example 2. This alternate method can be generalized to give a formula for the evaluation of

$$\binom{n}{m}$$

Observe that every *permutation* if the five objects taken three at a time could have been obtained in the following two steps:

(1) Choose a combination of the objects taken three at a time:

$$\binom{5}{3} \quad \text{ways.}$$

(2) Choose a permutation of the three objects chosen in (1): 3! ways.

[For example, if we had chosen the combination {a, e, i} in (1), we would get the 3! = 6 permutations:

$$
\begin{array}{lll}
\text{a, e, i} & \text{a, i, e} & \text{e, a, i} \\
\text{e, i, a} & \text{i, a, e} & \text{i, e, a.}]
\end{array}
$$

It follows from the Fundamental Counting Principle that

$$P_3{}^5 = \text{the number of permutations of five objects taken three at a time}$$

$$= \binom{5}{3} \cdot 3!$$

Thus,

$$\frac{5!}{2!} = \binom{5}{3} \cdot 6$$

$$\frac{5 \cdot 4 \cdot 3 \cdot \cancel{2} \cdot \cancel{1}}{\cancel{2} \cdot \cancel{1}} = \binom{5}{3} \cdot 6$$

$$\binom{5}{3} = \frac{5 \cdot 4 \cdot 3}{6} = \frac{60}{6} = 10 \qquad \text{(dividing both sides of the equation by 6)}$$

The argument used above can be used to establish a simple formula for the number of combinations of n objects taken m at a time. Every permutation of n objects taken m at a time can be obtained in two stages:

(1) Select a combination containing m objects: $\binom{n}{m}$ ways.

(2) Permute the m objects: $m!$ ways.

Thus,

$$P_m{}^n = \text{the number of permutations of } n \text{ objects taken } m \text{ at a time}$$

$$= \binom{n}{m} \cdot m!$$

so that

$$\frac{n!}{(n-m)!} = \binom{n}{m} \cdot m!$$

If we divide both sides of this last equation by $m!$, we obtain

$$\binom{n}{m} = \frac{n!}{m!(n-m)!}$$

This formula also is used when $m = 0$. For example,

$$\binom{7}{0} = \frac{7!}{0!7!} = \frac{7!}{1 \cdot 7!} = \frac{1}{1} = 1 \qquad \text{(Recall that } 0! = 1.\text{)}$$

By using the formula

$$\binom{n}{m} = \frac{n!}{m!(n-m)!}$$

we can calculate most of the numbers

$$\binom{n}{m}$$

without difficulty. If n and m are large, we use Table III at the end of the book to approximate the factorials.

Example 3. (a) The number of combinations of five things taken two at a time is

$$\binom{5}{2} = \frac{5!}{2!3!} = \frac{5 \cdot \overset{2}{\cancel{4}} \cdot \cancel{3!}}{1 \cdot \cancel{2} \cdot \cancel{3!}} = 10$$

(b) The number of combinations of seven things taken five at a time is

$$\binom{7}{5} = \frac{7!}{5!2!} = \frac{7 \cdot \overset{3}{\cancel{6}} \cdot \cancel{5!}}{\cancel{5!} \cdot \cancel{2} \cdot 1} = 21$$

(c) The number of combinations of four things taken one at a time is

$$\binom{4}{1} = \frac{4!}{1!3!} = \frac{4 \cdot \cancel{3!}}{1 \cdot \cancel{3!}} = 4$$

(d)
$$\binom{12}{12} = \frac{12!}{12!0!} = \frac{\cancel{12!}}{\cancel{12!} \cdot 1} = 1$$

(e)
$$\binom{25}{8} = \frac{25!}{8!17!} \approx \frac{(1.55)10^{25}}{(4.03)10^4(3.56)10^{14}} \quad \text{(by Table III)}$$

$$\approx \frac{1.55}{(4.03)(3.56)} 10^{25-4-14} \approx (0.108) \cdot 10^7$$

$$\approx (1.08) \cdot 10^6 \approx 1,080,000$$

EXERCISES

1. Evaluate the following numbers.

   (a) P_1^5 (d) $\binom{7}{1}$ (g) P_8^8

   (b) P_{11}^{13} (e) $\binom{12}{6}$ (h) $\binom{8}{8}$

   (c) P_4^8 (f) $\binom{10}{4}$ (i) $\binom{8}{0}$

2. Use Table III at the end of the book to evaluate the following numbers.

   (a) P_7^{20} (c) $\binom{20}{7}$

   (b) P_{10}^{23} (d) $\binom{23}{16}$

3. Calculate the number of permutations of n objects taken m at a time.
   (a) $n = 8, m = 2$ (c) $n = 5, m = 2$
   (b) $n = 12, m = 4$ (d) $n = 5, m = 5$
4. Calculate the number of combinations of n objects taken m at a time for the values of n and m in Exercise 3.
5. A five-member committee must be appointed from the 30 members of a class. In how many ways can this be done?

6. In how many ways can a seven-member committee be appointed from the 30 members of a class if either Billy or Beth must be on the committee?

7. The dial on a combination lock has eight numbers. The combination consists of three of these eight numbers which are dialed in sequence. A sneak thief dials four possible combinations at random. What is the probability that he will open the lock
    (a) if the three numbers in a combination must all be different?
    (b) if the numbers may be repeated in a combination?

8. The town council is composed of five Democrats, four Republicans, and two Independents. In how many ways can a six-member committee be formed that will consist of three Democrats, two Republicans, and one Independent?

10-8 APPLICATIONS

Combinations are used in a number of practical problems to calculate probabilities. They enable us to determine the number of outcomes of an event without actually making a tree diagram. Recall that

$$\binom{n}{m} = \frac{n!}{m!(n-m)!}$$

is the number of different subsets containing m objects that can be chosen from a set of n objects.

Example 1. A certain genetic defect occurs in one-half of the persons in a 100-person rural community. Five persons are chosen at random for a medical survey.

(a) In how many different ways could the five-person group be chosen from the 100 persons in the community?

(b) In how many different ways could the five-person group be chosen from the 50 persons in the community who lack the defect?

(c) What is the probability that a five-person group chosen at random will consist of individuals who do not have the defect?

(d) What is the probability that at least one person in the five-person group will have the defect?

Solution. (a) The total number of five-person groups that could be chosen in the community is

$$\binom{100}{5} = \frac{100!}{5!95!} = \frac{100 \cdot 99 \cdot 98 \cdot 97 \cdot 96 \cdot 95!}{5 \cdot 4 \cdot 3 \cdot 2 \cdot 1 \cdot 95!}$$

$$= \frac{\overset{20}{\cancel{100}}}{\cancel{5}} \cdot \frac{\overset{33}{\cancel{99}}}{\cancel{3}} \cdot \frac{\overset{49}{\cancel{98}}}{\cancel{2}} \cdot 97 \cdot \frac{\overset{24}{\cancel{96}}}{\cancel{4}} = 20 \cdot 33 \cdot 49 \cdot 97 \cdot 24$$

$$= 75,287,520$$

(b) The total number of five-person groups that could be chosen from the 50 persons who do not have the defect is

$$\binom{50}{5} = \frac{50!}{5!45!} = \frac{50 \cdot 49 \cdot 48 \cdot 47 \cdot 46}{5 \cdot 4 \cdot 3 \cdot 2 \cdot 1}$$

$$= \frac{\overset{5}{\cancel{50}}}{\cancel{5 \cdot 2}} \cdot 49 \cdot \frac{\overset{4}{\cancel{48}}}{\cancel{4 \cdot 3}} \cdot 47 \cdot 46$$

$$= 5 \cdot 49 \cdot 4 \cdot 47 \cdot 46 = 2{,}118{,}760$$

(c) The probability that a five-person group chosen at random will consist of individuals who do not have the defect is

$$p = \frac{\text{number of groups that could be formed from persons free of the defect}}{\text{total number of groups that could be formed}}$$

$$= \frac{2{,}118{,}760}{75{,}287{,}520} \approx 0.03$$

(d) Either a five-person group contains at least one person with the defect or all of the members of the group are free of it. These events are mutually exclusive. Therefore, the probability that at least one person in the group has the defect is

$$q = 1 - (\text{probability that no one in group has the defect})$$
$$\approx 1 - 0.03 \approx 0.97$$

LONG SHOTS
Many people make the error of assuming that an event with a very high probability must occur and that one with a very low probability cannot. These individuals should realize that long shots do occasionally pay off and that some very unlikely events happen every day. As we stated at the beginning of this chapter, probability theory is concerned with the likelihood of events. Only rarely can we assert that an event positively will or will not occur.

Thus, it is almost certain that at least one member of the five-person group will have the defect.

Example 2. In the game of *Showdown* in poker, each player is dealt a five-card hand.

(a) What is the total number of possible five-card hands?

(b) How many of the possible five-card hands will be composed entirely of *hearts?*

(c) What is the probability of being dealt a hand that is composed entirely of *hearts?*

(d) What is the probability of being dealt a hand composed of five cards in the same suit?

Solution. (a) The total number of five-card poker hands that can be dealt is equal to

$$\binom{52}{5} = \frac{52!}{5!47!} = \frac{52 \cdot 51 \cdot 50 \cdot 49 \cdot 48}{5 \cdot 4 \cdot 3 \cdot 2 \cdot 1}$$

$$= \frac{\overset{13}{\cancel{52}}}{\cancel{4}} \cdot \frac{\overset{17}{\cancel{51}}}{\cancel{3}} \cdot \frac{\overset{10}{\cancel{50}}}{\cancel{5}} \cdot 49 \cdot \frac{\overset{24}{\cancel{48}}}{\cancel{2}}$$

$$= 13 \cdot 17 \cdot 10 \cdot 49 \cdot 24 = 2{,}598{,}960$$

(b) The number of five-card hands composed entirely of hearts is

$$\binom{13}{5} = \frac{13!}{5!13!} = \frac{13 \cdot 12 \cdot 11 \cdot 10 \cdot 9}{5 \cdot 4 \cdot 3 \cdot 2 \cdot 1}$$

$$= 13 \cdot \frac{\cancel{12}}{4 \cdot 3} \cdot 11 \cdot \frac{\cancel{10}}{5 \cdot 2} \cdot 9$$

$$= 13 \cdot 1 \cdot 11 \cdot 1 \cdot 9 = 1{,}287$$

(c) The probability of being dealt a hand composed entirely of hearts is

$$p = \frac{\text{number of hands composed of hearts}}{\text{total number of hands}} = \frac{1{,}287}{2{,}598{,}960} \approx 0.0005$$

Thus, we would expect to be dealt a hand composed of hearts about 5 times out of every 10,000 hands dealt.

(d) There are four suits. The probability of being dealt a hand composed entirely of cards from any one of these is approximately equal to 0.0005. Since these events are mutually exclusive, the probability of being dealt a hand composed entirely of cards from the same suit is

$$p \approx \underbrace{0.0005}_{\substack{\text{probability of} \\ \text{five hearts}}} + \underbrace{0.0005}_{\substack{\text{probability of} \\ \text{five spades}}} + \underbrace{0.0005}_{\substack{\text{probability of} \\ \text{five diamonds}}} + \underbrace{0.0005}_{\substack{\text{probability of} \\ \text{five clubs}}}$$

$$\approx 0.002$$

We would expect to be dealt a hand consisting of five cards in the same suit twice out of every thousand hands.

Example 3. In the game of poker a five-card hand composed of cards in the same suit is called a *flush* unless the five cards are in numerical sequence. If the cards are in sequence, the hand is a *straight flush,* which ranks higher than a flush. (In determining the sequence, an ace can be considered as either the highest or the lowest card.)

For example, the hand

$$5\heartsuit,\ 7\heartsuit,\ 8\heartsuit,\ J\heartsuit,\ K\heartsuit$$

is a flush, while the hands

$$A\diamondsuit,\ 2\diamondsuit,\ 3\diamondsuit,\ 4\diamondsuit.\ 5\diamondsuit,$$
$$7\spadesuit,\ 8\spadesuit,\ 9\spadesuit,\ 10\spadesuit,\ J\spadesuit,$$

and

$$10\heartsuit,\ J\heartsuit,\ Q\heartsuit,\ K\heartsuit,\ A\heartsuit$$

are straight flushes.
(a) Find the probability of being dealt a straight flush.
(b) Find the probability of being dealt a flush.

The probability of being dealt a particular hand in bridge is

$$p = \frac{1}{635{,}013{,}559{,}600}$$

Once a hand has been dealt, however, the probability against it has no bearing on the situation. A similar principle holds for other events. Once an event has occurred, the probability against it can be ignored.

Solution. (a) A straight flush can have A, 2, 3, 4, 5, 6, 7, 8, 9, or 10 as its lowest-ranking card. Thus, there are 10 distinct straight flushes in each suit. Since there are four suits, then there are 40 possible straight flushes. The probability of being dealt a straight flush is

$$p = \frac{\text{number of straight flushes}}{\text{number of five-card hands}} = \frac{40}{2,598,960} \approx 0.000015$$

We would expect to be dealt a straight flush about 15 times out of every million hands.

(b) We found in Example 2 that there are 1,287 five-card hands composed of hearts. There also are 1,287 each composed entirely of spades, clubs, or diamonds. The total number of five-card hands composed of cards in the same suit is

$$4 \cdot 1,287 = 5,148$$

Forty of these hands are straight flushes. Thus, the total number of flush hands is

$$5,148 - 40 = 5,108$$

The probability of being dealt a flush is

$$p = \frac{\text{number of flush hands}}{\text{number of five-card hands}} = \frac{5,108}{2,598,960} \approx 0.002$$

Observe that this number (when rounded-off to three decimal places) is the same as the number we obtained in Example 2 for the number of poker hands composed of cards in the same suit. The number of straight flushes is so small that it makes the difference in the probabilities negligible.

Example 4. Four cakes are the prizes in a lottery at the church bazaar. Mrs. Goodbody, the minister's wife, purchased 10 of the 100 lottery tickets that were sold.

(a) What is the probability that she will not win anything?
(b) What is the probability that she will win at least one cake?
(c) What is the probability that she will win all four cakes?

Solution. The number of ways that four tickets can be drawn from 100 is

$$\binom{100}{4} = \frac{100 \cdot 99 \cdot 98 \cdot 97}{4 \cdot 3 \cdot 2 \cdot 1} = \frac{\overset{25}{\cancel{100}} \cdot \overset{33}{\cancel{99}} \cdot \overset{49}{\cancel{98}}}{\cancel{4} \cdot \cancel{3} \cdot \cancel{2}} \cdot 97$$

$$= 25 \cdot 33 \cdot 49 \cdot 97$$

$$= 3,921,225$$

The number of ways that four tickets can be drawn from the 90 tickets that Mrs. Goodbody did not buy is

$$\binom{90}{4} = \frac{90 \cdot 89 \cdot 88 \cdot 87}{4 \cdot 3 \cdot 2 \cdot 1} = \frac{\overset{30}{\cancel{90}}}{\cancel{3}} \cdot 89 \cdot \frac{\overset{11}{\cancel{88}}}{\cancel{4} \cdot \cancel{2}} \cdot 87 = 2,555,190$$

One of the first articles about probability to attract the attention of the general public was written by the mathematician *Jean-le-Rond d'Alembert* (1717–1783) for the French *Encyclopédie*. The main thrust of the article was directed toward the French state lottery, which was making large profits for the government. D'Alembert's article analyzed both the expectation of the average person to win and the probability of the government making a profit.

(a) The probability that Mrs. Goodbody will not win anything is

$$p = \frac{\text{number of outcomes in which she loses}}{\text{number of possible outcomes}} = \frac{2{,}555{,}190}{3{,}921{,}225} \approx 0.65$$

(b) The probability that she will win at least one cake is

$$q = 1 - p \approx 1 - 0.65 \approx 0.35$$

(c) The number of ways that all 4 tickets can be selected from the 10 bought by Mrs. Goodbody is

$$\binom{10}{4} = \frac{10 \cdot 9 \cdot 8 \cdot 7}{4 \cdot 3 \cdot 2 \cdot 1} = \frac{\overset{5}{\cancel{10}}}{\cancel{2}} \cdot \frac{\overset{3}{\cancel{9}}}{\cancel{3}} \cdot \frac{\overset{2}{\cancel{8}}}{\cancel{4}} \cdot 7 = 210$$

The probability that she will win all four cakes is

$$\frac{\text{number of successful outcomes}}{\text{number of possible outcomes}} = \frac{210}{3{,}921{,}225} \approx 0.00005$$

Trial and Conviction by Probability

In June of 1964, an elderly woman was mugged in San Pedro, California. Shortly thereafter, a witness saw a blonde woman, her hair arranged in a ponytail, run from the scene, get into a yellow car driven by a bearded black man, and speed away. Subsequently, the police arrested a married couple that fitted the description of the suspects—a blonde white woman, her hair in a ponytail, and a bearded black man, who owned a yellow automobile.

The prosecutor based his case on the unlikelihood of another couple that fitted the description and used the laws of probability to establish it. He estimated probability factors ranging from 1/4 that a girl would be blonde to 1/1000 that the couple would be black-white. He then multiplied the individual probability factors to obtain a resultant probability of less than 1/12,000,000 that any other couple could have robbed the woman. The jury agreed with the prosecutor and convicted the couple. Sentences of "not less than one year" and "one year to life" were given the woman and man, respectively.

The conviction was overturned in 1968 by the California Supreme Court when serious errors were established in the conclusions based on the probability argument: (1) The prosecutor estimated the individual probabilities rather than using statistics to prove them; (2) Certain of the individual events were not independent, thus the events could not be multiplied together to get the final probability; and (3) The most serious error was that the prosecutor had considered the unlikelihood of finding a couple that fitted the description rather than the relative frequency of such couples. A sophisticated argument based on advanced probability techniques was used to show that if at least one such couple was known to exist in the area, then there was a 41-percent chance that at least one other couple fitting the description lived there.

A more detailed report on the case can be found in *Time Magazine*, January 8, 1965, p. 42, and April 26, 1968, p. 41.

EXERCISES

1. A genetic defect is common to 20 persons in a 100-person community. What is the probability that the defect will show up in at least one person in the following size random samples?
   (a) 5 persons
   (b) 10 persons
   (c) 20 persons
   How many persons must be examined before it will be certain (probability of 1) that the defect will show up?

2. The Gasdrinker Engine Co. manufactures 1000 engines each day. Several of these engines, selected at random, are thoroughly tested. Assume that 300 of the engines are defective each day. What is the probability that at least one defective engine will be detected by this method
   (a) if 5 engines are tested?
   (b) if 10 engines are tested?

3. In the game of poker a *straight* is a hand composed of five cards, not all in the same suit, which are in numerical sequence. Examples of straights are

$$A\spadesuit,\ 2\diamondsuit,\ 3\diamondsuit,\ 4\heartsuit,\ 5\clubsuit$$

$$7\heartsuit,\ 8\heartsuit,\ 9\heartsuit,\ 10\heartsuit,\ J\diamondsuit$$

   and

$$10\diamondsuit,\ J\clubsuit,\ Q\heartsuit,\ K\clubsuit,\ A\spadesuit$$

   If five cards are in sequence *and* in the same suit (such as $3\diamondsuit$, $4\diamondsuit$, $5\diamondsuit$, $6\diamondsuit$, $7\diamondsuit$), the hand is a *straight flush,* which ranks much higher than a straight.
   (a) How many distinct poker hands are either straights or straight flushes?
   (b) How many of the hands in (a) are straight flushes? How many are straights?
   (c) What is the probability of being dealt a straight? A straight flush? Which is the more likely, a straight or a flush? (See Example 3.)

4. In the game of bridge, 13 cards are dealt to each person.
   (a) What is the total possible number of bridge hands?
   (b) What is the probability that a bridge hand will consist of 13 spades?
   (c) What is the probability that a bridge hand will consist of 13 cards in the same suit?
   (d) What is the probability that a bridge hand will contain all four aces, all four kings, and all four queens?

10-9 PASCAL'S TRIANGLE;
THE BINOMIAL THEOREM

Pascal, the discoverer of the formula

$$\binom{n}{m} = \frac{n!}{m!(n-m)!}$$

for the number of combinations of n objects taken m at a time, made a remarkable discovery about these numbers when he wrote them in a triangular array.

$n = 1$:
$$\binom{1}{0} \quad \binom{1}{1}$$

$n = 2$:
$$\binom{2}{0} \quad \binom{2}{1} \quad \binom{2}{2}$$

$n = 3$:
$$\binom{3}{0} \quad \binom{3}{1} \quad \binom{3}{2} \quad \binom{3}{3}$$

$n = 4$:
$$\binom{4}{0} \quad \binom{4}{1} \quad \binom{4}{2} \quad \binom{4}{3} \quad \binom{4}{4}$$

CHINESE VERSION OF
PASCAL'S TRIANGLE

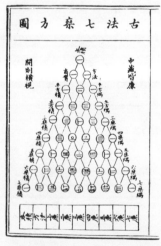

Photo from *Chu Shih-Chieh,* Ssu Yuan
Yü Chien, *1303.*

The Chinese knew about the Binomial Theorem hundreds of years before Pascal. This illustration is from a book published in 1303, 362 years before Pascal published his analysis of it. As far as we know, the Chinese did not connect the triangle with probability theory.

and so on. When he substituted the actual values for the symbols, he obtained the array:

$n = 1$: $\qquad$ 1 1
$n = 2$: $\qquad$ 1 2 1
$n = 3$: $\qquad$ 1 3 3 1
$n = 4$: $\qquad$ 1 4 6 4 1

and so on.

Pascal observed that every number in the interior of the triangle is the sum of the two numbers immediately above it. If we emphasize this by drawing diagonal lines from each interior entry to the two entries above it, we obtain the array shown in Figure 10-17. The array of numbers is known as *Pascal's Triangle.*

For example, the number 4 in the fourth row is the sum of the 1 and 3 above it; the 6 is the sum of the 3 and 3 above it, and so on.

Pascal's Triangle furnishes us a convenient way to calculate the numbers

$$\binom{n}{m}$$

for small values of n and m.

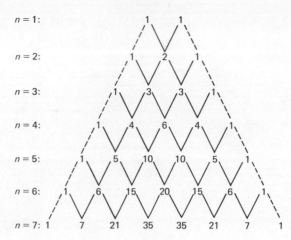

n = 1:

n = 2:

n = 3:

n = 4:

n = 5:

n = 6:

n = 7: 1 7 21 35 35 21 7 1

Figure 10-17 Pascal's Triangle.

Example 1. (a) List the eighth row of Pascal's Triangle.
(b) Identify each entry in the eighth row with the corresponding symbol

$$\binom{n}{m}$$

Solution. (a) The seventh row of Pascal's Triangle (Fig. 10-17) is

$$1 \quad 7 \quad 21 \quad 35 \quad 35 \quad 21 \quad 7 \quad 1$$

If we draw the diagonal lines down from this row, compute the sums, and adjoin the 1s to each end of the row, we obtain the eighth row:

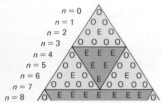

$n = 0$
$n = 1$
$n = 2$
$n = 3$
$n = 4$
$n = 5$
$n = 6$
$n = 7$
$n = 8$

The arithmetical properties of Pascal's Triangle have been studied extensively. The diagram shows the pattern of odd and even numbers in the first eight rows (The row for $n = 0$ was added at the top to complete the pattern.)

A different, but basic property of the triangle is related to prime numbers. If n is a prime, then all of the entries in the nth row except for the two 1s on the end are divisible by n.

$n = 7$: 1 7 21 35 35 21 7 1

$n = 8$: 1 8 28 56 70 56 28 8 1

(b) The symbols

$$\binom{n}{m}$$

in the eighth row are (in order):

$$\binom{8}{0}, \binom{8}{1}, \binom{8}{2}, \binom{8}{3}, \binom{8}{4}, \binom{8}{5}, \binom{8}{6}, \binom{8}{7}, \quad \text{and} \quad \binom{8}{8}.$$

Thus,

$$\binom{8}{0} = 1, \quad \binom{8}{1} = 8, \quad \binom{8}{2} = 28, \quad \binom{8}{3} = 56,$$

$$\binom{8}{4} = 70, \quad \binom{8}{5} = 56, \quad \binom{8}{6} = 28, \quad \binom{8}{7} = 8, \quad \binom{8}{8} = 1$$

The Binomial Theorem

The rows of Pascal's Triangle show up in another part of mathematics. In algebra we study the expression $(x + y)^n$. If we expand out the power $(x + y)^n$ and collect like terms, we find that the coefficients of the powers of x and y are exactly the numbers in the nth row of Pascal's Triangle.

For example,

$$(x + y)^2 = (x + y)(x + y) = x^2 + 2xy + y^2$$

$$= 1x^2 + 2xy + 1y^2$$

and the second row of Pascal's Triangle is

$$1 \quad 2 \quad 1$$

Similarly,

$$(x + y)^3 = x^3 + 3x^2y + 3xy^2 + y^3$$

$$= 1x^3 + 3x^2y + 3xy^2 + 1y^3$$

and the third row of Pascal's Triangle is

$$1 \quad 3 \quad 3 \quad 1$$

The relationship becomes striking if we list the powers of $(x + y)^n$ in rows opposite Pascal's Triangle as in Figure 10-18.

	Binomial Theorem	Pascal's Triangle
$n = 1$:	$(x + y)^1 = 1 \cdot x + 1 \cdot y$	1 1
$n = 2$:	$(x + y)^2 = 1 \cdot x^2 + 2xy + 1 \cdot y^2$	1 2 1
$n = 3$:	$(x + y)^3 = 1 \cdot x^3 + 3x^2y + 3xy^2 + 1 \cdot y^3$	1 3 3 1
$n = 4$:	$(x + y)^4 = 1 \cdot x^4 + 4x^3y + 6x^2y^2 + 4xy^3 + 1 \cdot y^4$	1 4 6 4 1
$n = 5$:	$(x + y)^5 = 1 \cdot x^5 + 5x^4y + 10x^3y^2 + 10x^2y^3 + 5xy^4 + 1 \cdot y^5$	1 5 10 10 5 1

Figure 10-18 The Binomial Theorem. The rows of Pascal's Triangle give the coefficients of the expansion of $(x + y)^n$.

The expression $x + y$ is called a *binomial* (bi = 2). The formula for the expansion of $(x + y)^n$ is called the *Binomial Theorem*. It follows from the

Binomial Theorem that

$$n = 1: \quad (x + y)^1 = \binom{1}{0}x + \binom{1}{1}y$$

$$n = 2: \quad (x + y)^2 = \binom{2}{0}x^2 + \binom{2}{1}xy + \binom{2}{2}y^2$$

$$n = 3: \quad (x + y)^3 = \binom{3}{0}x^3 + \binom{3}{1}x^2y + \binom{3}{2}xy^2 + \binom{3}{3}y^3$$

$$n = 4: \quad (x + y)^4 = \binom{4}{0}x^4 + \binom{4}{1}x^3y + \binom{4}{2}x^2y^2 + \binom{4}{3}xy^3 + \binom{4}{4}y^4$$

$$n = 5: \quad (x + y)^5 = \binom{5}{0}x^5 + \binom{5}{1}x^4y + \binom{5}{2}x^3y^2 + \binom{5}{3}x^2y^3$$
$$+ \binom{5}{4}xy^4 + \binom{5}{5}y^5$$

and so on.

Because the symbol

$$\binom{n}{m}$$

shows up as the coefficient of $x^{n-m}y^m$ in the expansion of $(x + y)^n$, this number is called a *binomial coefficient*.

Thus, we have several different interpretations of the symbol $\binom{n}{m}$:

$\binom{n}{m}$ = the coefficient of $x^{n-m}y^m$ in the expansion of $(x + y)^n$ by the Binomial Theorem

= the $(m + 1\text{st})$ entry in the nth row of Pascal's Triangle

= the number of combinations of n objects taken m at a time

= $\dfrac{n!}{m!(n - m)!}$

At this point the reader is probably wondering what connection the Binomial Theorem, which belongs in the realm of algebra, has with probability theory. This question will be answered in the next section. We shall see that the Binomial Theorem and Pascal's Triangle furnish us a quick and convenient way to solve a large number of important problems in probability—problems that would require much work by any other method.

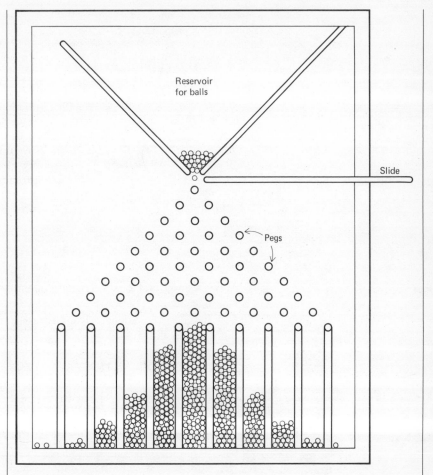

The Quincunx

Sir Francis Galton invented a simple device to illustrate the binomial coefficients that he called a *quincunx*. The device was named after an ancient Roman coin on which five spots were arranged in the pattern

Round pegs are driven into a board as shown above. When small balls are dropped from the reservoir at the top of the quincunx, they bounce from one peg to the next until they fall into the slots at the bottom. If the pegs are properly positioned with respect to the size of the balls, then a ball will have a probability of $\frac{1}{2}$ of falling to the right and a probability of $\frac{1}{2}$ of falling to the left when it strikes a peg.

The probability that a ball will fall into a given slot is proportional to the number of different paths that it can take from the first peg to that slot. It can be shown that this number is one of the binomial coefficients for the row $n = 10$ (the number of rows of pegs). The number of balls in the first slot is proportional to

$$\binom{10}{0}$$

the number in the second slot is proportional to

$$\binom{10}{1}$$

the number in the third slot is proportional to

$$\binom{10}{2}$$

and so on.

The distribution of balls in the slots is very much like a normal distribution, which is what Galton used the device to illustrate. As we shall see in the next section, the relative sizes of the binomial coefficients are very closely related to the normal-distribution curve.

EXERCISES

1. Make a copy of Pascal's Triangle. Complete the triangle through the row $n = 12$.
2. Use the Pascal's Triangle from Exercise 1 to find the values of the binomial coefficients.

   (a) $\binom{12}{7}$ (c) $\binom{11}{11}$ (e) $\binom{10}{5}$

   (b) $\binom{9}{3}$ (d) $\binom{12}{4}$ (f) $\binom{9}{0}$

3. Use the Pascal's Triangle from Exercise 1 to write out the expansion of $(x + y)^n$ for the following values of n:

   (a) $n = 6$ (c) $n = 8$ (e) $n = 10$
   (b) $n = 7$ (d) $n = 9$ (f) $n = 11$

4. Use the Pascal's Triangle from Exercise 1 to find
   (a) the number of combinations of 8 objects taken 5 at a time;
   (b) the number of combinations of 11 objects taken 3 at a time;
   (c) the number of 5-card poker hands that are composed entirely of aces, kings, and queens.
5. Use the formula

$$\binom{n}{m} = \frac{n!}{m!(n - m)!}$$

to calculate the coefficient of $x^{22}y^2$ in the expansion of $(x + y)^{24}$.
6. Use the relationship

$$\binom{n}{m} = \frac{n!}{m!(n - m)!}$$

to show that

(a) $\dbinom{8}{3} = \dbinom{8}{5}$ (b) $\dbinom{50}{17} = \dbinom{50}{33}$ (c) $\dbinom{27}{15} = \dbinom{27}{12}$

What general result would you conjecture? Can you make a general argument to show that it is true?

10-10 BINOMIAL DISTRIBUTIONS

The Binomial Theorem and Pascal's Triangle play an important role in probability theory. The following example illustrates the basic principles.

Example 1. We perform an experiment that consists of tossing a coin 10 times in a row. What is the probability that we shall toss exactly seven *heads* and three *tails*?

Solution. We first consider the probability of obtaining a specific sequence of seven *heads* and three *tails*, say

$$H\,H\,T\,T\,H\,H\,H\,H\,T\,H$$

On each toss, the probability is $\frac{1}{2}$ that we shall get *heads* and $\frac{1}{2}$ that we shall get *tails*. It follows that the probability of getting the particular sequence of seven *heads* and three *tails* is

$$\tfrac{1}{2} \cdot \tfrac{1}{2} \cdot \tfrac{1}{2} \cdot \tfrac{1}{2} \cdot \tfrac{1}{2} \cdot \tfrac{1}{2} \cdot \tfrac{1}{2} \cdot \tfrac{1}{2} \cdot \tfrac{1}{2} \cdot \tfrac{1}{2} = (\tfrac{1}{2})^{10}$$

Obviously, the above reasoning can be applied to any specific sequence of seven *heads* and three *tails*. Thus, we need to find the number of different ways we can write a sequence that consists of seven *heads* and three *tails*.

To find the number of different sequences of 7 *heads* and 3 *tails*, imagine that we have 10 empty slots in a row. We choose 7 of the 10 slots and label them H, then label the other 3 T. (See Fig. 10-19.) This gives us a particular sequence of seven *heads* and three *tails*. Since every such sequence can be obtained in this way, then the total number of such sequences is equal to the number of ways we can select 7 slots out of a set of 10. Thus, the number of sequences that consist of seven *heads* and three *tails* is equal to

$$\dbinom{10}{7} = \frac{10!}{7!3!} = 120$$

Since there are 120 different sequences of seven *heads* and three *tails* and each has a probability of $(\frac{1}{2})^{10}$ of being tossed, then the probability of tossing exactly seven *heads* and three *tails* is

$$p = 120 \cdot \left(\frac{1}{2}\right)^{10} = \frac{120}{1024} \approx 0.117$$

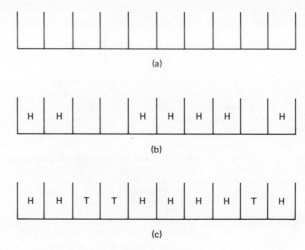

Figure 10-19 The number of different possible sequences of seven *heads* and three *tails* is

$$\binom{10}{7} = 120.$$

(a) Given 10 slots (b) select 7 slots, label them *heads*. This can be done in

$$\binom{10}{7} = \frac{10!}{7!3!} = 120 \text{ ways}$$

(c) Label the remaining 3 slots *tails*.

Thus, we can expect to toss 7 *heads* and 3 *tails* approximately 117 times out of every 1000 times we repeat the experiment.

The type of reasoning illustrated in Example 1 can be applied to any situation in which an event is repeated several times, provided there are only two possible outcomes for the event and the probabilities are known for each outcome.

Binomial Principle

Suppose an event has two possible outcomes: *success* and *failure*. Suppose also that successes or failures of previous events have no influence on future events. Let the probability of success on any event be p and the probability of failure be $q = 1 - p$. Then the probability of exactly m successes out of n events is

$$\binom{n}{m}p^m q^{n-m}$$

Example 2. Eight dice are tossed. What is the probability that a 6 is tossed on exactly four dice?

PROBABILITY
DISTRIBUTIONS

One of the most basic problems in statistics concerns the distribution of statistical data. Given an arbitrary bit of data, find the probability that its value is within a certain range, say between a and b. It can be shown that the probability of this event is

P = (area under the curve between a and b)/(total area under the curve).

Solution. For all practical purposes, this is the same problem as if one die were thrown eight times. The probability of a 6 being thrown on any one toss is $p = \frac{1}{6}$ and the probability of some other number being thrown is $q = \frac{5}{6}$. Thus, the probability that exactly four 6s will be tossed is

$$\binom{8}{4}p^4q^{8-4} = \binom{8}{4}\left(\frac{1}{6}\right)^4\left(\frac{5}{6}\right)^4 = 70 \cdot \frac{1}{6^4} \cdot \frac{5^4}{6^4}$$

$$= \frac{70 \cdot 1 \cdot 625}{6^8} = \frac{43,750}{1,679,616} \approx 0.026$$

Thus, we would expect exactly four 6s to be thrown approximately 26 times out of every 1000 times we repeat the experiment.

The Binomial Principle can be applied to any situation where there are just two possible outcomes. The following example shows an application to family planning.

Example 3. A couple plans to have six children.
(a) What is the probability that they all will be girls?
(b) What is the probability that three will be girls and three will be boys?
(Assume that the probability of a child being a boy is $\frac{1}{2}$ and the probability of being a girl is $\frac{1}{2}$.)

Solution. The probability that a child will be a girl is $p = \frac{1}{2}$, and the probability that it will be a boy is $q = \frac{1}{2}$.
(a) The probability that all six will be girls is

$$\binom{6}{6}p^6q^0 = 1 \cdot \left(\frac{1}{2}\right)^6\left(\frac{1}{2}\right)^0 = \frac{1}{2^6} = \frac{1}{64}$$

Thus, about 1 out of 64 six-child families will consist of all girls.
(b) The probability that three children will be girls and three will be boys is

$$\binom{6}{3}p^3q^3 = 20 \cdot \left(\frac{1}{2}\right)^3 \cdot \left(\frac{1}{2}\right)^3 = \frac{20}{64}$$

About 20 out of every 64 six-child families will consist of three girls and three boys. (See Fig. 10-20.)

Whenever possible, theoretical probabilities are checked against the actual frequencies of events. When the frequencies of large families having certain proportions of boys and girls were checked against the theoretical frequencies, an interesting fact was discovered. The theoretical frequencies (such as we computed in Example 3) are approximately correct except for large families where one child is of one sex and all of the others are of a different sex. For example, there is a higher frequency of six-child families with one boy and five girls than would be predicted by the theory. The researchers concluded that many of the parents had originally planned to have smaller families but, when all of the children had been born the same

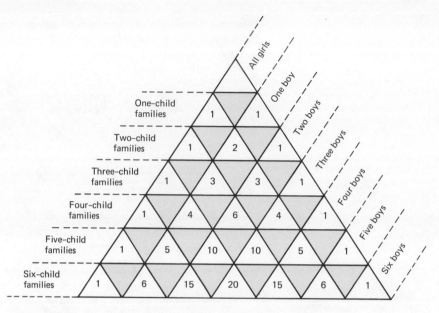

Figure 10-20 The rows of Pascal's Triangle give the expected distribution of boys and girls in families of a certain size. Out of every 64 six-child families, for example, we would expect 1 family to have all girls, 6 to have five girls and one boy, 15 to have four girls and two boys, 20 to have three girls and three boys, and so on.

sex, had decided to continue having children until they had one of the opposite sex.

Example 4. A man and his wife are both carriers for hereditary muscular atrophy. What is the probability that two of their three children will develop the defect?

Solution. As we saw in Section 10-5, the probability that any one child will develop the condition is $p = \frac{1}{4}$ and the probability that it will not develop it is $q = \frac{3}{4}$. It follows that the probability that exactly two children out of three will develop the condition is

$$\binom{3}{2}p^2 q = 3\left(\frac{1}{4}\right)^2\left(\frac{3}{4}\right) = 3 \cdot \frac{1}{16} \cdot \frac{3}{4} = \frac{9}{64}$$

Thus, we would expect exactly two children to have the defect in 9 out of 64 three-child families.

A close connection exists between the Binomial Principle and the Binomial Theorem. The probability of exactly m successful outcomes out of n events is

$$\binom{n}{m}p^m q^{n-m}$$

This is exactly the term for $p^m q^{n-m}$ in the expansion of $(p + q)^n$ using the

Binomial Theorem. Thus, we can identify the various terms of the Binomial Theorem with the probabilities of certain events.

Example 5. A couple plans to have four children. Use the Binomial Theorem to determine the probabilities for the different possible distributions of boys and girls.

Solution. Let $p = \frac{1}{2}$ be the probability of a boy and $q = \frac{1}{2}$ be the probability of a girl for any given child. If we expand $(p + q)^4$, we obtain

$$(p + q)^4 = \underbrace{1 \cdot p^4}_{\substack{\text{probability} \\ \text{of 4 boys}}} + \underbrace{4p^3q}_{\substack{\text{probability} \\ \text{of 3 boys} \\ \text{and 1 girl}}} + \underbrace{6p^2q^2}_{\substack{\text{probability} \\ \text{of 2 boys} \\ \text{and 2 girls}}} + \underbrace{4pq^3}_{\substack{\text{probability} \\ \text{of 1 boy} \\ \text{and 3 girls}}} + \underbrace{1 \cdot q^4}_{\substack{\text{probability} \\ \text{of 4 girls}}}$$

$$\frac{1}{16} + \frac{4}{16} + \frac{6}{16} + \frac{4}{16} + \frac{1}{16}$$

Hence

the probability of 4 boys is $\frac{1}{16}$;
the probability of 3 boys, 1 girl is $\frac{4}{16}$;
the probability of 2 boys, 2 girls is $\frac{6}{16}$;
the probability of 1 boy, 3 girls is $\frac{4}{16}$; and
the probability of 4 girls is $\frac{1}{16}$.

The Poisson Distribution

Because of its importance, we have concentrated on the normal distribution in our work with statistics. The reader should not, however, jump to the conclusion that this is the only distribution of importance. A number of distributions, including the binomial distribution, occur in nature and must be used when needed.

One of the most interesting of the distributions is the *Poisson distribution*, named after *Siméon Denis Poisson* (1781–1840), a French mathematician. The Poisson distribution, which can be obtained from the normal distribution, is concerned with the special case when the number of trials is very large, but the probability of "success" on any trial is very small. In this case, the number of possible "successes" is a small number.

In the nineteenth century, the Poisson distribution was used to predict the number of deaths that should occur each year in the German army from soldiers being kicked by horses. It also has been used to predict the number of deaths by drowning, the number of bomb hits in different areas of London during World War II, the number of wars that can be expected in a year, the number of strikes in industry each week, and the disintegration of radioactive matter.

Binomial Distributions

If we use the results of Example 5 in a frequency chart, we obtain the following expected distribution of 16 four-child families (Fig. 10-21):

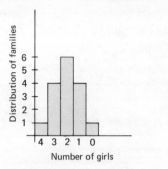
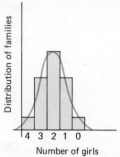

Figure 10-21 The expected distribution of four-child families according to the number of girls (Example 5). The distribution is very close to a normal distribution.

Class	Expected Frequency
All boys	1
3 boys, 1 girl	4
2 boys, 2 girls	6
1 boy, 3 girls	4
all girls	1
Sum:	16

Observe that the binomial coefficients

$$1 \quad 4 \quad 6 \quad 4 \quad 1$$

are the exact values of the frequencies. The sum of the coefficients is 16, the total number of families under consideration.

A distribution of this type, where the frequencies are the binomial coefficients, occurs whenever the probability of success in an event is $\frac{1}{2}$. Thus, this type of distribution occurs in family planning (probability of a boy is $\frac{1}{2}$; probability of a girl is $\frac{1}{2}$) in coin tossing (probability of *heads* is $\frac{1}{2}$; probability of *tails* is $\frac{1}{2}$), in genetics (if one parent is a carrier and the other is not, then the probability that a child will be a carrier is $\frac{1}{2}$), and in many other situations.

There is a very close relationship between the binomial distribution and the normal distribution. As an example, a normal-distribution curve is superimposed over the bar graph in Figure 10-21(b). In general, the larger the number of events, the closer the expected frequencies fit the normal-distribution curve.

Example 6.   An experiment consists of tossing eight coins. This experiment is repeated a large number of times. Use the Binomial Theorem to determine the expected frequencies with which each number of *heads* should occur.

Solution. Since we are tossing eight coins, then the coefficients of $(x + y)^8$ determine the frequencies. We use the eighth row of Pascal's Triangle:

Class	Expected Frequency
All *tails*	1
1 *heads*, 7 *tails*	8
2 *heads*, 6 *tails*	28
3 *heads*, 5 *tails*	56
4 *heads*, 4 *tails*	70
5 *heads*, 3 *tails*	56
6 *heads*, 2 *tails*	28
7 *heads*, 1 *tails*	8
All *heads*	1
Sum:	256

The frequency distribution obtained in Example 6 is a theoretical distribution that we expect to occur. We do not actually get exactly six *heads* and two *tails* 28 times out of every 256 experiments.

The total number of experiments necessary to get the expected distribution is 256.

The bar graph, with a normal-distribution curve superimposed over it, is shown in Figure 10-22.

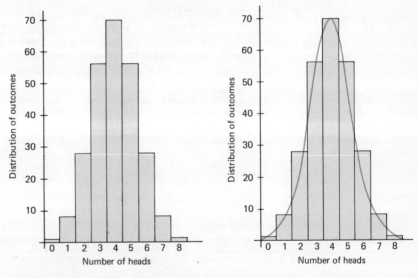

Figure 10-22 The expected distribution of outcomes of tossing eight coins a large number of times. (Example 6).

Observe in Figure 10-22 that the normal-distribution curve fits the bar graph very closely. As we increase the number of coins in the experiment, the bar graph more and more resembles the normal-distribution curve. For many applications in statistics, we simply assume that a binomial distribution is actually distributed normally and use the normal curve rather than the bar graph.

As we have seen in this chapter, there is a very close relationship between probability theory and statistics. It is clear that empirical probability must be based on statistics. It is surprising to learn, however, that the expected distributions of many events, such as tossing a large number of coins many times, are normal. This type of theoretical analysis, coupled with experiments and other forms of verification, is responsible for the great impact that probability and statistics have on our lives every day.

Gambler's Ruin

Probability theory can be used to show that extended betting over a long period of time against an opponent with superior resources is almost certain to lead to financial disaster.

Most gambling games in which the player bets against the house have almost equal odds with the house being a slight favorite. These odds are enough to ensure that the house will make a small profit over a long period of time. In actual practice, however, the house makes enormous profits.

The binomial distribution can show why the house makes such large profits. As an example, consider the simplest possible game—tossing a coin. It is easy to show that the probability of tossing four *heads* in a row is $\frac{1}{16}$. Suppose that *heads* have been tossed three times in a row. Many players would feel that it is almost

certain that *tails* will come up on the next toss. The actual probability of *tails*, however, is only $\frac{1}{2}$. It is just as likely that *heads* will occur a fourth time. If we examine the binomial distribution, we see that occasionally we should expect even longer runs of *heads*, perhaps as many as 50 *heads* in a row.

If we apply the binomial distribution to any game of chance, we see that we should expect runs of both good and bad luck. Because the house has almost infinite resources when compared to the player, then it can weather a long run of bad luck. The player, on the other hand, will be bankrupt after a relatively short run of bad luck and will be forced out of the game before he can recover his losses on a run of good luck.

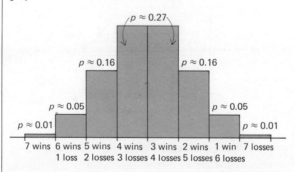

Expected distribution of wins and losses in seven plays of a fair game.

EXERCISES

1. A coin is tossed 10 times. What is the probability of each of the following events?
    (a) 5 *heads* and 5 *tails*
    (b) all *heads*
    (c) 4 *heads* and 6 *tails*
    (d) 1 *head* and 9 *tails*

2. A student makes an "intelligent" guess at each of the 10 problems on a multiple-choice test. Assume that the probability is 0.7 that he gets any particular problem correct.
   (a)   What is the probability that he gets exactly seven problems correct?
   (b)   What is the probability that he gets exactly three problems correct?
   (c)   What is the probability that he gets at least seven problems correct?
3. The genes for certain types of dwarfism are dominant. A man who has this condition marries a normal woman. They plan to have five children.
   (a)   What is the probability that all five will be normal?
   (b)   What is the probability that two will be dwarfs and three will be normal?
4. Use Pascal's Triangle to make a frequency chart for the expected outcome of tosses of 10 coins. (See Fig. 10-22.) How closely does the graph appear to fit a normal-distribution curve?
5. Calculate the mean and the standard deviation for the frequency chart in Exercise 4. What proportion of the data lies within one standard deviation of the mean? Within two standard deviations of the mean?

SUGGESTIONS FOR FURTHER READING

1.   Bell, E. T. "The Prince of Amateurs (Fermat)," and "Greatness and the Misery of Man (Pascal)," *Men of Mathematics*, Chaps. 4 and 5. New York: Simon & Schuster, 1937.
2.   David, F. N. *Games, Gods and Gambling*. New York: Hafner, 1962.
3.   Hogben, Lancelot. *Mathematics in the Making*. New York: Doubleday, 1960.
4.   Kahn, David. *The Codebreakers*. New York: Macmillan, 1967.
5.   Kline, Morris. *Mathematics for Liberal Arts*, Chap. 23. Reading, Mass.: Addison-Wesley, 1967.
6.   Ore, Øystein. *Cardano, the Gambling Scholar*. New York: Dover, 1965.
7.   Weaver, Warren. *Lady Luck*. Anchor Books. New York: Doubleday, 1963.

TABLES

Table I **Trigonometric Ratios**

Degrees	Sine	Cosine	Tangent	Cotangent
0	0	1.0000	0	—
1	.0175	.9998	.0175	57.290
2	.0349	.9994	.0349	28.636
3	.0523	.9986	.0524	19.081
4	.0698	.9976	.0699	14.301
5	.0872	.9962	.0875	11.430
6	.1045	.9945	.1051	9.5144
7	.1219	.9925	.1228	8.1443
8	.1392	.9903	.1405	7.1154
9	.1564	.9877	.1584	6.3138
10	.1736	.9848	.1763	5.6713
11	.1908	.9816	.1944	5.1446
12	.2079	.9781	.2126	4.7046
13	.2250	.9744	.2309	4.3315
14	.2419	.9703	.2493	4.0108
15	.2588	.9659	.2679	3.7321
16	.2756	.9613	.2867	3.4874
17	.2924	.9563	.3057	3.2709
18	.3090	.9511	.3249	3.0777
19	.3256	.9455	.3443	2.9042
20	.3420	.9397	.3640	2.7475
21	.3584	.9336	.3839	2.6051
22	.3746	.9272	.4040	2.4751
23	.3907	.9205	.4245	2.3559
24	.4067	.9135	.4452	2.2460
25	.4226	.9063	.4663	2.1445
26	.4384	.8988	.4877	2.0503
27	.4540	.8910	.5095	1.9626
28	.4695	.8829	.5317	1.8807
29	.4848	.8746	.5543	1.8040
30	.5000	.8660	.5774	1.7321
31	.5150	.8572	.6009	1.6643
32	.5299	.8480	.6249	1.6003
33	.5446	.8387	.6494	1.5399
34	.5592	.8290	.6745	1.4826
35	.5736	.8192	.7002	1.4281
36	.5878	.8090	.7265	1.3764
37	.6018	.7986	.7536	1.3270
38	.6157	.7880	.7813	1.2799
39	.6293	.7771	.8098	1.2349
40	.6428	.7660	.8391	1.1918
41	.6561	.7547	.8693	1.1504
42	.6691	.7431	.9004	1.1106
43	.6820	.7314	.9325	1.0724
44	.6947	.7193	.9657	1.0355
45	.7071	.7071	1.0000	1.0000
Degrees	*Sine*	*Cosine*	*Tangent*	*Cotangent*

Degrees	Sine	Cosine	Tangent	Cotangent
46	.7193	.6947	1.0355	.9657
47	.7314	.6820	1.0724	.9325
48	.7431	.6691	1.1106	.9004
49	.7547	.6561	1.1504	.8693
50	.7660	.6428	1.1918	.8391
51	.7771	.6293	1.2349	.8098
52	.7880	.6157	1.2799	.7813
53	.7986	.6018	1.3270	.7536
54	.8090	.5878	1.3764	.7265
55	.8192	.5736	1.4281	.7002
56	.8290	.5592	1.4826	.6745
57	.8387	.5446	1.5399	.6494
58	.8480	.5299	1.6003	.6249
59	.8572	.5150	1.6643	.6009
60	.8660	.5000	1.7321	.5774
61	.8746	.4848	1.8040	.5543
62	.8829	.4695	1.8807	.5317
63	.8910	.4540	1.9626	.5095
64	.8988	.4384	2.0503	.4877
65	.9063	.4226	2.1445	.4663
66	.9135	.4067	2.2460	.4452
67	.9205	.3907	2.3559	.4245
68	.9272	.3746	2.4751	.4040
69	.9336	.3584	2.6051	.3839
70	.9397	.3420	2.7475	.3640
71	.9455	.3256	2.9042	.3443
72	.9511	.3090	3.0777	.3249
73	.9563	.2924	3.2709	.3057
74	.9613	.2756	3.4874	.2867
75	.9659	.2588	3.7321	.2679
76	.9703	.2419	4.0108	.2493
77	.9744	.2250	4.3315	.2309
78	.9781	.2079	4.7046	.2126
79	.9816	.1908	5.1446	.1944
80	.9848	.1736	5.6713	.1763
81	.9877	.1564	6.3138	.1584
82	.9903	.1392	7.1154	.1405
83	.9925	.1219	8.1443	.1228
84	.9945	.1045	9.5144	.1051
85	.9962	.0872	11.430	.0875
86	.9976	.0698	14.301	.0699
87	.9986	.0523	19.081	.0524
88	.9994	.0349	28.636	.0349
89	.9998	.0175	57.290	.0175
90	1.0000	0	–	0
Degrees	Sine	Cosine	Tangent	Cotangent

Source: Adapted from Rinehart Mathematical Tables, Formulas and Curves, Holt, Rinehart and Winston, 1953.

Table II **Powers and Roots 1-100**

n	n^2	n^3	$\sqrt{n}$	$\sqrt[3]{n}$	n	n^2	n^3	$\sqrt{n}$	$\sqrt[3]{n}$
1	1	1	1.000	1.000	51	2,601	132,651	7.141	3.708
2	4	8	1.414	1.260	52	2,704	140,608	7.211	3.733
3	9	27	1.732	1.442	53	2,809	148,877	7.280	3.756
4	16	64	2.000	1.587	54	2,916	157,464	7.348	3.780
5	25	125	2.236	1.710	55	3,025	166,375	7.416	3.803
6	36	216	2.449	1.817	56	3,136	175,616	7.483	3.826
7	49	343	2.646	1.913	57	3,249	185,193	7.550	3.849
8	64	512	2.828	2.000	58	3,364	195,112	7.616	3.871
9	81	729	3.000	2.080	59	3,481	205,379	7.681	3.893
10	100	1,000	3.162	2.154	60	3,600	216,000	7.746	3.915
11	121	1,331	3.317	2.224	61	3,721	226,981	7.810	3.936
12	144	1,728	3.464	2.289	62	3,844	238,328	7.874	3.958
13	169	2,197	3.606	2.351	63	3,969	250,047	7.937	3.979
14	196	2,744	3.742	2.410	64	4,096	262,144	8.000	4.000
15	225	3,375	3.873	2.466	65	4,225	274,625	8.062	4.021
16	256	4,096	4.000	2.520	66	4,356	287,496	8.124	4.041
17	289	4,913	4.123	2.571	67	4,489	300,763	8.185	4.062
18	324	5,832	4.243	2.621	68	4,624	314,432	8.246	4.082
19	361	6,859	4.359	2.668	69	4,761	328,509	8.307	4.102
20	400	8,000	4.472	2.714	70	4,900	343,000	8.367	4.121
21	441	9,261	4.583	2.759	71	5,041	357,911	8.426	4.141
22	484	10,648	4.690	2.802	72	5,184	373,248	8.485	4.160
23	529	12,167	4.796	2.844	73	5,329	389,017	8.544	4.179
24	576	13,824	4.899	2.884	74	5,476	405,224	8.602	4.198
25	625	15,625	5.000	2.924	75	5,625	421,875	8.660	4.217
26	676	17,576	5.099	2.962	76	5,776	438,976	8.718	4.236
27	729	19,683	5.196	3.000	77	5,929	456,533	8.775	4.254
28	784	21,952	5.292	3.037	78	6,084	474,552	8.832	4.273
29	841	24,389	5.385	3.072	79	6,241	493,039	8.888	4.291
30	900	27,000	5.477	3.107	80	6,400	512,000	8.944	4.309
31	961	29,791	5.568	3.141	81	6,561	531,441	9.000	4.327
32	1,024	32,768	5.657	3.175	82	6,724	551,368	9.055	4.344
33	1,089	35,937	5.745	3.208	83	6,889	571,787	9.110	4.362
34	1,156	39,304	5.831	3.240	84	7,056	592,704	9.165	4.380
35	1,225	42,875	5.916	3.271	85	7,225	614,125	9.220	4.397
36	1,296	46,656	6.000	3.302	86	7,396	636,056	9.274	4.414
37	1,369	50,653	6.083	3.332	87	7,569	658,503	9.327	4.431
38	1,444	54,872	6.164	3.362	88	7,744	681,472	9.381	4.448
39	1,521	59,319	6.245	3.391	89	7,921	704,969	9.434	4.465
40	1,600	64,000	6.325	3.420	90	8,100	729,000	9.487	4.481
41	1,681	68,921	6.403	3.448	91	8,281	753,571	9.539	4.498
42	1,764	74,088	6.481	3.476	92	8,464	778,688	9.592	4.514
43	1,849	79,507	6.557	3.503	93	8,649	804,357	9.644	4.531
44	1,936	85,184	6.633	3.530	94	8,836	830,584	9.695	4.547
45	2,025	91,125	6.708	3.557	95	9,025	857,375	9.747	4.563
46	2,116	97,336	6.782	3.583	96	9,216	884,736	9.798	4.579
47	2,209	103,823	6.856	3.609	97	9,409	912,673	9.849	4.595
48	2,304	110,592	6.928	3.634	98	9,604	941,192	9.899	4.610
49	2,401	117,649	7.000	3.659	99	9,801	970,299	9.950	4.626
50	2,500	125,000	7.071	3.684	100	10,000	1,000,000	10.000	4.642

Source: Adapted from *Rinehart Mathematical Tables, Formulas and Curves*, Holt, Rinehart and Winston, 1953.

Table III   Factorials

$0! = 1$	$11! \approx (3.99)10^7$	$21! \approx (5.11)10^{19}$
$1! = 1$	$12! \approx (4.79)10^8$	$22! \approx (1.12)10^{21}$
$2! = 2$	$13! \approx (6.28)10^9$	$23! \approx (2.59)10^{22}$
$3! = 6$	$14! \approx (8.72)10^{10}$	$24! \approx (6.20)10^{23}$
$4! = 24$	$15! \approx (1.31)10^{12}$	$25! \approx (1.55)10^{25}$
$5! = 120$	$16! \approx (2.09)10^{13}$	$26! \approx (4.03)10^{26}$
$6! = 720$	$17! \approx (3.56)10^{14}$	$27! \approx (1.09)10^{28}$
$7! = 5040$	$18! \approx (6.40)10^{15}$	$28! \approx (3.05)10^{29}$
$8! = 40,320$	$19! \approx (1.22)10^{17}$	$29! \approx (8.84)10^{30}$
$9! = 362,880$	$20! \approx (2.43)10^{18}$	$30! \approx (2.65)10^{32}$
$10! = 3,628,800$		

Source: Adapted from *Rinehart Mathematical Tables, Formulas and Curves*, Holt, Rinehart and Winston, 1953.

Table IV Exponential Functions $(1 + r)^t$

t	$(1.01)^t$	$(1.02)^t$	$(1.03)^t$	$(1.04)^t$	$(1.05)^t$	$(1.06)^t$	t
1	1.0100	1.0200	1.0300	1.0400	1.0500	1.0600	**1**
2	1.0201	1.0404	1.0609	1.0816	1.1025	1.1236	2
3	1.0303	1.0612	1.0927	1.1249	1.1576	1.1910	3
4	1.0406	1.0824	1.1255	1.1699	1.2155	1.2625	4
5	1.0510	1.1041	1.1593	1.2167	1.2763	1.3382	5
6	1.0615	1.1262	1.1941	1.2653	1.3401	1.4185	**6**
7	1.0721	1.1487	1.2299	1.3159	1.4071	1.5036	7
8	1.0829	1.1717	1.2668	1.3686	1.4775	1.5938	8
9	1.0937	1.1951	1.3048	1.4233	1.5513	1.6895	9
10	1.1046	1.2190	1.3439	1.4802	1.6289	1.7908	10
11	1.1157	1.2434	1.3842	1.5395	1.7103	1.8983	**11**
12	1.1268	1.2682	1.4258	1.6010	1.7959	2.0122	12
13	1.1381	1.2936	1.4685	1.6651	1.8856	2.1329	13
14	1.1495	1.3195	1.5126	1.7317	1.9799	2.2609	14
15	1.1610	1.3459	1.5580	1.8009	2.0789	2.3966	15
16	1.1726	1.3728	1.6047	1.8730	2.1829	2.5404	**16**
17	1.1843	1.4002	1.6528	1.9479	2.2920	2.6928	17
18	1.1961	1.4282	1.7024	2.0258	2.4066	2.8543	18
19	1.2081	1.4568	1.7535	2.1068	2.5270	3.0256	19
20	1.2202	1.4859	1.8061	2.1911	2.6533	3.2071	20
21	1.2324	1.5157	1.8603	2.2788	2.7860	3.3996	**21**
22	1.2447	1.5460	1.9161	2.3699	2.9253	3.6035	22
23	1.2572	1.5769	1.9736	2.4647	3.0715	3.8197	23
24	1.2697	1.6084	2.0328	2.5633	3.2251	4.0489	24
25	1.2824	1.6406	2.0938	2.6658	3.3864	4.2919	25
26	1.2953	1.6734	2.1566	2.7725	3.5557	4.5494	**26**
27	1.3082	1.7069	2.2213	2.8834	3.7335	4.8223	27
28	1.3213	1.7410	2.2879	2.9987	3.9201	5.1117	28
29	1.3345	1.7758	2.3566	3.1187	4.1161	5.4184	29
30	1.3478	1.8114	2.4273	3.2434	4.3219	5.7435	30
31	1.3613	1.8476	2.5001	3.3731	4.5380	6.0881	**31**
32	1.3749	1.8845	2.5751	3.5081	4.7649	6.4534	32
33	1.3887	1.9222	2.6523	3.6484	5.0032	6.8406	33
34	1.4026	1.9607	2.7319	3.7943	5.2533	7.2510	34
35	1.4166	1.9999	2.8139	3.9461	5.5160	7.6861	35
36	1.4308	2.0399	2.8983	4.1039	5.7918	8.1473	**36**
37	1.4451	2.0807	2.9852	4.2681	6.0814	8.6361	37
38	1.4595	2.1223	3.0748	4.4388	6.3855	9.1543	38
39	1.4741	2.1647	3.1670	4.6164	6.7048	9.7035	39
40	1.4889	2.2080	3.2620	4.8010	7.0400	10.2857	40
41	1.5038	2.2522	3.3599	4.9931	7.3920	10.9029	**41**
42	1.5188	2.2972	3.4607	5.1928	7.7616	11.5570	42
43	1.5340	2.3432	3.5645	5.4005	8.1497	12.2505	43
44	1.5493	2.3901	3.6715	5.6165	8.5572	12.9855	44
45	1.5648	2.4379	3.7816	5.8412	8.9850	13.7646	45
46	1.5805	2.4866	3.8950	6.0748	9.4343	14.5905	**46**
47	1.5963	2.5363	4.0119	6.3178	9.9060	15.4659	47
48	1.6122	2.5871	4.1323	6.5705	10.4013	16.3939	48
49	1.6283	2.6388	4.2562	6.8333	10.9213	17.3775	49
50	1.6446	2.6916	4.3839	7.1067	11.4674	18.4202	50
t	$(1.01)^t$	$(1.02)^t$	$(1.03)^t$	$(1.04)^t$	$(1.05)^t$	$(1.06)^t$	t

Source: Adapted from *Rinehart Mathematical Tables, Formulas and Curves*, Holt, Rinehart and Winston, 1953.

ANSWERS TO SELECTED EXERCISES

Section 1-2, p. 10

3. (b) Part of the flow chart is shown:

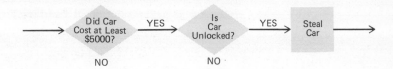

4. As stated the loop will be broken only if an immortal takes Prometheus' place.

Section 1-3, p. 16

1. Fig. 1-12 has one loop. Fig. 1-16 has one major loop that can be traversed in two different ways. 3. (a) They are necessary. 4. (b) $N = 3, A = 61$. 5. The loop never ends until the computer runs out of grades to read and stops work on its own. In that event the average is never calculated.

Section 1-4, p. 21

1. (b) $83\frac{1}{3}$ miles per hour. 2. (b) 990 feet per minute. 3. A little less than 6 cu. yd. 4. (a) After one payment he owes $850.

Section 2-1, p. 29

1. (b) ⌡⌡⌡⌡⌡ 999 ∩∩∩∩ / ∩∩∩∩ I , ४ ੨ ⟨ ⟩

3. Wires are needed for 1¢, 5¢, 10¢, 25¢, 50¢, $1.00 and $5.00. Four beads are needed on the $1.00 wire, only one is needed on the $5.00 wire. 4. Only 4 beads are needed on the I wire, only 1 is needed on the V wire, and so on. 6. 233.

Section 2-2, p. 34

3. (b) 397; (c) 897. 6. (b) It would be necessary to work an infinite number of particular examples.

Section 2-3, p. 38

1. (c) 11,403; (e) 467,022. 4. (a) It is necessary to keep track of both the units digit and the tens digit of each number. The product may have as many as three digits.

Section 2-4, p. 45

3. (a) True; (c) True. 4. Interpret ab and bc as the areas of two of the sides of the box and c and a, respectively, as the lengths of the perpendicular sides.

Section 2-5, p. 50

1. (b) 33; (e) 89; (f) 7808; (h) 119. 2. (a) Almost any three numbers will work as an example; (b) If a law is to be true it must be true in every case. A single example in which a proposition does not hold is sufficient to show that it does not hold in every case. 5. (b) Not necessarily true. Find an example in which the proposition does not hold; (c) Always true. Explain why!

Section 2-6, p. 55

1. (d) If we use 17 as a lower estimate for 17.31 and 14 as a lower estimate for 14.67 we obtain 238 as a lower estimate for the product. If we use 18 and 15 as upper estimates we obtain 270 as an upper estimate for the product. If 18 and 14 are used we obtain 252. **2. (d)** If we use the same estimates as in the solution of Exercise 1. (d) we find that

$$(17.31)(14.67) \approx \frac{238 + 270}{2} \approx 254$$

4. Approximately 5 cubic yards. **5.** The professor's number implies that the distance has been measured to within five miles of the true distance.

Section 2-7, p. 61

1. (c) The statement $a/b = n$ means that $a = bn$. Since $(9 \cdot 6) = 9 \cdot 6$ then it follows that $(9 \cdot 6)/6 = 9$. **2. (a)** 2 (approx.); **(c)** 200 (approx.). **3. (a)** The numerator is approximately equal to $2 \cdot 10^6$ and the denominator to $4 \cdot 10^4$. **4. (a)** 11 (approx.); **(c)** 18 (approx.). **5. (b)** 1, 2, 4, 8; **(f)** 1, 2, 4, 5, 10.

Section 2-8, p. 66

1. (a) $q = 41$, $r = 3$; **(c)** $q = 46$, $r = 5$; **(e)** $q = 597$, $r = 0$; **(f)** $q = 94$, $r = 87$. **3. (a)** $29 - 8 = 21$, $21 - 8 = 13$, $13 - 8 = 5$; $q = 3$, $r = 5$. **5. (b)** If the two even numbers are written as $2m$ and $2n$ then their sum is $2m + 2n = 2(m + n)$; **(d)** If the two even numbers are $2m$ and $2n$ then their product is $4mn = 2(2mn)$. **6.** *Hint:* There is one between 20 and 30. **7.** Three of these numbers are 3, 5, and 13. **8.** Show that an even integer greater than 2 has 2 as a proper divisor.

Section 3-1, p. 79

1. (a) $21 = 3 \cdot 7$, $36 = 2 \cdot 2 \cdot 3 \cdot 3$, $21/36 = 7/(2 \cdot 2 \cdot 3) = 7/12$. **2. (b)** $\frac{2}{3}$; **(e)** $\frac{8}{5}$. **3. (b)** $\frac{55}{104}$; **(e)** $\frac{13}{21}$. **8. (a)** 38 or 2280; **(d)** 43,593. **9. (a)** ▼▼▼▼ Symbol could also represent 7 or $7 \cdot 60 = 420$; **(d)** Symbol ▼▼▼ could also represent 257. **10. (b)** | | | | / | | |

Section 3-2, p. 84

1. (c) $\frac{5}{14}$; **(e)** $\frac{7}{6}$; **(i)** $\frac{2592}{7667}$.

Section 3-3, p. 88

1. (b) 113.52; **(c)** 87.0995; **(e)** 22.885; **(f)** 111.3628. **2. (b)** 92.532; **(d)** 1,842.2417; **(f)** 230.6318. **3. (c)** 1.955 (approx.); **(d)** 2.8 (approx.). **4. (a)** 20.32; **(b)** 68.58. **5. (b)** 4'8.3'' (approx.); **(d)** 7.559 (approx.). **7. (b)** 0.368; **(c)** 5.857. **8. (a)** 41.824; **(b)** 12.515. **10.** $\frac{12}{64}$.

Section 3-4, p. 93

1. (a) 10; **(c)** 133.33. **2.** Payroll deductions are approximately $800. **3. (b)** Plan A: Between 3760 and 3995 square feet. **4. (a)** $7\frac{1}{2}$% (approx.). **7.** $\frac{1}{6}$. **9.** Work the problem by considering what happens for each $1 of the original (pretax) bill.

Section 3-5, p. 102

1. (b) -10; **(d)** -25; **(f)** -3.3; **(h)** 6.645. **2. (b)** $-\frac{3}{8}$; **(d)** -7; **(f)** 1. **3. (b)** $-\frac{47}{60}$; **(d)** $-\frac{10}{3}$; **(f)** $-\frac{2}{9}$. **4. (d)** A shift of five units to the left followed by a shift of four units to the right.

Section 3-6, p. 109   **1.** (a) $\frac{3}{250}$; (d) $-\frac{1}{48}$.   **3.** (a) Irrational; (b) Rational.   **6.** (a) $31\frac{7}{16}$.
8. (a) Approx. 4.56.

Section 3-7, p. 117   **1.** (b) $\frac{4}{9}$; (d) $\frac{1}{4}$; (f) $5 \cdot 17^{10}$. **2.** (c) a^5b^2; (e) 1 (if $a \neq -b$). **3.** (b) $\frac{15}{41}$.
4. (a) Write 7^5 and 7^4 as products of the form $7 \cdot 7 \cdot 7 \dots$ **5.** (a) $(5^2)^3 =$
$(5 \cdot 5)^3 = ?$ **8.** (a) $(7.58123)10^3$. **9.** (b) 4.95.

Section 3-8, p. 124   **1.** (a) 0.222222; (d) 0.164286 (round-off); (f) 0.228915 (truncation).
2. (b) 3.1416. **4.** (d) 63. **5.** (b) $\frac{103}{999}$; (c) $\frac{314098}{999000}$.

Section 4-1, p. 135   **4.** (a) There are two such points; (d) There is only one such point.

Section 4-2, p. 142   **2.** (c) Each number must be positive and less than the sum of the other
two numbers. **5.** *Hint:* Make the line parallel to a base. **6.** *Hint:* Make
the line parallel to a base.

Section 4-3, p. 150   **1.** (b) 31.5 sq. units. **3.** See Exercise 2 for the definition of congruence;
(b) Neither; (c) Congruent; (f) Similar. **4.** $57\frac{1}{3}$ ft. **5.** *Hint:* Draw a figure.
Let x and y be the dimensions of one rectangle. Calculate the dimensions
of the other rectangle, then calculate their areas.

Section 4-4, p. 158   **1.** (b) $\beta \approx 64°$. **2.** (b) A is between $42°$ and $43°$, B is between $47°$ and
$48°$, b ≈ 29.5. **3.** 308,000 sq. ft. (approx.). **5.** 19 mi. (approx.). **6.** 1 mi.
(approx.).

Section 4-5, p. 167   **3.** 2.7 million (approx.).

Section 4-6, p. 175   **1.** (b) $m = \frac{2}{3}$; (d) $m = 2$. **2.** (b) 36 cm. (approx.); (d) 9 in.; (f) 50 in.
4. (a) and (b) are lines; (c) and (d) are cup-shaped curves called parabolas.

Section 4-7, p. 182   **1.** (b) 15 mo.: \$45. **2.** "Red Headed Screwball": 35 million (approx.),
"Down Home with the Folks": 18 million (approx.). **3.** (d) \$400.
4. m ≈ 1.7. **5.** Approx. 75 births per thousand. **6.** m $\approx .016$.

Section 5-2, p. 196   **1.** (b) $4x^2 - x + 6$; (e) $x^3 + x^2 + 3x + 8$. **2.** (c) $x^3 + 5x^2 - 2x - 24$;
(f) $2x^3 - x^2 - 5x - 2$. **3.** (a) Cancel $(3x + 1)$; (c) Cancel $x - 2 =$
$-(2 - x)$. **4.** (c) $(x - 1)/(x + 5)$; (f) $(x^2 + 3x + 2)/(2x^3 - 5x^2 + 14x -$
$35)$; (g) $(x^2 + 4x - 5)/(x^2 + 2x + 1)$.

Section 5-3, p. 201   **1.** (a) x_2 is a solution. **2.** (b) $\frac{1}{2}$; (d) 23; (f) 0. **3.** \$.29 per head. **4.** 49
quality points. **5.** Gross sales must be \$6300. **6.** 84.

Section 5-4, p. 208   **1.** (f) $-\frac{7}{3}$. **2.** (c) Any number x is a solution; (d) *Ans:* $x = -\frac{6}{5}$. **3.** (a) 3;
(c) -1; (e) No solution. **4.** (b) 2; (d) $-\frac{5}{2}$; (f) -3. **5.** (b) 19; (e) 79.
6. (b) (2): \$180,000. **7.** (b) \$4576. **8.** \$121,212.

Section 5-5, p. 213   **1. (b)** Conditional;   **(e)** Identity;   **(h)** Inconsistent.   **3. (b)** 208   gal. **4. (b)** 15.

Section 5-6, p. 222   **2. (b)** Yes; **(e)** Yes.   **3. (b)** $-\frac{7}{3}$; **(d)** $-\frac{31}{48}$.   **4. (b)** $m = -2$; **(d)** $m = \frac{3}{4}$; **(f)** $m = 0$; **(h)** $m = -\frac{8}{17}$.   **5. (d)** \$13.80; **(e)** 57 hundred gal. **6. (c)** \$4900.

Section 5-7, p. 229   **1. (b)** No.   **2. (a)** $x \approx -2.8$, $y \approx 0.8$;   **(d)** $x \approx -3.3$, $y \approx 2.7$; **(f)** $x \approx -0.1$, $y \approx -0.5$.   **3. (b)** $x = 4$, $y = 3$;   **(d)** $x = \frac{1}{14}$, $y = -\frac{31}{14}$; **(e)** $x = \frac{103}{36}$, $y = \frac{1}{36}$.   **4. (b)** $m = -5, 0, 0$;   **(e)** $m = \frac{13}{3}$, $-\frac{14}{13}$, $\frac{14}{3}$. **5. (c)** -40 degrees.

Section 5-8, p. 235   **1. (a)** $x = -2$, $y = 1$;   **(c)** $x = -2$, $y = -3$;   **(e)** $x = \frac{142}{73}$, $y = \frac{21}{73}$ **2. (a)** Unique; **(c)** Dependent.

Section 5-9, p. 238   **5.** \$1.65, 0.02.   **6.** \$2.51 and \$4.58 (approx.).   **7.** 16cc. of 15% alcohol.

Section 6-3, p. 254   **1.** The first line of the printout (Statement 25) should consist of the numbers 7, 42, 49, and 294. The fourth line should consist of the statements A = 5, B = 9.   **2. (a)** The first line of printout should consist of the statements 1, 2, C = 4.   **3.** A few of the errors are as follows: (1) First statement has no statement number; Statement 30 has an illegal expression to the left of the equals sign; Statement 25 is out of order (may not be illegal on some systems); Statement 25 is worthless since Statement 60 immediately fixes the value of N; There is no initial value of N.

Section 6-4, p. 258   **1. (a)** Legal; **(d)** Legal; **(g)** Illegal; **(j)** Illegal.   **2. (a)** The first time Statement 60 is executed the variables have the values A = 1, B = 2, C = 3, D = 6, K = 3; the second time it is executed the values are A = 1, B = 2, C = 4, D = 7, K = 5.

Section 6-5, p. 262   **1. (c)** $3/A + B\uparrow(5 * A)$; **(e)** $x + y/2 - z$.   **2. (d)** B is assigned the value 3.   **3. (a)** The computer prints six numbers: 4, 16, 256, and so on; **(b)** The computer prints the numbers 4, 16, 256, and so on, without stopping. The program has an infinite loop. **(c)** The computer prints the number 4 six times.   **4. (a)** The computer prints the numbers $-1, -2, -3, -4$, and so on, without stopping.

Section 6-6, p. 266   **1. (a)** Legal; **(b)** The format of the statement is legal on some computer systems, but it will cause infinite looping when the statement is executed; **(e)** Illegal on some systems.   **3.** Let S be the value of the smallest number. Set the initial value of S equal to the first number read. Test each number when it is read. If it is less than S replace S by the new number.

Section 6-7, p. 269   **1. (b)** 1,814,268,000; **(c)** .0000000978462.   **2. (a)** Either the word PRINT or the expression PRINT " " (literally "Print Nothing").   **3.** The semicolon causes the computer to close up the expressions, not leaving as much space between them.

Section 6-8, p. 277   **1. (b)** 1; **(d)** 5; **(g)** 3.5; **(i)** 6; **(k)** 5; **(l)** -4.   **3. (a)** $N = K/10$. **5. (b)** The computer continues to ask for input until you stop it.

Section 6-9, p. 282   **2.** Base your program on the one in Example 1.   **4.** Delete Statement 140, start the FOR . . . NEXT loop with Statement 150.

Section 7-1, p. 308   **1. (b)** Three of the points are $(-1, -\frac{1}{2})$, $(0,0)$, and $(2, -2)$; **(d)** Three of the points are $(1,1)$, $(2,\frac{1}{2})$, and $(4,\frac{1}{4})$; **(e)** Two of the points are $(-1,\frac{1}{2})$ and $(2,4)$. **4.** \$660 (approx.).   **5. (a)** 5 ft.   **6.** At 40 miles per hour the braking distance is 160 ft.

Section 7-2, p. 314   **1. (b)** Quartic, degree $= 4$; **(d)** Cubic, degree $= 3$.   **2. (b)** $(x + 1) \cdot (x + 2)$; **(e)** $(x - 4)(x + 3)$; **(h)** $(x - 7)(x + 9)$.   **3. (b)** $(x - 1)(x + 1)$. **(e)** $(y - \sqrt{7})(y + \sqrt{7})$; **(g)** $(t - \sqrt{8})(t + \sqrt{8})$.

Section 7-3, p. 319   **1. (a)** $0,4$; **(c)** $-1,5$; **(g)** $0,1,3$.   **2. (b)** $-1,-7$; **(d)** $\frac{1}{3},-\frac{3}{2}$; **(f)** 1; **(h)** $1,-5$. **3. (a)** $-2,2$; **(e)** $-\sqrt{18}, \sqrt{18}$; **(g)** $-1-\sqrt{8}$, $-1 + \sqrt{8}$; **(j)** $-5,0$.

Section 7-4, p. 327   **2. (b)** Symmetric point is $(2,5)$; **(d)** Symmetric point is $(-3,6)$.   **3.** The radical sign can only be used for nonnegative square roots.

Section 7-5, p. 339   **1. (b)** Vertex: $(-2,-8)$; **(c)** Line of symmetry: $x = -2$; **(f)** Vertex: $(3, -27)$; **(h)** Vertex: $(9,27)$.   **3.** The three parabolas have the same shape and orientation, but different vertices.   **4. (a)** Vertex: $(-1,-2)$; **(d)** Vertex $(3,0)$; **(f)** Vertex: $(-\frac{5}{2},\frac{37}{4})$.   **7. (a)** -4.38 and 0.38 (approx.); **(c)** $-.92$ and 4.92 (approx.); **(e)** 0.48 and 14.5 (approx.).   **8. (b)** None; **(d)** One; **(f)** Two.   **9. (a)** At the vertex $y = ax^2 + bx + c$ and $x = -b/2a$.

Section 7-6, p. 345   **1. (b)** 648; **(d)** 1562.5   **2. (c)** 150 m.   **3. (b)** 1800 m.   **4. (a)** $42\frac{1}{16}$ ft. **5. (c)** 64 ft.   **6.** 3 sec.   **7.** $x = 10$.   **8.** $x = \$3.75$.

Section 7-7, p. 352   **1. (b)** The positive number, which, when raised to the fifth power, is equal to 7.   **2. (d)** 8; **(f)** 16.   **3. (b)** 1.6 (approx.); **(d)** $-\frac{1}{7}$ (approx.).   **4. (b)** 1.86 (approx.); **(d)** 1.7 (approx.); **(f)** Not a real number.   **5. (b)** 14; **(d)** 1; **(h)** $\frac{1}{14}$.

Section 7-8, p. 359   **1. (a)** $12 - 2i$; **(d)** $-9 - 11i$.   **2. (b)** $58 + 44i$; **(d)** $56 + 213i$; **(f)** -5. **3. (c)** $12 + 5i$, 169; **(g)** $6i$, 36.   **4. (a)** $(1 - i)/2$; **(c)** $(41 - 11i)/34$; **(e)** $(143 + 24i)/145$; **(g)** $-i$.   **5. (a)** $x = \pm 6i$; **(d)** $x = \pm i\sqrt{17/3}$; **(f)** $x = (-1 \pm 5i)/2$; **(g)** $x = (2 \pm 7i)/3$.

Section 7-9, p. 367   **1. (a)** $-1 \pm 2i$; **(c)** $1,2$; **(e)** $-\frac{5}{3}$, -1; **(g)** $-1,2/5$; **(i)** $(-3 \pm i)/2$; **(k)** $-\frac{2}{3}$, -1.

Section 7-10, p. 373   **1.** (a) y varies jointly with x and z.   **2.** (b) $y = 5x^2/4$.   **3.** (b) 280 ft. at 80 mph.   **4.** (a) $6400; (d) $100.   **5.** (b) Slightly more than 1 cup.   **6.** (b) 120 lbs. per square inch.   **7.** (c) about 30 years.

Section 7-11, p. 380   **1.** (b) 5, 10, 160; (d) 6, 2, $\frac{2}{9}$; (h) 100, 1, $\frac{1}{10}$; (j) 1.6289, 2.6533.   **2.** (b) 58,000 (approx.); (c) 81,000 (approx.).   **3.** (a) $112,680; (b) 100,000(1.1268)^t.   **4.** (b) $32,600 (approx.).   **5.** (a) $4100 (approx.).

Section 8-1, p. 390   **7.** Use the method of Construction 5.   **9.** Let y be the length of a side of the large cube. Show that $y^3 = 2x^3$.   **10.** Let y be the side of the square. Show that $y^2 = \pi x^2$.

Section 8-2, p. 396   **1.** (a) No; (e) Yes; (f) No.   **2.** (b) 5; (d) $\sqrt{120}$; (f) $\sqrt{119}$.

Section 8-3, p. 403   **1.** (b) The distance can be measured to be $3 = |(-1) - 2|$.   **2.** (b) 13; (d) $\sqrt{53}$; (f) $\sqrt{13}$.   **3.** (b) Q is the point (6,1). $|P_1P_2|^2 = |P_1Q|^2 + |QP_2|^2 = 50$, $|P_1P_2| = \sqrt{50}$.   **5.** (a) Yes; (c) Yes; (e) No.   **6.** (a) 2.8 mi. (approx.); (c) 125 mi. (approx.).   **7.** (a) 1.4 mi. (approx.); (b) 28 mi. (approx.).

Section 8-4, p. 414   **1.** (a) 4.7 (approx.); (c) 29.5 (approx.).   **2.** (a) 12; (d) $3\sqrt{15}$.   **3.** (a) $A = 16\pi \approx 50.265$, $P = 8\pi \approx 25.133$; (d) $A \approx 103,277.9$ sq. cm. ≈ 10.32779 sq. m., $P \approx 1139.2$ cm. ≈ 11.392 m.   **4.** (a) The three sides measure 15, 20, and 25 units, area $= 150$ sq. units; (d) The three sides measure $5\sqrt{2}$, $8\sqrt{2}$ and $5\sqrt{2}$ units, area $= 24$ sq. units.   **5.** (a) Perimeter ≈ 13.6, area ≈ 9; (b) Perimeter ≈ 18, area ≈ 9.5; (c) Perimeter ≈ 12.5, area ≈ 8; (d) Perimeter ≈ 21.1, area ≈ 12.5.

Section 8-5, p. 425   **1.** Answers depend on the particular curves drawn through the points.   **2.** (a) Area ≈ 28.27; (b) Perimeter ≈ 18.85.   **5.** (b) 457.5, 1725; (c) Area $= 212,096$.   **7.** (a) 2.75; (b) 13.

Section 8-6, p. 436   **1.** (d) Surface area $= 26$, volume $= 12$.   **2.** (b) Surface area ≈ 314.16 sq. ft., volume ≈ 523.6 cu. ft. (d) Area ≈ 201 million sq. mi., Volume ≈ 268 billion cu. mi.   **3.** (a) Volume ≈ 56.55, lateral area ≈ 37.70, total surface area ≈ 94.25; (c) Volume $= 18$, lateral area $= 36$, total surface area $= 48$; (e) Volume ≈ 130.9, lateral area ≈ 111.1, total surface area ≈ 189.6.   **4.** $R = \frac{5}{2}$, $S = 10$, lateral area $= 25\pi$, Volume $= 125\pi\sqrt{15}/24 \approx 63.4$.   **5.** (a) 3.4 million cu. yds.; (b) 11.5 billion lbs.   **6.** 1850 gal. (approx.).   **7.** (a) 49 gal.; (b) $1\frac{2}{3}$ cu. ft.   **8.** The small piece has an approximate volume of 12.5 cu. ft.

Section 8-7, p. 443   **1.** The field should be square.   **3.** (a) *Triangle:* $s = 20$, area ≈ 173.2; *Square:* $s = 15$, area $= 225$; (b) *Triangle:* $s = P/3$, area $\approx .048P^2$; *Square:* $s = P/4$, area $\approx .063P^2$. Increase in area of square over area of triangle is approximately 31%; *Circle:* Radius $= P/2\pi$, area $= P^2/4\pi \approx .08P^2$. In-

crease in area of circle over area of triangle is approximately 67%.
4. *Second Hint:* Let the base be 100 ft., the area be 1 sq. in. $= \frac{1}{144}$ sq. ft. Calculate the height of the rectangle.

Section 9-2, p. 466

1. (a) $301 - 350$: None, $351 - 400$: two, $401 - 450$: 8, and so on. **3.** (a) 3A, 7B, and so on.

Section 9-3, p. 474

4. (a) Mean $= 6.6$, Median $= 7$. **5.** Based on the frequency chart, Mean $= 535.6$, Median class is $501 - 550$, the Median is the third element in this class ≈ 520, no Modal class.

Section 9-4, p. 480

1. $\bar{x} \approx 5.17$, $\sigma \approx 2.48$. **3.** $\bar{x} = 4.1$, $\sigma = 1.7$. **5.** $\bar{x} = 5.4$, $\sigma \approx 4.3$. **6.** (a) $\bar{x} \approx 536.7$, $\sigma \approx 100.4$. **6.** (b) $\bar{x} \approx 63.8$, $\sigma \approx 20.1$. **6.** (c) $\bar{x} \approx 551.6$, $\sigma \approx 95.2$. **7.** At least 90. **8.** At least 880.

Section 9-6, p. 492

2. Mean ≈ 27.2, $\sigma \approx .26$. **3.** (b) 160 (approx.). **4.** (c) 0.15%. **5.** (a) 90,000; (b) A bad law for several reasons.

Section 10-1, p. 506

1. (a) .97; (d) .40. **2.** (c) $\$1000/14.50 \approx \$68.97$ per year. **5.** About 58% of the persons who live to be at least one year old.

Section 10-2, p. 514

3. (b) $\frac{3}{4}$; (c) $\frac{2}{3}$. **4.** (b) $\frac{5}{12}$; (c) $\frac{2}{3}$. **5.** (b) $\frac{3}{4}$; (d) $\frac{1}{13}$; (e) $\frac{1}{26}$. **7.** The outcomes are not equally likely.

Section 10-3, p. 520

1. (a) Sixteen small branches on the diagram; (d) Eight small branches on the diagram. **2.** (a) $\frac{1}{4}$; (c) $\frac{1}{2}$. **3.** $\frac{1}{2}$. **4.** (b) $1/365^2$; (c) $\frac{1}{365}$. **5.** (b) $\frac{7}{8}$. **6.** (a) $\frac{1}{18}$; (d) $\frac{1}{9}$; (f) $\frac{1}{6}$. **7.** B. **8.** (b) $(.99)^2 \approx 98$. **9.** $1/10,000,000,000$.

Section 10-4, p. 525

1. (c) $\frac{1}{2}$; (d) $2/52^2$; (f) $\frac{1}{52}$. **2.** (c) $\frac{26}{51}$; (d) $\frac{2}{52} \cdot \frac{1}{51}$; (f) 0. **3.** $\frac{15}{36}$. **4.** (a) $\frac{8}{36} \approx .2222$; (b) $\frac{4}{36} \approx .1111$; (c) $\frac{24}{36} \cdot \frac{1}{6} \approx .1111$; (d) $\frac{4}{36} + \frac{24}{36} \cdot \frac{1}{6} \approx .2222$. **5.** $300 \cdot \frac{1}{6} = 50$. **6.** (b) $1 - (19/20)^5 \approx .23$.

Section 10-5, p. 533

1. (b) $\frac{2}{3}$. **2.** $\frac{2}{3} \cdot \frac{2}{3} \cdot \frac{1}{4} = \frac{1}{9}$. **3.** (b) All wrinkled. **4.** $\frac{3}{4}$ that it will be Rh$+$, $\frac{1}{4}$ that it will be Rh$-$. **5.** No.

Section 10-6, p. 539

1. Check your answer with Table III at the end of the book. **2.** (c) $n + 1$; (d) $9 \cdot 8 = 72$; (e) 10. **5.** (b) 36 possible outcomes. **6.** (c) $\frac{1}{12}$. **7.** $12 \cdot 11 \cdot 10 = 1320$. **9.** 25200. **10.** 17,643,225,600.

Section 10-7, p. 546

1. (b) $13!/2! = 3,113,510,400$; (d) 7; (e) $12!/6!6! = 924$; (g) $8! = 40,320$; (i) 1. **2.** (a) $20!/13! \approx (2.43)10^{18}/(6.28)10^9 \approx (3.87)10^8$; (c) $20!/7!13! \approx (7.68)10^4$. **3.** (b) $P_4^{12} = 11,880$; (d) $5! = 120$. **4.** (b) $12!/8!4! = 495$; (d) 1. **5.** $30!/25!5!$ **6.** $2 \cdot 29!/23!6!$

7. (a) $1 - \frac{335}{336} \cdot \frac{334}{335} \cdot \frac{333}{334} \cdot \frac{332}{333} \approx .012$. **8.** $\binom{5}{3} \cdot \binom{4}{2} \cdot \binom{2}{1} = 120$.

Section 10-8, p. 552

1. (b) $1 - \binom{80}{10} \Big/ \binom{100}{10} \approx .905.$ 2. (a) $1 - \binom{700}{5} \Big/ \binom{1000}{5} \approx .833.$

3. (a) 10,240; (c) Probability of a straight $= 10,200/2,598,960 \approx .004.$

4. (b) $p = 1 \Big/ \binom{52}{13} = 1/635,013,559,600;$ (d) $p = 40/635,013,559,600.$

Section 10-9, p. 558

2. (a) 792; (d) 495; (f) 1. 3. (e) The first four terms are $x^{10} + 10x^9y + 45x^8y^2 + 120x^7y^3.$ 4. (b) 165; (c) 792. 6. (a) Both binomial coefficients are equal to 8!/3!5!.

Section 10-10, p. 566

1. (a) $\binom{10}{5}\left(\frac{1}{2}\right)^{10} = \frac{252}{1024} \approx 0.25.$ 2. (a) $\binom{10}{7}(.7)^7(.3)^3 \approx .267;$

(c) $\binom{10}{10}(.7)^{10} + \binom{10}{9}(.7)^9(.3)^1 + \binom{10}{8}(.7)^8(.3)^2 + \binom{10}{7}(.7)^7(.3)^3 \approx 0.65.$

3. (b) $\frac{10}{32} \approx 0.31$ if the man is a carrier. 5. $\bar{x} = 5, \sigma = \sqrt{2.5} \approx 1.58.$

INDEX